Cambridge Studies in Biological and Evoultionary Anthropology 48

Feeding ecology in apes and other primates

Feeding Ecology in Apes and other Primates focuses on evolutionary perspectives of the complex interactions between the environment, food sources, physiology, and behavior in primates. This
highly interdisciplinary volume provides a benchmark to assess
dietary alterations that affected human evolution by putting the focus
on the diet of hominid primates. It also offers a new perspective
on the behavioral ecology of the last common ancestor by integrating
corresponding information from both human and non-human primates. The potential of innovations of applied biotechnology are
also explored to set new standards for future research on feeding
ecology, and new information on feeding ecology in humans, apes,
and other primates is synthesized to help refine or modify current
models of socioecology. By taking a comparative view, this book
will be interesting to primatologists, anthropologists, behavioral
ecologists, and evolutionary biologists who want to understand
better non-human primates, and the primate that is us.

GOTTFRIED HOHMANN is a Research Associate in the Department of Primatology at the Max Planck Institute for Evolutionary
Anthropology in Leipzig, Germany, researching bonobos. He has
also co-edited *Behavioural Diversity in Chimpanzees and Bonobos*
with Christophe Boesch and Linda Marchant (2002, ISBN
0521803543).

MARTHA M. ROBBINS is a Research Associate in the Primatology
department of the Max Planck Institute for Evolutionary Anthropology. She studies the behavioral ecology and reproductive strategies of
gorillas. She has also co-edited *Mountain Gorillas* (2001 ISBN
0521780047) with Pascale Sicotte and Kelly Stewart.

CHRISTOPHE BOESCH is Director of the Max Planck Institute for
Evolutionary Anthropology and Professor at Leipzig University. His
research priorities are: evolution of culture, socio-ecology,
evolutionary biology, and conservation biology.

Cambridge Studies in Biological and Evolutionary Anthropology

Series editors

HUMAN ECOLOGY
C. G. Nicholas Mascie-Taylor, University of Cambridge
Michael A. Little, State University of New York, Binghamton
GENETICS
Kenneth M. Weiss, Pennsylvania State University
HUMAN EVOLUTION
Robert A. Foley, University of Cambridge
Nina G. Jablonski, California Academy of Science
PRIMATOLOGY
Karen B. Strier, University of Wisconsin, Madison

Feeding ecology in apes and other primates

Ecological, physical, and behavioral aspects

GOTTFRIED HOHMANN,
MARTHA M. ROBBINS, and
CHRISTOPHE BOESCH

CAMBRIDGE
UNIVERSITY PRESS

CAMBRIDGE UNIVERSITY PRESS
Cambridge, New York, Melbourne, Madrid, Cape Town, Singapore, São Paulo

CAMBRIDGE UNIVERSITY PRESS
The Edinburgh Building, Cambridge CB2 2RU, UK
Published in the United States of America by Cambridge University Press, New York

www.cambridge.org
Information on this title: www.cambridge.org/9780521858373

First published 2006

Printed in the United Kingdom at the University Press, Cambridge

A catalog record for this publication is available from the British Library

ISBN-13 978-0-521-85837-3 hardback
ISBN-10 0-521-85837-2 hardback

Contents

Contributors

STUART A. ALTMANN Department of Ecology and Evolutionary Biology, Princeton University, Princeton, New Jersey, 08540, USA

DEAN ANDERSON Department of Zoology, University of Wisconsin-Madison, 430 Lincoln Drive, Madison, Wisconsin, 53706, USA

AUGUSTIN KANYUNYI BASABOSE Centre de Recherches en Sciences Naturelles, Lwiro, D.S. Bukavu, Democratic Republic of Congo

CHRISTOPHE BOESCH Department of Primatology, Max Planck Institute for Evolutionary Anthropology, Deutscher Platz 6, Leipzig, 04103, Germany

CAROLA BORRIES Department of Anthropology, SUNY at Stony Brook, Stony Brook, New York, 11794-4364, USA

BRENDA J. BRADLEY Department of Zoology, University of Cambridge, Downing Street, Cambridge, CB2 3EJ, UK

COLIN A. CHAPMAN Department of Anthropology and McGill School of Environment, McGill University, 855 Sherbrooke St West, Montreal, Quebec, H3A 2T7, Canada

NANCY LOU CONKLIN-BRITTAIN Department of Anthropology, Harvard University, Peabody Museum 55F, 11 Divinity Avenue, Cambridge, Massachusetts, 02138, USA

LISA DANISH Department of Zoology, University of Florida, Gainesville, Florida, 32611, USA

NATHANIEL J. DOMINY Department of Anthropology, University of California-Santa Cruz, 1156 High Street, Santa Cruz, California, 95064, USA

DIANE M. DORAN-SHEEHY Department of Anthropology, SUNY at Stony Brook, Stony Brook, New York, 11794, USA

ANDREW FOWLER Anthropology Department, University College London, Gower Street, London, WC1E 6BT, UK

JÖRG U. GANZHORN Biozentrum Grindel, Department of Animal Ecology and Conservation, University of Hamburg, Martin-Luther-King Platz 3, Hamburg, D-20146, Germany

MARTINE L. GEURTS Great Ape Trust of Iowa, 4200 SE 44th Ave, Des Moines, IA, 50320, USA

ZORO BERTIN GONÉ BI Wild Chimpanzee Foundation, Centre Suisse de Recherches Scientifiques, 01 BP 1303, Abidjan 01, Côte d'Ivoire

MARY BETH HALL U.S. Dairy Forage Lab, 1925 Linden Drive West, Madison, Wisconsin, 53706, USA

LISA A. HEIMBAUER Department of Anthropology, SUNY at Stony Brook, Stony Brook, New York, 11794, USA

GOTTFRIED HOHMANN Department of Primatology, Max Planck Institute for Evolutionary Anthropology, Deutscher Platz 6, Leipzig, 04103, Germany

CHARLES JANSON Department of Ecology and Evolution, SUNY at Stony Brook, Stony Brook, New York, 11794-4364, USA

CHERYL D. KNOTT Department of Anthropology, Harvard University, Peabody Museum, 11 Divinity Avenue, Cambridge, Massachusetts, 02138, USA

ANDREAS KOENIG Department of Anthropology, SUNY at Stony Brook, Stony Brook, New York, 11794-4364, USA

MARK LEIGHTON Department of Anthropology, Harvard University, Peabody Museum 26, 11 Divinity Avenue, Cambridge, Massachusetts, 02138, USA

PETER W. LUCAS Department of Anthropology, George Washington University, 2110 G Street Northwest, Washington, DC, 20052, USA

FRANK W. MARLOWE Department of Anthropology, Harvard University, Peabody Museum 26, 11 Divinity Avenue, Cambridge, Massachusetts, 02138, USA

ANDREW J. MARSHALL Department of Anthropology, Harvard University, Peabody Museum 26, 11 Divinity Avenue, Cambridge, Massachusetts, 02138, USA

ROBERT W. MAYES Macaulay Land Use Research Institute, Craigiebuckler, Aberdeen, AB15 8QH, UK

ALASTAIR MCNEILAGE Institute of Tropical Forest Conservation, Mbarara University of Science and Technology, and Wildlife Conservation Society, P.O. Box 44, Kabale, Uganda

KATHARINE MILTON Department of Environmental Science, Policy, and Management, Division of Insect Biology, University of California-Berkeley, 137 Mulford, Berkeley, California, 94720-3114, USA

TATANG MITRA SETIA Faculty of Biology, Nasional University, Jl. Sawo Manila, Pejaten, Pasar Minggu, Jakarta, Indonesia, 12520

DAVID MORGAN Goualougo Triangle Chimpanzee Project, Nouabalé-Ndoki National Park, B.P. 14537, Brazzaville, Republic of Congo

JOHN BOSCO NKURUNUNGI Makerere University Institute for Environment and Natural Resources, P.O. Box 7062, Kampala, Uganda

NUR SUPARDI NOOR Forest Ecology Unit, Forest Research Institute of Malaysia, Kepong, 52109, Kuala Lumpur, Malaysia

CEDRIC O'DRISCOLL WORMAN Department of Zoology, University of Florida, Gainesville, Florida, 32611, USA

JUAN CARLOS ORDOÑEZ JIMÉNEZ Junior Scientists Group on Cultural Phylogeny, Max Planck Institute for Evolutionary Anthropology, Bagaces, Guanacaste, Costa Rica

SYLVIA ORTMANN Institute for Zoo and Wildlife Research, Alfred-Kowalke-Str. 17, Berlin, 10315, Germany

SUSAN PERRY Junior Scientists Group on Cultural Phylogeny, Max Planck Institute for Evolutionary Anthropology, Deutscher Platz 6, Leipzig, 04103, Germany

JILL D. PRUETZ Department of Anthropology, Iowa State University, 324 Curtiss Hall, Ames, Iowa, 50011, USA

MARTHA M. ROBBINS Department of Primatology, Max Planck Institute for Evolutionary Anthropology, Deutscher Platz 6, Leipzig, 04103, Germany

KARYN D. RODE Alaska Department of Fish and Game, Division of Wildlife Conservation, 333 Raspberry Road, Anchorage, Alaska, 99518, USA

PETER S. RODMAN Department of Anthropology, University of California-Davis, One Shields Avenue, Davis, California, 95616, USA

CRICKETTE SANZ Department of Primatology, Max Planck Institute for Evolutionary Anthropology, Deutscher Platz 6, Leipzig, 04103, Germany

NATASHA F. SHAH Interdepartmental Doctoral Program in Anthropological Sciences, SUNY at Stony Brook, Stony Brook, New York, 11794-4364, USA

VOLKER SOMMER Anthropology Department, University College London, Gower Street, London, WC1E 6BT, UK

DANIEL STAHL Departments of Primatology and Developmental and Comparative Psychology, Max Planck Institute for Evolutionary Anthropology, Deutscher Platz 6, Leipzig, 04103, Germany

CAROLINE STOLTER Biozentrum Grindel, Department of Animal Ecology and Conservation, University of Hamburg, Martin-Luther-King Platz 3, Hamburg, D-20146, Germany

SRI SUCI UTAMI-ATMOKO Faculty of Biology, Nasional University, Jl. Sawo Manila, Pejaten, Pasar Minggu, Jakarta, Indonesia, 12520

ERIN VOGEL Department of Ecology and Evolution, SUNY at Stony Brook, Stony Brook, New York, 11794-4364, USA

SERGE A. WICH Behavioural Biology, Utrecht University, P.O. Box 80086, 3508 TB, Utrecht, The Netherlands

RICHARD W. WRANGHAM Department of Anthropology, Harvard University, Peabody Museum, 11 Divinity Avenue, Cambridge, Massachusetts, 02138, USA

JUICHI YAMAGIWA Laboratory of Human Evolution Studies, Graduate School of Natural Sciences, Kyoto University, Sakyo, Kyoto, 606-8502, Japan

Preface

GOTTFRIED HOHMANN, MARTHA M. ROBBINS, AND
CHRISTOPHE BOESCH

This book comes out of a conference on "Feeding Ecology in Apes and Other Primates," held August 17–19, 2004 at the Max Planck Institute for Evolutionary Anthropology in Leipzig. The meeting heard reports from 21 researchers. In addition to oral presentations by these senior researchers, poster sessions provided opportunities for students to present their most recent work.

Instead of publishing conference proceedings, we decided to organize a collection of integrated findings on key topics. Consequently, the chapters you will find here depart from the conference presentations in several ways. First, all chapters were greatly edited in order to include comments brought up during discussions at the meeting. Second, all chapters were reviewed by two or more external referees; modifications of the original text reflect the input of this process. Third, some authors who presented individually during the meeting agreed to prepare combined chapters for this book. Fourth, to expand the publication's scope, we invited others who were not present at the meeting to contribute.

The conference in Leipzig was supported by the Max Planck Society and by a generous donation from the Leipziger Stadtwerke. Claudia Nebel and Anja Herb took care of the logistics of the meeting, and their administrative efforts were much appreciated. Contributions to this book have benefited from the thoughtful comments made by external referees and the editors would like to thank the following people: Dean P. Anderson, Klaus Becker, Carola Borries, Colin A. Chapman, Marcus Clauss, Diane M. Doran-Sheehy, Anne Fischer, Klaus Hacklaender, Jean-Michel Hatt, Peter S. Henzi, Oliver Höhner, Heribert Hofer, Kevin D. Hunt, Lynne A. Isbell, Joanna E. Lambert, Mark Leighton, William C. McGrew, Katharine Milton, John C. Mitani, Melissa Remis, Dale A. Schoeller, Daniel Stahl, Emma D. Stokes, Karen B. Strier, Andrea B. Taylor, Elisabetha Visalberghi, Bettina Waechter, Peter Walsh, and Martin Wikelski.

We would also like to thank Tracey Sanderson and Betty Fulford from Cambridge University Press for their advice and assistance. Kevin Langergraber kindly donated the photograph on the front cover. Jeanette Hanby provided the cover photograph for chapter 14. This book would have never appeared without the combined efforts and skills of the following people: Alexander Burkhardt, Myriam Haas, Knut Finstermeier, Claudia Nebel, and Carolyn Rowney. Special thanks go to Paula Ross who edited the text, communicated with authors across the world, and coordinated the entire process of editing with great competence.

Primate feeding ecology: an integrative approach

MARTHA M. ROBBINS AND GOTTFRIED HOHMANN

No biologist would argue that ecology, "the scientific study of the interactions that determine the distribution and abundance of organisms," is not a complex topic. Feeding ecology is a central component of a species' biology, relating to its survival, reproduction, population dynamics, habitat requirements, and patterns of sociality. If one wants to best understand certain ecological processes, primates would not necessarily come to mind as the best organisms to study because of their long life-span and slow reproductive rates, the difficulties of getting sufficient sample sizes for statistical analysis, and the constraints on using wild primates for experiments. However, studying the feeding ecology of primates has been a major area of focus in primatological studies since field studies began because of the wide diversity of ecological niches occupied by primates, the heavy influence that ecology exerts on social behavior, and how it aids us in studying human evolution and behavior. Despite being a relatively small order, primates occupy a wide range of habitats and exhibit a huge diversity of grouping patterns and behavior. Studies of primate feeding ecology assist us in answering two main questions: Why do primates have the diets they do? Why do primates behave as they do? The aim of this book is to attempt to show the relationships among many of the aspects of the biology of primates to their environment.

Over the last four decades, there have been an ever-increasing number of field studies on apes and other primates that have focused on food acquisition, food processing, habitat utilization, foraging strategies, the relationship between ecology and sociality, and related topics. These studies have revealed that feeding ecology of apes and other primates is extremely diverse and complex. Perhaps more questions have been raised than answered concerning the relationship between feeding ecology and other variables such as habitat

Feeding Ecology in Apes and Other Primates. Ecological, Physical and Behavioral Aspects, ed. G. Hohmann, M. M. Robbins, and C. Boesch. Published by Cambridge University Press.

utilization and patterns of sociality. Additionally, newly developed techniques for laboratory analyses of primate foods are providing valuable tools that can be applied to answer questions on a finer scale than previously possible. To best understand the impact that feeding ecology has had on the evolution of the diversity of social systems that are observed in primates today, as well as human evolution, it is particularly useful to take an integrative approach and produce a synthetic volume that addresses these topics.

Organizational models of primate feeding ecology

To assist in orienting the reader to the topics discussed in this book, we present two frameworks in which to consider primate feeding ecology. First, we present a schematic model that shows the relationships and feedback among (A) the environment in which a primate lives, (B) the primate itself, and (C) the responses made by the primate to the environment, as constrained by the primate's morphology and physiology (Figure 1). The majority of primate ecology studies test hypotheses involving the relationship among at least two of these elements. Paradoxically, the primate itself may be the most difficult of these three elements to understand because what goes on internally in a primate is effectively a 'black box' due to the limits of invasive research on wild primates. The diet of a primate largely depends on the relationship between A and B. A primate can choose foods available in the environment and consume them within the limits of its ability to find them and process them (cognitively, manually, and digestively). A primate may in turn influence its environment through seed dispersal or how it influences the evolution of plant defences against consumption.

The relationship between A and C is the focus of the majority of primate behavioral ecology research. For example, patterns of diet, ranging, and habitat utilization are formed by food availability, within the constraints of a species' body size, digestive abilities, and abilities to detect and process foods (B). The socioecological model, which drives the majority of research on the social behavior of primates, rests on the assumption that the distribution and abundance of food resources has a strong influence on the type of relationships exhibited by female primates (Wrangham, 1980; van Schaik, 1989; Sterck *et al.*, 1997; Isbell & Young, 2002). Additionally, food availability will influence the density of primates that an area can support as well as modulate the population dynamics (e.g., Chapman *et al.*, 2004, Marshall & Leighton, Chapter 12, this volume). Conversely, the density of primates will determine the amount of food resources depleted from the environment.

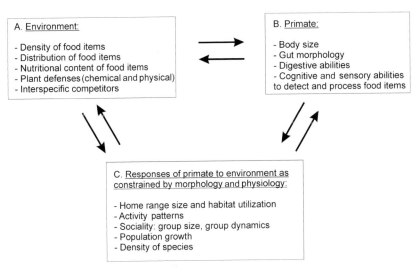

Figure 1. A schematic model of the relationships among a primate, its environment, and the responses of a primate to its environment.

Second, in an attempt to categorize complexity of the biological world, ecologists typically organize things according to a hierarchy or different levels of systems, specifically ecosystems, landscapes, communities, populations, and organisms (Goss-Custard & Sutherland, 1997; Barrett & Peles, 1999). In the case of social animals such as primates it is particularly important to insert 'social groups' in between populations and organisms. Interestingly, in the introductory chapter to the classic book on primate feeding ecology, Clutton-Brock stressed that it is the first volume to focus on social groups and not on population level processes (Clutton-Brock, 1977). Nearly 30 years later, this appears almost odd given that in recent decades the majority of primate field studies focus on social groups, because of the ecological influences on primate sociality. Primate feeding ecology is examined on the level of the organism, particularly when studying the relationship between the nutritional content and defenses of food resources and a species' food processing abilities. Populations of primates are the focus of research, particularly when addressing questions concerning what influences the density and distribution of primate populations, but such studies are often difficult to carry out because of the work involved in generating sufficient data on entire populations (longitudinally and/or cross-sectionally) and other variables such as food resources. Studying primates on the community level is most difficult because it involves studying the interactions of multiple species with their environment, but the topic has recently been the focus of a synthetic study in the book *Primate Communities* (Fleagle *et al.*, 1999).

A particularly important concept to keep in mind when studying ecology is the issue of scale (Brown *et al.*, 1995). For example, consider the difficulties of attempting to quantify the diet of a species. First there is the temporal scale. Not only is there likely to be daily and seasonal variation, but also variability between years. Second, there is spatial scale. Most primate studies are restricted to the study of only 1–2 social groups because of the logistical and financial constraints of habituating and monitoring more groups. Field observations of howler monkeys that were transferred to new habitats showed that individuals were able to adapt quickly to novel food sources (Silver & Marsh, 2003; see also Visalberghi *et al.*, 2003). These studies suggest that individuals tolerate major shifts in their overall diet composition. Meanwhile, an increasing number of studies are showing that there is remarkable variation in the diet of primate species, sometimes even among neighboring groups (Byrne *et al.*, 1993; Koenig & Borries, 2001; Chapman *et al.*, 2002; Ganas *et al.*, 2004; see also Boesch *et al.*, Chapter 7, this volume). Finally, at what level do we want to quantify the diet? According to the proportion of the diet in volume that consists of each type of plant part (e.g., fruit, seed, pith, leaves, etc.), of each species of plant consumed, time spent feeding on each resource, of each macronutrient (e.g., protein, sugars, etc.), or of total caloric intake? The scale at which a study is conducted is likely to be determined by the specific questions and hypotheses researchers are addressing, but also may be limited by the abilities of researchers to collect data (e.g., limited study period). However, a complete picture of primate feeding ecology will require looking 'upward' towards population and community level processes as well as looking 'inward' toward particular processes such as the nutritional composition of plants and digestive abilities of primates.

Understanding primate diet

The motivation of researchers to investigate the diet of a given primate species varies. Sometimes, data are obtained as a by-product of research on other topics. The value of this type of information is that it offers a bench mark against which data from other studies can be compared. For example, for a long time colobines were regarded as folivores (Davis & Oates, 1994). Equipped with a multichambered stomach, they seemed to be perfectly adapted to a diet dominated by leaves. Recent field studies made clear that colobines are seed-eaters that turn to leaves as fall-back food during times when other foods are in short supply (Koenig & Borries, 2001). In addition, to further our understanding of the intra-specific variability of dietary breadth, accounts of food lists have turned out to be a rich source of information for post-hoc comparison. Using information from a large number of field studies on great apes, Rodman (2002) analysed how the different species of African and Asian apes use the plant food sources available to them. The results of his study suggest that observed differences in food selection between the taxa reflect patterns of forest composition rather than preferences for certain subsets of plant food species. The integrative work by Rodman offers a unique source of information and has a large potential to expand in different ways. In addition, it highlights the significance of two parameters that are often neglected by primatologists, plant taxonomy and plant species diversity.

Primate diets appear to be complex in structure, diverse in content, and variable over time. Consequently, understanding a species' diet requires different perspectives. Identification of the nature of selected food items appears relatively easy but is complicated by the question of why individuals select a given food out of an almost uncountable number of potential resources. Depending on the dominant type of food, species have been classified as being frugivore, insectivore, or folivore. While this terminology has its legitimacy, it does not account for the traits that characterize diet selection by primates. Almost all primates exploit a large variety of different resources including fruit, seeds, leaves, flowers, bark, gum, insects, and meat and unlike many other omnivore vertebrates the diet of most primates combines a broad spectrum of plant foods with a narrow spectrum of animal food (Milton, 1987). There are different models explaining the intake of mixed diets (Westoby, 1978a; Freeland & Janzen, 1974a; for a review see Singer & Bernays, 2003) but so far few studies have applied the predictions to primates (e.g., Altmann, 1998). Given the wide spectrum of food items, primates are considered as being food-generalists (e.g., Harding, 1981). However, unlike other vertebrates that exploit the same sources, primates tend to process food dentally, manually, or technically (Boesch & Boesch, 1981; Taylor, 2002; Lambert *et al.*, 2003). Thus, food selected is not necessarily equal with food ingested and in spite of its potential significance in terms of nutrition, physical alteration of food items by primates has deserved surprisingly little attention.

"Hidden and consequently neglected keys to unlocking the mystery of why animals eat what they do is how food is processed after ingestion" (Levey & Martinez del Rio, 2001). While this quotation comes from a review on avian nutritional ecology, it applies equally well to primates. There is a rich literature on the content of macronutrients and anti-feedants of food items. However, what really matters in the context of nutrition is what individuals are able to extract from the ingested food. The efficiency of assimilation of nutrients depends, amongst others, on the time that food needs to pass through the digestive system (Lambert, 1998). Differences in food passage time are related to chemical and structural features of the food (Milton, 1981). From this, one may infer that information on dietary quality may provide considerable insight into the digestive strategy of the species in question. Findings from other mammals indicate a causal relationship between a species body size and diet (van Soest, 1996). Thus, in theory, data on body mass should predict dietary quality which, in turn, should allow us assign to the digestive strategy. Evidence suggests that primates violate the predictions of both models: For example, in spite of their large body size, chimpanzees consume a diet that is superior to the one eaten by sympatric Cercopithecines (Wrangham *et al.*, 1998). Another example is the unusually fast passage time of some medium-sized primates that appears to be related to the anatomy of certain parts of the digestive system rather than dietary quality (Milton, 1981; Lambert, 1998; see also Janson & Vogel, Chapter 11, this volume). While the studies cited above have advanced our understanding of the relationship between environment, food selection, and digestion, primate nutritional ecology is still in an early stage. Understanding dietary selection means to understand one of the most basic interactions between individuals and their environment: resource acquisition, and the external and internal processing of these resources. While previous studies have taught us much about how primates search, find, and compete for food, the modes of physical and physiological deployment of food sources are still largely unknown.

Understanding primate behavior

A major focus of primate field studies is the relationship between food availability, diet, movement patterns, and sociality. Three decades ago, Tim Clutton-Brock (1977) compiled a book on a number of comparative primate studies. Unlike earlier publications, the chapters of the book applied the same set of questions to a wide range of species including lemurs, monkeys, and apes. The book also contained some of the first attempts to test hypotheses that would show general trends in primate ranging and grouping patterns. For example, Clutton-Brock & Harvey (1977) showed the positive relationship between body size and home range size, which has stood the test of time and more modern statistical analysis (Nunn & Barton, 2000). Additionally, the prediction that as group size increases, the amount of food needed collectively by the group also increases, and daily travel distance and home range size should expand accordingly has been broadly supported and emphasizes the ecological constraints on group size (Clutton-Brock & Harvey, 1977; Janson & Goldsmith, 1995; Altmann, 1998; Chapman & Chapman, 2000). Various other aspects of movement patterns of primates have received considerable attention in the recent edited volume *On the Move* (Boinski & Garber, 2000). Understanding the cognitive abilities used by primates to locate food sources and to forage efficiently is a complex topic to tackle, but an increasing number of studies are elegantly addressing the issue (Boesch & Boesch, 1984; Byrne, 2000; Janson, 2000).

Interest in primate feeding ecology has been driven largely by questioning the adaptive significance of the highly variable social systems observed in primates. The major principle of the socioecological model is that the distribution, density, and quality of food resources will have an impact on the competitive interactions observed among female group members of diurnal primates, which in turn will influence the type of female social relationships exhibited (Wrangham, 1980; van Schaik, 1989; Sterck *et al.*, 1997; Koenig, 2002; Isbell & Young, 2002). In brief, the distribution and abundance of food resources will determine the strength of competitive interactions for access to food resources (e.g., if individuals exhibit scramble or contest competition), the structure of dominance relationships (e.g., despotic vs. egalitarian), the benefits derived from associating with kin (e.g., alliance formation), and female dispersal patterns. Testing the socioecological model has focused on comparing related species that vary in both ecological conditions and relevant traits (e.g., Barton *et al.*, 1996; Boinski *et al.*, 2002).

Without any doubt, the incorporation of socioecological theories into primate studies has significantly advanced our understanding of the function of primate social systems and social relationships within and between groups. However, like other theories, socioecological models are not unchallenged and several studies have identified some important constraints of current theories (e.g., Isbell & Young, 2002; Koenig & Borries, Chapter 10, this volume). For example, conventional models associate folivory with widely distributed food sources of relatively low quality and, as a consequence, predict that folivores do not gain by competing for access to resources. Field studies on common langurs in Nepal revealed that the leaves of major food sources are patchily distributed and of high nutritional quality. Consequently, female langurs gain by competing for access to these resources, leading to significant skews in food intake (Koenig *et al.*, 1998). More detailed studies of the nutritional value and distribution of food resources as well as behavioral studies of less-known species and populations of primates will provide further tests of the socioecological model, but we also need additional research that shows the link between social status, food intake, and fitness. To date, relatively few studies have shown a positive relationship between dominance rank, which is assumed to cause improved access to food resources, and female reproductive success, and even these are not able to show that reproductive success equals fitness (Pusey *et al.*, 1997; van Noordwijk & van Schaik, 1999; Altmann & Alberts, 2003).

Lastly, another link between feeding behavior and sociality is that of cultural variation in food processing (Whiten *et al.*, 1999; Fox *et al.*, 2000; van Schaik & Knott, 2001; Panger *et al.*, 2002) and even diet (Boesch *et al.*, this volume). Van Schaik *et al.* (1999) proposed a number of conditions that promote the evolution of tool use: extractive foraging, manual dexterity, intelligence, social tolerance. The complexity of cultural variations is thought to depend on opportunities for social learning, and social tolerance is seen as the key for the transmission of inventions. Data from chimpanzees, orangutans, and gorillas are consistent with this hypothesis (Byrne & Byrne, 1993; Boesch *et al.*, 1994; McGrew *et al.*, 1997; van Schaik, 2002).

Future directions

Perhaps one of the major things we have learned over the past decades about primate feeding ecology is that it is incredibly complex. We encourage researchers to take an integrative approach in the field, in the laboratory, and in the theoretician's armchair. While researchers strive to create elegant, parsimonious theoretical models to explain particular systems and drive future research, additional data from the field typically muddy the waters. However, field data are necessary to test the robustness of model assumptions and they should be used to create refinements to existing models, refute them, and/or lead to the creation of new models altogether. For example, the chapters by Koenig *et al.* and Janson and Vogel in this volume offer useful refinements to the socioecological model and, it is hoped, will stimulate additional research on other species. Comparative studies using similar methodologies, both of the same species in different habitats and different, closely related species in the same habitat, provide useful tests in the absence of experimental studies (Barton *et al.*, 1996; Sterk & Steenbeck, 1997; Pruetz & Isbell, 2000; Boinski *et al.*, 2002; Danish *et al.*, Chapter 18, this volume). Computer modeling could also be used to test predictions for which obtaining sufficient empirical data is difficult, for example examining the relationship between food availability, habitat utilization, and population dynamics.

Data collection needs to be expanded in all directions: more study sites, more long-term data, more detailed analysis of diet through nutritional analysis, etc. The key to understanding why primates eat the food they eat is hidden in the physical and chemical properties of food items on one hand, and the sensory skills and digestive strategies of primates on the other. Birds are perhaps the most important food competitors of primates and therefore, primatologists may benefit from looking at the findings from studies of avian nutritional ecology. Inter-specific differences in the ability to digest macronutrients, association patterns of secondary compounds, and differences in digestive anatomy are dimensions that have advanced the understanding of feeding ecology of birds but have been largely ignored in primate studies.

On a practical level, as an increasing number of primate species face extinction it is imperative to use research to assist with their conservation. Research should be focused to serve the dual purposes of answering questions relevant to understanding the evolution of primate feeding ecology and to conserving primates intact in their natural habitats. Studies of diet, ranging patterns, and habitat utilization are useful for understanding the habitat requirements to maintain viable populations and may also contribute to our comprehension of the population dynamics and carrying capacity of a particular area. Understanding the role that primates play in their community ecology as predators, prey, competitors, seed dispersers, etc. may assist in the conservation of entire ecosystems. As the habitats of primates shrink and become increasingly surrounded by human settlement, primates may be forced into marginal habitat and/or resort to crop raiding (Marsh, 2003). Knowledge of the dietary patterns of primates may assist in designing management strategies to reduce human–wildlife conflict. Studies of primate feeding ecology have revealed a great deal of flexibility in ecological patterns. While this knowledge of the variability may be used as an argument to conserve as many populations of a particular species as possible, conservationists also need to be aware that results from one population may not be appropriate for extrapolation to other populations of the same species. Only through concerted efforts of conservation and research will we be able to understand and appreciate the complexity of primates in their natural environments.

References

Altmann, J. & Alberts S. C. (2003). Variability in reproductive success viewed from a life-history perspective in baboons. *American Journal of Human Biology*, **15**, 401–9.

Altmann, S. A. (1998). *Foraging for Survival*. Chicago: The University of Chicago Press.

Barratt, G. W. & Peles, J. D. (1999). *Landscape Ecology of Small Mammals*. New York: Springer.

Barton, R. A., Byrne, R. W., & Whiten, A. (1996). Ecology, feeding competition and social structure in baboons. *Behavioral Ecology and Sociobiology*, **38**, 321–9.

Boinski, S. & Garber P. A. (2000). *On the Move: How and Why Animals Travel in Groups*. Chicago: The University of Chicago Press.

Boinski, S., Sughrue, K., Selvaggi, L. *et al.* (2002). An expanded test of the ecological model of primate social evolution: competitive regimes and female bonding in three species of squirrel monkeys (*Samiri oerstedii, S. boliviensis*, and *S. sciureus*). *Behaviour*, **139**, 227–61.

Boesch, C. & Boesch, H. (1981). Sex differences in the use of natural hammers by wild chimpanzees: a preliminary report. *Journal of Human Evolution*, **10**, 585–93.

    (1984). Mental map in wild chimpanzees: an analysis of hammer transports for nut cracking. *Primates*, **25**, 160–70.

Boesch, C., Marchesi, P., Marchesi, N., Fruth, B., & Joulian, F. (1994). Is nut cracking in wild chimpanzees a cultural behaviour? *Journal of Human Evolution*, **26**, 325–38.

Brown, J. H., Mehlman, D. W., & Stevens, G. C. (1995). Spatial variation in abundance. *Ecology*, **76**, 2028–43.

Byrne, R. W. (2000). How monkeys find their way: leadership, coordination, and cognitive maps of African baboons. In *On the Move: How and Why Animals Travel in Groups*, ed. S. Boinsky & P. A. Garber, pp. 491–518. Chicago: The University of Chicago Press.

Byrne, R. W. & Byrne, J. M. E. (1993). Complex leaf-gathering skills of mountain gorillas (*Gorilla g. beringei*): variability and standardization. *American Journal of Primatology*, **31**, 241–61.

Byrne, R. W., Whiten, A., Henzi, S. P., & McCulloch, F. M. (1993). Nutritional constraints on mountain baboons (*Papio ursinus*): implications for baboon socioecology. *Behavioral Ecology and Sociobiology*, **33**, 233–46.

Chapman, C. A. & Chapman, L. J. (2000). Determinants of group size in primates: the importance of travel costs. In *On the Move: How and Why Animals Travel in Groups*, ed. S. Boinsky & P. A. Garber, pp. 24–42. Chicago: The University of Chicago Press.

Chapman, C. A., Chapman, L. J., & Gillespie, T. R. (2002). Scale issues in the study of primate foraging: red colobus of Kibale National Park. *American Journal of Physical Anthropology*, **117**, 349–63.

Chapman, C. A., Chapman, L. J., Naughton-Treves, L., Lawes, M. J., & McDowell, L. R. (2004). Predicting folivorous primate abundance: validation of a nutritional model. *American Journal of Primatology*, **62**, 55–69.

Clutton-Brock, T. H. (1977). *Primate Ecology: Studies of Feeding and Ranging Behaviour in Lemurs, Monkeys, and Apes*. London: Academic Press.

Clutton-Brock, T. H. & Harvey, P. H. (1977). Primate ecology and social organization. *Journal of Zoology, London*, **183**, 1–39.

Davies, A. G. & Oates, J. F. (1994). *Colobine Monkeys: Their Ecology, Behaviour, and Evolution*. Cambridge: Cambridge University Press.

Fleagle, J. G., Janson, C., & Reed, K. E. (1999). *Primate Communities*. Cambridge: Cambridge University Press.

Fox, E. A., Sitompul, A. F., & van Schaik, C. P. (2000). Intelligent tool use in wild Sumatran orangutans. In *The Mentality of Gorillas and Orangutans*, ed S. T. Parker, L. Miles & R. Mitchell, pp. 99–116. Cambridge: Cambridge University Press.

Freeland, W. J. & Janzen, D. H. (1974). Strategies in herbivory by mammals: the role of plant secondary compounds. *The American Naturalist*, **108**, 269–89.

Ganas, J., Robbins, M. M., Nkurunungi, J. B., Kaplin, B. A., & McNeilage, A. (2004). Dietary variability of mountain gorillas in Bwindi Impenetrable National Park, Uganda. *International Journal of Primatology*, **25**, 1043–72.

Gros-Custard, J. D. & Sutherland, W. J. (1997). Individual behaviour, populations and conservation. In *Behavioural Ecology* (Fourth Edition), ed. J. R. Krebs & N. B. Davies, pp. 373–95. Oxford: Blackwell Science.

Harding, R. S. O. (1981). An order of omnivores: nonhuman primate diets in the wild. In *Omnivorous Primates: Gathering and Hunting in Human Evolution*, ed. R. S. O. Harding & G. Teleki, pp. 191–214. New York: Columbia University Press.

Isbell, L. A. & Young, T. P. (2002). Ecological models of female social relationships in primates: similarities, disparities, and some directions for future clarity. *Behaviour*, **139**, 177–202.

Janson, C. (2000). Spatial movement strategies: theory, evidence, and challenges, In *On the Move: How and Why Animals Travel in Groups*, ed. S. Boinsky & P. A. Garber, pp. 165–203. Chicago: The University of Chicago Press.

Janson, C. H. & Goldsmith, M. L. (1995). Predicting group size in primates: foraging costs and predation risks. *Behavioural Ecology*, **6**, 326–36.

Koenig, A. (2002). Competition for resources and its behavioral consequences among female primates. *International Journal of Primatology*, **23**, 759–83.

Koenig, A. & Borries, C. (2001). Socioecology of Hanuman langurs: the story of their success. *Evolutionary Anthropology*, **10**, 122–37.

Koenig, A., Beise, J., Chalise, M. K., & Ganzhorn, J. U. (1998). When females should contest for food – testing hypotheses about resource density, distribution, size, and quality with Hanuman langurs (*Presbytis entellus*). *Behavioural Ecology and Sociobiology*, **42**, 225–37.

Lambert, J. E. (1998). Primate digestion: interactions among anatomy, physiology, and feeding ecology. *Evolutionary Anthropology*, **7**, 8–20.

Lambert, J. E., Chapman, C. A., Wrangham, R. W., & Conklin-Brittain, N. L. (2003). Hardness of cercopithecine foods: implications for the critical function of enamel thickness in exploiting fallback foods. *American Journal of Physical Anthropology*, **126**, 363–8.

Levey, D. J. & Martinez del Rio, C. (2001). It takes guts (and more) to eat fruit: lessons from avian nutritional ecology. *The Auk*, **118**, 819–31.

Marsh, L. K. (2003). *Primates in Fragments: Ecology in Conservation*. New York: Kluwer Academic Press/Plenum Publishers.

McGrew, W. C., Ham, R. M., White, L. T. J., Tutin, C. E. G., & Fernandez, M. (1997). Why don't chimpanzees in Gabon crack nuts? *International Journal of Primatology*, **18**, 353–74.

Milton, K. (1981). Food choice and digestive strategies of two sympatric primate species. *The American Naturalist*, **117**, 476–95.

Milton, K. (1987). Primate diets and gut morphology: implications for hominid evolution. In *Food and Evolution: Toward a Theory of Human Food Habits*, ed. M. Harris & E. B. Ross, pp. 93–116. Philadelphia: Temple University Press.

Nunn, C. L. & Barton, R. A. (2000). Allometric slopes and interdependent contrasts: a comparative test of Kleiber's law in primate ranging patterns. *American Journal of Physical Anthropology*, **30** (suppl.), 239.

Panger, M. A., Perry, S., Rose, L. *et al.* (2002). Cross-site differences in foraging behavior of white-faced capuchins. *American Journal of Physical Anthropology*, **119**, 52–66.

Pruetz, J. D. & Isbell, L. A. (2000). Correlations of food distribution and patch size with agonistic interactions in female vervets (*Chlorocebus aethiops*) and patas monkeys (*Erythrocebus patas*) living in simple habitats. *Behavioral Ecology and Sociobiology*, **49**, 38–47.

Pusey, A., Williams, J., & Goodall, J. (1997). The influence of dominance rank on the reproductive success of female chimpanzees. *Science*, **277**, 828–31.

Rodman, P. S. (2002). Plants of the apes: is there a hominoid model for the origins of the hominoid diet? In *Human Diet: Its Origin and Function*, ed. P. S. Ungar & M. F. Teaford, pp. 77–109. Westpoint: Bergin & Harvey.

Singer, M. S. & Bernays, E. A. (2003). Understanding omnivory needs a behavioral perspective. *Ecology*, **84**, 2532–7.

Silver, S. C. & Marsh, L. K. (2003). Dietary flexibility, behavioral plasticity, and survival in fragments: lessons from translocated howlers, In *Primates in Fragments: Ecology in Conservation*, ed. L. K. Marsh. New York: Kluwer Academic Press/Plenum Publishers.

Sterck, E. H. M. & Steenbeck, R. (1997). Female dominance relations and food competition in the sympatric Thomas langur and longtailed macaques. *Behaviour*, **134**, 749–74.

Sterck, E. H. M., Watts, D. P., & van Schaik, C. P. (1997). The evolution of female social relationships in nonhuman primates. *Behavioral Ecology and Sociobiology*, **41**, 291–309.

Taylor, A. B. (2002). Masticatory form and function in the African apes. *American Journal of Physical Anthropology*, **117**, 133–56.

van Noordwijk, M. A. & van Schaik, C. P. (1999). The effects of dominance rank and group size on female lifetime reproductive success in wild long-tailed macaques, *Macaca fascicularis*. *Primates*, **40**, 105–30.

van Schaik, C. P. (1989). The ecology of social relationships amongst female primates. In *Comparative Socioecology*, ed. V. Standen & R. A. Foley, pp. 195–218. Oxford: Blackwell Scientific.

(2002) Fragile traditions: the disturbance hypothesis for the loss of local traditions in orangutans. *International Journal of Primatology*, **23**, 527–38.

van Schaik, C. P. & Knott, C. D. (2001). Geographic variation in tool use on Neesia fruits in orangutans. *American Journal of Physical Anthropology*, **114**, 331–42.

van Schaik, C. P., Deaner, R. O., & Merrill, M. Y. (1999). The conditions for tool use in primates: implications for the evolution of material culture. *Journal of Human Evolution*, **36**, 719–41.

van Soest, P. J. (1996). Allometry and ecology of feeding behavior and digestive capacity in herbivores: a review. *Zoo Biology*, **15**, 455–79.

Visalberghi, E., Sabbatini, G., Stammati, M., & Addessi, E. (2003). Preferences towards novel foods in *Cebus apella*: the role of nutrients and social influences. *Physiology and Behavior*, **80**, 341–9.

Westoby, M. (1978). What are the biological bases for varied diets? *American Naturalist*, **112**, 627–31.

Whiten, A., Goodall, J., McGrew, W. C. *et al.* (1999). Cultures in chimpanzees. *Nature*, **399**, 682–5.

Wrangham, R. W. (1980). An ecological model of female-bonded primate groups. *Behaviour*, **75**, 262–300.

Wrangham, R. W., Conklin-Brittain, N. L., & Hunt, K. D. (1998). Dietary response of chimpanzees and Cercopithecines to seasonal variation in fruit abundance. I. Antifeedants. *International Journal of Primatology*, **19**, 949–70.

Part I
Field studies

Introduction

PETER S. RODMAN

As an undergraduate student in 1966, I met Jane Goodall and Hugo van Lawick at a party at the apartment of my advisor, Irven DeVore. Because of Irv's generous hospitality, undergraduates and graduate students interested in primates and the !Kung San were rocking into the evening in honor of the visitors, who observed quietly. My conversation with Jane was one-sided because she was quiet and reserved, while I was wound up with the excitement of the event. I asked bluntly about her view of the criticism that her interpretations of chimpanzee behavior were too anthropomorphic and not real science. She was gracious as well as reserved and did not say much in response. At that point, I launched into a dissertation on the advantages of an anthropomorphic view of chimpanzees, because this was a useful way to interpret some of their behavior. We are, after all, quite a bit like them. After that, no doubt Jane and Baron van Lawick retired for a good rest, leaving a wild group of aspiring evolutionary anthropologists to keep the DeVores up far too late.

I mention this for two reasons. Goodall was the first to initiate a truly long-term field study of any non-human primate. Her years of observations of chimpanzees demonstrated the value of knowing individuals well, of knowing their genealogical relationships, and of seeing the development of lives longitudinally rather than simply seeing development in cross section. Her field work still contributes to important new knowledge about chimpanzees (Pusey *et al.*, 1997; Hill *et al.*, 2001) more than 40 years after she started.

Second, anthropomorphism was a pejorative term among biologists then, as it still is to some extent. It suggests that the anthropomorphist not only studies fuzzy animals, but also indulges fuzzy thoughts about them. Since 1966, chimpanzees have proven to be very closely related to us, not simply to be our closest living relatives. Shared characters of behavior must be present, and perhaps the best way to understand them is to allow anthropomorphism to guide hypotheses. Today even material and other cultural traits have been found in chimpanzees, and looking for these traits alone is a form of

Feeding Ecology in Apes and Other Primates. Ecological, Physical and Behavioral Aspects, ed. G. Hohmann, M.M. Robbins, and C. Boesch. Published by Cambridge University Press. © Cambridge University Press 2006.

anthropomorphism that may be quite productive if tempered thoroughly by rigorous methods of analysis.

The beginning of Goodall's study in 1959 was the dawn of a generation of field studies of animal behavior, of primate behavior, and particularly of behavior of the apes. There is normally little need to choose the most difficult location for studies of behavior – better to find a place and an animal to study that minimizes the costs of doing the work. Successful field studies of the great apes require heroic commitment, however. Gorillas, chimpanzees, and orangutans occupy remote forests that tend to be difficult to reach. Having found habitat occupied by one of these animals, the fieldworker may spend years attempting to habituate the subjects adequately to allow follows of more than a few minutes, perhaps accumulating only a few contacts with the wary animals each month. With luck, habituation evolves, and the fieldworker may be rewarded with the opportunity to follow individuals or groups more or less at will. Each grueling 12-hour day ends with supper of beans and rice or something similar, followed by a night in the tent. After cold rice for breakfast before sunrise, the next day is the same . . . and the next. Substantive results from such field research require not months, but years. There is seldom a "field season" corresponding to the calendar of life at home, like the temperate summer for paleontologists, archeologists, and other collectors, nor can the apes relate information about times in the past or about events that have transpired while the fieldworker has been away, as would be the case for studies of humans. In the midst of the labor, there are rewards as reported in the following chapters, but rewards come slowly for a long time after initiating a new study.

Attempts at field studies of apes date back to the 1920s (Bingham, 1932), but we are indebted to Schaller (1963), Goodall (1968), and MacKinnon (1974), for their early work that demonstrated such field research could be fruitful. Goodall (1986), Nishida (1969), Fossey (1974, 1983), Galdikas (1985, 1995), and their students and collaborators persisted through long-term studies of various apes at single sites. They showed examples of the fortitude necessary and provided basic information leading to complex and detailed studies of feeding ecology presented in the following chapters.

Authors of chapters in this section are part of a new generation of field workers. Most have initiated studies in remote locations with all the costs to time, health, and normal life. For some there have been important improvements associated with established research stations and the cooperation of government agencies in research. All provide detailed quantitative information on diets of African apes (or, in a departure from the apes, on capuchin monkeys of Central America; Perry & Jiménez, Chapter 8, this volume). The detail of their dietary analyses reflects the rising standards for rigor in studies of feeding ecology in two ways: first is the critical value of carefully identifying plant foods, which is a daunting task in forests of great diversity; second is the critical need for highly detailed information on feeding behavior and characteristics of plants to tease out subtle factors such as learned differences in patterns of behavior from ecologically determined variation.

Much has changed since the dawn of modern studies of African apes, and the changes may be addressed in two categories. The first is the study of the "new" apes, meaning lowland gorillas and bonobos. Most information on gorillas before the mid-1980s came from the long-term study of mountain gorillas of East Africa. They live at high altitude in a forest nearly devoid of edible fruit. Not surprisingly, mountain gorillas subsist on vegetation without fruit, feeding almost entirely on vegetative parts of plants (Watts, 1980). That they can subsist entirely on vegetative parts is extremely important for understanding gorilla ecology, but recently, studies of eastern and western lowland gorillas have revealed that gorillas feed on many different fruits. They may have greater diversity of fruits in the diet than sympatric chimpanzees, which are ripe fruit specialists, and the overlap in fruit species used is very high during fruiting seasons. The frugivory of gorillas has intrigued most investigators (e.g., Tutin *et al.*, 1991), and several chapters that follow give ample evidence of that interest.

Pruetz (Chapter 6, this volume) has been studying chimpanzees at Fongoli, Senegal, since April 2001. These chimpanzees live in one of a few dry sites where chimpanzees have been studied. Her results, based on direct observations of feeding and indirect assessment of diet by fecal analysis, confirm that chimpanzees remain fruit specialists even in what might be called marginal habitat at the margins of their geographical range. The apparently obligate reliance on succulent fruit by chimpanzees is important for understanding relationships of sympatric gorillas and chimpanzees.

How important fruit is for gorillas, how frugivory affects other variables of behavior and ecology, and how two large, frugivorous apes coexist are lingering questions. Robbins *et al.* (Chapter 1, this volume) provide results of continuous monitoring of one group of *Gorilla beringei*, the eastern gorilla, from September 1, 1998 to August 31, 2004, and discuss comparisons of the results with reported diets of other gorillas. They discuss the importance of fruit in the gorillas' diet, and report, among other things, that gorillas at Bwindi fed on fruit in 11% of observations of feeding.

Two chapters here recount studies of diets of sympatric chimpanzees and gorillas. Morgan and Sanz (Chapter 4, this volume) and Yamagiwa and Basabose (Chapter 3, this volume) used extensive fecal analyses to determine diets and dietary overlap of gorillas and chimpanzees in the Republic of Congo (*G. g. gorilla, Pan t. troglodytes*) and the Democratic Republic of the

Congo (*G. g. grauri*, *P. t. schweinfurthii*), respectively. Fecal analysis has advantageous consistency; all authors in this volume who have analyzed fecal samples have used the same procedure. Presence and absence of fruit parts and leaf parts are repeatable measures, and different studies in similar habitats give consistent results for chimpanzees and for gorillas, and for differences between the two apes. Diversity of fruits eaten should be accurately reflected in feces, although diversity of leaves should be more difficult to determine because of the effects of digestion. The relationship between actual time feeding on leaves or fruit and various measures in fecal analyses is somewhat elusive, but the elusive nature of the subjects and the time and heroic effort necessary to habituate them greatly increase the value of fecal analyses for comparisons.

The split of catarrhine primates into the two living subfamilies, Cercopithecoidea and Hominoidea sometime before the Upper Miocene was a major event in the history of the order and in the history of human origins (Temerin & Cant, 1983). Very few field studies have addressed this divergence by comparing ecology of old world monkeys and apes. In Chapter 2, Doran-Sheehy *et al.* compare the diet of western gorillas (*G. g. gorilla*) with two species of mangabeys (*Lophocebus albigena* and *Cercocebus agilis*) and discuss the detailed results in light of questions about the adaptive changes leading to the cercopithecoid–hominoid split. The chapter also provides more dietary information about frugivory in gorillas and important new results of direct observations of one adult male gorilla. This individual's diet suggests fecal analysis, while admirably consistent, may have over-estimated fruit consumption by western gorillas.

Nishida (1972) published the earliest report of a field study of the bonobo (*Pan paniscus*). Since then, there have been several long field studies of bonobos with gradually emerging understanding of their ecology and social behavior. Common chimpanzees (*P. troglodytes*) were well-known because of the work of Goodall and Nishida in the 1960s, and it was striking that early observations of bonobos revealed a number of characteristics that differed from their better known congener (Hohmann & Fruth, 2002). Females formed intense bonds incorporating homosexual behavior, and females largely dominated males, in contrast to common chimpanzees in which males, not females, have strong intrasexual bonds without homosexual behavior and females tend to be solitary and somewhat hostile to other females. Bonobos remained elusive, even during long studies, and questions persisted about the degree of difference from common chimpanzees. Hohmann *et al.* (Chapter 5, this volume) present results of analysis of a number of variables of dietary items for bonobos and chimpanzees, treating the data with elegant statistics to test several existing hypotheses about the difference in gregariousness between chimpanzees, which are not so gregarious, and bonobos, which are quite gregarious.

The African apes are our closest relatives, and much interest in them has to do with the desire to reconstruct the evolutionary history of humans. This anthropological approach is not necessary, of course, but even those trained as general animal behaviorists address questions about humans with reference to the behavior of apes. Evolved biological aspects of behavior are not the only avenues of interest in comparative studies of apes and humans, however. Recently, there has been important attention to cultural variation among populations of chimpanzees (Matsuzawa & Yamakoshi, 1996; Whiten *et al.*, 1999; Nakamura, 2002) and orangutans (van Schaik *et al.*, 2003). There are numerous small differences in the way chimpanzees and orangutans of different populations do the same things differently, which is one way Kluckhohn (1949) characterized cultural behavior, and the catalog of these learned differences is growing.

In Chapter 7, Boesch and his co-authors examine foraging by three neighboring communities of chimpanzees in the Taï National Park of Côte d'Ivoire, West Africa. They analyze very detailed observations of variation in abundance and synchrony of fruits, and their method is an interesting lesson in how to handle comparisons of feeding by groups taking into account different absolute abundances of foods in their respective ranges. This is another sophisticated statistical analysis that produces elegant measures extracted from data on food preferences. Results lead to the inference that there are subtle differences among the groups in their responses to increased relative abundance of foods and increased synchrony of foods.

Capuchin monkeys are the most encephalized of monkeys. Their index of cranial capacity is the largest among living non-human primates and approximately the same as for *Australopithecus africanus* (Martin, 1990). If encephalization indicates learning capacity, capuchins should be good models for learning abilities of human ancestors. Consequently, they fit well with the apes that are the subjects of the rest of the papers in this section. Primates might learn in several ways, and extended immaturity in the primates often is attributed to the advantages of an extended period of learning. One mode of learning is by observation, so we would expect that individuals would observe what is important to survival. In the last chapter of this section (Chapter 8) Perry and Jiménez report a detailed study of "food interest" (close observation of food and food processing) by white-faced capuchin monkeys (*Cebus capucinus*). They note that a capuchin has to learn how to identify foods, to catch or extract prey items, and to process foods that are protected by mechanical defenses, and their study aimed to determine whether capuchins might gain specific knowledge about foods and food processing techniques through socially biased learning ("socially biased learning" here means by close observation of another monkey that is feeding and of its food or its handling of the food).

Capuchins raise an important caution with regard to the anthropological significance of cultural elements of behavior in apes. They behave as though they will gain useful information by close observation (Perry & Jimenez, Chapter 8, this volume). Other studies of capuchin monkeys have already revealed interpopulational variation in processing the same foods, as noted by Boesch *et al.* (Chapter 7, this volume) citing the work of Panger *et al.* (2002).

The assertion that cultural differences among orangutans should push the origins of culture back to 14 million years (because of cultural variation in orangutans and their cladistic relationship to African apes and humans; van Schaik *et al.*, 2003) is dramatic, and it may be an overstatement of the significance of cultural variation in apes. It seems very likely that many primate and non-primate species, like capuchins, will reveal inter-popula-tional differences in learned behaviors when they are examined as closely and as long as chimpanzees and orangutans have been. If so, the origin of culture is much older than 14 million years. By the same logic as that above, the origin of culture should already be pushed back to 35 million years or more given the cladistic relationship of capuchin monkeys to humans, and if studies of other mammals reveal similar variations in cultural behavior, we may push the origin of culture back to the Mesozoic.

On the other hand, dramatic assertions about cultural behavior in primates will challenge others to undertake more intense study of learned behaviors in other animals, just as field studies of the primates have stimulated and led the way to field studies of other animals since 1959. The remarkable knowledge of natural behavior and ecology we now have, and are gaining because of work such as in this volume, makes the Order Primates an exceptional taxon within which to test models of evolution of diets and the relationship of diets to other socioecological variables.

References

Bingham, H. C. (1932). Gorillas in native habitat. *Carnegie Institute of Washington Publications*, **426**, 1–66.

Fossey, D. (1974). Observations on the home range of 1 group of mountain gorillas, *Gorilla gorilla beringei*. *Animal Behaviour*, **22**, 568–81.

(1983). *Gorillas in the Mist*. Boston: Houghton Mifflin.

Galdikas, B. M. F. (1985). Adult male sociality and reproductive tactics among orangutans at Tanjung-puting. *Folia Primatologica*, **45**, 9–24.

Galdikas, B. M. F. (1995). *Reflections of Eden: My Year with the Orangutans of Borneo*. Boston: Little Brown.

Goodall, J. (1968). Behaviour of free-living chimpanzees of the Gombe Stream area. *Animal Behaviour Monographs*, **1**, 163–311.

    (1986). *The Chimpanzees of Gombe: Patterns of Behavior*. Cambridge: Harvard University Press.

Hill, K., Boesch, C., Goodall, J. *et al.* (2001). Mortality rates among wild chimpanzees. *Journal of Human Evolution*, **40**, 437–50.

Hohmann, G. & Fruth, B. (2002). Dynamics in social organization of bonobos (*Pan paniscus*). In *Behavioural Diversity in Chimpanzees and Bonobos*, ed. C. Boesch, G. Hohmann, & L. F. Marchant, pp. 138–55. Cambridge: Cambridge University Press.

Kluckhohn, C. (1949). *Mirror for Man*. New York: Whittlesey House.

MacKinnon, J. R. (1974). *In Search of the Red Ape*. London: Collins.

Martin, R. D. (1990). *Primate Origins and Evolution*. Princeton: Princeton University Press.

Matsuzawa, T. & Yamakoshi, G. (1996). Comparison of chimpanzee material culture between Bossou and Nimba, West Africa. In *Reaching into Thought: The Minds of the Great Apes*, ed. A. E. Russon, K. A. Bard & S. T. Parker, pp. 211–34. Cambridge: Cambridge University Press.

Nakamura, M. (2002). Grooming-hand-clasp in Mahale M-group chimpanzees: implications for culture in social behaviours. In *Behavioural Diversity in Chimpanzees and Bonobos*, ed. C. Boesch, G. Hohmann, & L. F. Marchant, pp. 71–83. Cambridge: Cambridge University Press.

Nishida, T. (1969). The social group of wild chimpanzees in the Mahali Mountains. *Primates*, **9**, 167–224.

    (1972). Preliminary information of the pygmy chimpanzees, *Pan paniscus*, of the Congo Basin. *Primates*, **13**, 415–25.

Panger, M. A., Perry, S., Rose, L. *et al.* (2002). Cross-site differences in foraging behavior of white-faced capuchins (*Cebus capucinus*). *American Journal of Physical Anthropology*, **119**, 52–66.

Pusey, A., Williams, J., & Goodall, J. (1997). The influence of dominance rank on the reproductive success of female chimpanzees. *Science*, **277**, 828–31.

Schaller, G. B. (1963). *The Mountain Gorilla: Ecology and Behavior*. Chicago: University of Chicago Press.

Temerin, L. A. & Cant, J. G. H. (1983). The evolutionary divergence of old world monkeys and apes. *The American Naturalist*, **122**, 335–51.

Tutin, C. E. G., Fernandez, M., Rogers, M. E., Williamson, E. A., & McGrew, W. C. (1991). Foraging profiles of sympatric lowland gorillas and chimpanzees in the Lope Reserve, Gabon. *Philosophical Transactions of the Royal Society of London, B, Biological Sciences*, **334**, 179–86.

van Schaik, C. P., Ancrenaz, M., Borgen, G. *et al.* (2003). Orangutan cultures and the evolution of material culture. *Science*, **299**, 102–5.

Watts, D. (1980). Feeding ecology of mountain gorillas. *American Journal of Physical Anthropology*, **52**, 291.

Whiten, A., Goodall, J., McGrew, W. C. *et al.* (1999). Cultures in chimpanzees. *Nature*, **399**, 682–5.

1 Variability of the feeding ecology of eastern gorillas

MARTHA M. ROBBINS, JOHN BOSCO NKURUNUNGI, AND ALASTAIR MCNEILAGE

Introduction

Ecological conditions are predicted to have an impact on a multitude of behaviors and life history characteristics exhibited by a species, including diet, ranging patterns, reproduction, and social system. A useful way to test current models of socioecology is to make intraspecific comparisons of populations occurring in different ecological conditions (Barton *et al.*, 1996;

Feeding Ecology in Apes and Other Primates. Ecological, Physical and Behavioral Aspects, ed. G. Hohmann, M. M. Robbins, and C. Boesch. Published by Cambridge University Press. © Cambridge University Press 2006.

Sterck *et al.*, 1997; Koenig *et al.*, 1998; Boinski *et al.*, 2002). Gorillas are an appealing species for within-species comparative studies because they occur in widely varying habitats across Africa. Knowledge of the variability of feeding ecology in gorillas, particularly in areas where they are sympatric with chimpanzees, also contributes to understanding the evolution of the divergent ecological paths taken by African apes (see Yamagiwa & Basabose, Chapter 3; Morgan & Sanz, Chapter 4, this volume). A necessary first step to accurately characterize these variables is to gather sufficient data on the diet and ranging patterns of gorillas that live in different ecological circumstances. Second, long-term studies are important for understanding how temporal ecological variation affects a population. Recent short-term studies have shown that Bwindi mountain gorillas (*Gorilla beringei beringei*) have ecological patterns distinctive from other populations of gorillas (Goldsmith, 2003; Robbins & McNeilage, 2003; Ganas *et al.*, 2004; Nkurunungi, 2004; Ganas & Robbins, in press). The goal of this chapter is to use six years of data from one group of Bwindi mountain gorillas to address questions concerning long-term patterns of diet and ranging behavior, adding to our understanding of within and between population variability in the behavioral ecology of gorillas. Because the density of gorillas that an area can support is likely to be influenced by food availability and habitat utilization, we also discuss the ecological differences between Bwindi and other gorilla populations and the implications those differences carry for various aspects of their social structure and population dynamics, with particular focus on comparisons between the Virunga Volcanoes and Bwindi.

   Based on the pioneering studies conducted at the Karisoke Research Center in the Virunga Volcanoes (Fossey & Harcourt, 1977; Vedder, 1984; Watts, 1984), gorillas have traditionally been viewed as strict herbivores; however, their diet is limited because of low species diversity and the fact that almost no fruit suitable for consumption by gorillas is found at their high altitude (>2500 m) environment. In recent years we have seen an explosion of studies conducted at other field sites across Africa, and despite the difficulties in habituating gorillas, our view of their feeding ecology has changed greatly as a result of these studies. There is now little question that when available, fruit plays an important role in the gorilla diet, even though they rely heavily on fibrous foods (leaves, stems, and pith of herbs and other plants) as both staple (eaten on a daily/weekly basis throughout the year) and fallback (always available, but eaten only or mainly during fruit-scarce months) foods (Doran *et al.*, 2002; Ganas *et al.*, 2004; Rogers *et al.*, 2004; Yamagiwa *et al.*, unpublished data; Doran-Sheehy *et al.*, Chapter 2, this volume; Yamagiwa & Basabose, Chapter 3, this volume).

Table 1.1. *Percentages of areas covered by major altitudinal zones of the Virunga Volcanoes and Bwindi Impenetrable National Park*

	Virunga Volcanoes	Bwindi Impenetrable
Altitude	2227–4507 m	1160–2607 m
Mid-altitude forest (<1500 m)	–	15%
Lower montane forest (1500–2500 m)	–	33%
Upper montane/mixed forest (2000–2500 m)	37%	50%
Bamboo	26%	2%
Brush ridge/Hagenia/herbaceous (2500–3300 m)	25%	–
Alpine/subalpine (<3300 m)	12%	–

Modified from Sarmiento *et al.* (1996) and McNeilage (2001). Comparable data not available for comparison with Kahuzi-Biega.

Environmental conditions are apt to change significantly with altitude. Because eastern gorillas (*Gorilla beringei*) occupy a large altitudinal range (~500 to 3700 m) in Rwanda, Uganda, and Democratic Republic of Congo (DROC), the extensive variation in habitat types and food availability (Yamagiwa *et al.*, 1992, 1994, 1996; McNeilage, 2001; Ganas *et al.*, 2004; Nkurunungi *et al.*, 2004; Table 1.1) is not surprising. It is now apparent that findings from the well-studied Karisoke mountain gorillas, who live in one extreme of gorilla habitats, are not representative of all other eastern gorilla populations. Not only are there large differences between the diet of gorillas found in the Virunga Volcanoes, Bwindi Impenetrable National Park, and Kahuzi-Biega, but there are also notable differences in the diet of groups living relatively close to each other in each of these locations (see Ganas *et al.*, 2004 for summary). For example, gorillas at two sites in Bwindi that are located only about 15 km apart and with no altitudinal overlap, had less than 50% of the important foods in their diet (fibrous foods consumed > 5% of days; fruits consumed > 1% of days) in common (Ganas *et al.*, 2004). Furthermore these groups had remarkably few food species in common with gorillas located in the Virunga Volcanoes or Kahuzi-Biega. This study emphasized that Bwindi gorillas are intermediate in their frugivory patterns between the gorillas of the Virunga Volcanoes, who consume almost no fruit, and the more frugivorous Grauer's gorillas and western gorillas. However, previous research has shown that fruiting patterns vary significantly from year to year in Bwindi (Robbins & McNeilage, 2003), so long-term data on frugivory will assist us in understanding the importance of fruit in their diet by addressing the following questions: How variable are their frugivory patterns between years and is their fruit consumption seasonal? Is there a relationship between fruit and fibrous food consumption? Specifically, is there a decline in fibrous food consumption as fruit consumption increases?

Broad scale differences in diet, in particular the consumption of fibrous foods versus fruit, are predicted to lead to variability in ranging patterns (Clutton-Brock & Harvey, 1977; Janson & Goldsmith, 1995). A comparison of ranging patterns of gorillas across many locations has supported the prediction that as the degree of frugivory increases day journey length increases, presumably because of the greater degree of dispersion of high-quality fruit, with the Karisoke mountain gorillas having the shortest day journey length, western gorillas having the longest, and Bwindi gorillas being intermediate (Doran & McNeilage, 2001; Cipolletta, 2004; Doran-Sheehy *et al.*, 2004; Ganas & Robbins, in press). Home range size is also predicted to increase as fruit consumption increases. Karisoke mountain gorillas have the smallest annual home range size recorded, with larger values having been observed among Bwindi mountain gorillas, Grauer's gorillas, and western gorillas (Robbins & McNeilage, 2003; Bermejo, 2004; Cipolletta, 2004; Doran-Sheehy *et al.*, 2004; Yamagiwa & Basabose, Chapter 3, this volume; Ganas & Robbins, in press). Differences in the availability not only of fruit but of fibrous foods/terrestrial herbaceous vegetation (THV), should also influence ranging patterns (Watts, 1991; McNeilage, 1995, 2001). However, these observations of adjustments to ranging patterns when eating fruit emphasize that gorillas actively seek it out and should be considered fruit "pursuers" and not fruit opportunists (Rogers *et al.*, 2004).

If fruit is an important component in the diet of a species, the need to regularly monitor fruit trees in the habitat could lead to home ranges that are more stable than those of primates who feed predominantly on abundant, commonly available herbaceous vegetation (Milton, 1988). Watts (1998a) found that groups in the Virunga Volcanoes continued to add new areas to their range for several consecutive years and that the overlap of home ranges and core areas (area of the home range used most intensively) between years was generally low. This may allow the vegetation to regenerate in heavily used areas (Watts, 1998b) and it suggests that it may take several years to see the complete home range of a group. However, other variables such as mating competition may cause abrupt, large range shifts (Watts, 1998a, 2000; see also Cipolletta, 2004). The size of the core area may also be a larger proportion of the home range for frugivorous gorillas than for the gorillas in the Virunga Volcanoes if fruit resources are more dispersed and/or if THV density is lower. However, one would expect interannual core area overlap to be high if the gorillas need to regularly monitor all the fruit trees in these areas. Alternatively, if the species of trees that produce fruit vary greatly between years and these species are very spatially clumped in the habitat, core areas may shift between years based on which species produce fruit. Core area overlap would also be influenced by the regeneration time of the major herb species consumed. Long-term studies of the ranging behavior of frugivorous gorillas have yet to be conducted. Therefore, we ask how the consistency of the ranging patterns of Bwindi gorillas in the long term (home range size and area, core area stability) compares with results from the more herbivorous gorillas of the Virunga Volcanoes (Watts, 1998a) and other gorilla populations.

Methods

Study area and study group

Bwindi Impenetrable National Park is located in southwestern Uganda ($0°53'-1°08'N$ and $29°35'-29°50'E$). It covers approximately 331 km^2, characterized by steep hills and narrow valleys throughout. Continuous forest vegetation extends from 1160 m to 2607 m elevation, encompassing a rare ecological continuum of medium altitude and afromontane forest that make it a unique environment with high biodiversity (Butynski, 1984). The major vegetation zones include open forest, with a non-continuous canopy and dense herbaceous ground cover of herbs and vines, and mixed forest, with both trees and dense herbaceous understory, and mature forest, with a continuous canopy and little undergrowth (Nkurunungi *et al.*, 2004). Bwindi has two wet seasons (September–November and March–May) and two dry seasons (December–February and June–August), with an annual average rainfall between 1998 and 2002 of ~1300 mm.

The study group, Kyagurilo Group, ranges in the northeastern corner of the park at an altitude between 2100 and 2500 m. The group has been monitored since the 1980s by staff of the Institute of Tropical Forest Conservation (ITFC). Data included here were collected from September 1, 1998 to August 31, 2004. There was no monthly or seasonal bias in data collection (the only large gap was a 40-day halt in data collection in March and April 1999 due to a rebel invasion in the western region of the park). During this 6-year period the group has varied in size from 12 to 15 individuals, including three silverbacks (adult males), five to six adult females, and four to seven immature individuals ranging in age from 0–10 years of age (<12-year-old males, <8-year-old females). Demographic changes include the emigration of a young silverback in December 1999, the immigration of a young female in June 2000, three births (1999–2002), and the emigration of the older, previously dominant silverback in July 2004. None of the demographic changes are likely to have a significant impact on results presented here.

Data collection

The Kyagurilo Group has been habituated to human presence since the mid-1990s and was followed by field assistants on a nearly daily basis. Observations were restricted to approximately 4 hours per day (requirement of the Uganda Wildlife Authority), and while data from 7 am–6 pm were included, they were biased towards hours in the middle of the day (9 am–3 pm). Observations were not biased towards any particular age/sex class or individual. The group spent approximately 50% of observation time feeding (Robbins, unpublished data). During observations, field assistants recorded specific names of all plants eaten and the parts consumed (fruits, stems, leaves, pith, etc.). This method provides a basic measure of presence/absence of food species in the diet on a daily basis, probably underestimates the number of plants eaten in a day, and is biased towards excluding rarely eaten foods. It also does not give either an estimate of the time spent foraging nor the biomass of each food item.

We used the term "fibrous foods" to describe non-reproductive plant parts from herbs, shrubs, and trees. To facilitate direct comparisons of our results with those of other studies, we used the same definitions of important foods as previous gorilla diet studies. We defined "important" fibrous food species as those occurring on >5% of daily food trails or daily observations (following Doran *et al.*, 2002; Ganas *et al.*, 2004). Important fruit species were defined as those occurring in more than 1% of samples per group (following Remis, 1997; Ganas *et al.*, 2004).

To determine the location of the group, a GPS reading was taken at gorilla nest sites and/or upon first contact with gorillas, but only one data point per day, with preference given to nest site location, was used for the home range analysis (Robbins & McNeilage, 2003).

Diet analysis

For the analysis of frugivory patterns we used 6 years of data, totaling 1921 days of feeding observation (n = 6 years; $\bar{x} = 320$; range 248–360 days per year). We restricted our analysis of fibrous foods consumed to the last 3 years of the study (September 2001–August 2004), which include 1060 days of observation (n = 3 years; $\bar{x} = 353$; range = 337–360 days per year) because this was the point at which we were confident that the assistants' reports were accurate for all feeding observations. We used these data to calculate the proportion of days that each species was consumed on a monthly basis.

To test for seasonality of fruit eating and consumption of important herb species we performed analysis of variance (ANOVA). Using monthly consumption values for fruit (all species combined) and each important fibrous food species, linear regressions were performed to examine whether or not there was a relationship between fruit eating and consumption of each important fibrous food species.

Home range and core area analysis

GPS readings were entered into Arcview GIS software to calculate home range size in two ways, using the minimum convex polygon method (MCP; Southwood, 1966) and the grid cell method (500 m × 500 m grid cells are superimposed onto a map of the home range). Both methods have their advantages and disadvantages. The grid square method is highly sensitive to the size of the grid square used and the number of data points used. The minimum convex polygon method (MCP) eliminates the problem of grid squares within the range that are not entered and is more accurate when the number of data points is low, but peripheral data points may strongly influence home range size. Our use of the grid square method allows for direct comparisons with other studies of gorillas because this is the most commonly used method. Core areas were defined as the area where the gorillas spent 75% of their time. It was calculated by selecting grid squares that contained the highest frequency of group location points until 75% of the points were included (following Watts, 1998a; Robbins & McNeilage, 2003). Home range and core area were calculated on an annual basis and for the entire 6-year period. A total of 1851 daily data points were used ($\bar{x} = 309$ per year; range = 205–365).

Home range overlap and core area overlap between all pairs of years were calculated to examine site fidelity over time. Overlap was calculated as the percent of area used commonly between two time periods divided by the total area used during the two time periods. Home range overlap was calculated using both the MCP and 500-m grid square estimates, but due to our method of estimating core area (Watts, 1998a), core area overlap was calculated using only the 500-m grid square estimate.

Results

Patterns of frugivory

The Kyagurilo Group was observed feeding on fruit from 15 species of trees, 2 species of shrub, and 1 herb species over the 6-year period (range 9 to

16 species per year; Table 1.2). However, only 4 species were consumed on >1% of days for each of the 6 years (*Chyrsophyllum* sp, *Myrianthus holstii, Syzigium* sp., and *Rubus apetalus*). During any one of the 6 years, 13 species were consumed on >1% of observation days. Certain species were heavily fed upon during some years but others were not (e.g., *Drypetes gerrardii, Syzigium* sp.), which is likely due to variability in fruit production. Five species were fed on very rarely.

Within the 6-year period from which these data are taken, the percentage of days per year that fruit was observed being eaten varied considerably ($\bar{x} = 36.3\%$; range $= 15.6$–53.0%). There was no correlation between the number of observation days and the level of fruit consumption per year.

Fruit from trees was eaten during all months of the year, but with a significant decline during one of the wet seasons (September–November) compared with the other three seasons (ANOVA, F $= 5,734$, df $= 3$, p $= 0.001$; Figure 1.1.), suggesting that there is a moderate seasonality to fruit eating. Fruit was consumed for an average of 37% of days per month (n $= 71$ months; median $= 32\%$; range 0–100%).

Fibrous foods

A total of 67 species of fibrous foods was consumed from September 2001–August 2004, which is not greatly different from the 62 species noted during a single year of observation (Ganas *et al.*, 2004). The greater number is largely due to species that were observed to be eaten only rarely (<3 occurrences) and to the longer observation time.

A total of 16 species could be considered "important" over the 3-year period (consumed on >5% of observation days). This list of important fibrous food species is essentially the same as that observed for a 1-year study (September 2001–August 2002, and which is included in this 3-year dataset; Ganas *et al.*, 2004). However, although *Kosteletzkya grantii* was important in 2001–2002, across the 3-year period it was not. *Ficus* leaves and *Ganoderma australe* were not considered important in the 1-year study, but were across the 3 years. All three of these foods were eaten relatively infrequently (5–7% of days), which is probably what caused the variation in sampling periods.

Five of these important species showed seasonality in their consumption (Table 1.3). This is probably due to the seasonality of frugivory and not to the seasonal availability of these herbs since a negative correlation was observed between frugivory and consumption for three of the five important herb species and a trend for a fourth (Table 1.3; *Basella alba, Mimulopsis arborensens,*

Table 1.2. *Variability in consumption of fruit species across the 6 years of observation*

Species	Year 1	Year 2	Year 3	Year 4	Year 5	Year 6	Six years combined	Genus eaten in Kahuzi Biega
Allophyllus sp.	0.81	0.70	1.22	3.26	5.23	1.11	2.19	+
Chrysophyllum sp.	8.87	11.58	0.91	17.51	7.99	13.06	10.05	
Drypetes gerrardii	0.00	5.26	0.00	0.00	10.74	0.00	2.81	
Ficus spp.	0.40	2.46	0.30	0.00	1.10	1.11	0.88	+
Maesa lanceolata	0.00	5.26	3.96	4.45	16.53	0.56	5.47	+
Myrianthus holstii	8.06	9.82	6.71	6.53	12.95	8.89	8.90	+
Mystroxylon aethiopicum	0.00	4.56	0.30	1.19	1.10	0.00	1.15	
Olea capensis	6.85	1.05	0.00	8.61	0.00	0.28	2.60	
Olinia usambarensis	0.40	4.56	0.30	0.30	0.28	10.00	2.76	
Podocarpus milinjianus	0.00	0.35	0.61	0.00	0.55	0.00	0.26	
Rapennea rhodrodendoroides	0.00	0.00	0.00	0.00	0.55	0.00	0.10	
Rhytinginia kigeziensis	0.00	0.00	0.00	0.00	2.20	0.00	0.42	
Strombosia sp.	0.00	0.00	0.30	0.00	0.00	0.00	0.05	+
Rubus apetalus								
Symphonia globulifera	0.40	0.00	0.00	0.00	0.55	0.00	0.16	+
Syzigium sp.	2.42	5.61	1.22	7.42	4.13	11.39	5.57	+
Teclea nobilis	0.40	11.23	0.00	0.59	5.51	0.00	2.86	
Xylamos monospora	0.00	1.40	0.61	1.48	0.28	0.00	0.62	+
Total % fruit days	23.80	43.20	15.60	42.3	53.04	40.28	37.06	

Values are the percentage of observation days that the species was consumed. Genera eaten in Kahuzi-Biega from Yamagiwa *et al.* (2005). + indicates the genus is consumed by Grauer's gorillas in Kahuzi Biega (from Yamagiwa *et al.*, unpublished data). The only species recorded as eaten in the Virunga Volcanoes is *Rubus apetalus*.

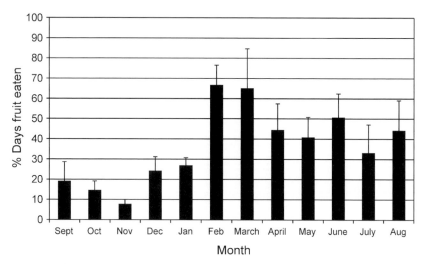

Figure 1.1. Monthly values of percentage of days that fruit was consumed. Bars represent the standard error for the monthly values averaged across 6 years (September 1998–August 2004).

Triumfetta sp., *Urera hypselodendron*). A negative correlation between frugivory and consumption of *M. holstii* leaves was also observed. While these relationships are significant, they are likely to be weak because all of the r^2 values were < 0.5. Qualitative observations have not shown any times when important herbaceous plants were conspicuously absent. Quantitative data collection on annual variation in herb availability is now under way to confirm these observations.

Home range and core area

Annual home range size varied from between 21–40 km^2 ($\bar{x} = 27.2$) using the MCP method and between 16–28 km^2 ($\bar{x} = 22$ km^2) using the 500-m grid square method (Table 1.4). The area used for the 6 years combined was 49.7 km^2 using the MCP method and 45.5 km^2 using the 500-m grid square method. The values obtained using the 500-m grid square method were between 77–94% of the values from the MCP method because the MCP method led to the inclusion of some areas that were not entered according to the grid square method.

Annual core area size varied between 7–12 km^2 ($\bar{x} = 9.3$). The total core area for the 6-year period was 18.8 km^2. Core area was between 30–41% of

Table 1.3. *Important fibrous foods consumed (eaten on > 5% of days), results of tests for seasonality in consumption of fibrous foods (ANOVAs), and correlations between consumption of fibrous foods and overall fruit consumption*

Species	Plant form	Part consumed	% days consumed across 3 years	Seasonality of consumption ANOVA F scores	Seasonality of consumption ANOVA p-values	Correlation with fruit consumption r^2 values	Correlation with fruit consumption p-values	Genus important ir Kahuzi Biega	Genus important in Virungas
Basella alba	herb	l, st	46.2	4.506	***0.010***	0.192	***0.007***	+	+
Cardus sp.	herb	l, st, fl	12.4	0.582	0.631	0.029	0.325		+
Epiphytes	herb	l, st, wp	7.5	1.464	0.243	0.010	0.554	+	
Ficus spp.	tree	b, l	5.7	0.596	0.622	0.001	0.895		
Ganoderma australe	fungus	wp	7.5	2.558	0.072	0.066	0.131		
Ipomea sp.	herb	b, l	46.6	1.209	0.322	0.070	0.119	−	
Mimulopsis arborescens	herb	l, p	42.2	4.873	***0.007***	0.105	0.054		
Mimulopsis solmsii	herb	b, l	45.9	1.680	0.191	0.051	0.183		
Momordica spp.	herb	b, l	71.1	0.984	0.413	0.092	0.073		

Species	Form	Plant parts						Kahuzi	Virunga
Myrianthus holstii	tree	b, l	26	2.264	0.100	0.107	0.051		+
Olea capensis	tree	l, dw	8.5	1.282	0.297	0.006	0.651		
Piper capense	herb	p	17.1	1.592	0.211	0.054	0.172		+
Rubus sp.	shrub	l	42.3	2.001	0.134	0.007	0.639		
Triumfetta sp.	herb	l	60.3	4.025	***0.015***	0.436	***0.000***		
Urera hypselodendron	herb	b, l	72.3	5.819	***0.003***	0.177	***0.011***		+
Vernonia calongensis	shrub	b, l	5.9	3.115	***0.040***	0.013	0.514	+	+

Plant parts: b = bark, l = leaves, st = stem, p = pith, wp = whole plant, fl = flower, dw = wood from dead tree. Epiphytes were not identified to genera level. *Ficus* were not identified to species level. *Mormodica* included three species: *M. solmsii*, *M. calantha*, and *M. foetida*. F values are for ANOVA tests for seasonality in consumption, df = 3. r^2 values are for a correlation between consumption of each fibrous food species and fruit consumption using monthly percentages, df = 34. Genus important in Kahuzi Biega and the Virunga Volcanoes are taken from Yamagiwa *et al.* (2005) and McNeilage (2001).

Table 1.4. *Summary of home range (HR) and core area for Kyagurilo Group*

	Year 1	Year 2	Year 3	Year 4	Year 5	Year 6	6 years combined	Average of 6 years
HR – MCP (km²)	21.7	21.1	40.1	31.3	23.2	25.9	49.7	27.2
HR – 500 m grid sq (km²)	17.0	16.3	28.0	25.5	21.3	24.3	45.5	22.1
Core area (km²)	8.5	7.0	12.0	10.5	8.5	9.3	18.8	9.3
% Core area / MCP range	39.2	33.2	29.9	33.5	36.7	35.7	37.7	
% Core area / 500 m range	50.0	42.9	42.9	41.2	40.0	38.1	41.2	

Home range overlap (%)

Year	1	2	3	4	5	6
1	–	70.1	54.1	66.0	56.7	54.2
2	–	–	52.6	66.2	69.1	49.0
3	–	–	–	61.5	47.7	45.2
4	–	–	–	–	63.6	68.2
5	–	–	–	–	–	53.4
6	–	–	–	–	–	–

Core area overlap (%)

Year	1	2	3	4	5	6
1	–	21.6	18.8	24.6	19.3	16.4
2	–	–	24.6	14.8	19.2	4.8
3	–	–	–	15.4	26.2	7.6
4	–	–	–	–	24.6	27.4
5	–	–	–	–	–	16.4
6	–	–	–	–	–	–

home range area using the MCP method for home range size and between 38–50% of home range size using the 500-m grid square method.

Home range overlap varied between 45–70% between all years using the MCP method (Table 1.4). There was no significant difference in the proportion of home range overlap between consecutive years and non-consecutive years (Mann–Whitney U test, $U = 22$, $p = 0.713$), suggesting that the gorillas do not shift between areas gradually, but probably move abruptly from one part of their overall range to another from one year to the next.

Core area overlap was less than home range overlap, varying between 5–27% (Table 1.4). Again, the amount of core area overlap between consecutive and non-consecutive years was not significantly different (Mann–Whitney U Test, $U = 21.5$, $p = 0.668$).

Discussion

Variability in frugivory and fibrous foods

The long-term observations of the Kyagurilo Group have revealed the large degree of interannual variability in fruit consumption and provided us with both a better understanding of the importance of particular species and of the overall frugivory patterns of mountain gorillas in Bwindi. Their fruit consumption tends to be strongly linked to fruit availability, as has been observed at other sites (Doran *et al.*, 2002; Rogers *et al.*, 2004; Yamagiwa *et al.*, 2005). Generally, the gorillas at this site in Bwindi consume relatively few species of fruit compared with a lower altitude site in Bwindi as well as Grauer's gorillas and western gorillas, probably due to decreasing fruit availability with increasing altitude (Doran *et al.*, 2002; Ganas *et al.*, 2004; Nkurunungi *et al.*, 2004; Yamagiwa *et al.*, 2005). At our site, very few important species are consumed every year (eaten on >1% of days; *Myrianthus holstii*, *Chrysophyllum* sp., *Syzigium* sp., *Rubus apetalus*). Other species may be consumed on 5–15% of the days of one year, and almost daily during the months consumed. However, for the next 2 years, consumption falls to almost zero, suggesting that many tree species are on multi-year fruiting cycles or only produce large quantities of fruit once every few years (*Drypetes gerrardii, Olea capensis, Olinia usambarensis, Teclea nobilis*). These results emphasize the value of long-term studies for determining patterns of frugivory, especially at high altitude sites.

As shown previously, the diet of Bwindi gorillas appears to be more similar to Grauer's gorillas in Kahuzi-Biega than to mountain gorillas in the Virunga Volcanoes, probably because of the similarity in food availability (Tables 1.2

and 1.3; Ganas *et al.*, 2004; Yamagiwa *et al.*, 2005). *Myrianthus holstii* appears to be one of the most important fruit species for eastern gorillas because it is heavily fed upon at two sites within Bwindi (Ganas *et al.*, 2004) and its seeds were found in one-third of the fecal samples analyzed across a 9-year study period in Kahuzi-Biega (Yamagiwa *et al.*, 2005). In contrast, figs appear to be a staple food in the diet of Grauer's gorillas at Kahuzi, and their seeds were found in more than 15% of the fecal samples of gorillas at a low altitude site in Bwindi (Ganas *et al.*, 2004), but they are rarely consumed by the Kyagurilo Group. This variation may be due to both the availability of figs and interspecific competition with chimpanzees (Stanford & Nkurunungi, 2003; Yamagiwa *et al.*, 2005).

How important is fruit to Bwindi gorillas? The methods used here and in other studies – calculating the percentage of days species were consumed, using either fecal analysis or direct observation for only part of the day – are limiting because they do not provide us with information on the time spent feeding on fruit nor the biomass of fruit consumed. They are, however, sensitive enough to detect seasonal differences in frugivory. The gorillas ate fruit, on average, 37% of observation days per year, suggesting that it is not an insignificant part of their diet. These values may also be underestimates of the daily occurrence of fruit eating because observations were limited to only part of the day; analysis of fecal samples for Year 4 showed evidence of fruit eating for 66% of days (Ganas *et al.*, 2004), compared with the 42% based on observations. Furthermore, a measure of presence/absence of fruit in the diet on a daily basis does not provide the same measure of frugivory as a measure of percent of time consuming fruit. Behavioral observations on the Kyagurilo Group show that approximately 11% of foraging time is spent feeding on fruit (range 0–65% per month; Robbins, unpublished data). This result is remarkably similar to values obtained for western gorillas for whom fruit availability tends to be much greater (Doran-Sheehy *et al.*, Chapter 2, this volume). Nonetheless, we still need intake rates and nutritional analyses to calculate the relative caloric and energy intake from fruit vs. fibrous foods. Nutritional analysis will also enable us to see how the nutritional gains from different fruit species varies, which is of particular importance given the large interannual variability in consumption of individual fruit species. Studies that address these issues are underway. Even if fruit consumption is relatively low in Bwindi, it still appears to influence the ranging patterns of the gorillas, with a positive relationship between frugivory and both daily travel distance and monthly home range size (Ganas & Robbins, 2005).

Fibrous foods clearly are the largest component of the diet of Bwindi gorillas, as has been observed at other gorilla sites. We have seen high consistency in consumption of important fibrous foods across years, which suggests that there is little variability in availability. By default, fibrous food consumption should decrease as fruit consumption increases. Even using our rough estimate of food consumption (presence/absence in daily diet), we found seasonality for the consumption of 4 of the 16 important fibrous food species and negative correlations between frugivory and their consumption. However, all of these species should still probably be considered as staple foods and not fallback foods since they are consumed frequently each month.

Given the large altitudinal range occupied by eastern gorillas and the distance between study sites, it is no surprise that few fibrous foods are found in common in the diets of gorillas of Bwindi, the Virunga Volcanoes, and Kahuzi-Biega (Ganas *et al.*, 2004). More genera are found in common between Bwindi and Kahuzi-Biega than with the Virunga Volcanoes (Table 1.3). The only fibrous food that occurs seasonally in all three areas is bamboo shoots (*Arundinaria alpina*), which has been shown to influence ranging patterns of gorillas in both the Virunga Volcanoes and Kahuzi Biega (Vedder, 1984; Watts, 1998a, 2000; Yamagiwa *et al.*, 2005). Bamboo is found in only a very small section of Bwindi and none is found in the home range of the Kyagurilo Group.

Ranging patterns

The results presented here show that gorillas in Bwindi consistently use an annual home range that is significantly bigger than the range observed in the Virunga Volcanoes (range 3–15 km^2 per year). The annual home range of the Kyagurilo Group was typically 20–25 km^2 per year, a total range area not reached until about 4 years had been spent observing the Virunga gorilla groups (Watts, 1998a, 2000). Annual core areas were also larger than that of Virunga gorillas, but represented similar proportions of the home range (30–40%; Watts, 1998a). Bwindi gorillas exhibited a slightly greater degree of home range overlap between years than did Virunga gorillas ($\sim$50–70% vs. <50%), but much less overlap of core areas between years (<25% vs. <40%). The total area used in 6 years of observation was also double that of one group of Virunga gorillas observed for 7 years (50 km^2 vs. 25 km^2; Watts, 1998a). Annual home ranges of three other groups in Bwindi were also larger than that observed in the Virunga Volcanoes (Ganas & Robbins, 2005). These comparisons suggest that Bwindi gorillas require more space to meet their nutritional needs than do Virunga gorillas. The lower overlap of core area suggests two things: either the herbaceous vegetation of Bwindi requires a longer regeneration time than that of the Virungas, or the core area is heavily influenced by fruit, which may be spatially clumped in the habitat according to which fruit species are heavily fed upon each year (see dietary variation above). To test these hypotheses, additional research is needed both on food availability in time and space as well as on regeneration rates.

Several studies of partially or fully habituated western gorillas have been published recently that now allow direct comparisons to be made among and between populations. Annual home range size of Bwindi gorillas is comparable to that observed in western gorillas and Grauer's gorillas (Cipolletta, 2003, 2004; Bermejo, 2004; Doran-Sheehy *et al.*, 2004; Yamagiwa & Basabose, Chapter 3, this volume; Ganas & Robbins, 2005). Core areas appear to occupy the same proportion (30–40%) of the home range in western gorillas, Grauer's gorillas, and both populations of mountain gorillas. Doran-Sheehy *et al.* (2004) state that western gorillas use a larger proportion of their annual home range on a monthly basis than do Virunga gorillas, which may be attributable to decreased availability of herbaceous vegetation and/or increased reliance on fruit, found there in more discrete and dispersed patches. They go on to argue that because of increased reliance on fruit and the benefits of regularly monitoring the same area, the home range of western gorillas is more stable over time than that of the gorillas of Virunga Volcanoes. However, their study was only 16 months in duration (with home range size stabilizing after 10 months) which is not directly comparable with the long-term studies done in the Virunga Volcanoes (Watts, 1998a) or the results reported here. Cipolletta (2004) found that home range size increased over time until it stabilized approximately 2 years into the 3.5-year period, but this may have been due to male mating competition. These results from several gorilla research sites are in strong contrast to the high site fidelity observed in chimpanzees (Lehmann & Boesch, 2003). Gorillas are not territorial and have large home range overlap with neighboring groups (Watts, 1998a; Doran-Sheehy *et al.*, 2004; Ganas & Robbins, 2005), whereas chimpanzees are strongly territorial and typically have only a small area of overlap with neighboring communities. These differences in sociality may be partially responsible for the variation in long-term stability of home ranges between species.

Implications of diet on gorilla density and sociality

Temporal and spatial variability in food availability appear to influence both diet and ranging patterns in gorillas, which should in turn influence the density of gorillas found in an area and may also cause variation in their social structure and behavior. Therefore, we can ask if the large differences between gorilla populations in diet and ranging lead to differences in population

Table 1.5. *Comparisons of Virunga and Bwindi population parameters*

	Virunga Volcanoes[a]	Bwindi Impenetrable[b]
Size of protected area	434km^2	330km^2
Size of "gorilla usable habitat"	310km^2	270km^2
Population size	380	320
Population density (total habitat)	0.90 gorillas/km^2	0.97 gorillas/km^2
Population density (gorilla usable habitat)	1.23 gorillas/km^2	1.19 gorillas/km^2
Number of social groups	32	27
Average group size (standard deviation)	11.4 (11.2)	11.3 (5.7)
Median group size	7.5	10
% of very large groups (>20 individuals)	15.6%	11%
% multimale groups	36%	44%

Notes:
Population density is based on the "gorilla usable habitat," which excludes areas in the alpine zone, meadow zone, and three-quarters of the bamboo zone (shoots only available approximately 3 months per year) in the Virunga Volcanoes and the northern sector of Bwindi that contains no gorillas, has high human encroachment, and markedly different habitat.
[a] Grey *et al.*, unpublished data.
[b] McNeilage *et al.*, unpublished data.

dynamics and social structure. Rogers *et al.* (2004) hypothesized that the density of gorillas is more strongly correlated to the abundance of their staple foods, which typically comprise herbaceous vegetation, than to other factors, such as the density of preferred seasonal fruit foods or competition from other large mammals, including chimpanzees and elephants. Stem density for herbaceous vegetation consumed by gorillas is comparable between the Virunga Volcanoes and Bwindi (Karisoke: 8.81 stems/m^2, Watts, 1984; Bwindi: 4.4–10.6 stems/m^2, Nkurunungi *et al.*, 2004), and significantly higher than found in western gorilla sites (0.78–2.25 stems/m^2, summarized in Doran *et al.*, 2002). The Virunga Volcanoes is the only location where gorillas are not sympatric with chimpanzees, and it also has the lowest availability of fruit.

The most useful comparisons of gorilla density are between the Virunga Volcanoes and Bwindi because these two areas have been censused in the same manner. The density of gorillas in these two populations is roughly comparable when using total habitat size, or "gorilla usable habitat" (which subtracts out vegetation zones unlikely to support gorillas in the long term; Table 1.5). However, it is likely that both of these populations are still increasing in size. Also they are both suffering from human disturbance, including the direct poaching of gorillas (Kalpers *et al.*, 2003; Gray *et al.*, unpublished data; McNeilage *et al.*, unpublished data), so these density estimates do not solely reflect the influence of food availability.

Comparisons with lowland gorillas are limited because of problems with the censusing methods used that may lead to large sampling errors (Walsh *et al.*, 2003). However, if one views with caution the density estimates of Grauer's and western gorilla populations summarized by Yamagiwa (1999), it appears that these gorilla populations typically occur at lower densities than mountain gorillas (<1 gorilla/km^2). The highest densities have been found in Odzala and Ndoki National Parks, Republic of Congo, both of which have the highest density of herbaceous vegetation recorded for western gorillas (>2 gorillas/km^2, Nishihara, 1995; Bermejo, 1999; Yamagiwa, 1999). While additional data are clearly needed for more accurate comparisons, including better controls for variables such as human encroachment and the influence of disease, there is no evidence yet to reject the hypothesis that staple foods (predominantly herbaceous vegetation) for gorillas heavily influence their density (Rogers *et al.*, 2004).

In addition to density, there are remarkable similarities in various parameters of social structure between the Bwindi and Virunga Volcano populations (Table 1.5). Average group size is similar and both populations contain groups that can be considered "very large" (>20 individuals). While average group size is roughly the same in all gorilla populations surveyed, very large groups have been observed in only one population of western gorillas in Odzala National Park, which, as already noted, has a high density of herbaceous vegetation (Bermejo, 1999; Magliocca *et al.*, 1999; Parnell, 2002; Robbins *et al.*, 2004). These results suggest that the availability and distribution of staple fibrous foods may limit group size, although fruit availability may also play a role (Doran & McNeilage, 2001). Both mountain gorilla populations have a significant number of multimale groups, which contrasts greatly with western gorillas, in which this group structure is only rarely observed (Robbins *et al.*, 2004). It is likely that any limits on group size may also influence male mating strategies, leading to greater rates of male emigration and a lower occurrence of multimale groups. Additional long-term research at several sites with different environmental conditions will help elucidate the role that fibrous foods and fruit play in the density, social structure, and social behavior observed in gorillas. Furthermore, an understanding of how gorillas respond to differing ecological conditions should assist in conservation efforts to protect these highly endangered primates.

Acknowledgments

We thank the Ugandan Wildlife Authority and the Uganda National Council for Science and Technology for their long-term support and permission to conduct this research. We thank all ITFC field assistants for data collection and gorilla monitoring, especially Tibenda Emmanuel, Twinomujuni Gaad, Mbabazi Richard, Ngambaneza Caleb, Kyamuhangi Narsis, Byaruhanga Gervasio, Twebaze Deo, Mayooba Godfrey, Tumwesigye Philimon, and Murembe Erinerico. Angela Higginson and Nick Parker assisted with data collection and data entry. Maryke Gray and Robert Barigira assisted with field assistant training, data collection, and plant identification. Robert Bitariho of the Ecological Monitoring Program of ITFC provided the rainfall data. This manuscript benefited from discussions with Jessica Ganas, statistical assistance from Andrew Robbins, and comments from Gottfried Hohmann and one anonymous reviewer. The project was funded by the Max Planck Society, Wildlife Conservation Society, with additional support to the Institute for Tropical Forest Conservation (ITFC) from the World Wide Fund for Nature.

References

Barton, R. A., Byrne, R. W. & Whiten, A. (1996). Ecology, feeding competition and social structure in baboons. *Behavioral Ecology and Sociobiology*, **38**, 321–9.

Bermejo, M. (1999). Status and conservation of primates in Odzala National Park, Republic of the Congo. *Oryx*, **33**, 323–31.

   (2004). Home-range use and intergroup encounters in western gorillas (*Gorilla g. gorilla*) at Lossi Forest, North Congo. *American Journal of Primatology*, **64**, 223–32.

Boinski, S., Sughure, K., Selvaggi, L. *et al.* (2002). An expanded test of the ecological model of primate social evolution: competitive regimes and female bonding in three species of squirrel monkeys (*Saimiri oerstedii*, *S. boliviensis* and *S. sciureus*). *Behaviour*, **139**, 227–61.

Butynski, T. M. (1984). *Ecological Survey of the Impenetrable (Bwindi) Forest, Uganda, and Recommendations for Its Conservation and Management*. Unpublished report to the Uganda Government.

Cipolletta, C. (2003). Ranging patterns of a western gorilla group during habituation to humans in the Dzanga-Ndoki National Park, Central African Republic. *American Journal of Primatology*, **24**, 1207–26.

   (2004). Effects of group dynamics and diet on the ranging patterns of a western gorilla group (*Gorilla gorilla gorilla*) at Bai Hokou, Central African Republic. *American Journal of Primatology*, **64**, 193–205.

Clutton-Brock, T. H. & Harvey, P. H. (1977). Species differences in feeding and ranging behavior in primates. In *Primate Ecology: Studies of Feeding and Ranging Behaviour in Lemurs, Monkeys, and Apes*, ed. T. H. Clutton-Brock, pp. 557–84. New York: Academic Press.

Doran, D. M. & McNeilage, A. (2001). Subspecific variation in gorilla behavior: the influence of ecological and social factors. In *Mountain Gorillas: Three Decades of Research at Karisoke*, ed. M. M. Robbins, P. Sicotte, & K. J. Stewart, pp. 123–49. Cambridge: Cambridge University Press.

Doran, D. M., McNeilage, A., Greer, D. *et al.* (2002). Western lowland gorilla diet and resource availability: new evidence, cross-site comparisons, and reflections on indirect sampling methods. *American Journal of Primatology*, **58**, 91–116.

Doran-Sheehy, D. M., Greer, D., Mongo, P., & Schwindt, D. (2004). Daily path length, home range and swamp use in one group of western gorillas: the influence of ecological and social factors on ranging. *American Journal of Primatology*, **64**, 207–22.

Fossey, D. & Harcourt, A. H. (1977). Feeding ecology of free ranging mountain gorillas (*Gorilla gorilla beringei*). In *Primate Ecology: Studies of Feeding and Ranging Behavior in Lemurs, Monkeys and Apes*, ed. T. H. Clutton-Brock, pp. 539–56. London: Academic Press.

Ganas, J. & Robbins, M. M. (2005). Ranging behavior of the mountain gorillas (*Gorilla beringei beringei*) in Bwindi Impenetrable National Park, Uganda: a test of the ecological constraints model. *Behavioral Ecology and Sociobiology*, **58**, 277–88.

Ganas, J., Robbins, M. M., Nkurunungi, J. B., Kaplin, B. A., & McNeilage, A. (2004). Dietary variability of mountain gorillas in Bwindi Impenetrable National Park, Uganda. *International Journal of Primatology*, **25**, 1043–72.

Goldsmith, M. L. (2003). Comparative behavioral ecology of a lowland and highland gorilla population: where do Bwindi gorillas fit? In *Gorilla Biology: A Multidisciplinary Perspective*, ed. A. B. Taylor & M. L. Goldsmith, pp. 358–84. Cambridge: Cambridge University Press.

Janson, C. H. & Goldsmith, M. L. (1995). Predicting group size in primates: foraging costs and predation risks. *Behavioral Ecology*, **6**, 326–36.

Kalpers, J., Williamson, E. A., Robbins, M. M. *et al.* (2003). Gorillas in the crossfire: population dynamics of the Virunga mountain gorillas over the past three decades. *Oryx*, **37**, 326–37.

Koenig, A., Beise, J., Chalise, M. K., & Ganzhorn, J. U. (1998). When females should contest for food – testing hypotheses about resource density, distribution, size, and quality with Hanuman langurs (*Presbytis entellus*). *Behavioral Ecology and Sociobiology*, **42**, 225–37.

Lehmann, J. & Boesch, C. (2003). Social influences on ranging patterns among chimpanzees (*Pan troglodytes verus*) in the Taï National Park, Côte d'Ivoire. *Behavioral Ecology*, **14**, 642–9.

Magliocca, F., Querouil, S., & Gautier-Hion, A. (1999). Population structure and group composition of western lowland gorillas in north-western Republic of Congo. *American Journal of Primatology*, **48**, 1–14.

McNeilage, A. (1995). *Mountain Gorillas in the Virunga Volcanoes: Ecology and Carrying Capacity*. Unpublished Ph.D. thesis, University of Bristol.

(2001). Diet and habitat use of two mountain gorilla groups in contrasting habitats in the Virungas. In *Mountain Gorillas: Three Decades of Research at Karisoke*, ed. M. M. Robbins, P. Sicotte, & K. J. Stewart, pp. 265–92. Cambridge: Cambridge University Press.

Milton, K. (1988). Foraging behaviour and the evolution of primate intelligence. In *Machiavellian Intelligence: Social Expertise and the Evolution of Intellect in Monkeys, Apes, and Humans*, ed. R. Byrne & A. Whiten, pp. 285–305. Oxford: Oxford Science Publications.

Nishihara, T. (1995). Feeding ecology of western lowland gorillas in the Nouable-Ndoki National Park, Congo. *Primates*, **36**, 151–68.

Nkurunungi, J. B. (2004). *The Availability and Distribution of Fruit and Non-fruit Resources in Bwindi: Their Influence on Gorilla Habitat Use and Food Choice.* Unpublished Ph.D. thesis, Makerere University.

Nkurunungi, J. B., Ganas, J., Robbins, M. M., & Stanford, C. B. (2004). A comparison of two mountain gorilla habitats in Bwindi Impenetrable National Park, Uganda. *African Journal of Ecology*, **42**, 289–97.

Parnell, R. J. (2002). Group size and structure in western lowland gorillas (*Gorilla gorilla gorilla*) at Mbeli Bai, Republic of Congo. *American Journal of Primatology*, **56**, 193–206.

Remis, M. J. (1997). Western lowland gorillas (*Gorilla gorilla gorilla*) as seasonal frugivores: use of variable resources. *American Journal of Primatology*, **43**, 87–109.

Robbins, M. M. & McNeilage, A. J. (2003). Home range and frugivory patterns of mountain gorillas in Bwindi Impenetrable National Park, Uganda. *International Journal of Primatology*, **24**, 467–91.

Robbins, M. M., Bermejo, M., Cipolletta, C., Magliocca, F., Parnell, R. J., & Stokes, E. (2004). Social structure and life history patterns in western gorillas (*Gorilla gorilla gorilla*). *American Journal of Primatology*, **64**, 145–59.

Rogers, M. E., Abernethy, K., Bermejo, M. *et al.* (2004). Western gorilla diet: a synthesis from six sites. *American Journal of Primatology*, **64**, 173–92.

Sarmiento, E. E., Butynski, T. M., & Kalina, J. (1996). Gorillas of Bwindi Impenetrable Forest and the Virunga Volcanoes: taxonomic implications of morphological and ecological differences. *American Journal of Primatology*, **40**, 1–21.

Southwood, T. R. E. (1966). *Ecological Methods*. London: Chapman & Hall

Stanford, C. B. & Nkurunungi, J. B. (2003). Behavioral ecology of sympatric chimpanzees and gorillas in Bwindi Impenetrable National Park, Uganda: diet. *International Journal of Primatology*, **24**, 901–18.

Sterck, E. H. M., Watts, D. P., & van Schaik, C. P. (1997). The evolution of female social relationships. *Behavioral Ecology and Sociobiology*, **41**, 291–309.

Vedder, A. L. (1984). Movement patterns of a free-ranging group of mountain gorillas (*Gorilla gorilla beringei*) and their relation to food availability. *American Journal of Primatology*, **7**, 73–88.

Walsh, P. D., Abernethy, K. A., Bermejo, M. *et al.* (2003). Catastrophic ape decline in western equatorial Africa. *Nature*, **422**, 611–14.

Watts, D. P. (1984). Composition and variability of mountain gorilla diets in the central Virungas. *American Journal of Primatology*, **7**, 323–56.

(1991). Strategies of habitat use by mountain gorillas. *Folia Primatologica*, **56**, 1–16.

(1998a). Long term habitat use by mountain gorillas (*Gorilla gorilla beringei*). 1. Consistency, variation, and home range size and stability. *International Journal of Primatology*, **19**, 651–80.

(1998b). Long term habitat use by mountain gorillas (*Gorilla gorilla beringei*). 2. Reuse of foraging areas in relation to resource abundance, quality and depletion. *International Journal of Primatology*, **19**, 681–702.

(2000). Mountain gorilla habitat use strategies and group movements. In *On the Move: How and Why Animals Travel in Groups*, ed. S. Boinski & P. A. Garber, pp. 351–74. Chicago: University of Chicago Press.

Yamagiwa, J. (1999). Socioecological factors influencing population structure of gorillas and chimpanzees. *Primates*, **40**, 87–104.

Yamagiwa, J., Mwanza, N., Yumoto, T., & Maruhashi, T. (1992). Travel distances and food habits of eastern lowland gorillas: a comparative analysis. In *Topics in Primatology, Volume 2: Behavior, Ecology and Conservation*, ed. N. Itoigawa, Y. Sugiyama, & G. P. Sackett, pp. 267–81. Tokyo: University of Tokyo Press.

(1994). Seasonal change in the composition of the diet of eastern lowland gorillas. *Primates*, **35**, 1–14.

Yamagiwa, J., Maruhashi, T., Yumoto, T., & Mwanza, N. (1996). Dietary and ranging overlap in sympatric gorillas and chimpanzees in Kahuzi-Biega National Park, Zaire. In *Great Ape Societies*, ed. W. C. McGrew, L. F. Marchant & T. Nishida, pp. 82–98. Cambridge: Cambridge University Press.

Yamagiwa, J., Basabose, A. K., Kaleme, K., & Yumoto, T. (2005). Diet of Grauer's Gorillas in the Montane Forest of Kahuzi, Democratic Republic of Congo. *International Journal of Primatology*, **26**, 1345–73.

2 Sympatric western gorilla and mangabey diet: re-examination of ape and monkey foraging strategies

DIANE M. DORAN-SHEEHY, NATASHA F. SHAH, AND LISA A. HEIMBAUER

Introduction

The two old world primate radiations, old world monkeys and apes, have distinct dental and gastrointestinal adaptations that indicate a divergence in foraging strategy. Old world monkeys have bilophodont dentition, which is thought to aid in the fracturing of tough seeds or cutting of leaves (Lucas &

Feeding Ecology in Apes and Other Primates. Ecological, Physical and Behavioral Aspects, ed. G. Hohmann, M.M. Robbins, and C. Boesch. Published by Cambridge University Press. © Cambridge University Press 2006.

Teaford, 1994), whereas apes have broader incisors and more generalized molars, more characteristic of frugivorous primates (Fleagle, 1999). One would assume that the larger-bodied apes would feed on lower-quality forage than smaller, old world monkeys because the ability to efficiently extract nutrients from low-quality resources (i.e., higher in fiber or anti-feedant content) has been thought to increase with increasing body size (Parra, 1978; Chivers & Hladik, 1980; Demment & Van Soest, 1983; Kay & Davies, 1994; Van Soest, 1996). However, the opposite appears to be true for apes and old world monkeys. Both old world monkey radiations – Colobines and cercopithecines – have gastrointestinal, anatomical, or physiological adaptations that allow them to feed on low-quality food. Colobines have a complex sacculated stomach that makes it possible for them to maintain bacterial colonies for fermentation (Fleagle, 1999). Cer-copithecines appear to have an unusually slow gut passage time that results in lengthy digestive retention times (Lambert, 2002). Given that the longer the digesta remains in the gut the greater the opportunity for fermentation and nutrient uptake, this permits the more efficient processing of lower-quality foods than found in primates of similar body size with faster gut passage rates (Maisels, 1993; Lambert, 1998, 2002). Apes, on the other hand, have a relatively unspecialized gut, and other than large body size, no diet-related gut modifications. Therefore, it has been argued that apes should be more selective in their diet, feeding on higher-quality, less readily available foods such as ripe fruits, compared with the more diverse, less selective consumption of more abundant and more heavily (chemically) defended foods of old world monkeys (Temerin & Cant, 1983; Wrangham *et al.*, 1998).

Wrangham and colleagues confirmed these ape/old world monkey dietary distinctions when they compared differences in diet and chemical content of foods consumed by sympatric chimpanzees and six species of old world monkeys (Conklin-Brittain *et al.*, 1998; Wrangham *et al.*, 1998). They found that compared with old world monkeys, chimpanzees had a higher-quality diet that included more ripe fruit (and total non-structural carbohydrates) and foods with lower fiber and tannin content. Chimpanzee diet composition fluctuated throughout the year, with ripe fruit consumption increasing with increasing ripe fruit availability. Old world monkeys, on the other hand, maintained greater dietary diversity (including more unripe fruit, seeds, and leaves) throughout the year, with consistently higher levels of antifeedants in the diet. Fruit consumption did not correlate significantly with changes in ripe fruit availability. Wrangham *et al.* (1998) concluded that chimpanzees are ripe fruit specialists, whereas cercopithecoids eat a lower-quality diet with higher amounts of antifeedants. They noted however, that chimpanzees are among the most frugivorous of all ape species, and that other apes are much less frugivorous and/or known to consume lower-quality forage at times. For example, non-reproductive plant parts make up >90% of the diet of Karisoke mountain gorillas (Watts, 1984), and bark consumption can, at times, account for greater than 50% of the diet in orangutans (Galdikas, 1988). Thus, it remains unclear to what extent chimpanzee/old world monkey dietary differences reflect distinct foraging adaptations of the ape and old world monkey radiations as a whole. Further study of ape/old world monkey dietary differences, and particularly of less frugivorous ape species in comparison to monkeys, is warranted in order to refine our understanding of ape and old world monkey dietary strategies.

Here, we compare the foraging strategies of sympatric western gorillas (*Gorilla gorilla*) with grey-cheeked (*Lophocebus albigena*) and agile (*Cercocebus agilis*) mangabeys. Mangabeys are medium-sized (males 7–12 kg; females 5–7 kg), frugivorous forest cercopithecines who have thick dental enamel and heavy jaw musculature, which enables them to feed on lignifed fruits and hard seeds (Kay, 1984; Fleagle & McGraw, 1999). The two mangabey taxa differ in their preference of forest strata: agile mangabeys are more terrestrial and use forest strata lower than that of grey-cheeked mangabeys (Shah, 2003). Western gorillas are large-bodied, lowland forest apes that are sexually dimorphic in body size (mean body weight: male = 170 kg, female = 71 kg; Smith & Jungers, 1997), and primarily terrestrial, although both males and females regularly climb as high as 30 m to forage (Doran-Sheehy, unpublished data). Western gorillas live in habitats with reduced herb and greater fruit availability compared with the well-studied Karisoke mountain gorilla (Doran & McNeilage, 2001), although the effect of these differences on diet is not yet clearly understood. Indirect studies (fecal samples and trail signs) of western gorilla diet have been conducted at more than six sites throughout Central Africa. In all cases, researchers have emphasized that western gorillas are much more frugivorous than their eastern counterparts (reviewed in Watts, 1996; Doran & McNeilage, 1998). Fruit species were the most diverse food category at all western sites studied, accounting for up to 70% of the food species identified (Rogers *et al.*, 2004). However, to date the amount of fruit western gorillas consume is unclear because they have been especially difficult to habituate to human observation (Tutin & Fernandez, 1991; Blom *et al.*, 2004), and as a result, no systematic, quantitative data on western gorilla diet based on direct observation are currently available.

The goal of this chapter is to provide the first (preliminary) quantification of western gorilla diet in order to directly compare it with that of a study of mangabeys conducted previously by one of us (NFS) at the same site. First, we will examine general dietary patterns and the degree of frugivory and diet diversity in the three species. Next, we will consider whether the pattern of fruit consumption in gorillas is more similar to that of chimpanzees versus old world monkeys. Specifically, we examine whether gorillas are more selective in their fruit choice than mangabeys and whether gorilla fruit consumption, unlike that of mangabeys, increases with ripe fruit availability. Finally, we examine how the radiations differ in their food choice, and whether these differences support the hypothesis that old world monkeys feed on a lower-quality diet.

Methods

Study site

Research was conducted at the Mondika Research Center (02°21′859″N, 016°16′465″E), located on the boundary of the Central African Republic

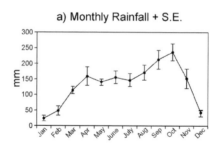

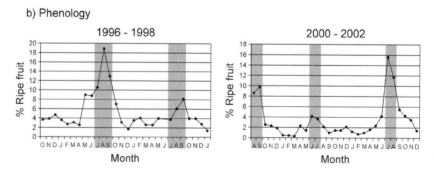

Figure 2.1. Rainfall and patterns of fruit availability at Mondika Research Center. (a) Mean monthly rainfall (in mm) + SE over 7 years. (b) The percentage of important gorilla food tree or liana individuals with ripe fruit over 56 months (n = 498 individuals). Gray shading indicates each annual fruiting peak.

(Dzanga-Ndoki National Park) and the Republic of Congo. For a complete description of the site, see Doran *et al.* (2002). There is monthly variation in rainfall (Figure 2.1a; ANOVA, F = 8.67, p < 0.0001, n = 7 years). Each year there is typically a 2 to 3-month period (Dec–Feb) in which less than 50 mm of rain falls per month, and a 1 to 2-month period (Sept–Oct) in which there is greater than average rainfall. There is also monthly and interannual variation in ripe fruit availability (Figure 2.1b), although each year there is a consistent (2–3 month) fruiting peak that occurs between June and September.

Sampling methods

Sampling time periods

We collected data during two study periods: September 1998–August 1999 (Study Period 1) and July 2003–June 2004 (Study Period 2). During Study Period 1, NFS conducted an average of 5 consecutive all-day follows per month of one habituated group each of agile and grey-cheeked mangabeys. During this time period, gorillas were not habituated to human observers and measures of gorilla diet based on direct observation were not possible. During Study Period 2, Doran and field assistants conducted daily (nearly) all-day follows (average 10 hours per day) of a habituated single adult silverback male. The other 12 group members were not sufficiently habituated to permit systematic sampling at that time, so results are based on a single individual.

To control for the possibility that apparent differences in gorilla and mangabey diet resulted from differences in fruit availability during the two time periods, we included indirect measures (fecal samples and trail signs) of gorilla diet during Study Period 1. This was possible because we were regularly tracking gorillas (with the aid of trackers) as part of a long-term study. Indirect measures of diet have been used in several previous studies of gorilla diet, both at this site, where a 3-year (July 95–June 98) dietary study was conducted (Doran *et al.*, 2002), and at several other sites in central Africa (LOPE, Gabon: Tutin *et al.*, 1991; BAI HOKOU, Central African Republic: Remis, 1997; Goldsmith, 1999; NDOKI, Congo: Nishihara, 1995), ranging in duration from 1–4 years. In general, fecal samples provide a good assessment of the diversity and frequency of gorilla fruit consumption because seeds of most species pass through the gut intact and are identifiable to species (Tutin & Fernandez, 1993; Doran *et al.*, 2002). Assessment of leaf and herb diversity from fecal samples is more problematic because individual leaf and pith species cannot be identified from macroscopic examination (Doran *et al.*,

2002; Rogers *et al.*, 2004). Generally, relative estimates of fruit and fiber (leaf and pith combined) intake are made from comparing the relative contribution of each to the fecal sample. Thus, fecal samples provide only a rough estimate of the quantity of fruit versus foliage in the diet. Feeding trail data, i.e., the gorilla feeding remains encountered while following a gorilla trail, add important additional dietary information missing from fecal samples, such as the type and frequency of leaf and pith remains, as well as an independent assessment of the number of fruit species consumed per day. However, it has been suggested that feeding trail data may underestimate the diversity of food items from trees, e.g., leaves (Rogers *et al.*, 2004).

Mangabey data sampling

During all-day follows of each mangabey group, 5-minute group scans (Altmann, 1974), were conducted at 20-minute intervals. During each scan the activity, food species, and part were recorded for all individuals in sight, an average of 7.1 and 5.4 individuals per scan for agile and grey-cheeked mangabeys, respectively. Data collection was more extensive for agile versus grey-cheeked mangabeys, with 12 vs. 10 months sampled during Study Period 1 (group scans: agile = 1977, grey = 1577). To improve the comparability of data, we subsampled the data and report results (with one exception) based on an average of 4.4 (of 5 consecutive) all-day follows per month for 10 of 12 months, excluding June and December, the 2 months where adequate data were not available for either species (agile: average number of days per month = 4.4, SD = 0.8, n = 10 months, total days sampled = 44, no. of group scans = 1337; grey-cheeked: average number of days per month = 4.4, SD = 0.5, n = 10 months, total days sampled = 45; no. of group scans = 1377). Results describing the pattern of fruit use, i.e., length of time agile mangabeys feed on a particular fruit species over the course of a year, are based on the larger 12-month agile data set.

Gorilla data sampling

Direct observation (Study Period 2): During daily (nearly) all-day gorilla follows, we recorded activity, food part, and species instantaneously at 1-minute intervals during each of four 10-minute sampling periods each hour. To produce a data set that was most comparable with that of the mangabeys, we randomly selected 4.5 (sd = 0.5) days per month for the same 10 months of the year sampled in the mangabey study.

Indirect measures of diet (Study Period 1)

Feeding trails

While following the trail of a group, we recorded the presence of each food item the first time we encountered it each day (n = 320 feeding trails). Food items were recognizable on trails by the characteristic manner with which each food type had been processed and the particular plant parts discarded (Williamson, 1989). We compiled a monthly list of all gorilla food items that occurred in >1% of feeding trails. Feeding trail data in this study provide a measure of how *often* a food item is consumed, but no measure of how *much* of an item is consumed.

Fecal samples

We analyzed a total of 60 fresh fecal samples (5 fecal samples per month for one year), but base the comparisons with mangabeys on 50 samples, excluding those from the months of June and December. Samples were weighed, dissociated in water, and rinsed through a 1-mm screen. We identified fruit species consumed in each sample and compiled a list of all fruit species consumed.

Phenology

Two types of phenological monitoring were conducted. During Study Period 1, NFS walked 7 km of transects monthly and recorded all fruits (unripe and ripe) observed on the ground. She then identified the tree, shrub, or liana from which the fruit fell, measured its diameter at breast height (DBH) (if a tree) and estimated the number of fruits present using a log 10 scale. For the gorilla study (which was of longer duration, beginning in 1995), each month for 56 months (October 96–January 99 and August 2000–December 2002) we monitored the presence or absence of ripe fruit and new and mature leaves in an average of 498 individuals of 57 species of trees and lianas previously determined to be important gorilla foods (Doran *et al.*, 2002). Phenological monitoring was not conducted during Study Period 2 due to shortages of staff, a result of reduced project financing.

Dietary diversity

Monthly dietary diversity was calculated for both mangabey species using Simpson's sample-size corrected index of diversity (Pielou, 1969):

$$D = 1 - \sum_{i}^{s} \left[\frac{n_i(n_i - 1)}{N(N - 1)} \right]$$

where s = total number of different foods eaten in the month, N = total number of feeding records in the month, and n_i = number of individuals feeding on this food i in the month.

Results

Western gorilla diet based on behavioral sampling

The first systematic sampling of western gorilla diet based on direct observation (n = 8643 feeding minutes) indicated that one male western gorilla ate 72 food items from 67 species, including terrestrial and aquatic herbs, fruits, leaves, termites, and bark (Table 2.1). The majority of gorilla food items consumed were non-reproductive, rather than fruits, contrary to all previous studies of western gorilla diet based on indirect sampling (Figure 2.2). Dicotyledonous leaves (51.3%) and bark (6.9%) and monocotyledonous stems and shoots (15.3%) accounted for 73.5% of all food items identified, whereas fruit accounted for only 22.2% of identified items (n = 72 food items). Additionally, gorillas consumed three species of insect, which accounted for the remaining 4.2% of species consumed.

All 72 foods were not eaten equally (Table 2.2). Two foods, *Celtis* sp. (Ulmaceae) leaves and *Aframomum limbatum* (Zingiberaceae) stem, were eaten much more frequently than others, accounting for 18% and 16% of overall feeding time, respectively (n = 8643 feeding minutes). The third most important food, *Duboscia macrocarpa* (Tiliaceae) fruit accounted for only 5.4% of the total diet. Together the 10 foods fed upon most frequently accounted for 62.7% of feeding time.

Not only were the majority of feeding items non-fruit, but the majority of feeding time was also spent feeding on non-reproductive plant parts; of total

Table 2.1. *Comparison of results based on differences in methods and time periods*

	# Food items	# Fruit species	# Leaf species	# Herb species	# Bark and plant structure species	# Insect species	# Flower species	Mushrooms
Gorilla								
Study period 1								
Fecal samples 10 months; n = 50 samples	NA	21	NA	NA	NA	2	NA	NA
Trail signs > 10 months; n = 320 days of trails	49	20	13	12	1	3	NA	NA
Study period 2								
Direct observation 10 months; (one male) n = 8643 feeding minutes	72	16	37	11	5	3	0	No
Mangabeys								
Study period 1								
Agile 10 months; (n = 1337 group scans)	58	52	0	3	0	Yes	1	Yes
Grey 10 months; n = 1377 group scans	68	50	8	0	5	Yes	4	No

NA indicates that data are not available. Mushroom species were not identified. Presence or absence of mushrooms in diet is noted as yes or no.

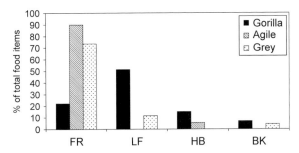

Figure 2.2. Gorilla and mangabey dietary components. The percentage of dietary items that are fruit (FR), leaf (LF), herb (HB), or bark (BK) identified for gorillas black, agile mangabeys shaded, and grey-cheeked mangabeys dotted during 10 months of behavioral sampling. Total number of food items identified = 72 (gorilla), 58 (agile mangabeys), and 68 (grey-cheeked mangabeys).

Table 2.2. *Foods eaten most frequently by gorillas, agile, and grey-cheeked mangabeys*

	Gorilla			Agile mangabey			Grey-cheeked mangabey		
Rank	Family	Species	%	Family	Species	%	Family	Species	%
1	Ulmaceae	*Celtis* sp.	18.4	Maranthaceae	*Haumania danckelmaniana* (pith)	15.8	Leg./Caesalpin	***Erythrophleum ivorense***	25.2
2	Zingiberaceae	*Aframomum limbatum*	15.8	Leg./Caesalpin	***Erythrophleum ivorense***	11.5		Insects	13.6
3	Tiliaceae	***Duboscia macrocarpa***	5.4	Ebenaceae	***Diospryros pseudomespilus***	8.5	Annonaceae	***Polyalthia suaveolens***	7.2
4	Hydrocharitaceae	*Hydrocharis chevalieri*	4.6		Insects	6.4	Moraceae	***Ficus* spp.**	4.2
5	Acanthaceae	*Thomandersia hensii*	4.0	Annonaceae	***Anonidium mannii***	5.8	Leg./Caesalpin	***Copaifera mildbraedii***	3.9
6	Commelinaceae	*Palisota brachythyrsa*	3.3	Flacourtiaceae	***Scotellia klaineana***	4.0	Annonaceae	***Xylopia* spp.**	3.8
7	Maranthaceae	*Megaphrynium macrostachyum*	3.2	Apocynaceae	***Tabernaemontana penduliflora***	3.4	Ulmaceae	*Celtis* sp.	3.7
8	Acanthaceae	*Whitfieldia elongata*	3.1	Annonaceae	***Polyalthia suaveolens***	3.3	Euphorbiaceae	***Croton cf haumanianus***	2.9
9	Zingiberaceae	*Aframomum subsericium*	2.5		Mushroom	3.2	Sapotaceae	***Manilkara mabokeensis***	2.6
10	Verbinaceae	***Vitex doniana* or *welwitschii***	2.4	Rubiaceae	***Nauclea diderrichii***	2.9		Unknown leaves	2.3
Sum			62.7			64.8			69.4

Foods are listed in descending order of rank based on the percent of feeding time during 8643 feeding minutes for gorillas and 1337 and 1377 group scans for agile and grey cheeked mangabeys respectively. **Bold** indicates the plant part is fruit (including seeds). All others indicate leaf, herb, or miscellaneous. Leg., Leguminosae, Caesalpin, Caesalpinioideae.

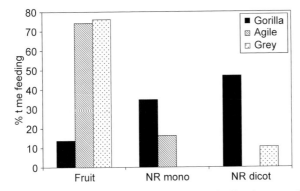

Figure 2.3. Gorilla and mangabey diet. Percent feeding time spent feeding on fruit, non-reproductive monocotyledonous stems and pith (NR mono) and dicotyledonous leaves (NR dicot) by gorillas, agile mangabeys, and grey-cheeked mangabeys. Data are from one-minute instantaneous sampling of gorillas (n = 8643 feeding minutes) and group scans of agile (n = 1337), and grey-cheeked mangabeys (n = 1377).

feeding time, leaves and bark accounted for 47.1% and herbs for 35.0%, whereas fruit and insects accounted for only an average of 13.9% and 3.9%, respectively (Figure 2.3; n = 8643 feeding minutes).

Comparison of western gorilla diet within and between Study Period 1 and 2

The total number of gorilla food items recorded during direct observation (n = 72; Study Period 2) was greater than that measured either from fecal samples (n = 24) or trail signs (n = 49) during Study Period 1. Nearly three times as many dicotyledonous leaf species were identified as food during direct observation compared with trail signs (Table 2.1). The trees on which these leaves occur are not deciduous, and consequently leaves are available year-round. Therefore, the limited number of leaf species identified as food in Study Period 1 compared with those in Study Period 2 was most likely an artifact of indirect sampling, consistent with previous findings that indirect sampling of diet underestimates leaf diversity.

Diversity in gorilla fruit consumption was more similar across the two periods, with 16 species identified as food in Study Period 2 and 20–21 identified in Study Period 1 (Table 2.1). A total of 31 gorilla fruit food species were identified over the two study periods by one of the three (fecal samples, trail signs, and observation) sampling methods. All but one of these species was previously identified as a fruit gorillas consumed at the site (Doran *et al.*, 2002), suggesting that previous indirect measures of diet sampling were largely effective in identifying the species of fruit consumed.

Mangabey diet in comparison to gorillas

Both mangabey species ate more than twice as many fruit species and far fewer herb and leaf species than did gorillas (Table 2.1). Fruit accounted for over 70% of food items identified (Figure. 2.2) and over 70% of feeding time (Figure. 2.3) for both mangabey species.

Mangabeys fed on a broader array of fruit species than did gorillas. This is true whether comparisons are made using similar methods (direct observation) at different time periods or different methods (indirect and direct sampling) during the same time period. A total of 81 fruit species were recorded for one of the three primate species based on similar amounts of direct observation (all foods identified from direct observation during 44 or 45 days of sampling in Study Period 1 (mangabeys) and 2 (gorillas). Gorillas ate only 20% of these fruit species, and shared only 12% and 6% in common with agile and grey-cheeked mangabeys, respectively, whereas the mangabeys ate 62% (grey-cheeked) and 64% (agile) of the total species, including the majority (69%) of gorilla fruit species. When compared during the same time period (Study Period 1), mangabeys fed on twice as many fruit species as did gorillas, in spite of more frequent gorilla sampling efforts (number of fruit species consumed: gorilla = 25, agile mangabey = 52, grey-cheeked mangabey = 50; sampling effort: gorillas, fecal = 50 samples and trail signs = 320 days of trail; mangabeys = 44 or 45 days of sampling).

When fruit species were shared in common by gorillas and mangabeys, mangabeys ate them over longer time periods than did gorillas. This extended use was possible because gorillas ate primarily ripe fruit, whereas mangabeys ate fruits and seeds that were unripe, ripe, or rotting, and dug up old seeds to feed on. For example, during Study Period 1, agile mangabeys fed on fruits/seeds of both *Anonidium manii* and *Polyalthia suaveolens* for 7 months, roughly twice as long as the 3–4 months of gorillas (Figure 2.4). Agile mangabeys began consuming both species 4 months earlier than did gorillas, consuming fruit while it was still unripe, and continued to feed on the pulp and seeds after gorillas ceased its consumption. Gorillas, on the other hand, nearly always consumed ripe fruit, and although they swallowed seeds, these passed largely undigested through the gut.

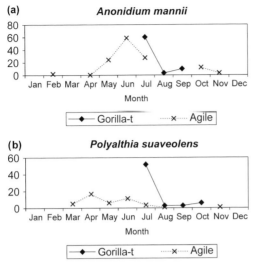

Figure 2.4. Months in which individual fruit species, (a) *Anonidium mannii* and (b) *Polyalthia suaveolens*, were eaten by gorillas and agile mangabeys during Study Period 1. Data are the percent of total feeding time for agile mangabeys (Agile) and (since direct observation of gorillas was not possible during that time period) the percent of days per month that the feeding remains of that species were recorded when following gorilla feeding trail (Gorilla-t). Gorillas and agile mangabeys were both sampled for 12 months, but zero values are omitted for clarity. Gorilla data are from 380 daily feeding trails and mangabey data are from 1977 group scans.

Mangabeys and gorillas differed in their fruit preferences as well. Both mangabey species relied more heavily on leguminous fruits than did gorillas. Mangabeys spent a greater proportion of their feeding time feeding on legumes and consumed a greater number of legume fruit species than did gorillas (percent of feeding time spent on legume fruits: gorilla = 1.1, agile = 13.6, grey = 22.0; number of legume fruit species consumed: gorilla = 2, agile = 5, grey = 8; n: gorilla = 8643 feeding minutes, agile = 1337 scans, grey = 1377 scans). This was particularly true for grey-cheeked mangabeys, for whom one species of legume, *Erythrophleum ivorense*, accounted for 25% of all feeding scans (Table 2.2), and was the most important food in 60% of months sampled, accounting for as much as 53% of the monthly diet (Table 2.3). This species was also the most important food species for agile mangabeys in 20% of months (Table 2.2), but was never recorded as a gorilla food during Study Period 1 or 2 or in the previous 3-year study at the site (Doran *et al.*, 2002).

Table 2.3. *Most important fruit species (percent of total monthly feeding time) consumed each month by gorillas and mangabeys*

	Gorilla			Agile mangabey			Grey-cheeked mangabey		
	Family	Species	%	Family	Species	%	Family	Species	%
Jan	Tiliaceae	*Duboscia macrocarpa*	6.6	Ebenaceae	*Diospryros pseudomespilus*	25.8	Leg./Caesalpin	*Erythrophleum ivorense*	26.8
Feb	Tiliaceae	*Duboscia macrocarpa*	1.0	Leg./Caesalpin	*Erythrophleum ivorense*	32.1	Moraceae	*Ficus sp.*	28.7
Mar	Tiliaceae	*Duboscia macrocarpa*	6.5	Flacourtiaceae	*Scotellia klaineana*	39.4	Leg./Caesalpin	*Erythrophleum ivorense*	37.0
Apr	Tiliaceae	*Duboscia macrocarpa*	4.0	Annonaceae	*Polyalthia suaveolens*	16.3	Leg./Caesalpin	*Erythrophleum ivorense*	31.6
May	Moraceae	*Treculia africana*	5.3	Annonaceae	*Anonidium mannii*	25.6	Annonaceae	*Xylopia spp.*	19.1
Jul	Tiliaceae	*Duboscia macrocarpa*	12.4	Annonaceae	*Anonidium mannii*	32.0	Annonaceae	*Polyalthia suaveolens*	51.5
Aug	Verbinaceae	*Vitex doniana or welwitschii*	19.0	Euphorbiaceae	*Drypetes polyantha*	10.3	Leg./Caesalpin	*Erythrophleum ivorense*	28.5
Sep	Verbinaceae	*Vitex doniana or welwitschii*	14.3	Maranthaceae	*Haumania danckelmaniana*	22.8	Sapotaceae	*Manilkara mabokeensis*	26.5
Oct	Tiliaceae	*Duboscia macrocarpa*	1.9	Leg./Caesalpin	*Erythrophleum ivorense*	25.7	Leg./Caesalpin	*Erythrophleum ivorense*	57.8
Nov	Tiliaceae	*Duboscia macrocarpa*	9.0	Ebenaceae	*Diospryros pseudomespilus*	45.6	Leg./Caesalpin	*Erythrophleum ivorense*	52.9

Leg.: Leguminoseae; Caesalpin: Caesalpinaceae.

Species differences in response to fluctuating fruit availability

Gorillas and mangabeys appear to differ in their response to changes in ripe fruit availability. Mangabeys' monthly dietary diversity was not significantly correlated with changes in fruit availability (all fruit species) during Study Period 1 (Spearman's rho: agile = 0.410, p = 0.27; grey = −0.555, p = 0.12). Thus, mangabeys do not appear to adjust the diversity of items in their diets in response to changing fruit availability, as also demonstrated in previous studies (Wrangham *et al.*, 1998). Although we could not test this directly for western gorillas due to the absence of phenological data during Study Period 2, there is evidence to suggest that gorillas decrease diet diversity in response to increasing fruit availability. In a previous study we documented that as fruit availability increased, the number of species of fruit consumed on a daily basis (as measured in fecal samples) increased and the number of herb and leaf species (measured from trail signs) in the diet decreased (Doran *et al.*, 2002), as in all previous western gorilla findings (Tutin *et al.*, 1991; Nishihara, 1995; Remis, 1997; Goldsmith, 1999). Here, we find that as gorilla time spent feeding on fruit increases, time spent feeding on leaves other than *Celtis* decreases significantly (Pearson correlation r = −0.605, p = 0.037; n = 12 months). Time spent feeding on fruit peaked (June–September) when fruit availability could reasonably be assumed to be highest, based on long-term phenological monitoring at the site (Figure 2.5). This suggests that as fruit availability increases, gorilla consumption of fruit increases, whereas that of leaves decreases. One leaf species (*Celtis* sp.), demonstrated at other sites to be high in protein (Bocian, 1997; Chapman & Chapman, 2002), is, however, not a fall-back species since its consumption did not decrease significantly in response to increased fruit consumption (Pearson correlation r = −0.258, p = 0.42, n = 12 months).

Mangabeys, on the other hand, did not vary fruit/seed consumption with changing ripe fruit availability. Rather, fruit consumption remained high throughout the year, accounting for 65–93% (agile) and 44–94% (grey-cheeked) of mangabey monthly feeding scans.

Discussion

Western gorilla diet

Results from previous studies of western gorilla diet, based on indirect sampling, have indicated that western gorillas have greater dietary breadth and eat more fruit than the well-studied Karisoke mountain gorilla, and these

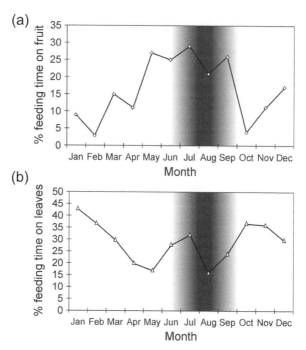

Figure 2.5. The percentage of total gorilla feeding time spent feeding on (a) fruit and (b) non-*Celtis* leaves during Study Period 2, shown relative to average ripe fruit availability at Mondika from 1996–2002. The grey bar indicates the time of the year when fruit is most abundant.

studies have generally emphasized the importance of this greater frugivory (Remis, 1997; Doran *et al.*, 2002). Results presented here, based on direct observation, albeit of a single male, confirm that the western gorilla diet was more diverse than that of mountain gorillas, routinely including fruit, dicotyledonous leaves, herbs, bark, and insects. However, contrary to expectations, we found the degree of frugivory was much lower than in any previous estimates. Fruit accounted for only 22% of food species identified and 14% of the annual diet, whereas previous estimates based on limited observation suggested that fruit consumption accounted for an average of 40–50% of western gorilla diet (Remis, 1997; Tutin *et al.*, 1997).

This discrepancy may result, in part, if ripe fruit availability was unusually low during this study, or if sampling a single individual failed to capture the range of fruit-eating in male and female gorillas. Phenological data are not available to test whether the former scenario was the case. However, although qualitative data suggest that overall fruit availability was reduced in Study Period 2 relative to Study Period 1, the majority of fruit species known to be important to gorillas were consumed during both study periods (Doran-Sheehy, unpublished data), indicating that interannual variation in fruit availability was likely not extreme enough to account for this discrepancy. Regarding the second possibility, it is noteworthy that previous studies have failed to detect any consistent sex difference in gorilla diet (Doran *et al.*, 2002), and although it has been hypothesized that males should be more folivorous than females due to their larger body size, it is equally plausible that males might be more frugivorous due to their ability to monopolize fruits (a preferred resource) through their dominance over females (Doran-Sheehy, unpublished data).

It is clear that further research on sex differences and interannual variation in gorilla diet will be necessary to provide a more complete understanding of western gorilla diet. However, these results suggest that frugivory in western gorillas is much lower than previously considered. This discrepancy may result from earlier inherent biases in indirect sampling or behavioral sampling prior to habituation, which tended to overemphasize frugivory. Although studies based on indirect sampling of diet can provide a good assessment of the diversity and frequency of fruit consumption, it provides, at best, a poor approximation of the actual quantity of fruit consumed (Tutin & Fernandez, 1985; Doran *et al.*, 2002). Additionally, previous results based on direct observation were based on relatively few observations and biased towards arboreal observations (since follows of gorillas on the ground were not possible), where fruit eating frequently occurs (Remis, 1997; Tutin *et al.*, 1997).

Differing patterns of frugivory in apes and mangabeys

In spite of rather limited frugivory, the pattern of fruit consumption in western gorillas was similar to that of chimpanzees studied previously (Wrangham *et al.*, 1998) and distinct from that of the mangabeys at this site. Gorillas were more selective in their fruit choice than mangabeys. First, gorillas ate a much smaller subset (roughly 25%) of available fruits than did mangabeys. Second, gorillas, unlike mangabeys, did not feed on species from the most commonly available trees. Mangabeys fed on fruits from 65–85% of the most commonly available trees, as measured by stem density or basal area (Shah, 2003), whereas gorillas only used 30% of them (Doran *et al.*, 2002). Third, gorillas largely restricted their fruit eating to ripe fruit, whereas mangabeys fed on fruit pulp or seeds from unripe, ripe, and rotting fruits, thereby feeding on any given fruit species over a much longer period of time than gorillas. Although we did not measure chemical content of food, it is likely that in their fruit and seed consumption, mangabeys regularly incorporated higher levels of antifeedants than gorillas since unripe fruits, seeds, and legumes are all known to be chemically defended (see Waterman & Kool, 1994 for review). Finally, gorilla fruit eating appeared to track ripe fruit availability, increasing at times when fruit availability is usually greatest. When fruit eating declined, gorilla dietary diversity increased as gorillas incorporated more leaves, herbs, and bark in the diet. Mangabeys, on the other hand, maintained high levels of frugivory and relatively constant diet diversity through the year, regardless of changes in fruit availability.

Thus, gorillas appear to adopt a dietary strategy of feeding on high quantities of leaves and herbs, and when possible, pursuing a diet of ripe, succulent fruit. This is further supported by previous studies of western gorilla ranging behavior. Western gorillas travel farther to add preferred ripe fruit to their diet when it is available, rather than subsist on lower-quality forage (Tutin, 1996; Remis, 1997; Goldsmith, 1999; Doran-Sheehy et al., 2004). Much of the variation in gorilla daily path length can be explained by variation in the degree of frugivory, with daily path length increasing with increased fruit availability and number of fruit trees visited (Doran-Sheehy et al., 2004). Mangabeys showed an opposite pattern, adopting a more generalized approach to diet, feeding on more readily available species of fruit, maintaining high levels of frugivory throughout the year, irrespective of availability. When fruit consumption increased, daily path length was either reduced, as in agile mangabeys, or not significantly altered, as in grey-cheeked mangabeys (Shah, 2003).

Ape and monkey dietary specializations

The distinct foraging strategies of gorillas and mangabeys seen in this study strongly resemble the pattern described by Wrangham et al. (1998) for chimpanzees and a broader array of old world monkeys. Finding a similar pattern of frugivory across ape species is especially interesting in light of the wide variation in the amount of fruit they consume and the relatively small amount of fruit consumed by gorillas in this study.

Previous studies of nutritional content of gorilla foods from the wild, as well as studies of gorilla taste preferences in captivity, indicate that gorillas, like other apes, are highly selective in food choice compared with old world monkeys, preferring plant parts that contain less fiber, more sugar, and lower levels of antifeedants (Remis et al., 2001; Remis & Kerr, 2002; Rogers et al., 2004) or relatively less fiber and more water-soluble sugar (Wrangham et al., 1998). Apes generally prefer ripe to unripe fruit (chimpanzees: Wrangham et al., 1998; orangutans: Knott, 1999), have strong preferences for particular and often rare species (Rogers *et al.*, 2004), incorporate ripe fruit in their diet on the basis of availability (Wrangham *et al.*, 1998; Knott, 1999), and travel farther to obtain it when available (orangutan: Galdikas & Teleki, 1981; Knott, 1999; chimpanzees: Doran, 1997; Wrangham *et al.*, 1998).

Old world monkeys, on the other hand, have a more generalized feeding strategy. They are less selective in fruit choice, feeding on a wide variety of commonly available species (Kaplin *et al.*, 1998; Kaplin & Moermond, 2000), showing no clear preference for ripe fruit (they don't incorporate it to greater degrees when available) (Kinnaird, 1990; Wrangham *et al.*, 1998), and feeding on fruits with higher amounts of antifeedants, including legumes, unripe fruits, and seeds (Wrangham *et al.*, 1998; Poulsen *et al.*, 2001).

The fact that the Cercopithecinae can subsist on a lower-quality diet, one that is higher in fiber and antifeedants, is perhaps surprising because of their small body size. However, given that their gut passage rates are much slower than most similarly sized new world monkeys (Maisels, 1993; Lambert, 1998), Lambert (1997, 1998) has suggested that selection acted to considerably slow gut passage time in Cercopithecines to allow for higher rates of fermentation and absorption. Furthermore, Lambert (2002) has suggested that small body size in Cercopithecines may provide a benefit when it comes to detoxification of plant metabolites since the process requires the activity of microsomal enzymes, and the rate of enzymatic activity may scale negatively with body size (Walker, 1978).

The other old world monkey radiation, the Colobinae, are similar to Cercopithecines in that they also subsist on a more generalized, lower-quality (but largely folivorous) diet. This is possible through modified stomach anatomy, which permits fermentation and efficient extraction of nutrients from readily available and heavily defended leaves and seeds (Kay & Davies, 1994). Although it is typical to see the two radiations as widely divergent in foraging strategy, the Cercopithecines as frugivorous and the Colobines as folivorous, comparison with apes show them to share a similar foraging strategy of less selective consumption of more abundant and more heavily defended foods, distinct from the apes' more selective feeding on less readily available fruits that are lower in fiber and antifeedant content (Temerin & Cant, 1983; Wrangham *et al.*, 1998). Given the great apes' larger body size, and resultant long gut passage times (Milton, 1984; Lambert, 1998, 2002; Caton, 1999; Remis & Dierenfeld, 2004), they may subsist on lower-quality forage when necessary, or even completely, as in the case of mountain gorillas. However, their taste preferences, feeding and ranging behavior all indicate that when possible, they prefer succulent, ripe fruit. This is consistent with earlier descriptions of the foraging adaptations that define the monkey/ape divergence (Temerin & Cant, 1983; Lambert, 1998, 2002; Wrangham *et al.*, 1998). This difference in the pattern of frugivory between apes and old world monkeys may have far-reaching implications for the evolutionary history of the two lineages, accounting for the greater taxonomic diversity and broader geographic distribution of old world monkeys compared with apes. Because monkeys' dental and gut adaptations allowed them to subsist on more common, lower-quality food sources, it may have permitted them to survive in a wider range of habitats and thus extend their biogeographic range during periods of climate change. Apes' preference for feeding more selectively on high-quality fruits may have limited their distribution to the ever-contracting forest patches where such foods could be found.

Acknowledgments

For permission to conduct research at Mondika, we gratefully acknowledge the Ministries of Eaux et Forêts and Recherches Scientifiques in Central African Republic and the Ministry of l'Enseignement Primaire, Secondaire et Supérieur Charge de la Recherche Scientifique in Republic of Congo. Financial support for the two studies is from the National Science Foundation (SBR-9729126), the office of the Vice President of Research of Stony Brook University, the Leakey Foundation, US Fish and Wildlife Great Ape Conservation Fund, and the Wildlife Conservation Society; IDPAS at Stony Brook University, Primate Conservation Inc., and the Wenner-Gren Foundation. We thank M. Alafei Abel, Directeur National of the Dzanga-Ndoki National Park, Conservateur Djoni of the Nouabale-Ndoki National Park, Emma Stokes, Mark Gately, Paul Elkan, and Bryan Curran of WCS, and Allard Blom, David Greer, and Chloe Cippoletta of WWF for support and logistical assistance. We would especially like to acknowledge the contribution of David Harris in identifying botanical specimens at Mondika. Finally, work at Mondika would not be possible without the efforts of many people in the field, including Tim Rayden, Roberta Salmi, Whitney Taylor DeSpain, and the skilled assistance and botanical lessons of many BaAka field colleagues, especially Mangombe, Mokonjo, Ndima, Mamandele, Bakombo, and Mbokola.

References

Altmann, J. (1974). Observational study of behavior: sampling methods. *Behaviour*, **49**, 227–65.

Blom, A., Cipolletta, C., Brunsting, A. M. H., & Prins, H. H. T. (2004). Behavioral responses of gorillas to habituation in the Dzanga-Ndoki National Park, Central African Republic. *International Journal of Primatology*, **25**, 179–96.

Bocian, C. (1997). *Niche Separation of Black and White Colobus Monkeys in the Ituri Forest (Zaire)*. Unpublished Ph.D. thesis, City University of New York.

Caton, J. (1999). A preliminary report on the digestive strategy of the western gorilla. *Australian Journal of Primatology*, **13**, 2–7.

Chapman, C. A. & Chapman, L. J. (2002). Foraging challenges of red colobus monkeys: influence of nutrients and secondary compounds. *Comparative Biochemistry and Physiology*, **133**, 861–75.

Chivers, D. & Hladik, C. (1980). Morphology of the gastrointestinal tract in primates: comparisons with other mammals in relation to diet. *Journal of Morphology*, **166**, 337–86.

Conklin-Brittain, N. L., Wrangham, R. W., & Hunt, K. D. (1998). Dietary response of chimpanzees and cercopithecines to seasonal variation in fruit abundance. II. Macronutrients. *International Journal of Primatology*, **19**, 971–98.

Demment, M. W. & Van Soest, P. J. (1983). *Body Size, Digestive Capacity, and Feeding Strategies of Folivores*. Morrilton: Winrock International.

Doran, D. M. (1997). Influence of seasonality on activity patterns, feeding behavior, ranging and grouping patterns in Tai chimpanzees. *International Journal of Primatology*, **18**, 183–206.

Doran, D. M. & McNeilage, A. (1998). Gorilla ecology and behavior. *Evolutionary Anthropology*, **6**, 120–31.

(2001). Subspecific variation in gorilla behavior: the influence of ecological and social factors. In *Mountain Gorillas: Three Decades of Research at Karisoke*, ed. M. M. Robbins, P. Sicotte, & K. J. Stewart, pp. 123–49. Cambridge: Cambridge University Press.

Doran, D. M., McNeilage, A., Greer, D., Bocian, C., Mehlman, P., & Shah, N. (2002). Western lowland gorilla diet and resource availability: new evidence, cross-site comparisons and reflections on indirect sampling methods. *American Journal of Primatology*, **58**, 91–116.

Doran-Sheehy, D. M., Greer, D., Mongo, P., & Schwindt, D. (2004). Ecological and social impact on ranging in western gorillas. *American Journal of Primatology*, **64**, 207–22.

Fleagle, J. G. (1999). *Primate Adaptation and Evolution*. New York: Academic Press.

Fleagle, J. G. & McGraw, W. S. (1999). Skeletal and dental morphology supports diphyletic origin of baboons and mandrills. *Proceedings of the National Academy of Sciences*, **96**, 1157–61.

Galdikas, B. M. F. (1988). Orangutan diet, range, and activity at Tanjung Putting, Central Borneo. *International Journal of Primatology*, **9**, 1–35.

Galdikas, B. M. F. & Teleki, G. (1981). Variations in subsistence activities of female and male pongids: new perspectives on the origins of hominid labor division. *Current Anthropology*, **22**, 241–56.

Goldsmith, M. L. (1999). Ecological constraints on the foraging effort of western gorillas (*Gorilla gorilla gorilla*) at Bai Hokou, Central African Republic. *International Journal of Primatology*, **20**, 1–23.

Kaplin, B. A. & Moermond, T. C. (2000). Foraging ecology of the mountain monkey (*Cercopithecus l'hoesti*): implications for its evolutionary history and use of disturbed forest. *American Journal of Primatology*, **50**, 227–46.

Kaplin, B. A., Munyaligoga, V., & Moermond, T. C. (1998). The influence of temporal changes in fruit availability on diet composition and seed handling in blue monkeys (*Cercopithecus mitis doggetti*). *Biotropica*, **30**, 56–71.

Kay, R. N. B. (1984). On the uses of anatomical features to infer foraging behavior in extinct primates. In *Adaptations for Foraging in Nonhuman Primates*, ed. P. S. Rodman & J. G. H. Cant, pp. 21–53. New York: Columbia University Press.

Kay, R. N. B. & Davies, A. G. (1994). Digestive physiology. In *Colobine Monkeys*, ed. A. G. Davies & J. F. Oates, pp. 229–49. Cambridge: Cambridge University Press.

Kinnaird, M. F. (1990). *Behavioral and Demographic Responses to Habitat Change by the Tana River Crested Mangabeys* Cercocebus galeritus galeritus. Unpubished Ph.D. thesis, University of Florida.

Knott, C. D. (1999). *Reproductive, Physiological, and Behavioral Responses of Orangutans in Borneo to Fluctuations in Food Availability*. Unpublished Ph.D. thesis, Harvard University.

Lambert, J. E. (1997). *Digestive Strategies, Fruit Processing, and Seed Dispersal in the Chimpanzees* (Pan troglodytes) *and Redtail Monkeys* (Cercopithecus ascanius) *of Kibale National Park, Uganda*. Unpublished Ph.D. thesis, University of Illinois-Urbana-Champaign.

(1998). Primate digestion: interactions among anatomy, physiology, and feeding ecology. *Evolutionary Anthropology*, **7**, 8–20.

(2002). Digestive retention times in forest guenons (*Cercopithecus* spp.) with reference to chimpanzees. *International Journal of Primatology*, **23**, 1169–85.

Lucas, P. W. & Teaford, M. F. (1994). Functional morphology of colobine teeth. In *Colobine Monkeys*, ed. A. G. Davies & J. F. Oates, pp. 173–203. Cambridge: Cambridge University Press.

Maisels, F. (1993). Gut passage rate in guenons and mangabeys: another indicator of a flexible feeding niche. *Folia Primatologica*, **61**, 35–7.

Milton, K. (1984). The role of food processing factors in primate food choice. In *Adaptations for Foraging in Nonhuman Primates*, ed. P. S. Rodman & J. G. H. Cant, pp. 249–79. New York: Columbia University Press.

Nishihara, T. (1995). Feeding ecology of western lowland gorillas in the Nouabale-Ndoki National Park, Congo. *Primates*, **36**, 151–68.

Parra, R. (1978). Comparisons of foregut and hindgut fermentation in herbivores. In *The Ecology of Arboreal Folivores*, ed. G. G. Montgomery, pp. 205–29. Washington, DC: Smithsonian Institution Press.

Pielou, E. C. (1969). *An Introduction to Mathematical Ecology*. New York: John Wiley & Sons, Inc.

Poulsen, J. R., Clark, C. J., & Smith, T. B. (2001). Seasonal variation in the feeding ecology of the grey-cheeked mangabey (*Lophocebus albigena*) in Cameroon. *American Journal of Primatology*, **54**, 91–105.

Remis, M. J. (1997). Western lowland gorillas (*Gorilla gorilla gorilla*) as seasonal frugivores: use of variable resources. *American Journal of Primatology*, **43**, 87–109.

Remis, M. J. & Dierenfeld, E. S. (2004). Digesta passage, digestibility and behavior in captive gorillas under two dietary regimes. *International Journal of Primatology*, **25**, 825–45.

Remis, M. J. & Kerr, M. E. (2002). Taste responses to fructose and tannic acid among gorillas (*Gorilla gorilla gorilla*). *International Journal of Primatology*, **23**, 251–61.

Remis, M. J., Dierenfeld, E. S., Mowry, C. B., & Carroll, R. W. (2001). Nutritional aspects of western lowland gorilla diet during seasons of fruit scarcity at Bai Hokou, Central African Republic. *International Journal of Primatology*, **22**, 807–36.

Rogers, E., Abernethy, K., Cipolletta, C., et al. (2004). Western gorilla diet: a synthesis from six sites. *American Journal of Primatology*, **64**, 173–92.

Shah, N. (2003). *Foraging Strategies in Two Sympatric Mangabey Species* (Cerco-cebus agilis *and* Lophocebus albigena). Unpublished Ph.D. thesis, Stony Brook University.

Smith, R. A. & Jungers, W. L. (1997). Body mass in comparative primatology. *Journal of Human Evolution*, **32**, 523–59.

Temerin, L. A. & Cant, J. G. H. (1983). The evolutionary divergence of old world monkeys and apes. *The American Naturalist*, **122**, 335–51.

Tutin, C. E. G. (1996). Ranging and social structure of lowland gorillas in the Lopé Reserve, Gabon. In *Great Ape Societies*, ed. W. C. McGrew, L. F. Marchant, & T. Nishida, pp. 58–70. Cambridge: Cambridge University Press.

Tutin, C. E. G. & Fernandez, M. (1985). Foods consumed by sympatric populations of *Gorilla g. gorilla* and *Pan t. troglodytes* in Gabon: some preliminary data. *International Journal of Primatology*, **6**, 27–43.

(1991). Responses of wild chimpanzees and gorillas to the arrival of primatologists: behaviour observed during habituation. In *Primate Responses to Environmental Change*, ed. H. O. Box, pp. 187–97. London: Chapman and Hall.

(1993). Composition of the diet of chimpanzees and comparisons with that of sympatric lowland gorillas in the Lope Reserve, Gabon. *American Journal of Primatology*, **30**, 195–211.

Tutin, C. E. G., Fernandez, M., Rogers, M. E., Williamson, E. A., & McGrew, W. C. (1991). Foraging profiles of sympatric lowland gorillas and chimpanzees in the Lope Reserve, Gabon. *Philosophical Transactions of the Royal Society of London, B*, **334**, 179–86.

Tutin, C. E. G., Ham, R. M., White, L. J. T., & Harrison, M. J. S. (1997). The primate community of the Lope Reserve, Gabon: diets, responses to fruit scarcity, and effects on biomass. *American Journal of Primatology*, **42**, 1–24.

Van Soest, P. J. (1996). Allometry and ecology of feeding behavior and digestive capacity in herbivores: a review. *Zoo Biology*, **15**, 455–79.

Walker, C. H. (1978). Species differences in microsomal monooxygenase activity and their relationship to biological half-lives. *Drug Metabolism Reviews*, **7**, 295–310.

Waterman, P. G. & Kool, K. M. (1994). Colobine food selection and plant chemistry. In *Colobine Monkeys*, ed. A. G. Davies & J. F. Oates, pp. 251–84. Cambridge: Cambridge University Press.

Watts, D. P. (1984). Composition and variability of mountain gorilla diets in the Central Virungas. *American Journal of Primatology*, **7**, 323–56.

(1996). Comparative socioecology of gorillas. In *Great Ape Societies*, ed. W. C. McGrew, L. F. Marchant, & T. Nishida, pp. 16–28. Cambridge: Cambridge University Press.

Williamson, E. A. (1989). *Behavioural Ecology of Western lowland Gorillas in Gabon*. Unpublished Ph.D. thesis, University of Stirling.

Wrangham, R. W., Conklin-Brittain, N. L., & Hunt, K. D. (1998). Dietary response of chimpanzees and cercopithecines to seasonal variation in fruit abundance. I. Antifeedants. *International Journal of Primatology*, **19**, 949–70.

3   *Effects of fruit scarcity on foraging strategies of sympatric gorillas and chimpanzees*

JUICHI YAMAGIWA AND AUGUSTIN KANYUNYI BASABOSE

Introduction

In tropical forests, most primates live sympatrically with other primate species. Several species form polyspecific associations, with large overlap

Feeding Ecology in Apes and Other Primates. Ecological, Physical and Behavioral Aspects, ed. G. Hohmann, M.M. Robbins, and C. Boesch. Published by Cambridge University Press.
© Cambridge University Press 2006.

in diet and ranging due to increase in benefits of predator detection, food resource location, and insect capture (Gautier-Hion *et al.*, 1983; Cord, 2000; Heymann & Buchanan-Smith, 2000). However, among most sympatric species, distinct differences are found in diet, ranging, foraging height, locomotion pattern, and activity period (Gautier-Hion *et al.*, 1983; Terborgh, 1983; Waser, 1987; Gebo & Chapman, 1995). Such differences may suggest niche divergence to mitigate interspecies competition, although it is difficult to elucidate the evolutionary effects of competition. Niche separation between sympatric primate species becomes more pronounced during periods of food shortage (Ungar, 1996; Tan, 1999; Powzyk & Mowry, 2003).

Gorillas and chimpanzees coexist sympatrically in equatorial Africa. Unlike most sympatric species, these apes rarely associate with other primate species. Unlike some small and medium-sized primates characterized by special diets and small ranges, both of these apes have large body size, diverse diets, and use a wide range of habitats, including most types of vegetation and all strata of tropical forests. It is still unclear how such wide-ranging species with large body size coexist sympatrically. In initial studies, the possible reasons for their coexistence were attributed to niche differentiation, for example, diet and ranging area (Schaller, 1963; Jones & Sabater Pi, 1971). Frugivorous chimpanzees tended to range in primary forests and to stay on dry ridges, while folivorous gorillas tended to range in secondary, regenerating forests and to stay in wet valleys. These differences in ecological features were considered to affect the apes' social features: fission-fusion grouping patterns caused by strong frugivory of chimpanzees and cohesive group formation by the consistent folivory of gorillas (Harcourt, 1978; Wrangham, 1987). These observations were, however, based on short-term studies on sympatric populations in Uganda and Rio Muni, or on an allopatric population of mountain gorillas in the Virungas and chimpanzees in Budongo and Gombe.

Recent studies on sympatric populations have shown extensive overlap between gorillas and chimpanzees in diet, foraging height, and ranging. Western and eastern lowland gorillas generally feed on fruits and insects and range in primary forests, as also observed for chimpanzees (Williamson *et al.*, 1990; Tutin & Fernandes, 1992, 1993; Yamagiwa *et al.*, 1994, 1996a; Kuroda *et al.*, 1996; Remis, 1997a; Doran & McNeilage, 1998; Rogers *et al.*, 2004). Fruit species eaten by western lowland gorillas at Lopé, Gabon overlapped by 79% those eaten by sympatric chimpanzees (Tutin & Fernandez, 1993).

The diets of gorillas and chimpanzees tend to diverge during periods of fruit scarcity. Gorillas rely heavily on vegetative foods when fruit availability is limited, while chimpanzees persistently seek fruits (Tutin & Fernandez,

1993; Yamagiwa *et al.*, 1994, 1996a; Nishihara, 1995; Remis, 1997a). Western and eastern lowland gorillas range in large open clearings or bamboo forests, where fruit production is very low throughout the year and chimpanzees rarely appear (Casimir & Butenandt, 1973; Doran-Sheehy *et al.*, 2004). Significant negative correlation is found between elevation of habitat and degree of frugivory for gorillas, while for chimpanzees no such relationship appears to exist between them (Stanford & Nkurunungi, 2003). It is assumed that gorillas and chimpanzees exhibit different foraging strategies when and where fruit availability is limited.

Which aspects of the dietary and ranging behaviors of these apes are influenced by fruit scarcity? How does interspecific feeding competition shape their socioecological features? In order to answer these questions, we conducted a research project on a sympatric population of eastern lowland gorillas (*Gorilla beringei graueri*) and eastern chimpanzees (*Pan troglodytes schweinfurthii*) inhabiting the montane forest of Kahuzi-Biega National Park, Democratic Republic of Congo (DRC). The study area is located in and below the bamboo forest, which constitutes the altitudinal upper limit of chimpanzee distribution; for both gorillas and chimpanzees the montane habitat is characterized by less diversity and productivity of fruit than are the lowland tropical forests (Yamagiwa *et al.*, 1992; Basabose, 2002). We have habituated a group of gorillas and a unit-group of chimpanzees and have monitored their daily movements and diet since 1994 through direct observations, by following their fresh trails, and using fecal analysis. We have also monitored the fruiting of their major plant foods by taking a fruit trail census. Gorillas exhibit a frugivorous diet during the dry season while consuming various kinds of vegetative food throughout the year (Casimir, 1975; Goodall, 1977; Yamagiwa *et al.*, 2005). Chimpanzees consume more kinds of fruits than gorillas and feed frequently on animal matter in a smaller home range than chimpanzees in other habitats (Basabose, 2002, 2005). In this paper, we use data on the diet and ranging of gorillas and chimpanzees, collected over a period of more than 8 years, to analyze their foraging strategies in relation to fruit scarcity. Dietary features and ranging patterns of gorillas and chimpanzees in the montane forest of Kahuzi are compared with those in other habitats.

Methods

Study area

The Kahuzi-Biega National Park is located to the west of Lake Kivu and covers an area of 6000 km^2 at an altitude of 600–3308 m. The park consists of highland (600 km^2) and lowland (5400 km^2) sectors, which are interconnected by a corridor of forest. The study area covers about 100 km^2 along the eastern border of the park at an altitude of 2050–2350 m, where a group of *G. b. graueri* and a unit-group of *P. t. schweinfurthii* range sympatrically (Yamagiwa *et al.*, 1996a; Basabose, 2002).

The study area is made up of bamboo *Arundinaria alpina* forest, primary montane forest, secondary montane forest, *Cyperus latifolius* swamp, and other vegetation, as described by Casimir (1975) and Yumoto *et al.* (1994). The mean annual rainfall during the study period (1994–2002) was 1658 mm (range: 1409–2180 mm) with a distinct dry season (June, July, and August) in which the mean rainfall was below 50 mm. The mean monthly temperature was 20.2 °C (mean maximum: 26.5 °C; mean minimum: 13.8 °C).

In August 1994, we set up a belt transect 5000 m long and 20 m wide in the study area (Basabose, 2002). Every tree, shrub, and strangling fig above 10 cm in diameter at breast height (DBH) was identified. A total of 2033 trees, including shrubs and strangling figs above 10 cm in DBH, of 49 species from 29 families and one unidentified species were recorded along the transect. The density (the number of individual trees/ha in the transect) and basal area [(1/2 DBH)2 × π] were calculated for each species. Until July 1996, when our census was interrupted by civil war, each tree was monitored twice per month to record fruiting. Among 49 species identified, we found 25 species whose fruit was eaten either by gorillas or chimpanzees. Since February 1998 we have focused on 28 plant species (22 woody plant species and 2 herb species whose fruit is frequently eaten by apes, and 4 species whose fruit is not eaten by apes). For each species, flowers, leaves, fruits (ripe and unripe distinguished by color, hardness, and odor) of at least 10 (range: 10–13) reproductively mature trees were monitored twice each month on the fruit trail.

To estimate fruit abundance (biomass and number) of tree species, we used diameter at breast height (DBH) (Chapman *et al.*, 1992). We calculated a monthly fruit index (Fm) as

$$F_m = \sum_{k=1}^{s} P_{km}B_k$$

where P$_{km}$ denotes the proportion of the number of trees in ripe fruit for species k in month m, and B_k denotes the total basal area/ha for species k. Data on ripe fruit were used for calculation because gorillas and chimpanzees rarely consumed unripe fruits. Among tree species monitored monthly, we used those ranked in the top three species by fecal analysis for calculation of the fruit index (seven for gorillas and 11 for chimpanzees, see Table 3.1). Shrubs (*Piper capense*), vines (*Rubus* spp. and *Landolphia owariensis*), and

Table 3.1. *Percentage of fecal samples out of the total samples collected during the study period that included each of the fruit species, and number of months in which each species was ranked in top three species*

	Percentage of fecal samples including each fruit species (%)		Number of months ranked in top 3	
	Gorilla	Chimpanzee	Gorilla	Chimpanzee
**Myrianthus holstii*	31.4	31.9	70	58
**Ficus* spp.	17.1	94.1	76	90
**Bridelia bridelifolia*	7	9.4	22	17
**Allophyllus* spp.	4.1	3.6	9	10
**Syzygium guineense*	3.7	6.5	19	16
Rubus spp.	2.5	11.9	20	22
**Psychotria palustris*	1.6	1.7	4	2
Piper capense	0.5	1.7	7	2
**Maesa lanceolata*	0.5	11.1	1	14
Landolphia owariensis	0.2	2.5	1	4
Aframomum sanguinium	0.2	2.9	1	2
**Ekebergia capensis*	0.1	7.7	1	16
Impatiens spp.	0.1	1	1	1
**Cassipourea ruwenzoriensis*	0.001	1.6	0	4
**Diospyros honleana*	0.001	3	0	6
**Newtonia buchananii*	0.001	3.3	0	4

Estimated to be preferred by both apes (**) or only by chimpanzees (*)

herbs (*Aframomum sanguinium* and *Impatiens* spp.), which were not monitored on the fruit trail, were excluded from calculation of the fruit index.

Data collection of ape diet and ranging

Since 1991 we have tried to habituate a group of gorillas and a single unit-group of chimpanzees in the study area. These groups had extensive overlapping ranging areas (Yamagiwa *et al.*, 1996a, 1996b). Until 1994, both groups had been semi-habituated and occasionally tolerated the presence of human observers when we stayed at a distance of 20 m. We formed two survey teams, each consisting of a field assistant and two trackers, who traced fresh trails (up to 1 day old) between consecutive nest sites, recording the locations and compositions of nests (height, material, diameter, and size of feces in each nest). Between 1994 and 2002 the total number of individuals in the study group of gorillas was 7–25 and that of chimpanzees was 22–23. The leading silverback and three females of the gorilla group were killed by poachers in 1998 and 1999, and the group split into two groups in 2000.

Comparison of diet between gorillas and chimpanzees was based mainly on fecal analysis. Discrimination of feces between gorillas and chimpanzees was based on fecal shape and odor. Fresh (up to 1-day-old) feces were collected mainly at nest sites, once from each individual's dung pile to avoid collection of multiple samples from the same individual. Fecal samples were washed in 1-mm mesh sieves, dried in sunlight, and stored in plastic bags. The contents of each sample were examined macroscopically and listed as seeds, fruit skins, fiber, leaves, fragments of insects, and other items. The volume percentage of each of these different items was estimated at 5% intervals. Fruit seeds and skin were identified at the species level macroscopically. Over the course of 92 months, from August 1994 to September 2002, 14 367 gorilla fecal samples, averaging 152 samples per month (range: 36–361), and 8070 chimpanzee fecal samples, averaging 87 samples per month (range: 17–427), were collected. The fruit species observed in more than 0.5% of total fecal samples and as one of the top three fruits in a month, as ranked by fecal analysis, were defined as being preferred by either gorillas or chimpanzees (Table 3.1). Because this definition did not take into account availability of each fruit, the term "preferred" may not precisely reflect an ape's preference but rather reflects their relative frequent consumption.

As used in the previous studies on mountain gorillas (Fossey & Harcourt, 1977; Vedder, 1984; Watts, 1998a), we plotted daily movements of the gorilla study group on a grid of 250×250-m quadrates superimposed on a 1:25 000 vegetation map of the study area. From 1994–2002, we collected complete day routes on 1440 days in 92 months, averaging 16 days per month (range: 5–29). Due to their fission-fusion nature and the difficulties of following trails on the ground, we could not record complete day routes of chimpanzees. Two teams of three people each walked simultaneously through the study area, searching chimpanzee signs. We usually used radios to communicate between teams to avoid recounting or missing a party. A team searched for chimpanzee signs and endeavored to cover all quadrates of the entire study site. On the map we recorded the locations of chimpanzee parties we encountered, their fresh trails we followed for at least 100 m, and their nest sites. We recorded these on 1122 days in 82 months, averaging 14 days per month (range: 5–27). Despite our efforts, we missed some parts of their day range, and our records may represent smaller than actual monthly ranges. Total home range size is the summed area of all quadrates that a group entered during the study period. The frequency of visits was counted once per quadrate per day, and multiple visits to the same quadrate within a day were counted as one visit. Core area is defined as the summed area of quadrates that contained the highest frequency of visits until 75% or 50% of total visits were included (Watts, 1998a; Robbins & McNeilage, 2003). The proportion of range overlap was calculated as the percentage of overlap in the total home range of both gorillas and chimpanzees during the same year. In order to avoid the bias of difference in sample size, monthly range size was calculated from the total number of quadrates visited by the apes per month divided by the number of sampling days.

We used regression coefficient analysis to examine the correlations between monthly variations, such as fruit index, consumption of fruits, and range size.

Results

Dietary overlap

During the 92-month study period from 1994–2002, we found 231 plant foods (116 species) of gorillas and 137 plant foods (104 species) of chimpanzees through direct observations, evaluation of feeding remains along fresh trails, and fecal analysis. Among these food items, 88 foods (38% of gorilla foods and 64% of chimpanzee foods) were eaten by both apes. More than half of fruit/seed food items were eaten by both apes (77% for gorillas and 59% for chimpanzees) (Figure 3.1). Gorillas fed on more kinds of vegetative foods than chimpanzees (especially bark). Most vegetative foods eaten by chimpanzees were also eaten by gorillas. Chimpanzees fed on more kinds of insects (honey bees, ants, and beetles) than gorillas, and preyed on mammals (*Cercopithecus* monkeys and giant forest squirrels), which gorillas were never seen to hunt or eat. Gorillas fed on several kinds of roots (trees, shrubs, and herbs) and rotten wood, on which chimpanzees were never seen to feed. Among 59 species of herbs and vines eaten by the apes, 33 species (56%) were eaten by both apes. In ten species of herbs and vines ranked in the top five food species for gorillas in a month by direct observations and feeding remains in their fresh trails, five species (50%) were eaten by chimpanzees. However, two species (*Urera hypselodendron* and *Basella alba*) most frequently eaten by gorillas were not used by chimpanzees as foods.

Fruit remains were found in 52% of gorilla fecal samples and 99% of chimpanzee fecal samples. The average monthly mean number of fruit species per gorilla fecal sample (0.8, n = 93, range: 0–2.9) was smaller than that per chimpanzee fecal sample (2.7, n = 92, range: 1.1–4.5). Monthly mean proportion of fruit remains per gorilla fecal sample fluctuated between 0 and 56%, and that per chimpanzee fecal sample fluctuated between 22% and 92%.

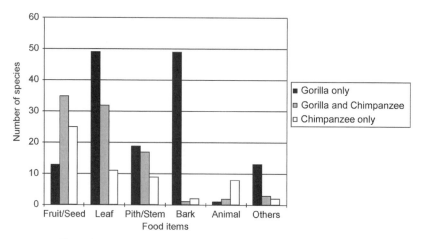

Figure 3.1. The number of food species eaten by gorillas and chimpanzees in each food category at Kahuzi during the study period.

Gorillas and chimpanzees showed similar preferences for most fruits. Thirteen fruit species were regarded as the species preferred by both apes because they were ranked as the top three (the first, second, or third highest frequency observed in fecal samples of both apes) in one or more months (Table 3.1). Among these, small differences between gorillas and chimpanzees were found for eight species in the number of months recorded as top three. Small differences between gorillas and chimpanzees were also found in a proportion of the fecal samples that included six fruit species. Four species (*Maesa lanceolata*, *Landolphia owariensis*, *Aframomum sanguinium*, and *Ekebergia capensis*), were more frequently eaten by chimpanzees than by gorillas. Fig fruits were ranked in the top three in most months for both gorillas and chimpanzees, but their seeds were found less frequently in gorilla fecal samples than in chimpanzee samples. Gorillas apparently consumed fig fruits over as long a period as did chimpanzees but did so less frequently during that period.

Among fruit species preferred by both apes, we selected two species (*Ficus* spp. and *Myrianthus holstii*) frequently consumed by them to examine the correlation between monthly proportion of fecal samples that included fruit remains (fruit consumption) and monthly proportion of trees bearing ripe fruits (fruit abundance). Neither gorillas nor chimpanzees consumed fig fruits according to their abundance ($r^2 = 0.003$, NS for gorillas; $r^2 = 0.0001$, NS, for chimpanzees), while both of them tended to consume fruits of *M. holstii* according to their availability ($r^2 = 0.482$, $p < 0.0001$ for gorillas, $r^2 = 0.281$, $p < 0.0001$ for chimpanzees) (Figure 3.2).

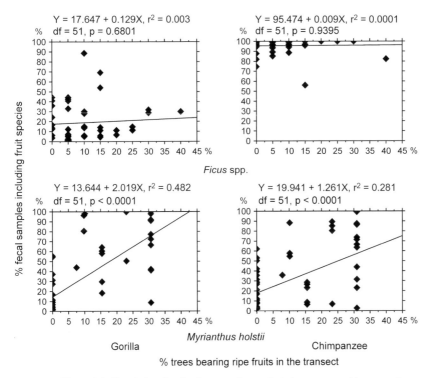

Figure 3.2. Correlations by simple regression analysis between monthly proportions of gorilla or chimpanzee fecal samples that included fruit remains and monthly proportion of trees bearing ripe fruits in the fruit trail. *Ficus* spp.: $r^2 = 0.003$, $p = 0.6801$ for gorillas; $r^2 = 0.0001$, $p = 0.9395$ for chimpanzees. *Myrianthus holstii*: $r^2 = 0.482$, $p < 0.0001$ for gorillas; $r^2 = 0.281$, $p < 0.0001$ for chimpanzees.

Monthly means of fruit consumption by chimpanzees were higher than those of gorillas (Figure 3.3). Fruit consumption of both apes increased with an increase in the ripe fruit index ($r^2 = 0.244$, $p < 0.001$ for gorillas; $r^2 = 0.187$, $p < 0.01$). The mean number of fruit species per gorilla fecal sample was also positively correlated with the fruit index ($r^2 = 0.436$, $p < 0.0001$), while no correlation was found between them for chimpanzees ($r^2 = 0.005$, NS).

Patterns of range use and range overlap

The differences in fruit consumption between gorillas and chimpanzees may have been influenced by both fruit availability and differences in ranging patterns of the apes. Based on the number of grid squares visited by the study

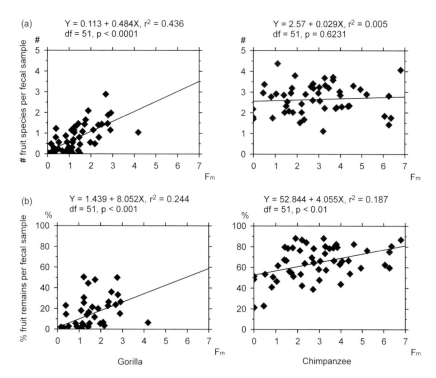

Figure 3.3. Correlations by simple regression analysis between the fruit index calculated from seven species preferred by gorillas or 11 species preferred by chimpanzees and mean number of fruit species per fecal sample (a), and mean proportion (%) of fruit remains per fecal sample (b). Significance level (a): $r^2 = 0.436$, $p < 0.0001$ for gorillas; $r^2 = 0.005$, $p = 0.623$ for chimpanzees. (b): $r^2 = 0.244$, $p < 0.001$ for gorillas; $r^2 = 0.197$, $p < 0.01$ for chimpanzees.

groups, the total home range of the gorilla group for 92 months (1440 days) was 42.3 km^2, and that of the chimpanzee unit-group for 82 months (1122 days) was 15.7 km^2 (Figure 3.4, Table 3.2). The maximum number of visits in a grid was 55 days by chimpanzees and 35 days by gorillas. Core area containing the highest frequency of visits up to 75% and 50% occupied 33% and 15% of gorilla total home range and 32% and 16% of chimpanzee total home range, respectively. The mean numbers of visits per grid in core areas up to 75% (27.9) and 50% (37.1) for chimpanzees were twice more than those (13.9 and 20.5, respectively) for gorillas. The overlap between total home ranges of gorillas and chimpanzees accounted for 26% of gorilla range and 70% of chimpanzee range during the entire study period. However, only small parts of core areas (especially core areas up to 50%) of gorillas and

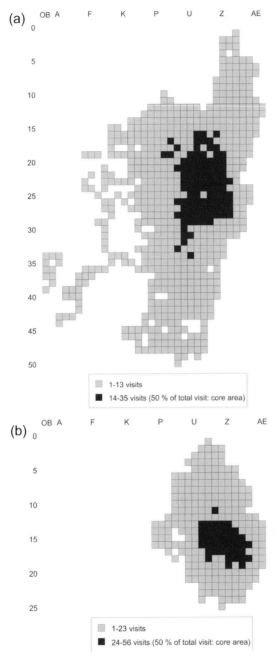

Figure 3.4. (a) Home range of the study group of gorillas. Core area is indicated by the dark grids. (b) Home range of the study group of chimpanzees. Core area is indicated by the dark grids.

Table 3.2. *Annual home range and core area (km²)*

Year	Gorilla Home range	Core area 75%	Core area 50%	Chimpanzee Home range	Core area 75%	Core area 50%	Overlap between gorilla and chimpanzee Area	% Gorilla	% Chimpanzee
1995	16.4	7.8	3.6	7.7	3.8	1.8	2.6	15.8%	33.8%
1998	13.2	6.6	3.0	5.8	2.1	1.1	3.9	29.5%	67.2%
1999	15.4	7.8	3.8	6.9	2.8	1.6	3.7	24.0%	53.6%
2000	14.4	7.6	4.1	7.6	3.5	1.8	2.3	16.0%	30.3%
2001	18.2	9.6	3.4	7.3	3.1	1.5	3.3	18.1%	45.2%
Average	15.5	7.9	3.6	7.1	3.0	1.5	3.2	20.6%	45.1%
Total	42.3	13.8 (32.6%)	6.3 (14.9%)	15.7	5.0 (31.9%)	2.6 (16.3%)	11.6 Core area 75%	25.9%	69.7%
							2.1 Core area 50%	14.9%	41.3%
							0.25	4.0%	9.8%

chimpanzees were overlapped. Annual home range and core areas (calculated from the 5 years in which data were collected for the whole 12 months) for the gorilla group were more than twice as large as those of the chimpanzee group in any year (Table 3.2). The proportion of overlapped area in the annual home range for chimpanzees was significantly larger than that for gorillas (Wilcoxon signed rank test, $Z = -2.0$, $p < 0.05$).

The number of new grid squares visited by gorillas tended to increase every year, while that of chimpanzees did not increase substantially after the first 2 years (Figure 3.5). More than half of chimpanzee total home range was covered by their visits for the first 5 months (Basabose, 2005). Gorillas tended to visit new areas in the later half of the year, when they changed diet from fruits to foliage.

The monthly range size divided by the number of observation days was not correlated with the monthly mean number of fruit species per fecal sample (gorillas: $r^2 = 0.003$, $p = 0.625$; chimpanzees: $r^2 = 0.008$, $p = 0.4258$), or with the monthly mean proportion of fruit remains per fecal sample for gorillas ($r^2 = 0.009$, $p = 0.3624$) but did show a weak negative correlation for chimpanzees ($r^2 = 0.067$, $p < 0.05$). Gorillas did not increase monthly range with increasing fruit consumption. Chimpanzees tended to increase monthly range when they decreased fruit consumption.

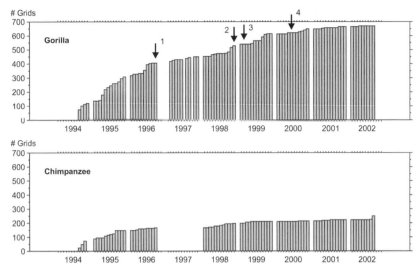

Figure 3.5. Cumulative number of grid squares visited per month. The numbers with arrows indicate the outbreak of civil war and occurrences of social changes in the study group of gorillas. 1: The occurrence of first civil war (October, 1996); 2: killing of a silverback by poachers (December, 1998); 3: killing of three females by poachers (February, 1999); 4: group fission by joining a solitary male (June, 2000).

Discussion

Preference for fruits and effects of fruit scarcity on diet of gorillas and chimpanzees

The strong preference for fruits by gorillas and chimpanzees in the lowland tropical forests may reflect their similar digestive capabilities (Yamagiwa *et al.*, 1992, 1994; Tutin & Fernandez, 1993; Kuroda *et al.*, 1996; Remis, 1997a). Although the diversity and amount of fruits are limited in the montane forest of Kahuzi, fruits accounted for the largest overlap in diet between gorillas and chimpanzees, as has also been observed for sympatric populations in the lowland tropical forests.

However, fruit diversity and fruit consumption of gorillas in montane forests are lower than for those in tropical forests. Collected over the course of several years, in the montane forests of Kahuzi and Bwindi averages of 0.8 and 1 species, respectively, of fruit remains were found per fecal sample of gorillas (this study; Goldsmith, 2003). Ganas *et al.* (2004) found no difference in monthly mean number of fruit species per daily dung sample between three gorilla groups (range: 0.85–1.11 species). These averages are far smaller than those reported for gorillas in the lowland forests: 3.2 and 3.4 species at Bai Hokou (Remis, 1997a; Goldsmith, 2003, respectively) and 3.5 species at Mondika (Doran *et al.*, 2002). By contrast, a similar number of fruit species was found in chimpanzee fecal samples in the montane forests of both Kahuzi (2.7 species, this study) and Bwindi (2.1 species, Stanford & Nkurunungi, 2003) and lowland forest of Lopé (2.7 species, Tutin & Fernandez, 1993). These differences are possibly caused by the chimpanzee's persistent pursuit of fruits irrespective of fruit abundance. Our study showed that the amount of fruit consumption by chimpanzees was positively correlated with the ripe fruit index, while the diversity of fruits consumed by them had no correlation with it. These results suggest that chimpanzees did not decrease the number of fruits consumed during the period of fruit scarcity.

Gorillas and chimpanzees may respond to fruit scarcity in different ways. During the lean period of fruits in tropical forests, gorillas tend to increase consumption of vegetative foods, while chimpanzees maintain a constant frugivorous diet (Tutin & Fernandez, 1993; Kuroda *et al.*, 1996). At Kahuzi, vegetative foods usually constitute the major part of gorilla diet regardless of fruit abundance, while chimpanzees increase consumption of herbaceous foods during lean fruit periods (Basabose, 2002; Yamagiwa *et al.*, in press). This study showed that in the montane forest of Kahuzi gorillas fed exclusively on bark, consuming some of most species. Although about half of the herb and vine species were eaten by both apes, the top two species (*U. hypselodendron* and *B. alba*) were eaten by gorillas exclusively. These observations suggest that fruit scarcity in montane forests may promote divergence of folivorous diets between gorillas and chimpanzees.

Fruit consumption during the lean fruit period was also different between gorillas and chimpanzees at Kahuzi. Chimpanzees usually consumed fig fruits more frequently than did gorillas, and neither gorilla nor chimpanzee fig fruit consumption was correlated with the ripe fruit index (Figure 3.2). Among five species of fig fruits identified in the study area, both gorillas and chimpanzees usually fed on two species (*Ficus oreadryadum* and *Ficus thonningii*; see Basabose, 2002; Yamagiwa *et al.*, in press). No correlation between gorillas and chimpanzees in monthly fig fruit consumption can possibly reflect the different periods or different stages of ripeness during which they consume these fruits. More precise observations are needed to examine these aspects.

The differences in the consumption frequency of the same fruit between gorillas and chimpanzees may reflect their preferences for fruits. Gorillas tend to avoid high-lipid foods at Lopé, while chimpanzees rely heavily on lipid-rich fruits of *Elaeis guineensis* (Rogers *et al.*, 1990; Tutin *et al.*, 1991). However, distinct differences between study sites in choice of plant food by gorillas and chimpanzees were also observed. This study showed that gorillas rarely ate fruit of *M. lanceolata*, which was frequently eaten by chimpanzees at Kahuzi. By contrast, this fruit was eaten frequently by gorillas but rarely by chimpanzees at Bwindi (Stanford & Nkurunungi, 2003). Large differences in frequency of fruit consumption (including *M. lanceolata*) were also reported between neighboring gorilla groups that overlapped ranging areas extensively at Bwindi (Ganas *et al.*, 2004). These observations suggest that differences in fruit consumption may be influenced not only by fruit availability but also by other factors, such as group or local traditions (Casimir, 1975; Goodall, 1986; Ganas *et al.*, 2004). In order to examine these possibilities, more research is needed to collect behavioral data on fruit consumption of gorillas and chimpanzees.

Ranging overlap and foraging strategies of sympatric gorillas and chimpanzees

Another factor influencing the frequency of fruit consumption by gorillas and chimpanzees is their ranging patterns. At Kahuzi, gorillas tend to range in the secondary forest with dense herbaceous undergrowth and frequently enter bamboo forest during the lean fruit period between September and December. Bamboo shoots may prompt gorillas to change their ranging patterns, resulting in a prominent reduction in their fruit consumption. By contrast, chimpanzees continuously stayed in the area covered by both primary and secondary forests. For at least 5 years, the study group of gorillas constantly moved in ranges more than twice as large as those of the study group of chimpanzees.

Gorillas at Kahuzi showed low site fidelity in order to expand home range yearly, as observed in mountain gorillas at Virungas (Watts, 1996, 1998b). In contrast, for more than 8 years chimpanzees showed high site fidelity in order to stay in a particular area, as also observed in western chimpanzees at Taï Forest (Lehmann & Boesch, 2003) and eastern chimpanzees at Gombe (Williams *et al.*, 2004). Mountain gorillas at Virungas exhibit folivorous strategy to seek high-quality herbaceous vegetation and to avoid previously used areas that contained trampled vegetation (Vedder, 1984; Watts, 1987, 1998b). Frugivorous primates harvest fruits efficiently to adjust search paths and patch revisitation (Waser, 1984; Garber, 1989). Differences in site fidelity between gorillas and chimpanzees are possibly anticipated in their folivorous and frugivorous foraging strategies. However, home range size of gorillas at Kahuzi over an 8-year period is relatively larger than those at other sites, while that of chimpanzees is smaller (Tables 3.3, 3.4). Complex foraging

Table 3.3. *Home range size and proportion of core area of gorillas in various habitats*

Gorilla population	Location	Habitat type	Home range size (# years)	Annual range size	Percentage of core area[a]	Source
Mountain	Virungas	Montane forest	21–25 km^2 (5–7 years)	9–12 km^2	24–27%	Watts, 1998a
Mountain	Bwindi	Montane forest	40.2 km^2 (3 years)	21–40 km^2	35%	Robbins & McNeilage, 2003
Eastern lowland	Kahuzi	Montane forest	42.3 km^2 (8 years)	13–18 km^2	33%	This study
Western lowland	Lope	Lowland forest	21.7 km^2 (10 years)	7–14 km^2	NP	Tutin, 1996
Western lowland	Bai Hokou	Lowland forest	23 km^2 (2.2 years)	NP	NP	Remis, 1997b
Western lowland	Bai Hokou	Lowland forest	18.3 km^2 (3.5 years)	8–13 km^2	31%	Cipolletta, 2004
Western lowland	Lossi	Lowland forest	11 km^2 (3.2 years)	NP	NP	Bermejo, 2004
Western lowland	Mondika	Lowland forest	15.8 km^2 (1.3 years)	15 km^2	NP	Doran-Sheehy et al., 2004

Notes:
[a] The area in which 75% of total visits or occupancy by gorillas recorded in their home range.
NP = Not provided

Table 3.4. *Home range size and proportion of core area of chimpanzees in various habitats*

Chimpanzee population	Location	Habitat type	Home range size (# years)	Annual range size	Percentage of core area	Source
Western	Mt. Assirik	Savanna	278–330 km² (4 years)	NP	NP	Baldwin et al., 1982
Western	Tai	Lowland forest	27 km² (10 years)	NP	NP	Boesch & Boesch, 1989
Western	Tai	Lowland forest	(10 years)	14–26 km²	35%	Lehmann & Boesch, 2003
Eastern	Kahuzi	Montane forest	15.7 km² (8 years)	6–8 km²	32%	This study
Eastern	Budongo	Forest*	20 km² (2 years)	NP	NP	Reynolds & Reynolds, 1965
Eastern	Budongo	Forest*	6.8 km² (1.5 years)	NP	NP	Newton-Fisher, 2003
Eastern	Kibale	Forest*	23–38 km² (1.4 years)	NP	NP	Chiglieri, 1984
Eastern	Mahale	Forest*	11–34 km² (5 years)	NP	NP	Nishida & Kawanaka, 1972
Eastern	Gombe	Forest*/woodland	(18 years)	6–14 km²	NP	Williams et al., 2004
Eastern	Kasakati	Woodland	122–124 km² (1 year)	NP	NP	Izawa, 1970
Eastern	Filabanga	Savanna	150 km² (1 year)	NP	NP	Kano, 1971
Eastern	Ugalla	Savanna	250–560 km² (1 year)	NP	NP	Kano, 1972

Notes:
*Tropical forest at medium altitudes around 1000–1500 m asl.
NP = Not provided

strategies used by gorillas at Kahuzi may result in their large home range. The abundance of fruits in tropical forests may enable gorillas there to subsist within smaller ranges, although they tend to extend daily path length when they exhibit frugivorous diet (Yamagiwa & Mwanza, 1994; Remis, 1997b; Cipolleta, 2004; Doran-Sheehy *et al.*, 2004). However, diversity of woody plants providing a large amount of fruits decreases in montane forests with decreasing canopy height at higher elevations (Moutsamboté *et al.*, 1994; Tutin *et al.*, 1994; Yumoto *et al.*, 1994; Lieberman *et al.*, 1996; Yamagiwa *et al.*, 2003). No clear relationship was found between daily path length or home range size and fruit eating by gorillas at Bwindi (Goldsmith, 2003; Robbins & McNeilage, 2003; Stanford & Nkurunungi, 2003). Although gorillas at Kahuzi tend to extend daily path length with increasing fruit consumption (Yamagiwa *et al.*, 2003), this study found no correlation between their monthly range size and fruit consumption. The low diversity and amount of fruits in montane forests may stimulate gorillas at Bwindi and Kahuzi to use both folivorous and frugivorous foraging strategies to expand their yearly home range.

Home range size of chimpanzees may be related to food availability, in particular to fruit availability. Chimpanzees need larger home ranges in woodlands or savannas with lower availability of fruits than they do in forests (Table 3.4) since they persistently seek fruits in any habitat (Yamagiwa, 1999; Stanford & Nkurunungi, 2003). A small range size (6.8 km^2) was reported for chimpanzees (38–46 individuals) at Budongo Forest, where high production of chimpanzee foods was expected (Newton-Fisher, 2003). This study reported a similar annual range size for chimpanzees (22–23 individuals) in the montane forest of Kahuzi. Why didn't the chimpanzees at Kahuzi expand their yearly home ranges to seek fruits widely, as observed for chimpanzees in drier habitats? The small community size and the small fragmentary distribution of primary forests providing plenty of fruit foods may constitute possible factors keeping them within a small range.

Various factors other than food distribution may also influence home range size of gorillas and chimpanzees. Human disturbances, social factors induced by changes in group size, female transfers, and encounters with other groups or solitary males may have a powerful effect on ranging patterns of gorillas across habitats (Fossey, 1974; Yamagiwa, 1983; Cipolletta, 2003, 2004; Robbins & McNeilage, 2003; Bermejo, 2004). Male chimpanzees defend a feeding territory for their resident females and sometimes attack and kill chimpanzees of other unit groups (Goodall *et al.*, 1979; Nishida *et al.*, 1985; Watts & Mitani, 2001; Williams *et al.*, 2004). The number of adult males was the best predictor of home range size for chimpanzees, and their ranging patterns were strongly influenced by intercommunity relations at Taï

Forest (Herbinger *et al.*, 2001; Lehmann & Boesch, 2003). At Kahuzi, home range of the chimpanzee study group did not overlap with home ranges of neighboring groups during the study period (Basabose, 2005). The study groups of gorillas experienced large disturbances during the civil wars in 1996 and 1998 (Yamagiwa, 2003). More analyses are necessary to examine not only human but also other factors such as presence or absence of competitors or neighboring groups.

In addition to these considerations, this study suggests that feeding competition with sympatric gorillas may stimulate chimpanzees to remain within a small range. Although incomplete sampling on ranging of chimpanzees may result in smaller home range and core area than those of gorillas, the percentage of overlap in core areas between gorillas and chimpanzees was far smaller than that in total home ranges between them. This implies that gorillas and chimpanzees avoided encounters with each other in the areas frequently visited by each of them. The overlapped core area included more primary than secondary forests, possibly due to the diversity and abundance of fruits. Chimpanzees tended to avoid nesting in trees bearing fruits preferred by gorillas in the secondary forests at Kahuzi (Basabose & Yamagiwa, 2002). Aggressive interactions or mutual avoidance of direct contacts between gorillas and chimpanzees were observed in montane forests, while more peaceful encounters were reported between them in lowland tropical forests (Suzuki & Nishihara, 1992; Tutin, 1996; Yamagiwa *et al.*, 1996a; Stanford & Nkurunungi, 2003). The sympatry observed during periods of prolonged fruit scarcity in montane forests may have promoted the distinct divergence in foraging strategies between gorillas and chimpanzees.

Acknowledgments

This paper was originally prepared for the International Symposium on "Feeding ecology in apes and other primates" held at the Max Planck Institute, August 16–19, 2004 in Leipzig, Germany. We would like to express our hearty thanks to Drs. Gottfried Hohmann and Christophe Boesch for giving us the opportunity to present our work on the sympatric apes at the symposium. This study was financed by a Grant for the Biodiversity Research of the 21st Century COE (A14) and by the International Scientific Research Program (No. 162550080 to J. Yamagiwa) sponsored by the Ministry of Education, Science, Sports and Culture, Japan. It was conducted in cooperation with CRSN (Centre de Recherches en Sciences Naturelles) and ICCN (Institut Congolais pour Conservation de la Nature). We thank Dr. S. Bashwira, Dr. B. Baluku, Mr. M. O. Mankoto, Mr. B. Kasereka,

Mr. L. Mushenzi, Ms. S. Mbake, Mr. B.I. Iyomi, and Mr. C. Schuler for their administrative help, and Dr. M. Matsubara for digitizing the daily travel routes of the study group onto a vegetation map. We are also greatly indebted to Mr. M. Bitsibu, Mr. S. Kamungu, and all of the guides, guards, and field assistants in the Kahuzi-Biega National Park for their technical help and hospitality throughout the fieldwork.

References

Baldwin, P.J., McGrew, W.C., & Tutin, C.E.G. (1982). Wide-ranging chimpanzees at Mt. Assirik, Senegal. *International Journal of Primatology*, **3**, 367–85.

Basabose, A.K. (2002). Diet composition of chimpanzees inhabiting the montane forest of Kahuzi, Democratic Republic of Congo. *American Journal of Primatology*, **58**, 1–21.

(2005). Ranging patterns of chimpanzees in a montane forest of Kahuzi, Democratic Republic of Congo. *International Journal of Primatology*, **26**, 31–52.

Basabose, A.K. & Yamagiwa, J. (2002). Factors affecting nesting site choice in chimpanzees at Tshibati, Kahuzi-Biega National Park: influence of sympatric gorillas. *International Journal of Primatology*, **23**, 263–82.

Bermejo, M. (2004). Home-range use and intergroup encounters in western gorillas (*Gorilla g. gorilla*) at Lossi Forest, North Congo. *American Journal of Primatology*, **64**, 223–32.

Boesch, C. & Boesch, H. (1989). Hunting behavior of wild chimpanzees in the Taï National Park. *American Journal of Physical Anthropology*, **78**, 547–73.

Casimir, M.J. (1975). Feeding ecology and nutrition of an eastern gorilla group in the Mt. Kahuzi region (République du Zaire). *Folia Primatologica*, **24**, 1–36.

Casimir, M.J. & Butenandt, E. (1973). Migration and core area shifting in relation to some ecological factors in a mountain gorilla group (*Gorilla gorilla beringei*) in the Mt. Kahuzi region (République du Zaire). *Zeitschrift für Tierpsychologie*, **33**, 514–22.

Chapman, C.A., Chapman, L.J., Wrangham, R.W., Hunt, K., Gebo, D., & Gardner, L. (1992). Estimates of fruit abundance of tropical trees. *Biotropica*, **24**, 527–31.

Cipolletta, C. (2003). Ranging patterns of a western gorilla group during habituation to humans in the Dzanga-Ndoki National Park, Central African Republic. *International Journal of Primatology*, **24**, 1207–26.

(2004). Effects of group dynamics and diet on the ranging patterns of a western gorilla group (*Gorilla gorilla gorilla*) at Bai Hokou, Central African Republic. *American Journal of Primatology*, **64**, 193–205.

Cord, M. (2000). Mixed-species association and group movement. In *On the Move*, ed. S. Boinski & P. Garber, pp. 73–99. Chicago: University of Chicago Press.

Doran, D. & McNeilage, A. (1998). Gorilla ecology and behavior. *Evolutionary Anthropology*, **6**, 120–31.

Doran, D.M., McNeilage, A., Greer, D., Bocian, C., Mehlman, P., & Shah, N. (2002). Western lowland gorilla diet and resource availability: new evidence, cross-site comparisons, and reflections on indirect sampling methods. *American Journal of Primatology*, **58**, 91–116.

Doran-Sheehy, D.M., Greer, D., Mongo, P., & Schwindt, D. (2004). Impact of ecological and social factors on ranging in western gorillas. *American Journal of Primatology*, **64**, 207–22.

Fossey, D. (1974). Observations on the home range of one group of mountain gorillas (*Gorilla gorilla beringei*). *Animal Behaviour*, **22**, 568–81.

Fossey, D. & Harcourt, A.H. (1977). Feeding ecology of free-ranging mountain gorilla (*Gorilla gorilla beringei*). In *Primate Ecology*, ed. T.H. Clutton-Brock, pp. 415–47. New York: Academic Press.

Ganas, J., Robbins, M.M., Nkurunungi, J.B., Kaplin, B.A., & McNeilage, A. (2004). Dietary variability of mountain gorillas in Bwindi Impenetrable National Park, Uganda. *International Journal of Primatology*, **25**, 1043–72.

Garber, P.A. (1989). Role of spatial memory in primate foraging patterns: *Saguinus mystax* and *Saguinus fuscicollis*. *American Journal of Primatology*, **19**, 203–16.

Gautier-Hion, A., Quris, R., & Gautier, J-P. (1983). Monospecific vs. polyspecific life: a comparative study of foraging and antipredatory tactics in a community of Cercopithecus monkeys. *Behavioral Ecology and Sociobiology*, **12**, 325–35.

Gebo, D.L. & Chapman, C.A. (1995). Positional behavior in five sympatric old world monkeys. *American Journal of Physical Anthropology*, **97**, 49–77.

Ghiglieri, M.P. (1984). *The Chimpanzees of Kibale Forest: A Field Study of Ecology and Social Structure*. New York: Columbia University Press.

Goldsmith, M.L. (2003). Comparative behavioral ecology of a lowland and highland gorilla population: where do Bwindi gorillas fit? In *Gorilla Biology*, ed. A.B. Taylor & M.L. Goldsmith, pp. 358–84. Cambridge: Cambridge University Press.

Goodall, A.G. (1977). Feeding and ranging behavior of a mountain gorilla group (*Gorilla gorilla beringei*) in the Tshibinda-Kahuzi region (Zaire). In *Primate Ecology*, ed. T.H. Clutton-Brock, pp. 450–79. New York: Academic Press.

Goodall, J. (1986). *The Chimpanzees of Gombe: Patterns of Behavior*. Cambridge: Belknap Press of Harvard University Press.

Goodall, J., Bandora, A., Bergmann, E., Busse, C., Matama, H., Mpongo, E., Pierce, A., & Riss, D. (1979). Intercommunity interactions in the chimpanzee population of the Gombe National Park. In *The Great Apes*, ed. D.A. Hamburg & E.R. McCown, pp. 13–53. Menlo Park: Benjamin/Cummings.

Harcourt, A.H. (1978). Strategies of emigration and transfer by primates, with particular reference to gorillas. *Zeitschrift für Tierpsychologie*, **48**, 401–20.

Herbinger, I., Boesch, C. & Rothe, H. (2001). Territory characteristics among three neighboring chimpanzee communities in the Taï National Park, Côte d'Ivore. *International Journal of Primatology*, **22**, 142–67.

Heymann, E.W. & Buchanan-Smith, H.M. (2000). The behavioral ecology of mixed-species troops of callitrichine primates. *Biological Reviews of the Cambridge Philosophical Society*, **75**, 169–90.

Izawa, K. (1970). Unit groups of chimpanzees and their nomadism in the savanna woodland. *Primates*, **11**, 1–45.

Jones, C. & Sabater Pi, J. (1971). Comparative ecology of *Gorilla gorilla* (Savage and Wyman) and *Pan troglodytes* (Blumenbach) in Rio Muni, West Africa. *Biblioteca Primatologica*, **13**, 1–96.

Kano, T. (1971). The chimpanzees of Filabanga, western Tanzania. *Primates*, **12**, 229–46.

(1972). Distribution and adaptation of the chimpanzee on the eastern shore of Lake Tanganyika. *Kyoto University African Studies*, **7**, 37–129.

Kuroda, S., Nishihara, T., Suzuki, S., & Oko, R.A. (1996). Sympatric chimpanzees and gorillas in the Ndoki Forest, Congo. In *Great Ape Societies*, ed. W.C. McGrew, L.F. Marchant & T. Nishida, pp. 71–81. Cambridge: Cambridge University Press.

Lehmann, J. & Boesch, C. (2003). Social influences on ranging patterns among chimpanzees (*Pan troglodytes verus*) in the Taï National Park, Côte d'Ivoire. *Behavioral Ecology*, **14**, 642–9.

Lieberman, D., Lieberman, M., Peralta, R., & Harshorn, G.S. (1996). Tropical forest structure and composition on a large-scale altitudinal gradient in Costa Rica. *Journal of Ecology*, **84**, 137–52.

Moutsamboté, J-M., Yumoto, T., Mitani, M., Nishihara, T., Suzuki, S., & Kuroda, S. (1994). Vegetation and list of plant species identified in the Nouabalé-Ndoki Forest, Congo. *Tropics*, **3**, 277–93.

Newton-Fisher, N.E. (2003). The home range of the Sonso community of chimpanzees from the Budongo Forest, Uganda. *African Journal of Ecology*, **41**, 150–6.

Nishida, T. & Kawanaka, K. (1972). Inter-unit-group relationships among wild chimpanzees of the Mahali Mountains. *Kyoto University African Studies*, **7**, 131–69.

Nishida, T., Hiraiwa-Hasegawa, M., Hasegawa, T., & Takahata, Y. (1985). Group extinction and female transfer in wild chimpanzees in the Mahale Mountains. *Zeitschrift für Tierpsychologie*, **67**, 284–301.

Nishihara, T. (1995). Feeding ecology of western lowland gorillas in the Nouabale-Ndoki National Park, Congo. *Primates*, **36**, 151–68.

Powzyk, J.A. & Mowry, C.B. (2003). Dietary and feeding differences between sympatric *Propithecus diadema diadema* and *Indri indri*. *International Journal of Primatology*, **24**, 1143–62.

Remis, M.J. (1997a). Western lowland gorillas (*Gorilla gorilla gorilla*) as seasonal frugivores: use of variable resources. *American Journal of Primatology*, **43**, 87–109.

(1997b). Ranging and grouping patterns of a western lowland gorilla group at Bai Hokou, Central African Republic. *American Journal of Primatology*, **43**, 111–33.

Reynolds, V. & Reynolds, F. (1965). Chimpanzees of the Budongo Forest. In *Primate Behavior*, ed. I. DeVore, pp. 368–424. New York: Holt, Rinehart & Winston.

Robbins, M.M. & McNeilage, A. (2003). Home range and frugivory patterns of mountain gorillas in Bwindi Impenetrable National Park, Uganda. *International Journal of Primatology*, **24**, 467–91.

Rogers, M.E., Abernethy, K., Bermejo, M., *et al.* (2004). Western gorilla diet: a synthesis from six sites. *American Journal of Primatology*, **64**, 173–92.

Rogers, M.E., Maisels, F., Williamson, E.A., Fernandez, M., & Tutin, C.E.G. (1990). Gorilla diet in the Lopé Reserve, Gabon: a nutritional analysis. *Oecologia*, **84**, 326–39.

Schaller, G.B. (1963). *The Mountain Gorilla: Ecology and Behavior.* Chicago: University of Chicago Press.

Stanford, C.B. & Nkurunungi, J.B. (2003). Behavioral ecology of sympatric chimpanzees and gorillas in Bwindi Impenetrable National Park, Uganda: diet. *International Journal of Primatology*, **24**, 901–18.

Suzuki, S. & Nishihara, T. (1992). Feeding strategies of sympatric gorillas and chimpanzees in the Ndoki-Nouabale Forest, with special reference to co-feeding behavior by both species. *Abstracts of the XIVth Congress of the International Primatological Society*, Strasbourg, France.

Tan, C.L. (1999). Group composition, home range size, and diet of three sympatric bamboo lemur species (Genus Hapalemur) in Ranomafana National Park, Madagascar. *International Journal of Primatology*, **20**, 547–66.

Terborgh, J. (1983). *Five New World Primates.* Princeton: Princeton University Press.

Tutin, C.E.G. (1996). Ranging and social structure of lowland gorillas in the Lopé Reserve, Gabon. In *Great Ape Societies*, ed. W.C. McGrew, L.F. Marchant & T. Nishida, pp. 58–70. Cambridge: Cambridge University Press.

Tutin, C.E.G. & Fernandez, M. (1992). Insect-eating by sympatric lowland gorillas (*Gorilla g. gorilla*) and chimpanzees (*Pan t. troglodytes*) in the Lopé Reserve, Gabon. *American Journal of Primatology*, **28**, 29–40.

   (1993). Composition of the diet of chimpanzees and comparisons with that of sympatric lowland gorillas in the Lopé Reserve, Gabon. *American Journal of Primatology*, **30**, 195–211.

Tutin, C.E.G., Fernandez, M., Rogers, M.E., Williamson, E.A., & McGrew, W.C. (1991). Foraging profiles of sympatric lowland gorillas and chimpanzees in the Lopé Reserve, Gabon. *Philosophical Transactions of the Royal Society of London, Series B*, **334**, 179–86.

Tutin, C.E.G., White, L.J.T., Williamson, E.A., Fernandez, M., & McPherson, G. (1994). List of plant species identified in the northern part of the Lope Reserve, Gabon. *Tropics*, **3**, 249–76.

Ungar, P.D. (1996). Feeding height and niche separation in sympatric Sumatran monkeys and apes. *Folia Primatologica*, **67**, 163–8.

Vedder, A.L. (1984). Movement patterns of a group of free-ranging mountain gorillas (*Gorilla gorilla beringei*) and their relation to food availability. *American Journal of Primatology*, **7**, 73–88.

Waser, P. (1984). Ecological differences and behavioral contrasts between two mangabey species. In *Adaptations for Foraging in Nonhuman Primates*, ed. P.S. Rodman & J.G.H. Cant, pp. 195–216. New York: Columbia University Press.

   (1987). Interactions among primate species. In *Primate Societies*, ed. B.B. Smuts, D.L. Cheney, R.M. Seyfarth, R.W. Wrangham, & T.T. Struhsaker, pp. 210–26. Chicago: University of Chicago Press.

Watts, D.P. (1987). The influence of mountain gorilla foraging activities on the productivity of their food species. *African Journal of Ecology*, **25**, 155–63.

   (1996). Comparative socioecology of gorillas. In *Great Ape Societies*, ed. W.C. McGrew, L.F. Marchant & T. Nishida, pp. 16–28. Cambridge: Cambridge University Press.

(1998a). Long-term habitat use by mountain gorillas (*Gorilla gorilla beringei*). 1. Consistency, variation, and home range size and stability. *International Journal of Primatology*, **19**, 652–79.

(1998b). Long-term habitat use by mountain gorillas (*Gorilla gorilla beringei*). 2. Reuse of foraging areas in relation to resource abundance, quality, and depletion. *International Journal of Primatology*, **19**, 681–702.

Watts, D.P. & Mitani, J.C. (2001). Boundary patrols and intergroup encounters in wild chimpanzees. *Behaviour*, **138**, 299–327.

Williams, J.M., Oehlert, G.W., Carlis, J.V., & Pusey, A.E. (2004). Why do male chimpanzees defend a group range? *Animal Behaviour*, **68**, 523–32.

Williamson, E.A., Tutin, C.E.G., Rogers, M.E., & Fernandez, M. (1990). Composition of the diet of lowland gorillas at Lopé in Gabon. *American Journal of Primatology*, **21**, 265–77.

Wrangham, R.W. (1987). Evolution and social structure. In *Primate Societies*, ed. B. B. Smuts, D.L. Cheney, R.M. Seyfarth, R.W. Wrangham, & T.T. Struhsaker, pp. 282–96. Chicago: University of Chicago Press.

Yamagiwa, J. (1983). Diachronic changes in two eastern lowland gorilla groups (*Gorilla gorilla graueri*) in the Mt. Kahuzi region, Zaire. *Primates*, **24**, 174–83.

(1999). Socioecological factors influencing population structure of gorillas and chimpanzees. *Primates*, **40**, 87–104.

(2003). Bushmeat poaching and conservation crisis in Kahuzi-Biega National Park, Democratic Republic of the Congo. *Journal of Sustainable Forestry*, **16**, 115–35.

Yamagiwa, J. & Mwanza, N. (1994). Day-journey length and daily diet of solitary male gorillas in lowland and highland habitats. *International Journal of Primatology*, **15**, 207–24.

Yamagiwa, J., Basabose, A.K., Kaleme, K., & Yumoto, T. (2005). Diet of Grauer's gorillas in the montane forest of Kahuzi, Democratic Republic of Congo. *International Journal of Primatology*, **26**, 1345–73.

Yamagiwa, J., Kahekwa, J., & Basabose, A.K. (2003). Intra-specific variation in social organization of gorillas: implications for their social evolution. *Primates*, **44**, 359–69.

Yamagiwa, J., Maruhashi, T., Yumoto, T. & Mwanza, N. (1996a). Dietary and ranging overlap in sympatric gorillas and chimpanzees in Kahuzi-Biega National Park, Zaire. In *Great Ape Societies*, ed. W.C. McGrew, L.F. Marchant & T. Nishida, pp. 82–98. Cambridge: Cambridge University Press.

Yamagiwa, J., Kaleme, K., Milynganyo, M., & Basabose, A.K. (1996b). Food density and ranging patterns of gorillas and chimpanzees in the Kahuzi-Biega National Park, Zaire. *Tropics*, **6**, 65–77.

Yamagiwa, J., Mwanza, N., Spangenberg, A. *et al.* (1992). Population density and ranging pattern of chimpanzees in Kahuzi-Biega National Park, Zaire: a comparison with a sympatric population of gorillas. *African Study Monographs*, **13**, 217–30.

Yamagiwa, J., Mwanza, N., Yumoto, Y., & Maruhashi, T. (1994). Seasonal change in the composition of the diet of eastern lowland gorillas. *Primates*, **35**, 1–14.

Yumoto, T., Yamagiwa, J., Mwanza, N., & Maruhashi, T. (1994). List of plant species identified in Kahuzi-Biega National Park, Zaire. *Tropics*, **3**, 295–308.

4 Chimpanzee feeding ecology and comparisons with sympatric gorillas in the Goualougo Triangle, Republic of Congo

DAVID MORGAN AND CRICKETTE SANZ

Introduction

Trends in feeding ecology that occur within and between populations over time have important implications for understanding the socioecology of a

Feeding Ecology in Apes and Other Primates. Ecological, Physical and Behavioral Aspects, ed. G. Hohmann, M. M. Robbins, and C. Boesch. Published by Cambridge University Press. © Cambridge University Press 2006.

particular species. Chapman *et al.* (2002) showed that primate foraging behavior and diets can vary considerably over small spatial and temporal scales. Such behavioral flexibility raises questions about both stereotypic characterization of primates and traditional comparative studies that do not incorporate intraspecific variation. Although chimpanzees have been studied at several long-term sites, questions still remain about the degree and nature of intraspecific variation in chimpanzee foraging ecology. Some general patterns such as frugivory and dietary diversity are consistent across all populations and habitats. Other aspects of feeding ecology differ between sites or even the same site in different years (Wrangham, 1977; Nishida & Uehara, 1983; Nishida *et al.*, 1983; Newton-Fisher, 1999). Assembling reports from different populations over a large geographic distribution provides the most complete depiction of the feeding ecology of this species and may elucidate patterns that are not apparent when comparing only a few populations in a restricted area of their range. Chimpanzees and gorillas are sympatric in certain regions, and descriptions of their dietary composition and overlap within a shared habitat contribute to understanding the behavioral ecology of these species and facilitate attempts at characterizing their coexistence. We report our preliminary observations of chimpanzee (*Pan troglodytes troglodytes*) and gorilla (*Gorilla gorilla gorilla*) feeding behavior at a new study site in the Goualougo Triangle, Republic of Congo (ROC). We specifically examine the diet composition and diversity of these chimpanzees in comparison to other sites and provide preliminary descriptions of dietary overlap with sympatric gorillas studied during the same time period.

Chimpanzees reside in a range of habitats from savanna to dense forests, and consume a wide variety of food items from several plant and animal species. Dietary composition is closely related to habitat, but has also been shown to vary over short time periods and increase with the length of the study at a particular site (Wrangham, 1977; Nishida & Uehara, 1983; Nishida *et al.*, 1983). For example, Hunt & McGrew (2002) reported that 33 and 60 food items were consumed by chimpanzees residing in the arid climates of Semliki and Assirik, respectively. Wrangham (1977) reported that this many food species could be consumed during a single month in the woodland forests of Gombe. In a similar habitat at the nearby Mahale site, researchers reported that after 7 years of study the cumulative diet composition of chimpanzees was 205 foods, which increased to 328 after 16 years (Nishida, 1974; Nishida & Uehara, 1983). Despite these differences in diet composition, chimpanzees in all habitat types have been characterized as persistent frugivores.

Chimpanzees thrive on a diverse, fruit-based diet and spend more time consuming fruits than any other food item (Wrangham, 1977; Nishida & Uehara, 1983; Tutin *et al.*, 1991; Newton-Fisher, 1999). Time spent feeding on fruits varies between populations, but comprises more than half of the feeding time at all sites (Table 4.1). Although it provides no indication of the importance of specific fruit species in the diet, the relative proportion of fruit species consumed of the total diet composition is also cited as an indication of frugivory. The overall mean proportion of fruit in the diet at 12 sites listed in Table 4.1 was 55%, with values ranging from 31%–88% at different sites. Although this range of values is large, the proportion of a particular food category in the chimpanzee diet can also vary to this extent over different time periods at the same site and may be attributable to fluctuating availability of preferred foods such as fruit. Hladik (1977) reported that fruit accounted for 68% of the food intake of chimpanzees in Gabon during a 1-year study period, but ranged from 40%–90% on a daily basis. Leaves normally constituted 28% of food intake, but increased to 50% during certain periods within Hladik's study (Hladik, 1977). Newton-Fisher (1999) reported from Budongo Forest that fruit was the most commonly consumed food item (65% of all observations), followed by leaves (20%) and flowers (8%), but that leaf consumption increased to 41% of feeding time during one month and flowers to 45% and 57% during two other months.

A combination of behavioral and ecological modifications allows chimpanzees to sustain this dietary pattern during periods of fruit scarcity (Wrangham, 1977). The flexibility of fission-fusion sociality enables chimpanzees to adjust their association patterns and ranging behavior to effectively forage during lean periods (Wrangham, 1977; Chapman *et al.*, 1995; Wrangham *et al.*, 1996; Sugiyama, 1999; Boesch & Boesch-Achermann, 2000; Fawcett, 2000; Anderson *et al.*, 2002). During such times, chimpanzees have also been shown to increase their dietary diversity, including consumption of less preferred fruits and foliage such as terrestrial herbaceous vegetation (THV) (Tutin *et al.*, 1991; Doran, 1997; Basabose, 2002). The amounts of leaves and THV consumed by different populations vary greatly. Wrangham reported that time spent eating leaves accounted for 21.2% of feeding time at Gombe, but only 2.6% of the monthly feeding time at Kibale in Uganda (Wrangham, 1977; Wrangham *et al.*, 1996). THV consumption for chimpanzees at Kibale ranged from 12.3%–19.9% of the monthly feeding time, but accounted for only 3.2% of the feeding time for chimpanzees at Budongo, Uganda (Wrangham *et al.*, 1996; Newton-Fisher, 1999). These reports clearly indicate the high degree of variation in chimpanzee feeding behavior.

The fruit consumption and dietary diversity of gorillas is also more variable across Africa than was previously thought from early studies of mountain gorilla feeding ecology (Doran & McNeilage, 1998). Gorillas are considered folivorous and possess anatomical specializations designed to efficiently digest large quantities of fiber such as THV and foods high in secondary

Table 4.1. *Comparison of chimpanzee diet composition and time allocated to feeding on particular food categories*

Diet composition Subspecies-study site	Methods	% Fruit	% Seeds	% Leaves	% Stems/pith	% Flowers	% Bark	% Other	Plant food items	Plant species
P. t. schweinfurthii										
Gombe, Tanzania[a]	D	43	7	27	8	10	–	7	201	–
Mahale, Tanzania[b]	D	31	5	36	11	9	4	5	328	198
Budongo, Uganda[c]	D	–	–	–	–	–	–	–	118	58
Budongo, Uganda[d]	D	–	–	–	–	–	–	–	91	49
Semliki, Uganda[e]	D	39	15	30	9	3	3	–	33	–
Bwindi, Uganda[f]	F	50	–	–	–	–	–	–	60	34
Kahuzi, D. R. C.[g]	F, T, D	38	–	31	19	–	6	5	99	75
Kahuzi, D. R. C.[h]	D, T, F	40	3	30	17	5	2	4	156	110
P. t. troglodytes										
Belinga, Gabon[i]	F, T	85	–	4	7	–	–	4	46	43
Lope, Gabon[j]	F, T	66	11	12	5	4	2	1	161	132
Goualougo, Rep. Congo[k]	D, F, T	56	7	16	8	8	3	2	158	116
Ndoki, Rep. Congo[l]	D, F, T	80	6	–	14	–	–	–	52	–
Ndoki, Rep. Congo[m]	D, F, T	88	–	3	5	2	–	3	114	108
P. t. verus										
Assirik, Senegal[n]	F, T, D	57	10	10	3	10	7	3	60	–
Bossou, Guinea[o]	D	52	7	18	13	5	3	2	246	–

Time spent feeding Study site	Methods	% Fruit	% Seeds	% Leaves	% Stems/pith	% Flowers
Budongo, Uganda[c]	15-min scans	65	–	20	3	9
Budongo, Uganda[d]	30-min focals	64	8	27	–	–
Gombe, Tanzania[a]	30-min scans	59	–	21	–	–
Kibale, Uganda[p]	15-min scans	79	–	3	17	–
Goualougo, Rep Congo[k]	20-min scans	57	–	32	2	4

Notes:

D = Direct observation, F = Fecal analysis, T = Trail Signs;

[a] Wrangham, 1977;
[b] Nishida & Uehara, 1983;
[c] Newton-Fisher, 1999;
[d] Fawcett, 2000;
[e] Hunt & McGrew, 2002;
[f] Stanford & Nkurunungi, 2003;
[g] Yamagiwa et al., 1996;
[h] Basabose, 2002;
[i] Tutin & Fernandez,, 1985;
[j] Tutin & Fernandez, 1993;
[k] This study;
[l] Kuroda, 1992;
[m] Kuroda et al., 1996;
[n] McGrew et al., 1988;
[o] Sugiyama & Koman, 1992;
[p] Wrangham et al., 1996.

compounds (Rogers *et al.*, 1990; Remis *et al.*, 2001). The degree of frugivory complementing their herbaceous diet has been shown to be positively correlated with tree species diversity at sites in east Africa. Tree species diversity is inversely related to altitude, and predictably, gorillas in the high altitude regions consumed fewer plant species than gorillas at low altitudes. (Stanford & Nkurungi, 2003; Ganas *et al.*, 2004). Researchers in the lowland forests of central Africa have reported that gorillas supplement their folivorous diet with many fruits (Williamson *et al.*, 1990; Tutin *et al.*, 1991; Remis, 1994; Fay, 1997; Tutin *et al.*, 1997; Doran *et al.*, 2002; Rogers *et al.*, 2004). The frugivorous tendencies of western lowland gorillas in areas where they are sympatric with chimpanzees have led researchers to suggest that resource competition may occur between these apes (Tutin & Fernandez, 1985; Tutin *et al.*, 1991; Yamagiwa *et al.*, 1996; Remis, 1997).

Chimpanzees and gorillas are sympatric throughout much of the Congo Basin and in some parts of eastern Africa. Several sites have reported a high degree of overlap in the diets of these sympatric apes (Table 4.2 summarizes these studies). Further examinations both of this dietary overlap and of specific foraging behaviors can lead to a better understanding of the ecological constraints and mechanisms governing the coexistence of these apes. Researchers at Kahuzi-Biega in the Democratic Republic of Congo (DRC) have suggested that dietary and ranging overlap of *P. t. schweinfurthii* and *Gorilla gorilla graueri* is high, but suggest that competition is avoided through mutual avoidance (Yamagiwa *et al.*, 1996). Studying *P. t. schweinfurthii* and *Gorilla gorilla beringei* at Bwindi Impenetrable Forest in Uganda, Stanford & Nkurungi (2003) have reported that gorillas show increased seasonal frugivory, which results in a high degree of dietary overlap with chimpanzees during particular months. However, it has been suggested that the degree of dietary overlap between these species may be most pronounced in the lowland forests of central Africa (Tutin & Fernandez, 1985, 1993; Tutin *et al.*, 1991; Nishihara, 1995; Kuroda *et al.*, 1996). Evidence of simultaneous exploitation of these shared resources was supported by several observations of co-feeding in the same tree canopy (J. M. Fay, pers. comm., 1998; Kuroda *et al.*, 1996; T. Nishihara, pers. comm., 1998). Although it is clear that dietary overlap occurs between these sympatric ape species, the ecological factors (such as forest productivity and seasonality) or demographic variables (such as ape abundance and group composition) that influence this phenomenon have been overshadowed by general methods and sampling bias. Cumulative food lists that do not provide an indication of the relative amount of food intake or temporal aspects of dietary convergence are often compared. Such comparisons are also based on the assumption that food lists compiled under different circumstances are relatively representative of the diets of these species.

In this study, we provide the first descriptions of chimpanzee feeding ecology derived from prolonged direct observations of *P. t. troglodytes* residing in an undisturbed, dense lowland forest in the Congo Basin. Based on previous reports of dietary overlap between apes in the Ndoki region and the relatively high density of apes in the Goualougo Triangle study area (1.53–2.23 chimpanzees/km^2 and 2.34 gorillas/km^2, respectively), we expected to see some degree of overlap between these sympatric apes (Morgan *et al.*, in press). We compare our results to previous studies to examine whether: (1) the diet composition or diversity of these chimpanzees is different from other sites; (2) the dietary overlap shown by chimpanzees and gorillas in the Goualougo Triangle resembles feeding overlap between sympatric apes at other sites; and (3) there is preliminary indication of niche differentiation by these apes as shown by divergent selectivity of the same species. In addition, preliminary data are provided on resource abundance in relation to foraging behavior. We assess whether our results support generalizations about chimpanzee feeding behavior, provide insights into the intraspecific behavioral flexibility of this species, and elucidate potential patterns in dietary overlap between these sympatric apes.

Methods

Study site

The Goualougo Triangle is located within the Nouabalé-Ndoki National Park (NNNP) (16°51′–16°56′N; 2°05′–3°03′E), Republic of Congo. The study area covers 30 000 ha of lowland forest and altitudes range between 330 and 600 m. Four habitat types occur in the Goualougo Triangle: monodominant *Gilbertiodendron* forest, *Gilbertiodendron* mixed species forest, mixed species forest, and swamp forest (based on Moutsambote *et al.*, 1994). The climate in the study area can be described as transitional between the Congo-equatorial and sub-equatorial climatic zones (White, 1983). The main rainy season is typically from August through November, with a short rainy season in May. Average monthly temperatures and rainfall were recorded at Mbeli Bai base camp, ROC (17 km from the study area). The annual rainfall averaged 1728 ± 47mm between 2000 and 2002 (E. Stokes, unpublished data). The average minimum and maximum temperatures during those years were 21.1 °C and 26.5 °C in 2000, 21.5 °C and 26.8 °C in 2001, and 21.9 °C and 26.5 °C in 2002, and showed little seasonal variation (E. Stokes, unpublished data).

Table 4.2. *Sites where sympatric chimpanzees and gorillas have been studied*

Site / Country	Altitude (m)	Vegetation classification	Sympatric apes	Ape density Indiv/km²	Method (# chimp; # gorilla)	Dietary diversity Parts (Spp.)	# Overlap species	% Dietary overlap
Belinga[a] Gabon	700–1000	Primary forest, secondary forest	P. t. troglodytes G. g. gorilla	0.49[i] 0.44[i]	**F** (25; 246) **T** (35; 165) **D** (14; 11)	46(43) 104(89)	46	100 60
Lope[b] Gabon	100–200	Semi-evergreen tropical rainforest, colonizing forest, Marantaceae forest, mixed species, closed canopy forest, Sacoglottis forest, savanna and forest savanna	P. t. troglodytes G. g. gorilla	0.2–1.1[j] 0.3–1.0[j]	**F** (1854; 3565) **D** (857; ?) **T** (277; ?)	161 (132) 213 (–)	123	73 57
Okorobiko[c] Eq. Guinea	<750	Dense and secondary forest, regenerating forest, Brachystegia, Landolphia, plantations	P. t. troglodytes G. g. gorilla	0.3–1.5		(43)		
Ndoki[d] Rep. Congo	300–600	Monodominant Gilbertiodendron forest, semi-deciduous mixed species forest, swamp forest	P. t. troglodytes G. g. gorilla	1.3 1.3	**F** (42; 29) **D** (7; 8) **T**	52 79	47	90 59
Ndoki[e] Rep. Congo	300–600	Monodominant Gilbertiodendron forest, semi-deciduous mixed species forest, swamp forest	P. t. troglodytes G. g. gorilla	2.7 1.9–2.6, 2.3–2.6	**F** (214; ?) **D** **T**	114 (108) 182 (152)	64	59 42

Goualougo[f] Rep. Congo	330–600	Monodominant *Gilbertiodendron* forest, semi-deciduous mixed species forest, swamp forest	*P. t. troglodytes*	1.5 (1.2, 1.9)[k]	**D** (650; 342)	158 (116)	67	58
			G. g. gorilla	2.3 (1.8, 3.0)[k]	**F** (497; 631)	107 (80)		84
					T			
Kahuzi[g] Dem. Congo	1800–3300	Bamboo forest, secondary montane forest, *Cyperus* swamp	*P. t. schweinfurthii*	0.1	**F** (394; 256)	99 (75)	55	73
			G. g. graueri	0.4–0.5	**D** (47; 39)	129 (79)		70
					T (11, 12)			
Bwindi[h] Uganda	2000–2300	*Parinari*-dominated forest, *Chrysophyllum*-dominated upland forest, *Newtonia*-dominated forest, swamp, bamboo zone	*P. t. schweinfurthii*		**F** (187; 264)	60 (>34)	32	94
			G. g. beringei		**D**	133 (>96)		33

Notes:

[a] Tutin & Fernandez, 1985;
[b] Tutin & Fernandez, 1993;
[c] Jones & Sabater Pi, 1971; Sabater Pi, 1979;
[d] Kuroda, 1992;
[e] Kuroda *et al.*, 1996;
[f] This study;
[g] Yamagiwa *et al.*, 1996;
[h] Stanford & Nkurunungi, 2003;
[i] Tutin & Fernandez, 1984;
[j] White, 1992;
[k] Morgan *et al.*, in press;

Methods: D = Direct observation, F = Fecal analysis, T = Trail signs. Dietary overlap is number of food items shared by apes divided by the total number of food items for each species.

Data collection

Direct observations

During 47 months between February 1999 and October 2004, we made contact with chimpanzees and gorillas, respectively, on 650 and 342 occasions for a total observation time of ~1500 hours and 54 hours, respectively. Data for this study were collected while conducting reconnaissance surveys of the semi-habituated chimpanzees in the Goualougo Triangle. Lack of full habituation may have biased our observations toward chimpanzees feeding and socializing in the canopy. While searching for chimpanzees, we frequently encountered gorillas but did not attempt to maintain prolonged contact. During each observation we recorded all food items (fruit, leaf, flower, seed, pith, bark, meat, insect, honey, soil) and species consumed. Interspecific co-feeding events with chimpanzees occurred when another primate species fed simultaneously upon food items from the same individual source.

Time spent feeding was measured by instantaneous group scans that were conducted at 20-minute intervals during chimpanzee encounters (Altmann, 1974). Scan data were collected during two field seasons (June 2000–June 2001, September 2001–December 2002), totaling 414 hours of direct observation (n = 195 contacts, average duration = 127 ± 116 min). Data collected included general activity patterns of all individuals present and any food item consumed.

Fecal analysis

A total of 497 chimpanzee and 631 gorilla fecal samples were analyzed for food content between June 2000 and March 2004. The feces of the two ape species were distinguishable by form, consistency, and associated traces. The contents of all fecal samples were examined macroscopically following the methods of Tutin *et al.* (1991). Fresh (<12 hours old) and recent (<24 hours old) samples were transported back to camp to be sluiced in a 1-mm mesh sieve. Older samples were examined on the path. All seeds were identified to species level and counted. The abundance of small seeds was ranked on a five-point scale of abundance (absent, rare, few, common, abundant). As it is often difficult to identify the species of foliage in feces, the abundance of green leaf fragments and fiber categories were independently ranked on the same scale with respect to the total mass of the fecal sample. When possible, the species of foliage was identified. The species and abundance of social insect remains, meat, bone, and soil found in fecal samples were also recorded.

Feeding traces

During 2001 and 2002, we recorded all feeding traces that were attributed with certainty to either chimpanzees or gorillas based on the presence of other verifying signs (presence, feces, odor, imprints, hair, etc.). These data were used to compile a more comprehensive food list for both chimpanzees and gorillas.

Tree diversity

Hall, Harris, & Finkral (unpublished data) conducted a floristic inventory of two mixed species forest sites in the Goualougo Triangle study area. The inventory was designed to gain an unbiased understanding of tree species diversity within the semi-deciduous forests of the southeastern sector of the NNNP. They chose a 2.25-km^2 area within which a sampling grid was laid out on the ground. A total of sixteen 30-m $\times$ 30-m plots were laid out at 500-m intervals along four parallel 1.5-km transects. Within each plot, diameters were measured and species identified for all trees > 10 cm diameter at breast height (DBH).

Data analysis

Food list

Food lists, including the species, life form of the species, and food item consumed, were compiled for chimpanzees and gorilla based on direct observations, fecal remains, and feeding traces. The life form of the plant was divided into five categories: tree, shrub, herb, liana, or other. The diet of chimpanzees and gorillas was broken into seven categories: fruit, seeds, leaves, stems/pith, flowers, bark, and other. In order to distinguish between fruit and seed categories we considered seed consumption to have taken place only if the apes actively sought to consume the seed component of a given fruit or pod. It was not necessary to consume the pulp of a fruit in order to feed on the associated seed(s). We report the five fruits and leaf species that were either most frequently directly observed being consumed or identified in fecal analysis.

Interspecific dietary overlap

Interspecific dietary overlap was calculated as the number of food species shared by gorillas and chimpanzees divided by the total number of food species consumed by each ape (Tutin & Fernandez, 1984, 1993). We also calculated interspecific dietary overlap in food items consumed.

Feeding analysis of fecal samples

The total number of fruit species was tallied for each fecal sample. These data were used to calculate the average number of fruit species consumed each month. Monthly averages of green leaf and fiber were also calculated. A foliage score for each fecal sample was calculated by summing the abundance of green leaf and fiber scores.

We analyzed an average of 13.43 ± 8.28 chimpanzee and 19.12 ± 12.34 gorilla fecal samples per month during the study period. We adopted the method of Doran *et al.* (2002) to determine if sample size had an effect on the amount of fruit and fiber in the fecal samples per month. We did not find a significant relationship between sample size and: mean number of fruit species per fecal sample (r_s chimp $= 0.05$, n $= 37$, p $= 0.77$; r_s gorilla $= -0.21$, n $= 34$, p $= 0.24$), or foliage score (r_s chimp $= -0.30$, n $= 37$, p $= 0.86$; r_s gorilla $= 0.21$, n $= 34$, p $= 0.25$). Only the months with more than five samples for each species were included in the comparison of chimpanzee and gorilla feeding patterns (Figures 4.1a,b).

Results

Chimpanzee diet

Chimpanzees consumed 158 vegetative food items of 116 species, which represented 41 families. The life form of these chimpanzee food items included 83 tree species, 11 strangler species, 10 lianas, 10 species of herbaceous vegetation, and 2 shrubs. Of all the feeding observations recorded for chimpanzee parties, 92% were of feeding on plant items. The food list composition and proportion of time spent consuming fruits, leaves, flowers, seeds, and bark/pith are shown in Table 4.1. The vegetative component of the chimpanzee diet consisted primarily of fruits, which accounted for 56% of feeding events and 57% of time spent feeding (Table 4.1). Leaves were the second-most prevalent food item in species representation (16%) and accounted for 32% of time spent feeding. Chimpanzees spent 4% of their feeding time consuming flowers and buds. The top five most frequently consumed fruit and leaf species are shown in Table 4.3.

An additional 18 non-plant food items were also identified, including meat, social insects, honey, soil, and water. In contrast to other sites with frequent monkey predation, chimpanzees in the Goualougo Triangle have thus far only been observed feeding on duikers (*Cephalophus monticola*, *Cephalophus callipygus*, *Cephalophus dorsalis*). Social insect predation included ants (*Dorylus* spp., *Oecophylla longinodo*), termites (*Macrotermes* spp.), and bees

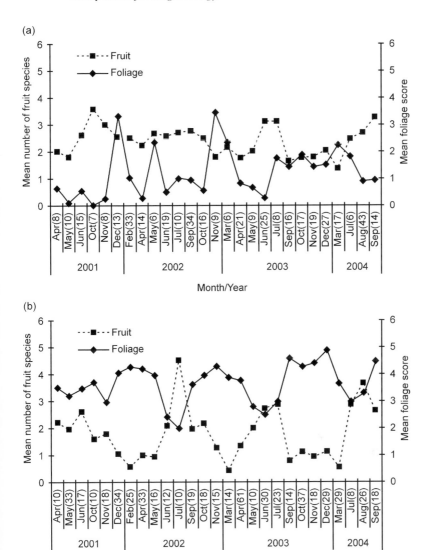

Figure 4.1. (a) Mean number of fruit species and foliage consumed by chimpanzees; abundance scores are per fecal sample per month. (b) Mean number of fruit species and foliage consumed by gorillas; abundance scores are per fecal sample per month.

(*Apis* sp., *Trigona* sp.). Chimpanzees frequently used tools to extract termites and ants from their nests (Sanz *et al.*, 2004). Tool use was also observed in honey gathering at the nests of at least three different bee species. Soil from termite mounds was also consumed. Similar to other long-term study sites, we

Table 4.3. *Top five fruit and leaf species consumed by chimpanzees in Goualougo as shown by percent of feeding observations, group scans, and fecal analysis*

	Chimpanzee			Gorilla	
Top fruit species	Feeding obsv. (n = 550)	Feces (n = 497)	Scans (n = 1390)	Feeding obsv. (n = 179)	Feces (n = 631)
Ficus spp.	**17.3**	**49.5**	**43.4**	3.4	4.8
Greenwayodendron suaveolens	**6.7**	**21.9**	0.7	**11.7**	**9.5**
Irvingia grandifolia	**5.6**	0.2	0.4	–	–
Duboscia sp.	**3.8**	**17.1**	3.4	**19.6**	**40.9**
Landolphia sp.	**3.8**	4.0	1.1	0.6	0.2
Mammea africana	3.3	2.2	**7.5**	1.7	–
Antiaris toxicaria	2.7	6.2	**6.7**	–	–
Celtis adolfi-friderici	2.5	7.8	**6.0**	–	1.7
Manilkara mabokeensis	2.0	–	**5.2**	–	–
Tetrapleura tetraptera	–	7.2	–	1.7	**13.3**
Chrysophyllum lacourtiana	1.6	**19.3**	–	**8.9**	**8.9**
Grewia sp.	1.1	3.8	1.5	**7.3**	**8.1**
Klainedoxa gabonensis	**2.7**	1.0	1.5	**6.1**	3.0
Total of top five	37.2	–	68.8	53.6	–

	Chimpanzee	
Top leaf species	Feeding obsv. (n = 143)	Scans (n = 1390)
Celtis mildbraedii	**62.9**	**87.8**
Celtis adolfi-friderici	**4.9**	**1.3**
Triplochiton scleroxylon	**4.2**	**5.1**
Liane sp.	**3.5**	**3.2**
Dalhousia africana	**2.1**	–
Milicia excelsa	0.7	**1.3**
Total of top five	77.6	98.7

Items in **bold** font are top five species for each column.

observed chimpanzees using leaf sponges to gather water from tree basins. Leaves used by local human populations for medicinal purposes were recovered intact in chimpanzee feces throughout this study.

The frugivorous tendency of chimpanzees and folivorous tendency of gorillas was evident in the results of fecal analyses. A greater proportion of chimpanzee fecal samples contained at least one fruit item (fecal samples with fruit: chimpanzee – 98.8%; gorilla = 79.9%). The average number of fruit species per fecal sample was higher for chimpanzees than gorillas (mean number of fruit species per fecal sample: chimpanzees = 2.33 ± 1.22, n = 497, range = 0, 7; gorillas = 1.63 ± 1.42, n = 631, range = 0, 8). A paired comparison of the monthly averages of these values showed a significant difference between chimpanzees and gorilla fruit consumption (t = 3.51, df = 26, p < 0.01). In contrast, gorillas consumed more foliage than chimpanzees. Foliage was present in a greater proportion of gorilla (94.1%) versus chimpanzee (46.1%) fecal samples and in greater amounts per sample (abundance score of fecal samples: gorilla = 3.74 ± 1.54, n = 631, mode = 4, range = 0, 8; chimpanzee = 1.07 ± 1.57, n = 497, mode = 0, range = 0, 8). The monthly averages of foliage scores also showed a significant difference between chimpanzees and gorilla (t = −13.35, df = 26, p < 0.01).

In mixed forest, the 25 most common tree species and their exploitation by apes are listed in Table 4.4. Four important chimpanzee fruit species (*Greenwayodendron suaveolens, Celtis adolfi-friderici, Manilkara mabokeensis, Duboscia* sp.) and three leaf species (*Celtis mildbraedii, Celtis adolfi-friderici, Triplochiton scleroxyn*) were represented in this list of trees. This list does not include a single representative of the *Ficus* genus, which was identified by all measures as an important food for chimpanzees (Table 4.3). Of the species listed above that were both important in chimpanzee diet and prevalent in the study area, four were also consumed by gorillas.

Dietary overlap between chimpanzees and gorillas

Of the total number of plant species eaten by apes, 52% of food species overlapped in the chimpanzee and gorilla diets (n = 67 shared species). Our data indicated that 84% of gorilla food species were consumed by chimpanzees and 58% of chimpanzee food species were consumed by gorillas. A lower degree of overlap was found in specific food items (30.1% of chimpanzee food items shared with gorillas, 54.7% of gorilla food items shared with chimpanzees), a finding which is most likely affected by skewed sampling effort toward chimpanzees. The highest degree of dietary overlap

Table 4.4. *Most common tree species in mixed species forest to the*
Goualougo Triangle, ranked by stem density

Rank	Genus	Stems/ha	M/ha	C/G*
1	*Diospyros bipindensis*	29.17	0.41	
2	*Diospyros canaliculata*	18.75	0.36	
3	**Celtis mildbraedii**	**13.89**	**1.25**	**C/G**
4	**Greenwayodendron suaveolens**	**13.89**	**0.68**	**C/G**
5	*Strombosia pustulata*	12.50	0.96	C
6	*Pancovia* sp.	11.81	0.30	C/G
7	*Strombosia nigropunctata*	10.42	0.54	
8	*Entandrophragma cylindricum*	8.33	2.21	
9	*Nesogordonia papaverifera*	8.33	1.31	
10	*Anonidium mannii*	8.33	0.55	C/G
11	*Petersianthus macrocarpus*	7.64	1.91	C/G
12	*Sterculia oblonga*	7.64	1.46	
13	**Celtis adolfi-friderici**	**7.64**	**0.50**	**C/G**
14	*Camptostylus mannii*	7.64	0.14	
15	*Strombosiopsis tetandra*	6.94	0.41	
16	*Guarea vel* sp. *aff. Thompsonii*	6.25	0.26	
17	*Cola lateritia*	5.56	0.52	C
18	*Strombosia grandifolia*	4.86	0.69	
19	*Trichilia* sp.	4.86	0.06	
20	*Terminalia superba*	4.17	1.62	
21	*Cleistanthus mildbraedii*	4.17	0.62	
22	*Diospyros crassiflora*	4.17	0.36	C/G
23	*Barteria* sp.	4.17	0.09	
24	*Angylocalyx pynaertii*	3.47	0.18	C/G
25	*Xylopia chrysophylla*	3.47	0.16	

Notes:
*Consumed by chimpanzees (C), gorillas (G), or both (C/G).
Items in **bold** font are important foods consumed by chimpanzees. Rank indicates species position
in relation to all other species with regards to density. Area is the percent each species contributes
to the total basal area per hectare.

occurred in fruit items, followed by leaves and stems. Dietary overlap did not
occur with meat species or social insects. Chimpanzees and gorillas employ
different foraging strategies to feed on social insects. Chimpanzees used a
complex tool set to extract two different species of *Macrotermes* from epigeal
and subterranean nests, whereas gorillas gathered *Cubitermes* by breaking the
nest with their hands (Sanz *et al.*, 2004). Gorillas have not yet been observed
eating ants or honey, but ant traces have been recovered in their feces.

There was overlap in the five fruit species most frequently consumed by
chimpanzees and gorillas (Table 4.3). All measures (direct observation of

feeding events, fecal analysis, group scans) indicated that *Ficus* was the most frequent food item for chimpanzees. They were observed consuming 13 species of this genus, whereas only 4.8% of gorilla feces contained *Ficus* seeds and direct observations of gorillas revealed only three species. *Duboscia* was the most common fruit item consumed by gorillas during direct observations and recovered in feces. This was also identified as an important food item for chimpanzees. *G. suaveolens* was the second most common item for chimpanzees (21.9% of feces) and third for gorillas (9.5% of feces). Overlap also occurred in the consumption of *Chrysophyllum lacourtiana*. Although our observation time was biased toward chimpanzees, all but one of the top foods for gorillas identified in the study were listed as important foods by Doran *et al.* (2002), who studied feeding in habituated western lowland gorillas.

Over the duration of the study, chimpanzees and gorillas showed different patterns in fruit and foliage consumption (Figure 4.1). Chimpanzees consistently consumed a greater variety of fruit species (as measured by the average number of fruit species per fecal sample) than gorillas, but this was not related to their foliage consumption each month. Gorillas consumed more foliage than chimpanzees in all months surveyed. Foliage scores consisted of both green leaf fragments and fiber. The level of consumption and relative contribution of these components clearly differed between chimpanzees and gorillas (Figures 2a, b). There was not a significant difference in the average monthly green leaf scores between chimpanzees and gorillas ($U = 51.5$, $n = 24$, $p = 0.25$), but average monthly fiber scores were different ($U = 0.0$, $n = 24$, $p < 0.01$). Our results indicated a trend rather than a negative correlation between the monthly fruit and foliage intake of chimpanzees ($r_s = -0.36$, $n = 29$, $p = 0.06$). In contrast, there was an inverse relationship between the average number of fruit species and foliage scores for gorillas over the same time period ($r_s = -0.62$, $n = 29$, $p < 0.01$).

Chimpanzees and gorillas were observed co-feeding on eight occasions in six different species of fruit tree. This included co-feeding in the same tree crown ($n = 4$) and co-feeding in the same tree but separated by vertical distance, with a solitary silverback gorilla feeding on fallen fruits on the ground while chimpanzees foraged in the canopy of the same tree ($n = 4$). There were six additional occasions when gorillas approached a tree where chimpanzees were feeding, but fled after detecting human presence. Co-feeding parties could be lengthy, lasting an average of 46 minutes (sd = 41, range = 18–150 minutes; $n = 8$) and ranging in size from 3–22 individuals. Average chimpanzee and gorilla party size during co-feeding events was 7.0 ± 6.2 ($n = 8$, range = 2–18) and 2.8 ± 1.8 ($n = 6$, range = 1–6),[1] respectively. Most of these intersections occurred in two species of *Ficus*

(a)

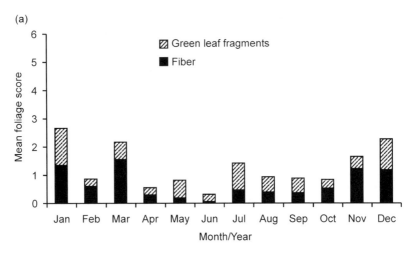

(b)

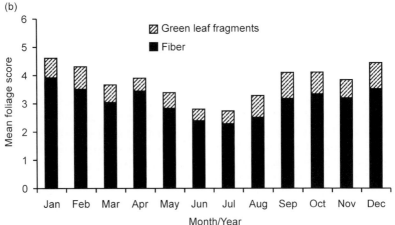

Figure 4.2. (a) Average abundance of foliage consumed each month by chimpanzees. (b) Average abundance of foliage consumed each month by gorillas.

trees (*Ficus calyptrate* and *Ficus recurvata*), but co-feeding was also observed at *C. lacourtiana*, *Mammea africana*, and *Treculia africana*.

Discussion

The chimpanzees of the Goualougo Triangle are frugivores who regularly supplement their diet with vegetative plant parts, meat, and social insects.

They have a diverse diet that currently encompasses more than 176 food items and continues to expand. As documented at other sites with sympatric chimpanzees and gorillas, we observed a high degree of dietary overlap and occasional interspecific encounters at food sources (Tutin & Fernandez, 1985, 1993; Kuroda, 1992). Food list overlap showed that chimpanzees and gorillas consume many of the same plant foods, but fecal analysis indicated that their patterns of fruit and foliage consumption differed over the same time period. Divergences in dietary selectivity were indicated by differences in fruit consumption and exploitation of common tree species in the study area.

Chimpanzees in the Goualougo Triangle were persistent and selective in their fruit consumption throughout the year. We found that this central subspecies community of chimpanzee allocated more than half of their feeding time to fruits, a finding similar to that found for other subspecies residing in east and west Africa (Hladik, 1977; Wrangham, 1977; Wrangham *et al.*, 1991; Wrangham *et al.*, 1996; Newton-Fisher, 1999; Fawcett, 2000). We observed that 56% of the food list comprised fruits, which was within the range of other sites. However, given that chimpanzee diets vary considerably over different temporal and spatial scales and that the chimpanzees were not completely habituated during the course of this investigation, continued research will be required to determine if our findings are representative of the feeding ecology of the chimpanzees in this population.

Researchers in central and east Africa have suggested that leaves are an important resource for chimpanzees (Hladik, 1977; Nishida *et al.*, 1983; Tutin *et al.*, 1991; Fawcett, 2000). We found that the time spent feeding on leaves by chimpanzees in this study was higher than observed in other populations. The chimpanzees regularly fed upon several leaf species from different life forms such as trees and lianas as well as leaves in different stages of maturity. Specifically, our observations confirm suggestions that *C. mildbraedii* leaves may be a key resource for apes in northern Congo (Nishihara, 1995; Kuroda *et al.*, 1996). Direct observations showed that chimpanzees and gorillas in the Goualougo Triangle consumed *C. mildbraedii* leaves throughout the year, with significant increases when new leaves were available (Morgan, unpublished data). The apparent preference of chimpanzees for *C. mildbraedii*, based on direct observations and its abundance in the study area (13.9 stems/ha) make this a good candidate for being a staple food item in the diet of these apes. Important foods and their abundance in the environment have been suggested as playing a key role in ape social behavior (Wrangham, 1986). Investigating the importance of *Celtis m.* leaves in the diet of the Goualougo chimpanzees and the relationship of overall foliage intake to ape social behavior in this region will be an important avenue of future research.

In a synthesis of the diet of western lowland gorillas from six sites, Rogers *et al.* (2004) showed this species to have a diverse feeding repertoire while remaining a selective fruit pursuer. Fruit consumption of western gorillas in the Goualougo and other sites in the region is less consistent than for chimpanzees, which could be the result of a selective fruit-eating strategy based on particular fruit species (Williamson *et al.*, 1990; Tutin *et al.*, 1991; Kuroda *et al.*, 1996; Fay, 1997; Remis, 1997; Remis *et al.*, 2001; Doran *et al.*, 2002; Rogers *et al.*, 2004). Foraging intersections and dietary overlap with chimpanzees occurred most often in the consumption of fruit. The overlap in fruits was followed by shared leaf and stem food items. Researchers have consistently reported a divergence in non-food plant items, such as insects and mammals. However, the degree of habituation and the sampling effort of observing gorillas were much less than for chimpanzees in this study, and this may have affected our overlap results. A higher degree of overlap is reported from all sites where the food list for one ape species is more complete than for the other species (Table 4.2). The gorilla food lists have been more complete at several of these sites (Belinga, Lopé, Ndoki, Bwindi), resulting in high or complete overlap for chimpanzees. Goualougo and (recently) Kahuzi (Basabose, 2002) are the only sites where chimpanzees are studied primarily by direct observation rather than indirect evidence (feces, trails signs). If an earlier estimate of diet composition is used to calculate overlap at Kahuzi, then it is 73% with gorillas, but if the more recent estimate of 110 food species is used, then overlap decreases to 54% (Yamagiwa *et al.*, 1996; Basabose, 2002). The proportion of dietary overlap in food lists is obviously influenced by skewed sampling effort, so interpretations of the degree of dietary overlap from short-term studies of unhabituated groups should be interpreted with caution. For comparisons of interspecific feeding ecology, it is more informative to depict feeding patterns between species at specific time intervals.

In the Goualougo Triangle, chimpanzee and gorilla diets clearly diverged in their patterns and degrees of foliage consumption during the same sampling period. Gorilla foliage scores were correlated with fruit consumption. This was in contrast to chimpanzees, who maintained relatively low foliage scores and stable fruit consumption over the same period. Chimpanzees consumed relatively low but equally attributable amounts of green leaf and fiber as indicated in their monthly foliage scores. In contrast, the stems and leaves of terrestrial herbaceous vegetation from the *Marantaceae* and *Zingiberaceae* families largely comprised the high levels of year-round foliage consumed by gorillas, which is similar to other findings in central Africa (Tutin & Fernandez, 1985, 1993; Williamson *et al.*, 1990; Remis, 1994; Nishihara, 1995; White *et al.*, 1995; Kuroda *et al.*, 1996; Fay, 1997; Doran *et al.*, 2002; Rogers *et al.*, 2004). Although stem densities have been shown to vary across sites, gorillas consistently consume large amounts of terrestrial herbaceous vegetation, a reflection of their specialized digestive physiology (Doran *et al.*, 2002; Rogers *et al.*, 2004).

It is well documented that figs are a key resource for many chimpanzee populations, but members of the *Ficus* genus have not been particularly prevalent in reports of gorilla feeding ecology (Conklin & Wrangham, 1994; Wrangham *et al.*, 1996; Newton-Fisher, 1999; Basabose & Yamagiwa, 2002; Stanford & Nkurungi, 2003; but see Ganas *et al.*, 2004). The representation of *Ficus* spp. as locations for co-feeding events with gorillas, despite the relatively low stem density of this tree species (<1 stem/ha) suggests that this is also an important resource for apes in the Goualougo Triangle. Although gorillas in the Goualougo Triangle overlapped with chimpanzees in their fig consumption, their exploitation of this genus was less frequent and more selective. The fig species eaten by gorillas tend to be characterized by relatively large syconium that may provide more efficient nutritional intake for the foraging effort compared with smaller figs. Conklin & Wrangham (1994) found that there was significant variation in the nutritional content of different *Ficus* species. This is an indication of dietary selectivity that could be an important aspect of the niche differentiation between these sympatric apes.

In this chapter we also report on prolonged observations of co-feeding and found that these interspecific encounters may be linked to particular food resources. Associations between chimpanzees and gorillas occurred when both species were mutually attracted to a specific fruiting tree. There was no evidence that these species were otherwise attracted to each other. Chimpanzees in the Goualougo Triangle are amenable to co-feeding with gorillas at large fruiting tree species for relatively long periods. We did not observe any of the indications of mutual avoidance or competitive behavior between chimpanzees and gorillas that have been reported from other sites (Tutin & Oslisly, 1995; Basabose & Yamagiwa, 2002; Stanford & Nkurungi, 2003). Chimpanzees were also observed co-feeding with *Lophocebus albigena* and *Cercopithecus nictitans*. Their responses to mangabeys and guenons were more variable and included neutral co-feeding, displacement, threat, and active eviction from tree crowns. There were no indications of chimpanzee hunting behavior or apparent hesitation by monkeys to feed in proximity to chimpanzees. More detailed data will be necessary to determine the mechanisms governing interactions between chimpanzees and other members of the primate community.

Researchers in central Africa have just begun to document the behavior and ecology of apes residing in this region. A better understanding of the flexibility in chimpanzee diets and their relationships with sympatric gorillas will enable us to proceed with more valid comparisons among species. Although several primate taxa in central Africa have been studied independently, little research has focused on the complex relationships within primate communities. Such research will only be possible if we ensure the immediate protection and long-term preservation of the remaining intact tropical forest ecosystems within the forests of the Congo Basin.

Acknowledgments

We are deeply appreciative of the opportunity to work in the Nouabalé-Ndoki National Park and especially the Goualougo Triangle. This work would not be possible without the continued support of the Government of the Republic of Congo and Wildlife Conservation Society-Congo. Special thanks are due to J.M. Fay, P. Elkan, S. Elkan, M. Gately, E. Stokes, B. Djoni, and J.R. Onononga. Comments from M. Robbins, G. Hohmann, D. Doran, C. Boesch, P. Lee, and anonymous reviewers improved the quality of this manuscript. Invaluable assistance in the field was provided by R. Mokanga, J.P. Koba, G. Djokin, J. Mbio, N. Ewouri, and L. Oessembati. Grateful acknowledgment of funding is due to the U.S. Fish and Wildlife Service, National Geographic Society, Wildlife Conservation Society, Columbus Zoological Park, Brevard Zoological Park, and Lowry Zoological Park.

References

Altmann, J. (1974). Observational study of behavior: sampling methods. *Behaviour*, **49**, 227–65.

Anderson, D.P., Nordheim, E.V., Boesch, C., & Moermond, T.C. (2002). Factors influencing fission-fusion grouping in chimpanzees in the Tai National Park, Côte d'Ivoire. In *Behavioural Diversity in Chimpanzees and Bonobos*, ed. C. Boesch, G. Hohmann, & L.F. Marchant, pp. 90–101. Cambridge: Cambridge University Press.

Basabose, A.K. (2002). Diet composition of chimpanzees inhabiting the montane forests of Kahuzi, Democratic Republic of Congo. *American Journal of Primatology*, **58**, 1–21.

Basabose, A.K. & Yamagiwa, J. (2002). Factors affecting nesting site choice in chimpanzees at Tshibati, Kahuzi-Biega National Park: influence of sympatric gorillas. *International Journal of Primatology*, **23**, 263–82.

Boesch, C. & Boesch-Achermann, H. (2000). *The Chimpanzees of the Tai Forest: Behavioural Ecology and Evolution.* Oxford: Oxford University Press.

Chapman, C.A., Chapman, L.J., & Gillespie, T.R. (2002). Scale issues in the study of primate foraging: red colobus of Kibale National Park. *American Journal of Physical Anthropology*, **117**, 349–63.

Chapman, C.A., Chapman, L.J., & Wrangham, R.W. (1995). Ecological constraints on group size: an analysis of spider monkey and chimpanzee subgroups. *Behavioral Ecology and Sociobiology*, **36**, 59–70.

Conklin, N.L. & Wrangham, R.L. (1994). The value of figs to a hind-gut fermenting frugivore: a nutritional analysis. *Biochemical Systematics and Ecology*, **22**, 137–51.

Doran, D. (1997). Influence of seasonality on activity patterns, feeding behaviour, ranging and grouping patterns in Taï chimpanzees. *International Journal of Primatology*, **19**, 183–206.

Doran, D. & McNeilage, A. (1998). Gorilla ecology and behavior. *Evolutionary Anthropology*, **6**, 120–31.

Doran, D.M., McNeilage, A., Greer, D., Bocian, C., Mehlman, P., & Shah, N. (2002). Western lowland gorilla diet and resource availability: new evidence, cross-site comparisons, and reflections on indirect sampling. *American Journal of Primatology*, **58**, 91–116.

Fawcett, K. (2000). *Female Relationships and Food Availability in a Forest Community of Chimpanzees.* Unpublished Ph.D. thesis, University of Edinburgh.

Fay, J.M. (1997). *The Ecology, Social Organization, Populations, Habitat and History of the Western Lowland Gorilla (Gorilla gorilla gorilla Savage and Wyman 1847).* Unpublished Ph.D. thesis, Washington University.

Ganas, J., Robbins, M.M., Nkurunungi, J.B., Kaplin, B.A., & McNeilage, A. (2004). Dietary variability of mountain gorillas in Bwindi Impenetrable National Park, Uganda. *International Journal of Primatology*, **25**, 1043–72.

Hladik, C.M. (1977). Chimpanzees of Gabon and chimpanzees of Gombe: some comparative data on diet. In *Primate Ecology*, ed. T.H. Clutton-Brock, pp. 481–501. London: Academic Press.

Hunt, K.D. & McGrew, W.C. (2002). Chimpanzees in the dry habitats of Assirik, Senegal and Semliki Wildlife Reserve, Uganda. In *Behavioural Diversity in Chimpanzees and Bonobos*, ed. C. Boesch, G. Hohmann, & L.F. Marchant, pp. 35–51. Cambridge: Cambridge University Press.

Jones, C. & Sabater Pi, J. (1971). Comparative ecology of *Gorilla gorilla* (Savage and Wyman) and *Pan troglodytes* (Blumenbach) in Rio Muni, West Africa. *Bibliotheca Primatologica*, **13**, 1–96.

Kuroda, S. (1992). Ecological interspecies relationships between gorillas and chimpanzees in the Ndoki-Nouabalé reserve, northern Congo. In *Topics in Primatology, Volume 2: Behavior, Ecology, and Conservation*, ed. N. Itoigawa, Y. Sugiyama, G.P. Sackett & R.K.R. Thompson, pp. 385–94. Tokyo: University of Tokyo Press.

Kuroda, S., Nishihara, T., Suzuki, S. & Oko, R.A. (1996). Sympatric chimpanzees and gorillas in the Ndoki Forest, Congo. In *Great Ape Societies*, ed. W.C. McGrew, L.F. Marchant & T. Nishida, pp. 71–81. Cambridge: Cambridge University Press.

McGrew, W.C., Baldwin, P.J., & Tutin, C.E.G. (1988). Diet of wild chimpanzees (*Pan troglodytes verus*) at Mt. Assirik, Senegal. I. Composition. *American Journal of Primatology*, **16**, 213–26.

Morgan, D., Sanz, C., Onononga, J.R., & Strindberg, S. (in press). Ape abundance and habitat use in the Goualougo Triangle, Republic of Congo. *International Journal of Primatology*.

Moutsambote, J.-M., Yumoto, T., Mitani, M., Nishihara, T., Suzuki, S., & Kuroda, S. (1994). Vegetation and plant list of species identified in the Nouabale-Ndoki Forest, Congo. *Tropics*, **3**, 277–94.

Newton-Fisher, N.E. (1999). The diet of chimpanzees in the Budongo Forest Reserve, Uganda. *African Journal of Ecology*, **37**, 344–54.

Nishida, T. (1974). The ecology of wild chimpanzees. In *Human Ecology*, ed. R. Ohtsuka, J. Tanaka, & T. Nishida, pp. 15–60. Tokyo: Kyoritsu-Shuppan.

Nishida, T. & Uehara, S. (1983). Natural diet of chimpanzees (*Pan troglodytes schweinfurthii*): long-term record from the Mahale Mountains, Tanzania. *African Study Monographs*, **3**, 109–30.

Nishida, T., Wrangham, R.W., Goodall, J., & Uehara, S. (1983). Local differences in plant-feeding habits of chimpanzees between the Mahale Mountains and Gombe National Park, Tanzania. *Journal of Human Evolution*, **12**, 467–80.

Nishihara, T. (1995). Feeding ecology of western lowland gorillas in the Nouabalé-Ndoki National Park, northern Congo. *Primates*, **36**, 151–68.

Remis, M.J. (1994). *Feeding Ecology and Positional Behavior of Lowland Gorillas in the Central African Republic*. Unpublished Ph.D. thesis, Yale University.

(1997). Western lowland gorillas (*Gorilla gorilla gorilla*) as seasonal frugivores: use of variable resources. *American Journal of Primatology*, **43**, 87–109.

Remis, M.J., Dierenfeld, E.S., Mowry, C.B., & Carroll, R.W. (2001). Nutritional aspects of western lowland gorilla (*Gorilla gorilla gorilla*) diet during seasons of fruit scarcity at Bai Houkou, Central African Republic. *International Journal of Primatology*, **22**, 807–36.

Rogers, M.E., Abernathy, K., Bermejo, M. *et al.* (2004). Western gorilla diet: a synthesis from six sites. *American Journal of Primatology*, **64**, 173–92.

Rogers, M.E., Maisels, F., Williamson, E.A., Fernandez, M., & Tutin, C.E.G. (1990). Gorilla diet in the Lope Reserve, Gabon: a nutritional analysis. *Oecologia*, **84**, 326–39.

Sabater Pi, J. (1979). Feeding behavior and diet of chimpanzees (*Pan troglodytes troglodytes*) in the Okorobiko Mountains of Rio Muni (West Africa). *Zeitschrift fur Tierpsychologie*, **50**, 265–81.

Sanz, C., Morgan, D., & Gulick, S. (2004). New insights into chimpanzees, tools, and termites from the Congo Basin. *The American Naturalist*, **164**, 567–81.

Stanford, C.B. & Nkurunungi, J.B. (2003). Behavioral ecology of sympatric chimpanzees and gorillas in Bwindi Impenetrable National Park, Uganda: diet. *International Journal of Primatology*, **24**, 901–18.

Sugiyama, J. (1999). Socioecological factors influencing population structure of gorillas and chimpanzees. *Primates*, **40**, 87–104.

Sugiyama, Y. & Koman, J. (1992). The flora of Bossou: its utilization by chimpanzees and humans. *African Study Monographs*, **13**, 127–69.

Tutin, C.E.G. & Fernandez, M. (1984). Nationwide census of gorilla and chimpanzee populations in Gabon. *American Journal of Primatology*, **6**, 313–36.

(1985). Foods consumed by sympatric populations of *Gorilla gorilla* and *Pan troglodytes* in Gabon: some preliminary data. *International Journal of Primatology*, **6**, 27–43.

(1993). Composition of the diet of chimpanzees and comparisons with that of sympatric lowland gorillas in the Lope Reserve, Gabon. *American Journal of Primatology*, **30**, 195–211.

Tutin, C.E.G. & Oslisly, R. (1995). *Homo, Pan,* and *Gorilla*: coexistence over 60,000 years at Lope in central Gabon. *Journal of Human Evolution*, **28**, 597–602.

Tutin, C.E.G., Fernandez, M., Rogers, M.E., Williamson, E.A., & McGrew, W.C. (1991). Foraging profiles of sympatric lowland gorillas and chimpanzees in the Lopé Reserve, Gabon. *Philosophical Transactions of the Royal Society of London*, **334**, 178–86.

Tutin, C.E.G., Ham, R.M., White, L.J.T., & Harrison, M.J.S. (1997). The primate community of the Lope Reserve, Gabon: diets, responses to fruit scarcity and effects on biomass. *American Journal of Primatology*, **42**, 1–24.

White, F. (1983). *The Vegetation of Africa*. Paris: UNESCO.

White, L. (1992). *The Effects of Mechanized Selective Logging on the Flora and Mammalian Fauna of the Lope Reserve, Gabon*. Edinburgh: University of Edinburgh.

White, L.J.T., Rogers, M.E., Tutin, C.E.G., Williamson, E.A., & Fernandez, M. (1995). Herbaceous vegetation in different forest types in the Lope Reserve, Gabon: implications for keystone food availability. *African Journal of Ecology*, **33**, 124–41.

Williamson, E.A., Tutin, C.E.G., Rogers, M.E., & Fernandez, M. (1990). Composition of the diet of lowland gorillas at Lope in Gabon. *American Journal of Primatology*, **21**, 265–77.

Wrangham, R.W. (1977). Feeding behaviour of chimpanzees in Gombe National Park, Tanzania. In *Primate Ecology*, ed. T. H. Clutton-Brock, pp. 503–38. London: Academic Press.

(1986). Ecology and social relationships in two species of chimpanzee. In *Ecology and Social Evolution: Birds and Mammals*, ed. D.I. Rubenstein & R.W. Wrangham, pp. 352–78. Princeton: Princeton University Press.

Wrangham, R.W., Chapman, C.A., Clark-Arcadi, A.P., & Isabirye-Basuta, G. (1996). Social ecology of Kanyawara chimpanzees: implications for understanding the costs of great ape groups. In *Great Ape Societies*, ed. W.C. McGrew, L.F. Marchant, & T. Nishida, pp. 45–57. Cambridge: Cambridge University Press.

Wrangham, R.W., Conklin, N.L., Chapman, C.A., & Hunt, K.D. (1991). The significance of fibrous foods for Kibale chimpanzees. *Philosophical Transactions of the Royal Society of London*, **334**, 171–8.

Yamagiwa, J., Maruhashi, T., Yumoto, T., & Mwanza, N. (1996). Dietary and ranging overlap in sympatric gorillas and chimpanzees in Kahuzi-Biega National Park, Zaire. In *Great Ape Societies*, ed. W.C. McGrew, L.F. Marchant, & T. Nishida, pp. 82–98. Cambridge: Cambridge University Press.

Notes

1. The sample size for gorillas at co-feeding events is smaller because on two occasions it was not possible to accurately count all gorilla group members present.

5 Frugivory and gregariousness of Salonga bonobos and Gashaka chimpanzees: the influence of abundance and nutritional quality of fruit

GOTTFRIED HOHMANN, ANDREW FOWLER, VOLKER
SOMMER, AND SYLVIA ORTMANN

Feeding Ecology in Apes and Other Primates. Ecological, Physical and Behavioral Aspects, ed.
G. Hohmann, M.M. Robbins, and C. Boesch. Published by Cambridge University Press.
© Cambridge University Press 2006.

Introduction

The exceptional diversity of social systems in non-human primates has stimulated a strong interest in explaining this variation in an evolutionary context (Kummer, 1971; Wrangham, 1980). Abundance and distribution of food resources are key elements of theories explaining interspecific variation of social systems (Wrangham, 1979). Socioecological models suggest that these two parameters determine the type of competition within and between groups (van Schaik, 1989; Sterck *et al.*, 1997). It has been suggested that competition style affects other variables, such as group size (Chapman & Chapman, 2000), the expression of dominance (Sterck *et al.*, 1997), reproduction (Koenig & Borries, 2001), migration, and as a consequence of the latter, kin relationships (Vigilant *et al.*, 2001). While data from many primate species support the major predictions made in these models, it has become increasingly clear that multiple factors modulate competition and, consequently, social relations (van Schaik, 1996; Sterck *et al.*, 1997; Isbell & Young, 2002; Koenig, 2002; Janson & Vogel, Chapter 11, this volume; Koenig & Borries, Chapter 10, this volume).

Food choice does not always reflect availability, i.e., individuals may select food items from relatively rare species instead of exploiting those that are more abundant. Optimal Diet theory predicts that food selection is triggered by maximization of net energy (Emlen, 1966). Field studies and experimental work show that food selection and foraging efficiency is likely to be affected by a number of parameters, including nutritional quality and distribution and abundance of sources (Carlo *et al.*, 2003; Saracco *et al.*, 2004). It appears reasonable to assume that individuals select sources with high nutrient and energy levels, and at the same time, avoid the intake of chemical components that impede digestion. However, recent studies suggest that avoidance of antifeedants may be the driving force of food selection (Alm *et al.*, 2000; Clauss *et al.*, 2003). Moreover, since species differ in their preference for different nutrients, general predictions about food selection and foraging strategies require a working knowledge of digestive physiology (Witmer & van Soest, 1998; Schaefer *et al.*, 2003). In addition, temporal and spatial fluctuations of food availability are also known to affect the quality of ingesta (Owen-Smith, 1994; Dearing *et al.*, 2000). Finally, nutritional requirements vary with age, sex, reproductive status, and other demographic parameters, and the design of an optimal diet is likely to vary in relation with a number of life history parameters (Altmann, 1998). Taken together, understanding observed patterns of food choice requires input from various information sources and a solid knowledge of how the different parameters interact with each other. Few attempts have been made by primate researchers to integrate quantitative measures of resource abundance, parameters of resource quality, and digestive physiology. Among the notable exceptions are the studies on howler monkeys and other primates by Milton (1978, 1981, 1986; Milton & McBee, 1983), the longitudinal study on baboons by Altmann (1998), and Leighton's (1993) field studies on foraging strategies of wild orangutans. Given the challenges imposed by a frugivorous lifestyle (Levey & Martinez del Rio, 2001), nutritional ecology is a promising field for primate research that is still waiting to be explored.

In this study we investigated the availability, and the chemical quality of fruit produced at two sites in central and western Africa, respectively. One site is inhabited by bonobos (*Pan paniscus*), the other by chimpanzees (*Pan troglodytes*). The two *Pan* species provide an interesting model for studying the relationship between resource distribution, food competition, and social relations.

Bonobos typically occupy the moist evergreen lowland forests south of the Congo river, although in the southern range where the forest gives way to savannas, the habitat may be a mosaic combining primary forest, secondary growth, and grassland. Chimpanzees inhabit the forests north of the Congo river and in some regions, they coexist with gorillas. Chimpanzees inhabit a variety of different habitats across equatorial Africa, ranging from dry forests and woodland savanna to moist evergreen forests of variable altitude. Independent of habitat, the diet of both species is dominated by fruit from trees and climbers. Leaves, flowers, tubers, and pith are minor components of a plant diet that is complemented by insects and the meat of vertebrates. Both species live in fission-fusion societies that allow them to vary group size, and by so doing, mitigate food competition. In both species, females tend to transfer from their natal groups whereas males are philopatric. Therefore, resident males are expected to be more closely related to each other than are females (Morin *et al.*, 1994; Gerloff *et al.*, 1999).

While some populations of the two species inhabit similar environments and may therefore exploit similar resources, others inhabit distinct types of habitat and are expected to exploit different types of resources. From this one would predict that populations living under similar conditions would resemble each other in terms of social patterns, while populations living under different environmental conditions would not. While the notion that each *Pan* species lives in a different social environment remains widespread, recent work on both species has identified similarities of social patterns that were once thought to be disparate. While some of the gaps separating the two species have therefore seemed to disappear, recent field studies still maintain that regardless of ecological conditions, the two species differ markedly in terms of social relationships between group members and between neighboring groups (for reviews see Wrangham, 1986). Female bonobos are highly gregarious and associate regularly with other adult females and with adult males (Idani, 1991; Hohmann *et al.*, 1999). By comparison, social bonds between males are neutral or weak (Furuichi & Ihobe, 1994). Females are co-dominant with males and, under certain circumstances, may even be able to dominate males due to female–female coalitions (Parish, 1996). In chimpanzees, relations between females are more variable, ranging from affiliative and cooperative (e.g., at Taï Wittig & Boesch, 2003) to solitary and uncooperative (e.g., at Kanyawara; Wrangham *et al.*, 1992). Intraspecific variation in female social relations covaries with reproductive strategies (Wrangham, 2002; Lehmann & Boesch, 2004), alliance formation (Parish, 1996; Vervaecke *et al.*, 2000), and food sharing (Fruth & Hohmann, 2002). The differences between *Pan* species are even more pronounced when looking at males: Male chimpanzees are the social sex and are invariably dominant over females. Males associate with each other and form strategic alliances, cooperating in social affairs with community members and in interactions with members of other communities (Goodall, 1986; Nishida, 1990; Boesch & Boesch-Acherman, 2000).

The gregariousness of female bonobos has been associated with a reduced level of competition, which, in turn, has been linked to the (1) abundance of terrestrial herbs (Badrian & Malenky, 1984); (2) high quality of terrestrial herbs (Wrangham *et al.*, 1991); and (3) large food patch size (White & Wrangham, 1988). While comparison of data from single, distinct populations of bonobos and chimpanzees provided some support for these hypotheses, reevaluation of the same data showed that none of it could sufficiently explain the differences in gregariousness and social behavior of the two *Pan* species (Chapman *et al.*, 1994).

Studies on the feeding ecology of bonobos and chimpanzees have produced a wealth of information on the general type of food ingested, its temporal and spatial distribution, and abundance (Nishida & Uehara, 1983; Malenky & Wrangham, 1994; White & Lanjouw, 1994; Wrangham *et al.*, 1996). By comparison, studies on the nutritional ecology of these apes are still rare and are restricted to single communities that represent few regional populations. Preliminary information about bonobos comes from Malenky's (1990) work at Lomako. The results of his study suggested that the diet of bonobos consisted of high levels of macronutrients and remarkably low levels of antifeedants. In the absence of corresponding information about non-food items, the high quality of food was explained by strong selectivity for ripe food (Malenky, 1990). Data on the availability and use of terrestrial herbs from the same site showed that herbs provided an essential component of the bonobo plant food diet (Malenky & Stiles, 1991).

Information on qualitative parameters of the food of wild chimpanzees comes from different sites representing a range of habitats that vary in terms of geography, climate, forest cover, and flora: Ipassa, Gabon (Hladik, 1977), Gombe, Tanzania (Wrangham & Waterman, 1983), Kanyawara, Uganda (Wrangham *et al.*, 1991), Budongo, Uganda (Reynolds *et al.*, 1998), Mahale, Tanzania (Matsumoto-Oda & Hayashi, 1999), and Bossou, Guinea (Takemoto, 2003). The information emerging from these studies has answered a number of questions about the nutritional ecology of chimpanzees and will be very useful in designing future studies in this field. Comparing the diet of chimpanzees and sympatric Cercopithecines, it was found that the nutritional quality of the diet of chimpanzees was higher than predicted by body mass (Conklin-Brittain *et al.*, 1998). The same study showed that only chimpanzees were able to maintain the high levels of macronutrients and relatively low levels of antifeedants throughout the year (Wrangham *et al.*, 1998). Information from other studies suggests that food choices of chimpanzees are guided by multiple cues. While specific avoidance of unripe fruit seems to be triggered by high levels of antifeedants (Wrangham & Waterman, 1983), selection of food species in general was found to be based on the content of macronutrients (Reynolds *et al.*, 1998; Takemoto, 2003). Reports from chimpanzee study sites indicate substantial variation in floristic composition: In an extensive review of the taxonomy of food plants consumed by Asian and African Pongides, Rodman (2002) found that the use of food plants reflects geographic variation of forest composition rather than preferences for certain plant taxa. Phytochemical studies provide evidence for variation in the chemistry of the taxa that are common elements in the diet of *Pan* species. For example, *Ficus populifolia* contains a toxic substance that is used to poison arrows and that is not found in other species of this genus (Neuwinger, 1998, and citations therein). Pharmacological studies suggest variation of the content of saponines and alkaloids in the leaves of Dialium species (Neuwinger, 1998, and citations therein). Studies on livestock have shown that tannin activity increases with ambient temperature (Makkar & Becker, 1998), and that it is therefore reasonable to assume that the capacity of tannin to bind with protein may vary with altitude. Plants that are typical for young, regenerating forests have been reported to contain relatively low amounts of secondary compounds (Harbourne, 1993), while vegetation that grows under harsh conditions contains higher levels of defensive compounds (Mueller-Harvey & McAllan, 1992). Accordingly, patterns of forest composition can also be expected to affect the nutritional quality of plant foods. Taken together, given the wide range of habitats occupied by chimpanzees, one would predict that differences in the forest composition are reflected not only in patterns of food availability but also in a number of parameters that define the nutritional quality of plant foods.

The goal of the study reported on below was to collect corresponding data from bonobos and chimpanzees for the purpose of comparing the abundance of food sources and the quality of plant foods, and to relate measurements of food abundance to group size. Here we used data on the frugivore section of the diet to (1) identify annual variation of fruit production in both habitats and relate this to annual changes in gregariousness; (2) evaluate the nutritional quality of fruits eaten compared with fruits that were not eaten; and (3) compare the proportions of functionally different chemical components (nutrients, fiber, antifeedants) in food items.

Methods

Study sites and populations

Fieldwork on bonobos was conducted at Lui Kotal, Salonga National Park, Democratic Republic of Congo (DRC) (Hohmann & Fruth, 2003a). Data from chimpanzees came from the Kwano community at Gashaka Gumti National Park, Nigeria (Sommer *et al.*, 2004). The two sites belong to different biomes (Figure. 5.1). Salonga is situated in the southern part of the Cuvette Central. The terrain is flat and covered by a closed canopy forest that is occasionally

Figure 5.1. Geographic location of the two study sites.

interspersed by small, circular savanna patches. Gashaka belongs to the Guinea savanna belt, the terrain is undulated, and the habitat is more diverse, encompassing savanna woodland and gallery forests. At Salonga, data were collected during an early phase of the project when information on home range size, demography of the study community, and other aspects were fragmentary or missing. Initial information on the ecology and demography of the Kwano community was published by Sommer *et al.* (2004).

Data collection

Methodology and data collection protocols were identical at both sites and streamlined in consultations between GH, VS, and AF prior to commencement of the study.

Assessments of fruit production

Data were collected from the beginning of March 2002 through April 2003. The diversity and productivity of the two forests were assessed by monitoring the phenology of about 1000 trees (DBH > 10 cm) along line transects (8000 × 2 m) at each site. Corresponding information was collected from woody climbers attached to those transects trees. Trees and climbers were tagged and efforts were made to identify them taxonomically. Transects were monitored twice per month basis and scores were given for the presence/ absence of leaves, flowers, and fruit. In the absence of data on fruit chemistry, the distinction between "ripe" and "unripe" fruit tends to be arbitrary. Phenology records include comments on the quality of fruit, using size, color, taste, smell, condition of seeds, and other descriptive criteria. However, reports from other sites suggest that at certain times of the year, chimpanzees ingest substantial amounts of unripe fruit (e.g., Wrangham & Waterman, 1983; Reynolds *et al.*, 1998). Therefore, assessments of food abundance included all sources producing fruit of mature size. Crop size was estimated on an exponential scale (1, 10, 100, 1000 . . . fruit/plant).

Assessment of diet composition

At both sites, direct observations of feeding helped to identify food sources, mode of processing, and times of consumption in relation to the state of ripeness. However, direct observations were restricted in time and space because individuals were not fully habituated to the close proximity of observers. Therefore, systematic dietary assessments were based on macroscopic inspections of fecal samples (see below). Results obtained with this method are likely to create a bias towards food sources that contain undigestible components visible in the feces. Being aware of this limitation, we decided to use fecal inspection as an indirect source of information for between-species comparison to assure a higher degree of interobserver reliability and to increase detection rate of food items. In fact, while traces of all plants that were observed being eaten were also found in the remains, a large number of food item remains that appeared in the feces but did not have corresponding records from direct observations.

For evaluation of diet composition, about 40 samples of fresh feces were collected from nest sites or other places each month at each site. The number of nest groups providing samples ranged from 1–8 and attempts were made to spread sample collection evenly across the month. Samples were weighted and then washed with a fine sieve (1-mm mesh). The rough volume proportions of seed, fiber, unstructured matrix, and other components (e.g., termites) were estimated in the field using the following categories: <10 %, 11–25%, 26–50%, 51–75%, $>75\%$. Seeds were counted and identified. Samples of fiber were removed and stored in liquid nitrogen for identification. Results from macroscopic inspection of feces were pooled for each month.

Assessment of fruit abundance

Data from the monthly phenology walks were used to assess fruit abundance. Monthly fruit abundance indices (Am) were calculated as

$$AM \sum (Dk \times Bk \times Pkm)$$

where Dk refers to the density of species k in the transect area; Bk is the average basal area of this species; and Pkm is the proportion of species k on all fruit-bearing trees (Anderson *et al.*, 2002).

Since trees and climbers growing along phenology trails were selected only by metric criteria (DBH), phenology samples consisted both of food plants and plants that were not consumed by *Pan*.

Assessment of party size

The size of daytime parties tends to be more variable than the size of nest groups (Fruth, 1995). We therefore took the size of night nest groups as an indirect measure to assess seasonal fluctuations in gregariousness within populations. The assignment of nests to the same group was based on four criteria: (1) direct observation of nest construction; (2) close spatial aggregation of nests; (3) freshness of nest materials (i.e., consisting of green, fresh leaves, and twigs); and (4) traces of fresh feces and/or urine underneath the nest. Most nest counts were made immediately after the site had been abandoned. Using the above criteria, nest counts yielded two figures: The minimum number of nests refers to all nests that consisted of fresh leaves, were well-spaced (i.e., a clearly evident distance between a nest and its neighbors), clearly visible from the ground, and showed traces of fresh urine/feces spatially discernible from those of neighboring nests. This figure excludes fresh nests without droppings, and nests showing physical association (e.g., integrated and sharing nest materials) or signs of decomposition. The maximum number of nests refers to all nests with fresh leaves. Re-use of nest sites and multiple nest construction by the same individual means that nest counts overestimate the number of individuals who have nested together. Therefore, the minimum nest group size was used to calculate monthly group sizes.

Collection of food samples for chemical analyses

Efforts were made to collect samples from fruit of all taxa of trees and woody climbers on line transects. Fruit actually eaten by *Pan* but not found on line transects or plots was also collected. Samples were preferably collected from individual trees that had been visited by *Pan*. In these cases, samples were collected when the fruit were of mature size and when the physical properties such as coloration, softness of texture, and smell indicated a stage of ripeness. In total, sample collection at Salonga included 47 food species and 39 non-food species. The majority (72%) of food species could be identified at least to the family level while only half (49%) of the non-food species could be identified. At Gashaka, samples represented 53 food species and 46 non-food species. Here taxonomic identification included 81% of the food species and 57% of the non-food species. Fresh samples collected in the forest were brought back to the camp where we took weight and size measurements. For chemical analyses, we tried to remove seeds and hard shells from the parts that were actually consumed by *Pan*. If reports on the consumption did not exist, we took samples from parts that were likely to be ingested: the soft meso- and exocarp of fruit. To prevent changes in the chemical content of the sampling material by mold and other biochemically active sources, all samples tubes were stored in liquid nitrogen.

Assessment of phytochemistry

To identify intra- and interspecific variation in phytochemical composition of the diet of both species, samples were assigned to the following two categories: (1) *food* refers either to fruits actually seen to be eaten during the course of this study or to remains of fruit obtained from fecal samples; (2) *non-food* refers to species not eaten.

Macronutrients and energy

Samples were freeze-dried and ground prior to analysis. Dry matter content was determined by drying a portion of the sample at 105 °C overnight, crude protein by using Dumas-Combustion. Total nitrogen (N) of the sample provides an estimate of crude protein (protein level = N × 6.25). Crude fat was assessed by petroleum ether extraction (Soxhlett), starch and mono-/disaccharides enzymatically, and energy content of the sample by bomb calorimetry. Detergent Fiber Analysis was performed after van Soest (1994) and provided a rapid stepwise procedure for determining soluble cellular components as well as the insoluble cell wall matrix and its major subcomponents: hemicellulose, cellulose, and lignin. Cell contents and soluble components were estimated by boiling the sample in neutral detergent solution. The residue Neutral Detergent Fiber (NDF) contains hemicellulose, cellulose, and lignin. Hemicellulose was extracted by boiling the sample in acid detergent solution. The residue Acid Detergent Fiber (ADF) only contains cellulose and lignin. The last step extracts cellulose by acid hydrolysis and burns the sample to ash at 550 °C. The residue Acid Detergent lignin (ADL) only contains lignin. Hemicellulose and cellulose contents are calculated by weighing and subtracting residues with: hemicellulose = NDF−ADF and cellulose = ADF−ADL.

Antifeedants

For extraction of simple phenolics and tannins, 100 mg of plant material was dried and finely ground (<0.18mm particle size). The sample was mixed with exactly 5 ml aqueous acetone (70% v/v) and sonicated for 20 min at room temperature. The mixture was then centrifuged for 10 min at 3000 g at 4 °C. The supernatant was collected in a fresh tube and used for the following steps as "sample phenolic extract".

The total phenolics were determined essentially according to Makkar *et al.* (1993). Suitable quantities of the aliquots (determined by trial and error so that the absorbance value below was within the range of the calibration curve) of the above sample phenolic extract were placed in a test tube and made up to 500 μl with distilled water; 250 μl of the Folin Ciocalteu reagent followed by 1.25 ml of the sodium carbonate solution was added to this and the mixture was stirred on a vortex machine. Absorbance of the solution was recorded at 725 nm after a 40-min incubation period in the dark. The total amount of phenols was then calculated as tannic acid equivalent from a calibration curve prepared with tannic acid. The results were then expressed as g/100 g tannic acid equivalent on a dry matter basis.

Estimation of total tannins from total phenolic extracts: 100 mg of polyvinyl polypyrrolidone (PVPP) was weighed into a 100×12-mm test tube; 1.0 ml distilled water followed by 1.0 ml of the sample phenolic extract was added to the test tube (100 mg PVPP is sufficient to bind 2 mg of total phenols; if total phenolic content of the sample is more than 10 % on a dry matter basis, the extract has to be diluted). The contents of the tube were shaken on a vortex machine and kept at 4 °C for 15 min, after which it was shaken again and centrifuged (3000 g for 10 min) to collect the supernatant, which only contains simple phenolics other than tannins (the tannins are precipitated with the PVPP). The phenolic content of the supernatant was measured as described above. The content of the non-tannin phenolics was expressed on a dry matter basis. The tannin content of the sample was then calculated as follows: total phenolics (%) – non-tannin phenolics (%) = tannins (%). The result is then expressed as tannic acid equivalent on a dry matter basis.

Determination of condensed tannins followed Porter *et al.* (1986). Phenolic extract (0.50 ml) was diluted with 70 % acetone and pipetted into a 100 mm × 16-mm glass test tube. We added 3.0 ml of the butanol-HCI reagent and 0.1 ml of ferric reagent. The sample was heated on a heating block set at 95 °C for 60 min. After cooling, the absorbance of the mixture was recorded at 550 nm.

Condensed tannins (% in dry matter) as leucocyanidin equivalent are calculated using the formula: (A 550nm × 78.26)/ (% dry matter).

Statistics

All statistical analyses were done with SPSS (version 11.5). A General Linear Model (GLM) was used to investigate two aspects of food abundance, first, whether the variation in food abundance correlates with party size, and second, whether the correlation is the same in both species. To fulfill the assumptions of parametric tests, nest counting data were square root transformed (Zar, 1999). For comparison of chemical components between species and plant types, we used 2-way Analyses of Variance (ANOVA). Due to the positive skew of the data, the measurements of the content of chemical components and energy were Ln-transformed. The assumptions of both parametric tests were controlled by visual inspection of the residuals and tests for equality of variance (Howell, 2002). To detect structure between variables of different chemical components, we applied a Principal Component Analysis (PCA). To fulfil assumptions of the PCA, Ln-transformed data were used.

Results

Environmental features

Climatic conditions at the two sites were different (Figure 5.2). Mean temperatures in Salonga were relatively low (min 20.7 °C, max 26.9 °C) and stable over the year. Annual rainfall exceeded 2000 mm and there were no months without rain. In Gashaka, average temperature was higher (min 23.2 °C, max 32.2 °C) and showed more variation between wet and dry seasons. Annual rainfall was below 2000 mm and was concentrated in the period between April and October. The dry season was pronounced, with 4 months without rain.

Preliminary comparison of the floristic diversity at the two sites suggests that plant diversity at Gashaka is likely to exceed that of Salonga. Gashaka is situated in a zone of transition between two biogeographical regions, and is therefore assumed to have an exceptionally high biodiversity (Fjeldsa & Lovett, 1997). Moreover, because of the altitudinal changes at Gashaka, the diversity of microhabitats is likely to be higher than at Salonga. Finally, the flora of Gashaka is characterized by plant forms that can withstand the pronounced dry season and therefore, may be absent from Salonga.

Overall, the abundance of trees of different size categories at both sites was similar (Figure. 5.3). However, the basal area of trees at Salonga was significantly smaller than the basal area at Gashaka (Salonga: M = 218 cm^2, range min 79 cm^2, max 3.34 m^2, n = 959 trees, Gashaka: M = 283 cm^2, range min 79 cm,^{2} max 21.05 m^2, n = 1000 trees, Mann-Whitney U-test, z = −3.831, p < 0.001). Climbers were more abundant at Salonga than at Gashaka (Salonga mean = 1.8 climbers/tree, range 0–6, Gashaka mean = 0.8, range 0–4).

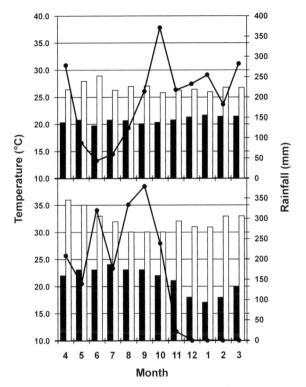

Figure 5.2. Climatic profiles of Salonga (top) and Gashaka (bottom). Black bars refer to means of minimum and unshaded bars to mean of maximum temperatures. Solid lines indicate monthly rainfall.

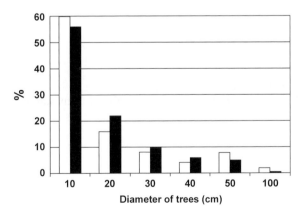

Figure 5.3. Percentage of transect trees of different size classes ranging from 10 cm to > 100 cm dbh at Salonga (unshaded) and Gashaka (black bars).

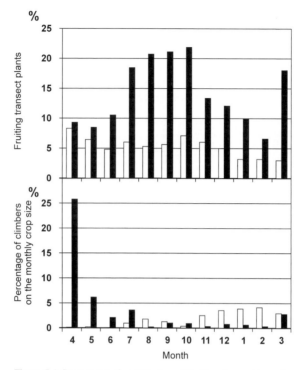

Figure 5.4. Percentage of contribution of fruiting transect trees (upper section) and percentage of woody climbers (lower section) to total monthly fruit crop from Salonga (unshaded bars) and Gashaka (black bars) over 12 months. Figures on the x-axis refer to the different months: 4 = April 2002, 3 = March 2003.

Fruit production

The proportion of trees producing fruit in each of the 12 months was consistently higher at Gashaka than at Salonga (Figure. 5.4). There was also substantial monthly variation of fruit abundance at Gashaka and the peak of fruit production coincided with the end of the wet season in October (Figure 5.4 top). In contrast to fruit crops produced by trees, the proportion of fruit from woody climbers within a given month was small, with the exception of one month when fruit from climbers accounted for 25% of the monthly crop at Gashaka (Figure. 5.4 bottom).

Diet composition

To explore the diversity of diet over time, fresh feces were screened for feeding remains. In the analysis of fruit remains such as seeds, fibers, and skin fragments, monthly fecal samples also showed variance in the other types of plant foods as well as the number of different items/species. The content of fecal samples was always skewed towards relatively few species accounting for the majority of structured remains. In this study we only scored remains that accounted for 10% or more of all structured fruit remains within a given sample. In the bonobo samples, remains of 22 different fruit species were found (M = 5.5 species per month, min = 3, max = 10). In the chimpanzee samples, fruit from 25 different species were found (M = 4.5 species per month, min = 1, max = 7). To test if the number of fecal samples collected for each month had an impact on the number of fruit species obtained, we performed a Spearman rank correlation. In the data set from Salonga bonobos the two parameters were not related (r = 0.002, p = 0.996, n = 12 months). In the data set from Gashaka chimpanzees, the two parameters showed a positive correlation (r = 0.781, p = 0.008, n = 10 months). At both sites a few food species clearly dominated the monthly diet as assessed by fecal inspection. Table 5.1 lists those species that appeared in more than 50% of the monthly fecal samples, and therefore were considered to be important fruit species. Table 5.1a shows that at Salonga remains of single food species appeared in the feces for relatively long periods (up to 6 months). At Gashaka (Table 5.1b), the type of feeding remains in feces tended to fluctuate more rapidly.

Nest-group size

Comparing the minimum number and the maximum number for nest group size for each site revealed a significantly strong correlation (Salonga: Spearman's rho max # of nests 0.935, p < 0.001, N = 119; Gashaka: Spearman's rho max # of nests 0.960, p < 0.001, N = 44). The following calculations are all based on the minimum number of nest group sizes. In both species, nest-group size fluctuated from month to month, but average monthly nest group sizes were almost always larger in bonobos (Figure. 5.5). Comparing the median of monthly nest group size revealed significant differences between the two species (Mann-Whitney U-test, z = −2.457, p = 0.014, for sample sizes see Figure. 5.5).

Nest-group size and fruit abundance

To test the relationship between food abundance and party size, we used the median of monthly nest group size and the calculated monthly fruit index of the two sites. A General Linear Model analysis with party size as dependent variable, and fruit index and species as independent variable revealed a

Table 5.1. *Dominant fruit obtained from visual inspection of fresh fecal samples from bonobos (a) and chimpanzees (b)*

(a)

	Apr	May	Jun	Jul	Aug	Sep	Oct	Nov	Dec	Jan	Feb	Mar	# months with 50%
N	15	24	43	36	45	35	37	24	36	63	29	46	
Meliaceae	87	0	0	0	0	0	0	0	0	0	0	0	1
No id	100	0	0	0	0	0	0	0	0	0	0	0	1
Dialium	67	96	100	94	100	86	14	4	0	0	0	0	6
Drypetes	53	0	0	0	0	0	0	0	0	0	0	0	1
Cissus	0	0	0	3	64	49	30	0	0	0	0	0	1
Ficus	0	0	0	14	0	0	57	63	31	2	59	46	3
Polyathia	0	0	0	0	0	9	89	100	44	2	0	0	2
Grewia	0	0	0	0	0	0	0	0	25	75	100	100	3

(b)

	Apr	May	Jun	Jul	Aug	Sep	Oct	Nov	Dec	Jan	Feb	Mar	# months with 50%
N	33	27	2	23	19	11	2	2	0	27	3	0	
Landolphia	27	96	100	17	0	0	0	0		11	0		2
Cola m.	0	4	100	9	0	0	0	0		0	0		1
Afromomum	0	0	0	52	37	0	0	0		0	0		1
Canarium	0	0	0	43	63	27	0	0		0	0		1
Leea	0	0	0	0	11	9	50	100		19	0		2
Bridelia	0	0	0	0	0	81	0	0		0	0		1
Malacantha	0	0	0	0	0	36	50	100		4	0		2
No id	0	0	0	0	0	55	0	0		0	0		1
Lasianthus	0	0	0	0	0	0	50	100		0	0		2
Anthiaris	0	0	0	0	0	0	0	0		56	100		2

The tables only show food species that appeared in 50% or more of the monthly fecal sample collection.

significant interaction between species and fruit abundance (species: df = 1, 18, F = 16.212, p = 0.001, fruit index: df = 1, 18, F = 2.518, p = 0.13, species* fruit index: df = 1, 18, F = 10.502, p = 0.005, r^2 = 0.57). This meant that the effect of the fruit index was different at the two sites. To further analyze the interaction, a linear regression between fruit index and party size was performed separately for each site (Figure 5.6). Data from Gashaka chimpanzees revealed a significant positive relationship between nest group size and fruit production (df = 1, 8, F = 14.834, standardized beta = 0.806, p = 0.005). The data from Salonga bonobos did not show a significant

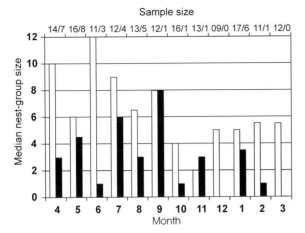

Figure 5.5. Median of nest group size at Salonga (unshaded) and Gashaka (black bars). The x-axis indicates the months, the y-axis shows the median values for different months. Figures on top of the bars refer to the sample size of monthly nest groups for Salonga/Gashaka.

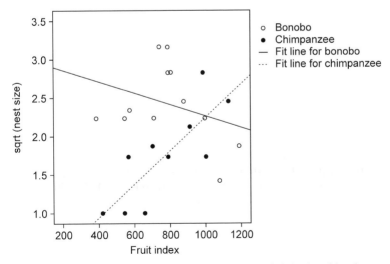

Figure 5.6. Linear regression between monthly fruit index (x-axis) and square root transformed count data on minimum nest group size (y-axis).

relationship between the two parameters (df = 1, 10, F = 1.223, standardized beta = −0.33, p = 0.295).

Macronutrients and antifeedants

Table 5.2 gives the mean values of phytochemical components of food plants and non-food for both species. Inspecting the data from chimpanzees on a descriptive level revealed that with one exception, concentration of all macronutrients tended to be higher in food plants compared with non-food, while the concentration of fiber components and antifeedants showed the opposite tendency. In bonobos, the trend was similar for macronutrients but less consistent in terms of fiber components and antifeedants (see Table 5.2).

Table 5.2. *Mean values and standard deviation of phytochemical components in plants collected in the habitats of bonobos and chimpanzees during the course of this study*

	Bonobo		Chimpanzee	
	Food (N = 47)	Non-food (N = 39)	Food (N = 53)	Non-food (N = 46)
EN	17.5 ± 2.5 (30)	18.1 ± 5.4 (17)	21.7 ± 5.4 (12)	18.7 ± 3.3 (9)
DM	23.9 ± 12.8 (8)	25.9 ± 15.2 (13)	26.0 ± 16.3 (38)	29.5 ± 18.5 (23)
CP	8.3 ± 4.7 (42)	11.4 ± 4.7 (29)	7.3 ± 3.1 (53)	7.6 ± 3.7 (46)
SU	9.8 ± 6.9 (43)	7.0 ± 6.9 (33)	8.4 ± 7.6 (52)	5.1 ± 5.5 (36)
ST	3.7 ± 9.1 (43)	3.9 ± 9.0 (33)	2.2 ± 6.1 (52)	1.8 ± 4.3 (36)
CF	6.6 ± 8.9 (31)	3.3 ± 4.6 (20)	8.8 ± 12.7 (52)	5.4 ± 10.0 (35)
NDF	26.8 ± 16.3 (43)	32.1 ± 15.3 (32)	26.9 ± 13.1 (53)	32.3 ± 16.7 (46)
ADF	16.6 ± 11.1 (43)	20.8 ± 12.0 (32)	18.1 ± 9.7 (53)	20.2 ± 10.8 (46)
ADL	5.2 ± 5.1 (43)	6.9 ± 5.6 (32)	6.5 ± 4.7 (53)	5.8 ± 4.2 (46)
CE	11.5 ± 7.2 (43)	13.9 ± 7.0 (32)	11.6 ± 6.4 (53)	14.4 ± 7.8 (46)
HC	10.1 ± 8.1 (43)	11.3 ± 6.4 (32)	8.8 ± 6.6 (53)	12.1 ± 12.0 (46)
TP	4.9 ± 5.1 (28)	4.7 ± 5.7 (9)	3.5 ± 3.1 (16)	11.6 ± 12.3 (18)
TT	2.8 ± 3.0 (28)	3.0 ± 4.0 (9)	2.0 ± 1.5 (16)	8.8 ± 9.9 (18)
CT	4.7 ± 10.1 (28)	3.6 ± 5.4 (9)	8.7 ± 11.5 (16)	7.3 ± 9.8 (18)
CP/ADF	0.81 ± 0.7 (42)	0.87 ± 0.79 (29)	0.53 ± 0.35 (53)	0.5 ± 0.4 (46)

Figures in pasentheses are sample sizes of analyzed plants. All values are percent of chemical components in relation to dry matter weight. The figures for energy are kj/g dry mass. Key for abbreviations of phytochemical components: EN = energy, DM = dry mass, CP = crude protein, SU = sugar, ST = starch, CF = crude fat, NDF = neutral detergent fiber, ADF = acid detergent fiber, ADL = acid detergent lignin, CE = cellulose, HC = hemicellulose, TP = total phenol, TT = total tannin, CT = condensed tannin, CP/ADF = ration of crude protein/acid detergent fiber.

Table 5.3. *Test of between-subjects effects from Univariate Analyses of Variances (SPSS) of phytochemical components between species (bonobo vs. chimpanzee), plant types (food versus non-food), and the interaction between species and plant type*

Component	Test	F	Df, Error	p
FN	species	6.136	1, 64	0.016
	plant	1.426		0.237
	species* plant	3.547		0.064
CP	species	14.680	1, 165	<0.001
	plant	7.285		0.008
	species* plant	5.447		0.021
CP/ADF	species	11.745	1, 165	0.001
	plant	0.384		0.682
	species* plant	0.204		0.815
ST	species	7.580	1, 159	0.007
	plant	3.786		0.533
	species* plant	0.289		0.863
SU	species	2.080	1, 159	0.151
	plant	9.231		0.003
	species* plant	0.073		0.787
CF	species	1.991	1, 1.76	0.161
	plant	4.157		0.043
	species* plant	0.003		0.953
NDF	species	0.231	1, 0.347	0.631
	plant	5.521		0.02
	species* plant	0.056		0.813
ADF	species	1.718	1, 0.427	0.192
	plant	3.412		0.066
	species*plant	0.628		0.429
ADL	species	5.936	1, 169	0.016
	plant	1.954		0.164
	species*plant	2.467		0.118
CE	species	0.83	1, 0.361	0.363
	plant	6.155		0.014
	species*plant	0.04		0.841
HC	species	0.048	1, 0.675	0.826
	plant	4.396		0.038
	species*plant	0.07		0.792
TP	species	2.514	1, 67	0.118
	plant	0.353		0.554
	species*plant	3.653		0.60
TT	species	3.043	1, 67	0.086
	plant	0.031		0.86
	species*plant	3.495		0.066
CT	species	1.865	1, 338	0.177
	plant	0.014		0.906
	species*plant	0.016		0.901

Statistical comparison of differences in variation between species (bonobo vs. chimpanzee) and between plant types (food vs. non-food) revealed significant differences on both levels (Table 5.3). The content of sugar and fat was significantly higher in food plants than in non-food. Similarly, the contents of three fiber fractions (NDF, CE, and HC) were lower in food plants compared with non-food. Differences between species appeared to be less pronounced: chimpanzees, as compared with bonobos, had access to resources offering more energy, while bonobos had access to resources with a higher concentration of starch and a more favorable ratio between crude protein and ADF. Similarly, phytochemical analyses of the samples from Salonga indicate that bonobos had access to sources that contained significantly more protein than did chimpanzees (F = 14.68, df = 1, 165, p < 0.001). However, there was also a significant interaction between species and plant type (F = 5.447, df = 1, 165, p = 0.021). Posthoc pairwise comparison revealed that this difference in protein was caused by high concentrations of protein in non-food plants of Salonga bonobos (bonobo food plants: $\bar{X}$ ± sd = 8.3 ± 4.0, n = 42 species, bonobo non-food: $\bar{X}$ ± sd = 11.4 ± 4.7, n = 29 species).

To explore how the different phytochemical components related to each other, we compared measurements of macronutrients, fiber fractions, and antifeedants in food of bonobos and chimpanzees, respectively (Tables 5.4 and 5.5): In the food of both species, strong positive correlations were found between all fiber fractions (n = 5) on the one hand, and antifeedants (n = 3) on the other. In the food of bonobos, positive correlations existed between crude protein and two fiber fractions (ADL, HC). The content of sugar and starch showed negative correlations with some fiber fractions while concentrations of crude fat appeared to be independent of other chemical components. In the food of chimpanzees, some fiber fractions (NDF, ADL, HC) showed positive correlations with some antifeedants (TP, CT). Sugar showed negative correlations with the other macronutrients, most fiber fractions, and condensed tannin. Two macronutrients, starch and crude fat, showed positive correlations with fiber fractions and antifeedants.

To identify patterns in the data, we applied PCA to the 12 phytochemical variables for each site. This method identifies variables or subsets of variables that are correlated with one another, but largely independent of other variables, and combines these variables into components. In the case of bonobo foods, the information from the 12 variables can be compressed into four components, which explain 80% of the original data. The rotated component matrix (Varimax with Kaiser normalization) shows the following: The five fiber fractions loaded high and positively with the first component (r_{min} = 0.81, r_{max} = 0.96), and the three parameters of antifeedants loaded high and

Table 5.4. *Pearson correlation between chemical components in the food of Salonga bonobos*

PP		CP	SU	ST	CF	NDF	ADF	ADL	CE	HC	TP	TT	CT
CP	r		−0.076	0.081	0.195	0.254	0.226	0.336	0.158	0.34	0.152	0.038	−0.266
	p		0.632	0.612	0.294	0.105	0.15	0.029	0.316	0.028	0.439	0.849	0.172
	N		42	42	31	42	42	42	42	42	28	28	28
SU	r			−0.044	−0.308	−0.447	−0.392	−0.415	−0.273	−0.343	−0.009	−0.035	−0.091
	p			0.777	0.091	0.003	0.009	0.006	0.076	0.024	0.966	0.86	0.644
	N			43	31	43	43	43	43	43	28	28	28
ST	r				−0.041	−0.303	−0.325	−0.124	−0.376	−0.159	0.179	0.19	−0.067
	p				0.826	0.048	0.033	0.427	0.013	0.309	0.363	0.332	0.737
	N				31	43	43	43	43	43	28	28	28
CF	r					−0.025	0.05	0.104	−0.132	−0.037	−0.145	0.031	0.007
	p					0.896	0.788	0.577	0.478	0.841	0.52	0.891	0.974
	N					31	31	31	31	31	22	22	22
NDF	r						0.926	0.783	0.863	0.762	−0.236	−0.097	−0.157
	p						<0.001	<0.001	<0.001	<0.001	0.226	0.623	0.425
	N						43	43	43	43	28	28	28
ADF	r							0.794	0.925	0.636	−0.216	−0.093	−0.215
	p							<0.001	<0.001	<0.001	0.27	0.639	0.271
	N							43	43	43	28	28	28
ADL	r								0.579	0.576	−0.044	0.154	0.015
	p								<0.001	<0.001	0.826	0.433	0.941
	N								43	43	28	28	28
CE	r									0.598	−0.279	−0.211	−0.245
	p									<0.001	0.15	0.282	0.209
	N									43	28	28	28
	r										−0.371	−0.273	−0.314

Table 5.4. (*cont.*)

	PP	CP	SU	ST	CF	NDF	ADF	ADL	CE	HC	TP	TT	CT
HC p											0.052	0.16	0.103
N											28	28	28
TP r												0.851	0.543
p												<0.001	0.003
N												28	28
TT r													0.665
p													<0.001
N													28
CT p													
N													

Within each cell, the first (top) value represents the correlation coefficient (Pearson), the second (middle) the corresponding p-value, and the third (bottom) the sample size. Shaded areas indicate groups of chemical components showing high intercorrelation. For abbreviations of chemical components, see Table 5.2.

Table 5.5. *Pearson correlation between chemical components in the food of Gashaka chimpanzees*

PT		CP	SU	ST	CF	NDF	ADF	ADL	CE	HC	TP	TT	CT
CP	r		−191	0.133	0.265	0.074	0.021	−0.1	0.101	0.097	−0.49	−0.212	−0.25
	p		0.176	0.348	0.057	0.597	0.884	0.478	0.474	0.491	0.054	0.432	0.351
	N		52	52	52	53	53	53	53	53	16	16	16
SU	r			−0.57	−0.557	−0.502	−0.388	−0.377	−0.264	−0.538	−0.074	−0.467	−0.529
	p			<0.001	<0.001	<0.001	0.004	0.006	0.058	<0.001	0.792	0.079	0.042
	N			52	52	52	52	52	52	52	15	15	15
ST	r				0.429	0.377	0.34	0.388	0.108	0.415	0.193	0.422	0.555
	p				0.002	0.006	0.014	0.004	0.444	0.002	0.49	0.117	0.032
	N				52	52	52	52	52	52	15	15	15
CF	r					0.417	0.333	0.308	0.203	0.467	0.622	0.775	0.902
	p					0.002	0.016	0.026	0.15	<0.001	0.013	0.001	<0.001
	N					52	52	52	52	52	15	15	15
NDF	r						0.895	0.806	0.694	0.863	0.52	0.33	0.635
	p						<0.001	<0.001	<0.001	<0.001	0.039	0.211	0.008
	N						53	53	53	53	16	16	16
ADF	r							0.805	0.866	0.568	0.426	0.208	0.47
	p							<0.001	<0.001	<0.001	0.1	0.44	0.067
	N							53	53	53	16	16	16
ADL	r								0.436	0.649	0.542	0.23	0.613
	p								0.001	<0.001	0.03	0.33	0.012
	N								53	53	16	16	16
CE	r									0.305	0.033	−0.041	0.033
	p									0.026	0.903	0.879	0.902
	N									53	16	16	16

Table 5.5. (*cont.*)

PT	CP	SU	ST	CF	NDF	ADF	ADL	CE	HC	TP	TT	CT
HC r										0.528	0.411	0.72
p										0.035	0.114	0.002
N										16	16	16
TP r											0.57	0.757
p											0.021	0.001
N											16	16
TT r												0.845
p												<0.001
N												16
CT r												
p												
N												

Within each cell, the first (top) value represents the correlation coefficient (Pearson), the second (middle) the corresponding p-value, and the third (bottom) the sample size. Shaded areas indicate groups of chemical components showing high intercorrelation. No id = not identified. For abbreviations of chemical components, see Table 5.2.

positively with the second component ($r_{min} = 0.78$, $r_{max} = 0.87$). This was followed by sugar ($r = -0.82$) and crude fat ($r = 0.75$) for the third and fourth component, respectively. Data from chimpanzees are slightly different: Three components explain 85% of the observed variance. From the five fiber components, four loaded high and positively with the first component ($r_{min} = 0.75$, $r_{max} = 0.96$). The three antifeedants and crude fat loaded high for the second component ($r_{min} = 0.68$, $r_{max} = 0.88$) and starch loaded high with the third component ($r = 0.92$).

To explore relationships between four macronutrients (CP, SU, ST, CF) and antifeedants in more detail, another PCA was run that compressed the measures of phenol, total tannin, and condensed tannin into one component. The resulting component explained more than 85% of the variance of each antifeedant (phenol $r = 0.85$, total tannin $r = 0.89$, condensed tannin $r = 0.96$). The results of this PCA suggest that with one exception, macronutrients vary independent of antifeedants. Only in the food of chimpanzees did the content of crude fat covary significantly with the antifeedant component ($r = 0.857$, $p < 0.001$, $n = 15$).

Discussion

Fruit abundance and gregariousness

Comparison of the records from phenology transects showed consistent differences in annual variation of fruit abundance between both sites. Compared with Salonga, scores of fruit production by trees and climbers were always higher at Gashaka. The proportion of fruit-bearing trees and climbers at Salonga appears very low but resembles findings of other studies. For example, phenology work at Kanyawara revealed that the average number of fruit-bearing trees per month was about 3% (Chapman *et al.*, 1994). Still, we cannot exclude the possibility that the protocol used here to assess monthly fruit production underestimated fruit abundance. One potential source for sampling errors is the exclusion of fruit crops produced by smaller trees. The more likely reason, however, is that the low fruit abundance results from low productivity of small transect trees. Figure 5.3 shows that small trees (dbh 10–20 cm) accounted for the majority of transect trees. In the absence of information on productivity patterns of tropical forest trees, we can only speculate that our sample of transect trees included a substantial proportion of immature individuals that had yet to reproduce. The difference in fruit production may be due to differences in the synchrony in fruit production of individuals of the same tree/climber species. Pronounced shifts between wet and dry seasons are known to be synchronized with flowering and fruit production (Leigh & Windsor, 1982). High synchrony is likely to result in a short period of fruit availability for a given species, while asynchrony makes the fruit of a given species available for longer periods (Poulin *et al.*, 1999). Alternatively, fruit productivity may also decrease with the numbers of species that do not produce fruit annually. The results of this study suggest that the differences in terms of fruit availability between Salonga and Gashaka reflect a combination of both effects. Forest productivity at Gashaka is likely to be influenced by two annual shifts in climate, one at the beginning and one at the end of the dry season. The low dynamics in fruit production at Salonga are in line with the modest fluctuations in temperature and rain. Phenology records show that single species produced fruit for up to six months, and fecal inspections (Table 5.1) revealed that the fruit of these species dominated the diet of bonobos for long periods of time. This indicates not only a considerable lack of synchrony in fruiting between individual trees of the same species but also a tendency of bonobos to focus on single species for extended periods. Meanwhile, phenology records from Lui Kotal cover more than 40 months, and it became obvious that a substantial number of species have long fruiting cycles. In fact, some species produced no fruit during the entire period of data collection. During the period of this study, the contribution of climbers to the total fruit crop was marginal except for one month when climbers at Gashaka made up 25% of the fruit crop. Direct observations and inspection of fecal samples made during other years showed that fruit from climbers may dominate the diet of bonobos in Salonga (Berghoudt *et al.*, 2005; authors' unpublished data). Thus, the small proportion of fruit from climbers reported here is not representative for Salonga.

Studies on wild tree species suggest substantial interannual variation in fruit production (Poulin *et al.*, 1999). In a study at Lopé (Gabon), Tutin & Fernandez (1993) reported that in a given year 98% of sample trees did not produce any fruit. At other times, high synchrony within and between species provided peaks of fruit availability. Given the differences in fruit abundance during this study, one would predict that bonobos may have relied on non-fruit food sources like herbs, leaves, or animal matter. The use of alternative food sources in relation to fruit availability is the topic of an ongoing project and will shed light on the diversity of the diet of the two species.

Counts of night nests at the two sites showed a trend that did not match with predictions based on fruit production: In Salonga, the number of individuals nesting together was always higher than in Gashaka. This supports previous findings on species differences in nest group size (Fruth & Hohmann, 1996) and is in line with the general view of bonobos being more gregarious than chimpanzees. During this study, records of the size of travel parties and feeding parties observed during the daytime were rare, and we cannot exclude the possibility that bonobo nest group size deviates from day group size. Records from Lomako showed that bonobos indeed have the tendency to aggregate in the evening in order to nest close together (Fruth, 1995). In this study, most nest sites were detected by vocalizations, and because large parties are more likely to vocalize than small ones, the results are probably biased towards larger groups. However, assuming that the detectibility of the distance calls of both species is similar (Hohmann & Fruth, 1995), this bias is unlikely to account for the consistent difference of nest group size between the two species.

An unexpected result of this study was the asymmetry of the variation of nest group size in relation to fruit abundance. In Gashaka chimpanzees, the significant relation between party size and fruit index suggests that fruit abundance did constrain gregariousness. In contrast, variation in bonobo nest group size appeared to be independent of fruit availability. Comparing data from a number of East African chimpanzee populations, Hashimoto *et al.* (2004) found that at some sites party size varied with food abundance, whereas at other sites the two parameters did not correlate. The authors proposed that food abundance is likely to affect party size only when re-sources are scarce for most of the time. In populations living in more productive locations, party size varies independently from food abundance. The lack of correlation between fruit index and group size suggests that the higher gregariousness of Salonga bonobos is not caused by fruit abundance. Previous studies suggested a positive correlation between party size and size of food patches (White, 1992), but comparison of data from bonobos and forest chimpanzees did not support this idea (White & Wrangham, 1988). Preliminary studies on feeding party size from Salonga also found that party size was independent of food patch size (Berkhoudt *et al.*, 2005). Therefore, the failure to detect a correlation between fruit abundance and gregariousness could in part be due to the fact that our phenology protocol did not include small trees and other small-sized woody plants.

Multivariate analyses of data from chimpanzees at Taï and Ngogo revealed that party size varied with the number of estrous females rather than food abundance (Anderson *et al.*, 2002; Mitani *et al.*, 2002). Data from Lomako bonobos showed that estrous female bonobos tended to travel in the same party, and parties with estrous females were attractive for both males and females; as a consequence, party size increased with the number of estrous females (Hohmann & Fruth, 2003b). Thus, while between-species differences in gregariousness may reflect differences in food resource availability, reproductive strategies of males and females may cause the temporal dynamics of party size of both species. The data on nest group size obtained in this study support previous findings suggesting that bonobos are more gregarious than chimpanzees. On the other hand, the results on fruit production do not correspond to the widespread view that food is more abundant for bonobos. What our study showed was a difference in seasonal *fluctuation* of fruit availability, as well as a tendency for differences in the synchrony of fruit production. From this we infer that the diet of Gashaka chimpanzees is more diverse than that of bonobos. How this affects foraging strategies depends on, among other things, the spatial distribution of trees bearing fruit simultaneously. As direct observations of bonobos and chimpanzees at the two sites improve, future studies will address questions of foraging and ranging patterns of the two populations.

Chemical composition of food and non-food

Analyses of the phytochemical composition revealed both differences between plant categories (food vs. non-food) and differences between the two species. Judging from measurements of the different chemical components, it appears that both species ate fruit of similar quality. Compared with non-food, food had higher levels of water, sugar, and crude fat, and lower levels of fiber components. The combination of relatively high levels of macronutrients and relatively low levels of fiber in food items matches predictions of Optimal Diet theory and is in line with findings from other studies (Rogers *et al.*, 1990; Reynolds *et al.*, 1998). However, one should keep in mind that the samples analyzed in our study came from fruit that had physical properties indicating ripeness. Although bonobos and chimpanzees might prefer a relatively narrow range of fruit ripeness, there is clear evidence that they also consume substantial amounts of unripe fruit (Reynolds *et al.*, 1998). Therefore, the patterns of chemical composition of food analyzed in this study are likely to represent a diet that was biased towards above-average quality.

An interesting result, albeit not at the level of statistical significance, was the difference in average concentration of antifeedants in non-food of chimpanzees. The levels of phenol and hydrolysable tannin ingested by chimpanzees differed from the concentrations found in non-food items. In contrast, bonobo food had concentrations of antifeedants similar to those of non-food, and the values of both components were on the same level as that of chimpanzee food. The content of tannin in plants varies with soil, availability of water, and other environmental factors (Mueller-Harvey & McAllan, 1992). There is evidence that tannin levels and tannin activity increase with ambient temperature (Makkar & Becker, 1998). Thus, in addition to the higher values of tannin content in plants at Gashaka, it is reasonable to assume that the activity of tannins might also be higher compared with the relatively cool climate of Salonga.

The densities of different nutrients in food items are not symmetric, and in wild plants, nutrients are regularly associated with components that reduce the palatability and/or digestibility of food (Dearing & Schall, 1992). Therefore, one important aspect for understanding the observed patterns of food choice is the identification of association patterns of different chemical food components. Well-known examples of the functional links between chemical composition of food and food choice come from studies on insects, birds, and herbivore mammals (Belovsky, 1978; Lotz & Nicolson, 1996; Raubenheimer & Simpson, 1997). Because of differences in digestive strategies, the same type of macronutrient may have a different value for different consumer species (Milton, 1981; Witmer & van Soest, 1998). Food experiments with herbivore mammals revealed that food choice was triggered by tannin content rather than concentrations of macronutrients (Alm *et al.*, 2000; Clauss, 2003).

Previous studies have revealed important features of dietary composition for both *Pan* species. Here we expanded the topic by asking how the different components of macronutrients, fiber, and antifeedants relate to each other. The data obtained from Salonga bonobos and Gashaka chimpanzees show some interesting asymmetries in the geometry of macronutrients, fiber fractions, and antifeedants: unlike bonobos, food items of chimpanzees showed a positive correlation between two macronutrients (CF, ST) on the one hand, and some fiber fractions and antifeedants on the other. Whether or not the association constrains the intake of starch and crude fat remains a challenging topic for future studies. Looking at the relation of the different chemical components in the food of bonobos, the intake of all macronutrients appeared not to be affected by antifeedants.

While the differences in chemical properties of food plants of bonobos and chimpanzees found in this study do not allow firm conclusions concerning features of dietary composition, they can be used to generate testable predictions. If chimpanzees ingest, along with macronutrients, higher amounts of fiber and/or antifeedants, they may need to adopt strategies to reduce the negative impact of these components. Following the detoxification constraints hypothesis (Freeland & Janzen, 1974) one would predict that chimpanzees (1) may have shorter feeding bouts; (2) visit more food patches per day; and (3) in order to avoid ingesting too many of the same toxic substances, do not exploit food patches of the same species in close succession. Gut passage time is thought to facilitate detoxification (Lambert, 1998), and therefore, chimpanzees may also have longer retention times than bonobos. In addition, chimpanzees may reduce tannin activity by eating soils that are rich in clay (Krishnamani & Mahaney, 2000). Another way to cope with high levels of tannin is to regulate the synthesis of salivary proline in relation to the tannin levels of food. Proline aggregates with tannins and reduces the capacity of tannin to bind other proteins. Detailed socioecological and experimental studies are required to explore the behavioral and physiological responses of the two *Pan* species to fluctuating levels of tannins.

General patterns of feeding ecology of the African Great Apes seem to be influenced by geographic variation in forest composition (Rodman, 2002). How this variation affects the availability and accessibility of nutrients remains an open question. Information on some components of fruit eaten by bonobos is available from the study on bonobos at Lomako. According to this study, fruit eaten by Lomako bonobos had much lower values of tannin (0.28% dry mass in Lomako vs. 2.8% dry mass in Salonga) while protein and lipid was similar (Malenky, 1990, Appendix II). Based on these findings, it was proposed that in Lomako bonobos, food selection may be driven by tannin avoidance. The results from studies on chimpanzees addressing this question are mixed: Analysing a large sample of fruit eaten by chimpanzees at Gombe, Wrangham & Waterman (1983) found that chimpanzees did not select species with lower tannin levels. Reports from Kanyawara and Budongo suggest that chimpanzees consume fruit containing more condensed tannins and total tannins, respectively, than other ripe fruit (Wrangham *et al.*, 1998, Table V; Reynolds *et al.* 1998, Table 5.2). At Bossue, one food category (mature leaves) was found to contain low levels of condensed tannins but another (young leaves) did not differ (Takemoto, 2003). Likewise, in feeding experiments, chimpanzees did not avoid tannin-enriched foods (Remis, 2002). Taken together the results suggest that food selection by chimpanzees may not be driven by tannin levels.

The study reported here has various limitations and it would be premature to draw firm conclusions on differences in food selectivity between bonobos and chimpanzees. Both species complement their diet with a selection of herbs and animal food, and future analyses of the nutritional ecology should incorporate these dietary components. Considering the variety of habitats occupied by chimpanzees, it is likely that populations use different strategies to solve the dual problem of avoiding unsuitable food while increasing nutrient intake. Understanding the patterns of nutrient geometry and associations between nutrients and antifeedants may help to explain differences in foraging strategies, resource competition, and grouping patterns.

Acknowledgments

The Institut Congolaise pour la Conservation de la Nature kindly granted permission to conduct fieldwork at Salonga National Park. Funding for the bonobo project came from The Leakey Foundation for Anthropological Research, the Max Planck Society and the Volkswagen Foundation. Barbara Fruth gave access to unpublished data on phenology and plant taxonomy. Daniel Stahl provided assistance and advice for statistical analyses. Klaus Becker lent his expertise in nutritional analyses. Verena Brauer, Julia Gessner, Alex Gregoriev, Etienne Mabonzo, Esperance Miezi, Musuyu Muganza, Valentin Omasombo, Veronica Vecellio, and Leilani Zeumer helped to collect, identify, and process samples in the field.

The Nigeria National Park Service kindly granted a research permit to the Gashaka Primate Project. Gashaka-Gumti National Park, the Nigerian Conservation Foundation, and the WWF-UK provided vital logistic support. The Chester Zoo Nigeria Biodiversity Programme provided generous corefunding. Additional support came from the Leakey Fund/London (to AF) and the Dean's travel Fund of UCL (to VS) and Leventis Ltd. & Pro Natura International/Lagos. Field work at Gashaka would have been impossible without local field assistants (Hammaunde Guruza, Felix Vitalis, Ali Tappare, Yakubu Ahmadu, Sam Yusufu), co-researchers and conservationists (Jeremiah Adanu, Ymke Warren, Aaron Nicholas, Susannah Garcia, Taiwo Oviasuyi). This is Gashaka publication Nr. 2.

Dean Anderson, Christophe Boesch, Daniel Stahl, Martha Robbins, and two anonymous reviewers made helpful comments on earlier drafts of this manuscript.

References

Alm, U., Birgersson, B., & Leimar, O. (2000). The effect of food quality and relative abundance on food choice in fallow deer. *Animal Behaviour*, **64**, 439–45.

Altmann, S.A. (1998). *Foraging for Survival*. Chicago: The University of Chicago Press.

Anderson, D.P., Nordheim, E.V., Boesch, C., & Moermond, T.C. (2002). Factors influencing fission-fusion grouping in chimpanzees in the Taï National Park, Côte d'Ivoire. In *Behavioural Diversity in Chimpanzees and Bonobos*, ed. C. Boesch, G. Hohmann, & L.F. Marchant, pp. 90–101. Cambridge: Cambridge University Press.

Badrian, N.L. & Malenky, R.K. (1984). Feeding ecology of *Pan paniscus* in the Lomako Forest, Zaire. In *The Pygmy Chimpanzee: Evolutionary Biology and Behavior*, ed. L.S. Susman, pp. 275–99. New York: Plenum Press.

Belovsky, G.E. (1978). Diet optimization in a generalist herbivore: the moose. *Theoretical Population Biology*, **14**, 105–34.

Berghoudt, K., Fruth, B., & Garber, P.A. (2005). *Food Patch Choice of Bonobos* (Pan paniscus) *in Lui Kotal, the Democratic Republic of Congo*. Poster presented at the American Association of Physical Anthropologists, April 6–9, 2005.

Boesch, C. & Boesch-Achermann, H. (2000). *The Chimpanzees of the Taï Forest: Behavioural Ecology and Evolution*. Oxford: Oxford University Press.

Carlo, T.A., Collazo, J.A., & Groom, M.J. (2003). Avian fruit preferences across a Puerto Rican forest landscape: pattern consistency and implications for seed removal. *Oecologia*, **134**, 119–31.

Chapman, C.A. & Chapman, L.J. (2000). Determinants of group size in primates: the importance of travel costs. In *On the Move: How and Why Animals Travel in Groups*, ed. S. Boinsky & P.A. Garber, pp. 24–42. Chicago: The University of Chicago Press.

Chapman, C.A., White, F.J., & Wrangham, R.W. (1994). Party size in chimpanzees and bonobos: a reevaluation of theory based on two similarly forested sites. In *Chimpanzee Cultures*, ed R.W. Wrangham, W.C. McGrew, F.B.M. de Waal, & P.G. Heltne, pp. 41–57. Cambridge, MA: Harvard University Press.

Chapman, C.A., Wrangham, R., & Chapman, L.J. (1994). Indices of habitat-wide fruit abundance in tropical forests. *Biotropica*, **26**, 160–71.

Clauss, M. (2003). Tannins in the nutrition of wild animals: a review. In *Zoo Animal Nutrition, Volume 2*, ed. A. Fidgett, M. Clauss, U. Gansloser, J.M. Hatt, & J. Nijboer, pp. 53–89. Fürth: Filander Verlag.

Clauss, M., Lason, K., Gehrke, J., *et al.* (2003). Captive roe dee (*Capreolus capreolus*) select for low amounts of tannin acid but not quebracho: fluctuation of preferences and potential benefits. *Comparative Biochemistry and Physiology, B*, **136**, 369–82.

Conklin-Brittain, N.L., Wrangham, R.W., & Hunt, K.D. (1998). Dietary response of chimpanzees and Cercopithecines to seasonal variation in fruit abundance. II. Macronutrients. *International Journal of Primatology*, **19**, 971–98.

Dearing, M.D. & Schall, J.J. (1992). Testing models of diet assembly by the generalist herbivore lizard, *Cnemidophorus murinus*. *Ecology*, **73**, 845–67.

Dearing, M.D., Mangione, A.M., & Karasov, W.H. (2000). Diet breadth of mammalian herbivores: nutrient versus detoxification constraints. *Oecologia*, **123**, 397–405.

Emlen, J.M. (1966). The role of time and energy in food preferences. *The American Naturalist*, **100**, 611–7.

Fjeldsa, J. & Lovett, J.C. (1997). Geographical patterns of old and young species in African forest biota: the significance of specific montane areas as evolutionary centres. *Biodiversity and Conservation*, **6**, 325–46.

Freeland, W.J. & Janzen, D.H. (1974). Strategies in herbivory by mammals: the role of plant secondary compounds. *The American Naturalist*, **108**, 269–89.

Fruth, B. (1995). *Nests and Nest Group in Wild Bonobos* (Pan paniscus): *Ecological and Behavioural Correlates*. Aachen: Verlag Shaker.

Fruth, B. & Hohmann, G. (1996). Nest building behavior in the great apes: the great leap forward? In *Great Ape Societies*, ed. W.C. McGrew, L.F. Marchant, & T. Nishida, pp. 225–40. Cambridge: Cambridge University Press.

(2002). How bonobos handle hunts and harvests: why share food? In *Behavioural Diversity in Chimpanzees and Bonobos*, ed. C. Boesch, G. Hohmann, & L.F. Marchant, pp. 231–43. Cambridge: Cambridge University Press.

Furuichi, T. & Ihobe, H. (1994). Variation in male relationships in bonobos and chimpanzees. *Behaviour*, **130**, 211–28.

Gerloff, U., Hartung, B., Fruth, B., Hohmann, G., & Tautz, D. (1999). Intracommunity relationships, dispersal pattern and paternity success in a wild living community of bonobos (*Pan paniscus*) determined from DNA analysis of faecal samples. *Proceedings of the Royal Society of London, B*, **266**, 1189–95.

Goodall, J. (1986). *The Chimpanzees of Gombe: Patterns of Behavior*. Cambridge: The Belknap Press of Harvard University Press.

Harbourne, J.B. (1993). *Introduction to Ecological Biochemistry*. San Diego: Academic Press.

Hashimoto, C., Suzuki, S., Takenoshita, Y., Yamagiwa, J., Basabose, A.K., & Furuichi, T. (2004). How fruit abundance affects the chimpanzee party size: a comparison between four study sites. *Primates*, **44**, 77–81.

Hladik, C.M. (1977). Chimpanzees of Gabon and chimpanzees of Gombe: some comparative data on the diet. In *Primate Feeding Ecology: Studies of Feeding and Ranging Behaviours in Lemurs, Monkeys, and Apes*, ed. T.H. Clutton-Brock, pp. 481–501. London: Academic Press.

Hohmann, G. & Fruth, B. (1995). Loud calls in great apes: sex differences and socialcorrelates. In *Current Topics in Primate Vocal Communication*, ed. E. Zimmermann, J.D. Newman, & U. Jürgens, pp. 161–84. New York: Plenum Press.

(2003a). Lui Kotal – a new site for field research on bonobos in the Salonga National Park. *Pan African News*, **10**, 25–7.

(2003b). Intra- and inter-sexual aggression by bonobos in the context of mating. *Behaviour*, **140**, 1389–413.

Hohmann, G., Gerloff, U., Tautz, D., & Fruth, B. (1999). Social bonds and genetic ties: kinship, association and affiliation in a community of bonobos (*Pan paniscus*). *Behaviour*, **136**, 1219–35.

Howell, D.C. (2002). *Statistical Methods for Psychology*. Duxbury: Thomas Learning.

Idani, G. (1991). Social relationships between immigrant and resident bonobo (*Pan paniscus*) females at Wamba. *Folia Primatologica*, **57**, 83–95.

Isbell, L.A. & Young, T.P. (2002). Ecological models of female social relationships in primates: similarities, disparities, and some directions for future clarity. *Behaviour*, **139**, 177–202.

Koenig, A. (2002). Competition for resources and its behavioral consequences among female primates. *International Journal of Primatology*, **23**, 759–83.

Koenig, A. & Borries, C. (2001). Socioecology of Hanuman langurs: the story of their success. *Evolutionary Anthropology*, **10**, 122–37.

Krishnamani, R. & Mahaney, W.C. (2000). Geophagy among primates: adaptive significance and ecological consequences. *Animal Behaviour*, **59**, 899–915.

Kummer, H. (1971). *Primate Societies*. Chicago: Aldine.

Lambert, J.E. (1998). Primate digestion: interactions among anatomy, physiology, and feeding ecology. *Evolutionary Anthropology*, **7**, 8–20.

Lehmann, J. & Boesch, C. (2004). To fission or to fusion: effects of community size on wild chimpanzee (*Pan troglodytes verus*) social organization. *Behavioral Ecology and Sociobiology*, **56**, 207–16.

Leigh, E.G. & Windsor, D.M. (1982). Forest production and regulation of primary consumers on Barro Colorado Island. In *The Ecology of a Tropical Forest: Seasonal Rhythms and Long-term Changes*, ed. A. Stanley Rand, D.M. Windsor, & E.G. Leigh, pp. 111–22. Washington, DC: Smithsonian Institution Press.

Leighton, M. (1993). Modeling dietary selectivity by Bornean orangutans: evidence for integration of multiple criteria in fruit selection. *International Journal of Primatology*, **14**, 257–313.

Levey, D.J. & Martinez del Rio, C. (2001). It takes guts (and more) to eat fruit: lessons from avian nutritional ecology. *The Auk*, **118**, 819–31.

Lotz, C.N. & Nicolson, S.W. (1996). Sugar preference of a nectarivorous passerine bird, the lesser double-collared sunbird (*Nectarinia chalybea*). *Functional Ecology*, **10**, 360–5.

Makkar, H.P.S. & Becker, K. (1998). Do tannins in leaves of trees and shrubs from African and Himalayan regions differ in level and activity? *Agroforestry Systems*, **40**, 59–68.

Makkar, H.P.S., Bluemmel, M., Borowy, N.K., & Becker, K. (1993). Gravimetric determination of tannins and their correlations with chemical and protein precipitation methods. *Journal of the Science of Food and Agriculture*, **61**, 161–4.

Malenky, R. (1990). *Ecological Factors affecting Food Choice and Social Organization in* Pan paniscus. Unpublished Ph.D. thesis, State University of New York-Stony Brook.

Malenky, R.K. & Stiles, E.W. (1991). Distribution of terrestrial herbaceous vegetation and its consumption by *Pan paniscus* in the Lomako Forest, Zaire. *American Journal of Primatology*, **23**, 153–69.

Malenky, R.K. & Wrangham, R.W. (1994). A quantitative comparison of terrestrial herbaceous food consumption by *Pan paniscus* in the Lomako Forest, Zaire, and *Pan troglodytes* in the Kibale Forest, Uganda. *American Journal of Primatology*, **32**, 1–12.

Matsumoto-Oda, A. & Hayashi, Y. (1999). Nutritional aspects of fruit choice by chimpanzees. *Folia Primatologica*, **70**, 154–62.

Milton, K. (1978). Behavioral adaptations to leaf-eating by the mantled howler monkey (*Alouatta palliate*). In *The Ecology of Arboreal Folivores*, ed. G.G. Montgomery, pp. 535–49. Washington, DC: Smithsonian Institution Press.

(1981). Food choice and digestive strategies of two sympatric primate species. *The American Naturalist*, **117**, 476–95.

(1986). Digestive physiology in primates. *News in Physiological Sciences*, **1**, 76–9.

Milton, K. & McBee, R.H. (1983). Rates of fermentative digestion in the howler monkey (*Alouatta palliate*) (Primates: Cebidae). *Comparative Biochemistry and Physiology*, **7**, 4–29.

Mitani, J.C., Watts, D.P., & Lwanga, J.S. (2002). Ecological and social correlates of chimpanzee party size and composition. In *Behavioural Diversity in Chimpanzees and Bonobos*, ed. C. Boesch, G. Hohmann, & L.F. Marchant, pp. 102–11. Cambridge: Cambridge University Press.

Morin, P.A., Moore, J.J., Chakraborty, R., Jin, L., Goodall, J., & Woodruff, D. (1994). Kin selection, social structure, gene flow, and the evolution of chimpanzees. *Science*, **265**, 1193–201.

Mueller-Harvey, I. & McAllan, A.B. (1992). Tannins, their biochemistry and nutritional properties. *Advances in Plant Cell Biochemistry and Biotechnology*, **1**, 151–217.

Neuwinger, H.D. (1998). *Afrikanische Arzneipflanzen und Jagdgifte: Chemie, Pharmakologie, Toxikologie*. Stuttgart: Wissenschaftliche Verlagsgesellschaft mbh.

Nishida, T. (1990). *The Chimpanzees of the Mahale Mountains: Sexual and Life History Strategies*. Tokyo: University of Tokyo Press.

Nishida, T. & Uehara, S. (1983). Natural diet of chimpanzees (*Pan troglodytes schweinfurthii*): long-term record from Mahale Mountains, Tanzania. *African Study Monographs*, **3**, 109–30.

Owen-Smith, N. (1994). Foraging responses of kudus to seasonal changes in food resources: elasticity in constraints. *Ecology*, **75**, 1050–62.

Parish, A.R. (1996). Female relationships in bonobos (*Pan paniscus*): evidence for bonding, cooperation, and female dominance in a male-philopatric species. *Human Nature*, **7**, 61–96.

Porter, L.J., Hrstich, L.N., & Chan, B.G. (1986). The conversion of procyanadins and prodelphinidins to cyanidin and delphinidin. *Phytochemistry*, **25**, 223–30.

Poulin, B., Wright, S.J., Lefebvrev, G., & Calderon, O. (1999). Interspecific synchrony and asynchrony in the fruiting phenologies of congeneric bird-dispersed plants in Panama. *Journal of Tropical Ecology*, **15**, 213–27.

Raubenheimer, D. & Simpson, S.J. (1997). Integrative models of nutrient balancing: application to insects and vertebrates. *Nutritional Research Reviews*, **10**, 151–79.

Remis, M.J. (2002). Food preferences among captive western gorillas (*Gorilla gorilla gorilla*) and chimpanzees (*Pan troglodytes*). *International Journal of Primatology*, **23**, 231–49.

Reynolds, V., Plumptre, A.J., Greenham, J., & Harborne, J. (1998). Condensed tannins and sugars in the diet of chimpanzees (*Pan troglodytes schweinfurthii*) in the Budongo Forest, Uganda. *Oecologia*, **115**, 331–6.

Rodman, P.S. (2002). Plants of the apes: is there a hominoid model for the origins of the hominoid diet? In *Human Diet: Its Origin and Function*, ed. P.S. Ungar & M. F. Teaford, pp. 77–109. Westpoint: Bergin & Harvey.

Rogers, M.E., Maisels, F., Williamson, E.A., Fernandez, M., & Tutin, C.E.G. (1990). Gorilla diet in the Lopé Reserve, Gabon: a nutritional analysis. *Oecologia*, **84**, 326–39.

Saracco, J.F., Collazo, J.A., & Groom, M.J. (2004). How frugivores track resources? Insight from spatial analyses of bird foraging in a tropical forest. *Oecologia*, **139**, 235–45.

Schaefer, H.M., Schmidt, V., & Baierlein, F. (2003). Discrimination abilities for nutrients: which difference matters for choosy birds and why? *Animal Behaviour*, **65**, 531–41.

Sommer, V., Adanu, J., Faucher, I., & Fowler, A. (2004). Nigerian chimpanzees (*Pan troglodytes vellerosus*) at Gashaka: two years of habituation efforts. *Folia Primatologica*, **75**, 295–316.

Sterck, E.H.M., Watts, D.P., & van Schaik, C.P. (1997). The evolution of female social relationships in nonhuman primates. *Behavioral Ecology and Sociobiology*, **41**, 291–309.

Takemoto, H. (2003). Phytochemical determination for leaf food choice by wild chimpanzees in Guinea, Bossou. *Journal of Chemical Ecology*, **29**, 2551–73.

Tutin, C.E.G. & Fernandez, M. (1993). Relationships between minimum temperature and fruit production in some tropical forest trees in Gabon. *Journal of Tropical Ecology*, **9**, 241–8.

van Schaik, C.P. (1989). The ecology of social relationships amongst female primates. In *Comparative Socioecology*, ed. V. Standen & R.A. Foley, pp. 195–218. Oxford: Blackwell Scientific Publications.

  (1996). Social evolution in primates: the role of ecological factors and male behaviour. *Proceedings of the British Academy*, **88**, 9–31.

van Soest, P.J. (1994). *Nutritional Ecology of the Ruminant*. Ithaca, NY: Cornell University Press.

Vervaecke, H., de Vries, H., & van Elsacker, L. (2000). Function and distribution of coalitions in captive bonobos (*Pan paniscus*). *Primates*, **41**, 249–65.

Vigilant, L., Hofreiter, M., Siedel, H., & Boesch, C. (2001). Paternity and relatedness in wild chimpanzee communities. *Proceedings of the National Academy of Sciences*, **98**, 12890–5.

White, F.J. (1992). Pygmy chimpanzee social organization: variation with party size and between study sites. *American Journal of Primatology*, **26**, 203–14.

White, F.J. & Lanjouw, A. (1994). Feeding competition in Lomako bonobos: variation in social cohesion. In *Topics in Primatology, Volume 1, Human Origins*, ed. T. Nishida, W.C. McGrew, P. Marler, M. Pickford, & F.B.M. de Waal, pp. 67–79. Tokyo: University of Tokyo Press.

White, F.J. & Wrangham, R.W. (1988). Food competition and patch size in the chimpanzee species *Pan paniscus and Pan troglodytes*. *Behaviour*, **105**, 148–64.

Witmer, M.C. & van Soest, P.J. (1998). Contrasting digestive strategies of fruit-eating birds. *Functional Ecology*, **12**, 728–41.

Wittig, R. & Boesch, C. (2003). The choice of post-conflict interactions in wild chimpanzees (*Pan troglodytes*). *Behaviour*, **140**, 1527–59.

Wrangham, R.W. (1979). On the evolution of ape social systems. *Social Science Information*, **18**, 335–68.

  (1980). An ecological model of female-bonded primate groups. *Behaviour*, **75**, 262–300.

(1986). Ecology and social evolution in two species of chimpanzees. In *Ecology and Social Evolution: Birds and Mammals*, ed. D.I. Rubenstein & R.W. Wrangham, pp. 352–78. Princeton: Princeton University Press.

(2002). The cost of sexual attraction: is there a trade-off in female *Pan* between sex appeal and received coercion? In *Behavioural Diversity in Chimpanzees and Bonobos*, ed. C. Boesch, G. Hohmann, & L.F. Marchant, pp. 204–15. Cambridge: Cambridge University Press.

Wrangham, R.W. & Waterman, P.G. (1983). Condensed tannins in fruits eaten by chimpanzees. *Biotropica*, **15**, 217–22.

Wrangham, R.W., Chapman, C.A., Clark-Arcadi, A.P., & Isabirye-Basuta, G. (1996). Socio-ecology of Kanyawara chimpanzees: implications for understanding the costs of great ape groups. In *Great Ape Societies*, ed. W.C. McGrew, L.F. Marchant, & T. Nishida, pp. 45–57. Cambridge: Cambridge University Press.

Wrangham, R.W., Clark, A., & Isabirye-Basuta, G. (1992). Female social relationships and social organization of Kibale Forest chimpanzees. In *Topics of Primatology, Volume 1, Human Origins*, ed. T. Nishida, W.C. McGrew, P. Marler, M. Pickford, & F.B.M. de Waal, pp. 81–9. Tokyo: University of Tokyo Press.

Wrangham, R.W., Conklin, N.L., Chapman, C.A., & Hunt, K.D. (1991). The significance of fibrous foods for Kibale Forest chimpanzees. *Philosophical Transactions of the Royal Society of London, B*, **334**, 171–8.

Wrangham, R.W., Conklin-Brittain, N.L., & Hunt, K.D. (1998). Dietary response of chimpanzees and Cercopithecines to seasonal variation in fruit abundance. I. Antifeedants. *International Journal of Primatology*, **19**, 949–70.

Zar, J.H. (1999). *Biostatistical Analysis*, 4th edn. Upper Saddle River: Prentice Hall.

6 Feeding ecology of savanna chimpanzees (Pan troglodytes verus) at Fongoli, Senegal

JILL D. PRUETZ

Introduction

Chimpanzees are commonly known as ripe fruit specialists (Goodall, 1968, 1986; Hladik, 1977, 1977, 1979; Nishida, 1990; Matsumoto-Oda & Hayashi, 1997; Tutin *et al.*, 1997; Wrangham *et al.*, 1998; Newton-Fisher, 1999; Balcomb *et al.*, 2000; Basabose, 2002), and this dietary emphasis is thought to be a major factor influencing their fission–fusion social organization

Feeding Ecology in Apes and Other Primates. Ecological, Physical and Behavioral Aspects, ed.
G. Hohmann, M.M. Robbins, and C. Boesch. Published by Cambridge University Press.
© Cambridge University Press 2006.

161

(Wrangham, 1979; Sugiyama & Koman, 1992; Wrangham, 2000; Newton-Fisher *et al.*, 2000; Mitani *et al.*, 2002; Lehmann & Boesch, 2004). In order to maximize their utilization of ripe fruit resources, which are generally described as patchy and variable in size (e.g., Ghiglieri, 1984), chimpanzee subgroups or parties fluctuate in size and individual make-up in response to resources. Presence of estrous females also influences the size and composition of chimpanzee parties, and this effect has been found to equate with food availability at some sites or even to surpass it (Goodall, 1986; Sakura, 1994; Boesch, 1996; Newton-Fisher *et al.*, 2000; Anderson *et al.*, 2002; Mitani *et al.*, 2002).

Models of primate socioecology, such as that of Wrangham (1980), which has since been modified by van Schaik (1989), Isbell (1991), and Sterck *et al.* (1997), have attempted to predict the effects of food availability, as well as other variables, on primate social behavior and organization. Wrangham (1980) and Isbell (1991) see food availability as the main factor influencing social grouping in non-human primates, while van Schaik (1989) views predation pressure as the significant determining factor. Sterck *et al.* (1997) consider additional variables such as infanticide risk as potentially significant factors influencing social grouping in primates. However, each of the models interprets food availability as a key influence on the social behavior and organization of primate species.

A chimpanzee's diet is constituted largely of fruit, which accounts for more than 60% of the food for most chimpanzee populations that have been studied (Conklin-Brittain *et al.*, 2001; Pruetz & McGrew, 2001). During times of fruit scarcity, chimpanzees may range more widely in search of ripe fruit (Sugiyama & Koman, 1992; Wrangham *et al.*, 1996; Matsumoto-Oda & Kasagula, 2000) or switch to alternative food sources (Isabirye-Basuta, 1989; Doran, 1997). For chimpanzees living in savanna habitats, the hypothesized scarcity of foods in this environment may have a greater effect on their behavior and, subsequently, aspects of their social organization compared with chimpanzees living in more forested environments. It is generally assumed that open or savanna habitats present a problem for apes because foods such as fruit are scarce and widely distributed (e.g., Isbell & Young, 1996). However, as Moore (1996) points out, this assertion has not yet been demonstrated. Both ecological measures of food availability and behavioral measures of feeding are needed to understand the selective pressures acting upon chimpanzees in a savanna environment. Such an understanding is especially relevant to our ability to decipher the selective pressures operating on fossil apes in an open environment. The importance of fallback foods (Wrangham *et al.*, 1998) in particular may be key to understanding evolutionary changes in the diet of early hominids (Teaford & Ungar, 2000). In order to examine the effects that savanna food resources have on chimpanzees' behavior, I will present information on the feeding ecology of the Fongoli community of chimpanzees living in southeastern Senegal as a step towards a better understanding of the behavior of this species in an open habitat.

Methods

Study site

The Fongoli community's home range in southeastern Senegal is at the margin of chimpanzees' geographical range in Africa, in a semi-arid, open environment (Carter *et al.*, 2003). The study site (12°39′N 12°13′W) lies approximately 10 km NE of the town of Kedougou and approximately 45 km SE of the Assirik site in the Parc National du Niokolo Koba (PNNK). Fongoli is roughly 35 km from the Guinea border to the south and 85 km from Mali to the east. Rainfall averages 900–1100 mm annually, with June–September constituting the rainy season (Ba *et al.*, 1997). Average annual temperature in Kedougou from 1961–1990 was 28.2 °C, with an average monthly minimum of 25 °C in December and an average monthly maximum of 33 °C in May (Ba *et al.*, 1997).

This region of Senegal can be characterized as a transition between the Sudanian and Sudo-Guinean vegetative belts, with the former predominating. The topography is composed of valleys, plateaus, and hills, with the highest peak in the region being 426 m (Baniomba Mountain). Common woody tree species include those indicative of the Sudanian zone (*Zizyphus mauritiana, Combretum glutinosum*) and the Sudo-Guinean zone (*Pterocarpus erinaceus, Piliostigma thonningii*) (Ba *et al.*, 1997). The habitat is dominated by woodland and wooded grassland interspersed with areas of bamboo woodland, plateau, and thicket. Small areas of gallery forest provide chimpanzees with important food and shelter.

Subjects

The Fongoli community of chimpanzees has been studied continuously since April 2001. Based on estimates from nest surveys, chimpanzees occur here at a density of 0.09 individuals per km^2 (Pruetz *et al.*, 2002). Chimpanzees at Fongoli are semihabituated in that certain individuals are habituated to researchers' presence, with 29 being identified (17 males, 12 females). Based on party size counts and including offspring of known females, a total of 32 individuals have been identified as belonging to this community. Over 200 contact hours have been spent with these chimpanzees in more than 400 encounters (April 2001–August 2004). Contacts vary but often last from 3–7 hours. After 3 years of habituation, average contact time with a chimpanzee party is 98 minutes. Fongoli chimpanzees are estimated to have a minimum community range of 63 km^2, predicated on the following evidence: nesting patterns (i.e., based on high density "core" nesting areas and peripheral areas where nests occur but at low density), and estimated core ranges of identified individuals. Community range size estimates have increased yearly, as chimpanzees become more habituated to observer presence and new areas are recorded in which chimpanzees range. The nearest neighboring chimpanzee community may be that of Baniomba Mountain, approximately 15 km NW of the Fongoli site and between the Assirik site and Fongoli. Chimpanzees in this region of Senegal are sympatric with humans of the Malinke, Bassari, Diahanke, Puhlar, and Bedik groups. Fongoli chimpanzees have not been found to raid the area's crops, which include peanuts, maize, cotton, and millet, although they are known to raid local humans' beehives.

Data collection

From May 2001–August 2004 data on chimpanzees' diet were collected based on direct observation and analyses of chimpanzee feces. Only fresh feces were collected, and these were sieved by washing the samples through 1-mm wire mesh. Data collected included species and number of seeds in feces, presence or absence of animal prey, and species of entire leaves, if present. Seeds were identified by Senegalese field assistants and checked with a reference collection of seeds from known plants. Botanical specimens were identified according to Arbonnier (2000) and with the assistance of S. Johnson-Fulton, Department of Botany, Miami University, Oxford, OH. Sample size analyzed per month varied from 36 to 186, with most fecal samples collected during the late dry–early wet season and late wet season.

Beginning in May 2003, feeding traces left by chimpanzees were also considered dietary corroboration since by then researchers were familiar with evidence left by chimpanzees as opposed to baboons (*Papio hamadryas papio*), patas (*Erythrocebus patas*), green monkeys (*Cercopithecus aethiops sabeus*), and other animals. Criteria followed McGrew *et al.* (1981), and associated evidence such as chimpanzee sightings, foot and knuckle prints, feces, and other features particular to chimpanzee feeding (e.g., tool use) were used to confirm certain items in the chimpanzees' diet. Most dietary data, however, are based on fecal analyses and are supplemented with direct observation of feeding.

When chimpanzees were observed, feeding bouts were recorded as independent feeding events associated with separate trees, shrubs, or climbers. In order to compensate for the bias (i.e., larger trees, more fruit, more chimpanzees) of multiple individuals feeding in large food sources, one or more chimpanzees feeding in one tree was considered a single feeding bout. In certain cases, one chimpanzee was observed feeding sequentially on one food from different plants of the same species (*Bassia multiflora* vine leaves, for example) and these were considered a single bout. Data collected during feeding bouts included date, time, and food species and part eaten.

Data collection on habitat structure and food availability at Fongoli was patterned after McBeath & McGrew (1982) and Pruetz *et al.* (2002). Habitat structure was estimated using 5-m radius circular plots placed every 100 m along 7.5 km of transects. Transects were placed randomly within the home range (1.5 km) or bisected the core home range EW (3 km) and NS (3 km). Phenological data on over 150 trees, shrubs, or climbers with dbh > 10 cm were recorded monthly from July 2002–July 2004 (excluding September and December 2002). Plants were distributed along a 1.5-km by 10-m transect that bisected the Fongoli chimpanzee community's core range. The number of large trees, shrubs, or climbers monitored per month varied from approximately 100 during the initial 14 months to approximately 165 after the transect had been lengthened in July 2003. Results presented here are for chimpanzee feeding species only. An average of 82 such individuals were monitored monthly at the same time each month. The abundance of these plants along this transect was used as a rough estimate of the common or rare occurrence of large trees, shrubs, and climbers within the home range, as well as their periods of flower, fruit, and leaf production.

Results

Dietary diversity

Data from fecal analyses, feeding observations, and feeding traces revealed that Fongoli chimpanzees fed on at least 47 different plant species in 27 families (Table 6.1). The majority of species were trees (57%), but shrubs or shrub-like trees accounted for 30% of species eaten, while liana, vines, or herbaceous-level vegetation, including grasses, accounted for 13% of the remaining species. Of the 60 different plant items eaten, most were fruit (53%). Termites and other insect prey, as well as vertebrate prey, could also be identified in feces, and these accounted for another five species. Leafy vegetation and flowers were rarely identifiable in feces, however, so these

Table 6.1. *Foods eaten by Fongoli chimpanzees (2001–2004)*

Genus and species Plant foods	Family	Life form	Part eaten	J	F	M	A	M	J	J	A	S	O	N	D	Criteria
Acacia ehrenbergiana	Mimosaceae	T	b	+	+		+									R
Adansonia digitata	Bombacaceae	T	uf, rf, fl, b	+	+	+	+	+	+	+	+	+	+	+	+	O, F, R
Afzelia africana	Cesalpiniaceae	T	f	+								+		+	+	F
Allophyllus africanus	Sapindaceae	T	f									+	+			F
Andropogon chevalieri	Graminaceae	H	l±						+	+						F
Asparagus sp.	Liliaceae	H	p													R
Baissea multiflora	Apocynaceae	C	l					+	+	+	+					O
Bombax costatum	Bombacaceae	T	fl, uf	+	+									+	+	O
Borassus aethiopum[a]	Palmae	T	p		+											R
Ceiba pentandra	Bombacaceae	T	fl		+											O
Cissus populnea[a]	Vitaceae	S	f	+									+	+		F
Cissus rufescens[a]	Vitaceae	S	f	+									+	+		F
Cola cordifolia[a]	Sterculiaceae	T	f				+	+			+					F, R
Cordyla pinnata[a]	Papilionaceae	T	f				+	+	+	+						F, R
Daniellia olivera	Cesalpiniaceae	T	f, fl	+	+	+	+	+	+	+	+	+				O
Diospyros mespiliformis[a]	Ebenaceae	T	f, l, b	+	+				+	+	+	+	+	+	+	O, F, R
Elaeis guineensis[a]	Palmae	T/S	p	+							+		+	+	+	R
Ficus sp.	Moraceae	T/S	uf, rf	+					+	+	+		+			F
Ficus asperifolia[a]	Moraceae	S	l±		+					+					+	F
Ficus ingens	Moraceae	T	uf, l	+					+				+			O
Ficus sycomorus?[a]	Moraceae	T	b				+									R
Ficus trichopoda[a]	Moraceae	T	l					+								O, R
Ficus sur[a]	Moraceae	S	p			+										O
Gardenia erubescens[a]	Rubiaceae	S	f	+	+											F, R

All months in which foods were eaten

Species	Family	Habit	Parts												Uses
Grewia lasiodiscus[a]	Tiliaceae	S	f							+			+	+	F
Hannoa undulata[a]	Simaroubaceae	T	f							+	+			+	F
Hexalobus monopetalus[a]	Annonaceae	T	f, l						+			+			O, F, R
Hymenocardia acida	Euphorbiaceae	S	fl				+								O
Landolphia heudelotii[a]	Apocynaceae	C	f					+	+						O, F, R
Lannea sp.	Anacardiaceae	T	f					+	+						F
Lannea microcarpa	Anacardiaceae	T	f						+					+	O, F
Nauclea latifolia[a]	Rubiaceae	S	p				+	+	+						R
Oncoba spinosa[a]	Flacourtiaceae	S	f				+				+		+		F
Parinari excelsum[a]	Chrysobalanaceae	T	f								+		+		F
Parkia biglobosa	Mimosaceae	T	f, fl	+	+	+	+	+							F, R
Piliostigma thonningii	Cesalpiniaceae	T	uf, rf	+	+	+	+	+	+		+	+	+		F
Pterocarpus erinaceus	Papilionaceae	T	b, f, fl	+	+	+	+	+	+	+					O, F, R
Saba senegalensis[a]	Apocynaceae	C	uf, rf, p					+						+	O, F, R
Sclerocarya birrea[a]	Anacardiaceae	T	f				+		+						F
Spondias monbin[a]	Anacardiaceae	T	uf, rf	+	+	+	+		+		+		+		O, F
Sterculia setigera	Sterculiaceae	T	f	+	+										F
Strychnos spinosa[a]	Loganiaceae	S	f	+	+			+							F, R
Tamarindus indica	Cesalpiniaceae	T	f	+		+			+			+	+		F
Vitellaria paradoxa[a]	Sapotaceae	T	f												F
Vitex madiensis	Verbinaceae	T/S	f	+	+				+		+	+	+		F
Zizyphus sp.[a]	Rhamnaceae	S	f					+							F
"Petit minkon"	?	T	l, uf					+		+					O
"Mora"	?	S	l			+									O
Herbaceous vine	?	H	l			+									O

Table 6.1. (*cont.*)

Animal foods	Family or Order								
Unknown vertebrate			a					+	F
Cercopithecus aethiops sabeus	Cercopithecidae	M	a			+	+?		O, F, R
Galago sp.	Galagidae	M	a			+	+	+	F
Oecophylla longinoda	Hymenoptera	I	a	+					O, R
Apis mellifera	Hymenoptera	I	a	+	+	+			O, R
Dorylus sp.	Hymenoptera	I	a	+	+	+	+	+	F, R
Macrotermes sp.	Isoptera	I	a	+	+	+	+	+	O, F, R

Notes:

T = tree, C = climber, H = herbaceous level/grass, S = shrub, M = mammal, I = insect. f = fruit (u = unripe, r = ripe), l = leaves/shoots, fl = flower/inflorescence, p = pith/stem, a = animal food. ± Leaves swallowed whole, possibly for medicinal purposes. O = direct observation, F = fecal analyses, R = feeding traces.

[a] Qualifies as large, fleshy fruit (fleshy drupes, berries or figs > 1.5 cm in length) according to Balcomb *et al.*, 2000.

Table 6.2. *Number of plant species and plant parts eaten by chimpanzees at various study sites*

Study site	Number of plant species	Number of plant parts	Reference
Dry sites[a]			
Fongoli, Senegal	47	60	This study
Semliki, Uganda	–	36	Hunt & McGrew, 2002
Assirik, Senegal	43	60	McGrew *et al.*, 1988
Average at dry sites	45	52	
Wet sites			
Budongo, Uganda	56	87	Newton-Fisher, 1999
Bossou, Guinea	200	246	Sugiyama & Koman, 1992
Gombe, Tanzania	–	201	Wrangham, 1977
Ipassa, Gabon	–	141	Hladik, 1977
Kahuzi-Biega, DRC	114	171	Basabose, 2002
Lope, Gabon	–	182	Tutin *et al.*, 1997
Mahale, Tanzania	198	–	Matsumoto-Oda & Kasagula, 2000
Kanyawara, Uganda	112	–	Wrangham *et al.*, 1991
Average at wet sites	36	171	
Overall chimpanzee average	110	132	

Note:
[a] Dry sites are defined as those receiving, on average, less than 1500 mm annually (see Hunt & McGrew, 2002).

food types are under-represented in the diet of Fongoli chimpanzees. Indirect evidence exists for geophagy by Fongoli chimpanzees. Knuckle prints, hair, and tooth marks indicated that chimpanzees eat soil at small caves, substantiating data from fecal analyses indicative of this behavior (i.e., consistency and coloration). Fongoli chimpanzees fed on a similar number of plant species compared with chimpanzees studied at Assirik during the 1970s (McGrew *et al.*, 1988) and at the Semliki, Uganda site (Hunt & McGrew, 2002), but on fewer species than chimpanzees living in wetter areas (Table 6.2).

Frugivory

Fruit was consumed during each month of the year. Analyses of 1007 fecal samples collected from May 2001 through July 2004 (38 months) revealed that Fongoli chimpanzees dispersed seeds of over 32 fruiting species. Foods that were represented in >50% of fecal samples in a single month were considered most important for Fongoli chimpanzees and included fruits of

Saba senegalensis, Diospyros mespiliformis, Ficus sp., *Adansonia digitata, Lannea* sp., *Spondias mombin,* and *Hexalobus monopetalus.* Four foods were found in >50% of feces in more than one month: *A. digitata, S. senegalensis, D. mespiliformis,* and *Ficus* species. At most, only two fruit species were represented in >50% of samples per month (n = 3 months), and most months were similarly characterized by only a single fruit species (n = 7). Only March and August were excluded using this criterion.

Fruits of the genus *Ficus* were eaten in 11 out of 12 months. However, seeds from feces could not be identified according to species for this genus. Fongoli chimpanzees also fed on leaves of at least three and the inner bark/ cambium or pith of two additional *Ficus* species, so that at least some part of this plant was eaten monthly. Based on the data collected thus far at Fongoli, fruits of *Ficus* appear to be staple foods (Newton-Fisher, 1999) for chimpanzees here, eaten in periods of both high and low overall fruit availability. The next most utilized fruit was *S. senegalensis,* eaten in 9 months out of the year. Again, chimpanzees also fed on the pith of *Saba,* so that this species was eaten monthly as well. Chimpanzees at Fongoli appear to utilize the fruit of *A. digitata* at different times of the year. *A. digitata* was eaten heavily (i.e., found in >50% of fecal samples) both during the early dry season (November–January) and in the late dry season (May), and this species was fed on to a lesser extent in every other month. This particular fruit has a chalky pulp enclosed in a woody husk that remains on trees for months without rotting.

Behavioral observations of Fongoli chimpanzees indicate that this community relies heavily on fruit, similar to what has been observed at all other sites where chimpanzees have been studied (Figure 6.1). A total of 144 feeding observations were analyzed regarding food type. The majority of feeding was on fruit (62.5%), followed by leaves (16%), flowers or inflorescence (11%), invertebrates (5%), pith (3%), and cambium/bark (2.5%) (Figure 6.1). When fruits could be categorized according to maturity, unripe fruit eating accounted for 55% (n = 16 bouts) and ripe fruit eating accounted for 45% of bouts (n = 13 bouts).

Animal prey

Animal prey was found in most fecal samples analyzed. In addition to invertebrate prey, consisting of termites (*Macrotermes*), driver ants (*Dorylus*), and weaver ants (*Oecophylla*), Fongoli chimpanzees also fed on vertebrates identified as a Lorisoid species, most likely a *Galago,* and on green monkeys (*Cercopithecus aethiops sabeus*) (Table 6.1). In comparison to other sites

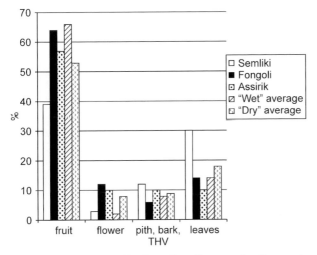

Figure 6.1. Observations of feeding at Fongoli compared to diet at various chimpanzee study sites. Fongoli data = feeding bouts. Semliki and Assirik data = percent of plant parts in diet based on fecal analyses, traces, and observation (McGrew *et al.*, 1988; Hunt & McGrew 2002). Dry site (<1500 mm annual rainfall) values summarized from Assirik, Semliki, and Fongoli. Wet site values summarized from: Gombe, Tanzania: Wrangham, 1977; Rodman, 1984 (observation); Mahale, Tanzania: Nishida & Uehara, 1983; Matsumoto-Oda *et al.*, 1998; Matsumoto-Oda & Kasagula, 2000 (observation); Ngogo, Uganda: Ghiglieri, 1984 (observation); Kanyawara, Uganda: Chapman *et al.*, 1994; Wrangham *et al.*, 1996 (observation); Bwindi, Uganda: Stanford & Nkurunungi, 2003 (feces); Bossou, Guinea: Sugiyama & Koman, 1987, 1992 (observation); Lope, Gabon: Tutin *et al.*, 1997 (feces); Budongo, Uganda: Newton-Fisher, 1999 (observation).

such as Gombe and Kibale, however, Fongoli chimpanzees exhibit low levels of meat eating (Figure 6.2). Savanna chimpanzees appear to exhibit low levels of meat eating in general. A lower density of prey species and individuals is a likely explanation for this difference.

Fongoli chimpanzees may utilize invertebrate prey more than chimpanzees at other sites. At most sites where chimpanzees have been studied extensively, termites are included in the diet mainly during wet season months (McGrew *et al.*, 1979; McBeath & McGrew, 1982; Goodall, 1986), although chimpanzees at Ndoki Forest, Congo and Okorobiko, Rio Muni, like the Fongoli chimpanzees, also exhibited a similar lack of seasonality in feeding on termites (McGrew *et al.*, 1979; Kuroda *et al.*, 1996). At Fongoli, termites were eaten monthly, with at least two and perhaps three species consumed (Bogart *et al.*, 2005). Additionally, some type of non-termite insect prey was eaten in most months of the year, while bees and honey were eaten seasonally, during the late dry season (Table 6.1). In general, while Fongoli

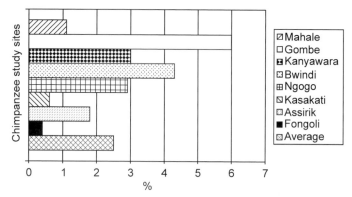

Figure 6.2. Percentage of vertebrate prey in the diet at various chimpanzee study sites based on fecal analyses. Data from Fongoli (this study) and Assirik (McGrew, 1983), Senegal and Kasakati Basin, Tanzania (Suzuki, 1966) represent drier, more open habitat chimpanzee study sites. Data from: Gombe, Tanzania: McGrew, 1983; Goodall, unpublished data; Bwindi, Uganda: Stanford & Nkurunungi, 2003; Mahale, Tanzania: Takahata *et al.*, 1984; Kanyawara and Ngogo, Uganda: Wrangham *et al.*, 1991 represent more closed, wetter chimpanzee study sites.

chimpanzees may not have access to an abundance of the vertebrate prey usually included in the chimpanzee diet (e.g., Colobus monkeys do not occur in this arid region), invertebrate prey and opportunistic feeding on vertebrates appear to characterize their diet. Invertebrate feeding, especially, may allow Fongoli chimpanzees to compensate for the lack of vertebrate prey characteristic of the diet of chimpanzees at other sites. If the percentage of all animal foods in the diet of Fongoli chimpanzees is taken into account, it is more similar to that of chimpanzees at other sites than if only vertebrate prey is considered (Figure 6.2). However, it is still much less than at Gombe and Mahale (Figure 6.3).

Food availability and habitat structure

Habitat structure at Fongoli is similar to the Assirik site in the PNNK but substantially different from other sites where chimpanzees have been studied. Analyses of the vegetation structure at the Assirik site and at Fongoli reveal similarities, even though humans are present at Fongoli (Pruetz *et al.*, unpublished data). The Fongoli site is characterized by a small percentage (2%) of closed habitat (gallery forest and ecotone), similar to Assirik's 5% (McGrew *et al.*, 1981; McBeath & McGrew, 1982). Woodland accounts for the greatest percentage of habitat types at each site, with 46% of Fongoli chimpanzees'

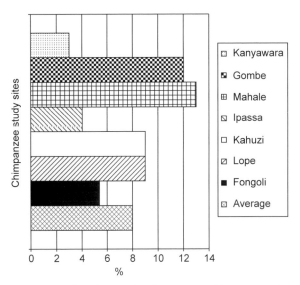

Figure 6.3. Animal foods in the diet at various chimpanzee study sites. Data from Fongoli (percent feeding bouts); Gombe, Tanzania: Wrangham, 1977 (percent dietary items); Ipassa, Gabon: Hladik, 1977 (percent feeding time); Mahale, Tanzania: Matsumoto-Oda *et al.*, 1998 (percent feeding time); Lope, Gabon: Tutin *et al.*, 1997 (percent fecal analyses); Kanyawara and Ngogo, Kibale Forest, Uganda: Wrangham *et al.*, 1991 (percent fecal analyses); Kahuzi-Biega, DRC: Basabose, 2002 (percent fecal analyses).

range consisting of woodland and 37% of available habitat at Assirik being woodland. The remaining habitat at Fongoli comprises grassland (16%), plateau (20%), bamboo woodland (12%), and horticultural fields (a minimum of 4%). The Fongoli site is characterized by a density of approximately 130 large (>5 m in height) stems/ha based on point-center-quarter measures (Pruetz *et al.*, unpublished data).

The percentage in fruit per month of large (>10 cm dbh) trees, shrubs, or climbers that were also chimpanzee feeding species is presented in Figure 6.4. Sixty-one percent of individual plants monitored along the phenological transect were chimpanzee food species (17 of 28 species). On average, food species were represented by five plants (range 1–14, median 7). Transect species, on average, were represented by four plants. The most common feeding species monitored included *P. erinaceus, P. thonningii*, and *S. senegalensis*. Overall, the most common genus was *Combretum*. Virtually all species included in the diet of Fongoli chimpanzees could be found in woodland or other open habitat. Rarely were food species associated only with a closed habitat (e.g., *Cola cordifolia, Ceiba pentandra*). The greatest percentage of fruiting plants was

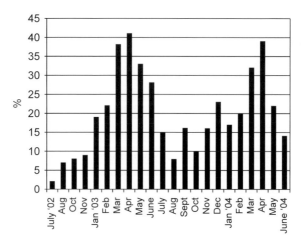

Figure 6.4. Percentage of feeding tree, shrub, and climber species in fruit per month.

available during the late dry season (Figure 6.4). Both ripe and unripe fruit are considered here since it is difficult to determine from analyses of feces the degree to which chimpanzees fed on each.

Discussion

As Wrangham *et al.* (1991) note, dietary data on chimpanzees that are not well habituated may be biased toward certain foods, given the visibility of feeding subjects. These authors note that during initial years of study, Kanyawara chimpanzees were more easily located at tree fruit sources, which they repeatedly used and at which they frequently gave loud calls. Similar behavior has been seen at Fongoli, and observers often located chimpanzees after hearing pant-hoots (Pruetz, unpublished data). As Isabirye-Basuta (1989) notes, the proportion of the diet devoted to arboreal fruits is likely to be over-estimated relative to terrestrial foods such as terrestrial herbaceous vegetation (THV), and this is especially the case for semihabituated subjects.

In order to strengthen the data set on chimpanzee diet, indirect evidence, e.g., fecal analyses and feeding traces, was also considered. These data are limited in specific ways as well. While fecal analyses may give a less biased view of chimpanzee fruit feeding than earlier observational data from studied communities, these data too are biased towards fruits since most fruits' seeds were dispersed by chimpanzees after passing through their gut intact. With these limitations in mind, I provide a preliminary examination of the diet of Fongoli chimpanzees in comparison to chimpanzees living at other sites.

Are Fongoli chimpanzees limited in terms of ripe, fleshy fruit?

It might be expected that chimpanzees living in such an open environment like Fongoli would be limited in terms of the amount of fruit available to them compared with chimpanzees living in more forested habitats. Both fecal analyses and direct observations of chimpanzees feeding illustrate the importance of fruit in the diet. Fruits accounted for 53% of plant parts eaten as evidenced by fecal analyses and 63% of foods eaten during observed feeding bouts. Most plant parts eaten at Fongoli were fruits characteristic of a woodland habitat, which might be predicted to offer less available food than a more forested environment. The habitat at Fongoli is dominated by woodland and grassland. In terms of the relative contribution of the different types of habitat to their home range, chimpanzees at Fongoli resemble Assirik chimpanzees (McGrew *et al.*, 1981) but differ from East African chimpanzees, for example, in the mosaic of woodland and other habitats used by chimpanzees in Tanzania (Kano, 1972). Balcomb *et al.* (2000) found a trend at Kibale, Uganda between the density of chimpanzee nests and the number of trees producing large, fleshy fruits, defined by these authors as fleshy drupes, berries, or figs >1.5 cm in length. At Kibale, chimpanzees had a total of 34 tree species available that produced large, fleshy fruits (Balcomb *et al.*, 2000). At Fongoli, using these criteria, chimpanzees are known to have 27 such species available to them (*Ximenia, Detarium*, and *Annona* species qualify as large, fleshy fruits but are not yet known to be eaten by Fongoli chimpanzees), and they are known to feed on the fruit of at least 17 of them (Table 6.1). Thus, even though savanna chimpanzees are hypothesized to live in an environment significantly different ecologically from forest-dwelling chimpanzees, examination of foods available to these apes reveals some similarities. Chimpanzees' most important fruit species were almost exclusively those that could be characterized as having large, fleshy fruit. Of the seven foods that were found in >50% of samples in more than one month, five could be identified according to species. Of these, all but baobab were characterized by fleshy drupes >1.5 cm in length. Similar to what has been reported at other sites (Wrangham *et al.*, 1993; Newton-Fisher, 1999; Tweheyo & Obua, 2001; Basabose, 2002; Mitani *et al.*, 2002), Fongoli chimps relied heavily on *Ficus* fruits throughout the year.

Further comparison of the distribution and abundance of fruits for savanna versus forest chimpanzees is needed in order to fully understand the ecological pressures associated with these respective environments. Data on the phenology of large trees showed fewer trees were in fruit during wet season months, while the greatest number of fruiting trees occurred during late dry season months. Mitani *et al.* (2002) also found that dry season months for Ngogo chimpanzees at Kibale National Park, Uganda were not those characterized by greatest fruit scarcity. The proportion of trees in fruit per month at Fongoli is similar to the site at Kahuzi Biega, DRC (Basabose, 2002). This montane site is also considered relatively limited in food resources compared with other sites where chimpanzees have been studied over the long term (Basabose, 2002). Stanford & Nkurunungi (2003) reported a monthly average of 23–57% of trees in fruit per month for chimpanzees and gorillas (*Gorilla gorilla beringei*) at Bwindi Impenetrable National Park, Uganda. Apes at this site inhabit a montane forest at an altitude of 2000–2300 meters. Although these sites differ significantly from the Fongoli site in terms of plant communities, similar selective pressures in the form of fruit scarcity may affect chimpanzees at all of them. As Boesch (1996) notes, however, the abundance of fruit rather than of fruiting trees (or species) may be most relevant to understanding chimpanzee feeding and grouping behavior.

If Fongoli chimpanzees are limited by their dry, open habitat in terms of overall fruit abundance, during periods of limited fruit availability they might be predicted to fall back on lower-quality foods such as bark and pith (Nishida, 1976; Wrangham *et al.*, 1991). Based on limited observational data, chimpanzees at Fongoli fed on bark or cambium of at least five species and on the pith of an additional six. Nishida (1976) found that Mahale chimpanzees included the bark or cambium of at least 21 species in their diet and described this food type as an emergency food. Overall, bark accounted for 3.4% of all feeding bouts observed at Mahale during six separate time periods from 1966–1974. In comparison, Fongoli chimpanzees fed on bark during 2.5% of observed feeding bouts, similar to what has been reported for Semliki chimpanzees (Hunt & McGrew, 2002). At Assirik, bark of four different plant species accounted for 7% of plant parts eaten (Hunt & McGrew, 2002), and chimpanzees fed on two species of what Wrangham *et al.* (1991) term 'terrestrial piths and leaves' (McGrew *et al.*, 1981). At Kibale, Uganda chimpanzees fed on 42 such species (Wrangham *et al.*, 1991). As Fongoli chimpanzees become better habituated to human observers and observations of feeding increase, the dietary repertoire of Fongoli chimpanzees is also expected to expand, especially in terms of bark, terrestrial piths, and leaves. In terms of other non-fruit foods, Fongoli chimpanzees feed on termites monthly, although this behavior is normally associated with the wet season at other sites. Chimpanzees here also appear to hunt opportunistically.

Insect feeding may allow chimpanzees to compensate for the scarcity of vertebrate prey at Fongoli.

Dietary diversity and frugivory

As is typical of chimpanzees elsewhere, Fongoli chimpanzees include some fruit in their diet monthly. Similar findings have been reported at other sites, even where fleshy fruit is thought to be relatively scarce. At Kahuzi-Biega, Democratic Republic of Congo, chimpanzees included some fruit in the diet during each month of study, although this site is characterized as having a low diversity of fruit species compared with other sites (Basabose, 2002). Additionally, one fruit species was usually the focus of the chimpanzee diet at any one time. Based on fecal analyses, Fongoli chimpanzees also relied heavily on a few fruits each month, similar to reports from Kanyawara and Ngogo, Uganda and at Gombe, Tanzania (Wrangham, 1977; Wrangham *et al.*, 1991). Five fruit species were found in >50% of all samples in a single month (*A. digitata, S. senegalensis, D. mespiliformis, H. monopetalus, S. mombin*). *Ficus* and *Lannea* fruits also accounted for >50% of samples in a single month, but these seeds were not distinguishable to the species level. The fruit *S. senegalensis* may be the most important food resource for Fongoli chimpanzees, being the focus of the diet from May–June, which is often the period when water is least available and when the availability of fruit begins to decline (Figure 6.4). Humans in the area intensively gather this fruit during this time and are estimated to extract more than 450 000 of them from the study area annually (Pruetz & Knutsen, 2003). The harvest of *Saba* fruit by humans in the Kedougou region has increased fivefold over the past several years, from 90 000 metric tons in 2000 to 470 000 metric tons in 2002 (P. Knutsen, pers. comm., 2003). The ramifications of this major increase in harvesting is not yet known, but it is anticipated that significant changes in chimpanzee diet may be apparent within the next decade if steps are not taken to enforce a sustainable level of extraction. Humans consume or use medicinally at least 60% of plant foods that occur in the diet of chimpanzees at Fongoli (Johnson-Fulton & Pruetz, unpublished data). The narrow diet characteristic of Fongoli chimpanzees may exacerbate the effects that mutual consumption of species by chimpanzees and humans at this site has on the future of the chimpanzee community. In contrast, for example, Sugiyama & Koman (1992) report that only 24% of the 246 different plant food items eaten by Bossou chimpanzees are also eaten by local humans.

Fongoli chimpanzees exhibit a narrow diet not only in terms of total species and parts consumed but also in the number of fruits included in the diet monthly. For example, Humle & Matsuzawa (2004) recorded between 11 and 23 different fruit species per month during seven different rainy season months at Bossou, Guinea. Additionally, at most other study sites the number of plant species included in the chimpanzee diet usually averages over 100 (Table 6.2). The least diverse diet appears to be that of chimpanzees at Semliki, Uganda who are recorded to feed on only 36 different food items (Hunt & McGrew, 2002). Although the dietary diversity of the Fongoli chimpanzees appears low, lack of habituation may explain this in part. Direct observation of chimpanzee feeding behavior at Fongoli has resulted in the addition of nine food items and four plant species utilized for leaf and flower eating, neither of which is easily discerned based on fecal analyses alone. This possibility demonstrates the need for the ecological study of habituated communities of open habitat chimpanzees.

Acknowledgments

Research was carried out with permission from and thanks to the Republic of Senegal and Department du Eaux et Forêts. Funding provided by Iowa State University, National Geographic Society, National Science Foundation (BCS 0122518), Primate Conservation Inc., and U.S.F.W.S. Great Ape Conservation Grant. Thanks to O. and D. Pruetz for support and to D. Kante, M. Camara, S. Bogart, M. Cook, H. Davis, M. Gaspersic, S. Johnson-Fulton, W. McGrew, A. Piel, F. Stewart, P. Stirling, and M. Waller for assistance in data collection. Special thanks to organizers of the Great Ape Feeding Ecology Symposium at Max Planck Institute for Evolutionary Anthropology, and especially to Gottfried Hohmann and four anonymous reviewers for comments on this manuscript.

References

Anderson, D.P., Nordheim, E.V., Boesch, C., & Moermond, T.C. (2002). Factors influencing fission-fusion grouping in chimpanzees in the Tai National Park, Côte d'Ivoire. In *Behavioural Diversity in Chimpanzees and Bonobos*, ed. C. Boesch, G. Hohmann & L.F. Marchant, pp. 90–101. Cambridge: Cambridge University Press.

Arbonnier, M. (2000). *Arbres, Arbustes et Lianes des Zones Séches d'Afrique de l'Ouest*. Paris: Cirad/MNHN.

Ba, A.T., Sambou, B., Ervik, F., Goudiaby, A., Camara, C., & Diallo, D. (1997). *Végétation et Flore. Parc Transfrontalier du Niokolo Badiar*. Niokolo Badiar: Union Européenne-Niokolo Badiar.

Balcomb, S.R., Chapman, C.A., & Wrangham, R.W. (2000). Relationship between chimpanzee (*Pan troglodytes*) density and large, fleshy-fruit tree density: conservation implications. *American Journal of Primatology*, **51**, 197–203.

Basabose, A.K. (2002). Diet composition of chimpanzees inhabiting the montane forest of Kahuzi, Democratic Republic of Congo. *American Journal of Primatology*, **58**, 1–21.

Boesch, C. (1996) Social grouping in Tai chimpanzees. In *Great Ape Societies*, ed. W.C. McGrew, L.F. Marchant, & T. Nishida, pp. 101–13. Cambridge: Cambridge University Press.

Bogart, S.L., Pruetz, J.D., & McGrew, W.C. (2005). Termite du jour? Termite fishing by west African chimpanzees (*Pan troglodytes verus*) at Fongoli, Senegal. *American Journal of Physical Anthropology, Supplement*, **126**, 75.

Carter, J., Ndiaye, S., Pruetz, J., & McGrew, W.C. (2003). Senegal. In *West African Chimpanzees: Status Survey and Conservation Action Plan*, ed. R. Kormos, C. Boesch, M.I. Bakarr, & T.M. Butynski, pp. 31–9. Washington, DC: IUCN/SSC Primate Specialist Group.

Chapman, C.A., White, F.J., & Wrangham, R.W. (1994). Party size in chimpanzees and bonobos: a reevaluation of theory based on two similarly forested sites. In *Chimpanzee Cultures*, ed. R.W. Wrangham, W.C. McGrew, F.B.M. deWaal, & P.G. Heltne, pp. 41–58. Cambridge, MA: Harvard University Press.

Conklin-Brittain, N.L., Knott, C.D., & Wrangham, R.W. (2001). The feeding ecology of apes. In *The Apes: Challenges for the 21st Century, Conference Proceedings of the Brookfield Zoo*, pp. 167–74. Chicago: Chicago Zoological Society.

Doran, D. (1997). Influence of seasonality on activity patterns, feeding behavior, ranging, and grouping patterns in Tai chimpanzees. *International Journal of Primatology*, **18**, 183–206.

Ghiglieri, M.P. (1984). Feeding ecology and sociality of chimpanzees in Kibale Forest, Uganda. In *Adaptations for Foraging in Nonhuman Primates*, ed. P.S. Rodman & J.G.H. Cant, pp. 161–94. New York: Columbia University Press.

Goodall, J. (1968). The behaviour of free-ranging chimpanzees in the Gombe Stream Reserve. *Animal Behavior Monographs*, **1**, 161–311.

(1986). *The Chimpanzees of Gombe: Patterns of Behavior*. Cambridge, MA: Harvard University Press.

Hladik, C.M. (1977). Chimpanzees of Gabon and chimpanzees of Gombe: some comparative data on the diet. In *Primate Ecology: Studies of Feeding and Ranging Behaviour in Lemurs, Monkeys and Apes*, ed. T.H. Clutton-Brock, pp. 481–501. London: Academic Press.

Humle, T. & Matsuzawa, T. (2004). Oil palm use by adjacent communities of chimpanzees at Bossou and Nimba Mountains, West Africa. *International Journal of Primatology*, **25**, 551–81.

Hunt, K.D. & McGrew, W.C. (2002). Chimpanzees in the dry habitats of Assirik, Senegal and Semliki Wildlife Reserve, Uganda. In *Behavioural Diversity in Chimpanzees and Bonobos*, ed. C. Boesch, G. Hohmann, & L.F. Marchant, pp. 35–51. Cambridge: Cambridge University Press.

Isabirye-Basuta, G. (1989). Feeding ecology of chimpanzees in the Kibale Forest, Uganda. In *Understanding Chimpanzees*, ed. P.G. Heltne & L.A. Marquardt, pp. 116–27. Cambridge, MA: Harvard University Press.

Isbell, L.A. (1991). Contest and scramble competition: patterns of female aggression and ranging behavior among primates. *Behavioral Ecology*, **2**, 143–55.

Isbell, L.A. & Young, T.P. (1996). The evolution of bipedalism in hominids and reduced group size in chimpanzees: alternative responses to decreasing resource availability. *Journal of Human Evolution*, **30**, 389–97.

Kano, T. (1972). Distribution and adaptation of the chimpanzee on the eastern shore of Lake Tanganyika. *Kyoto University African Studies*, **7**, 37–129.

Kuroda, S., Nishihara, T., Suzuki, S., & Oko, R.A. (1996). Sympatric chimpanzees and gorillas in the Ndoki Forest, Congo. In *Great Ape Societies*, ed. W.C. McGrew, L.F. Marchant, & T. Nishida, pp. 71–81. Cambridge: Cambridge University Press.

Lehmann, J. & Boesch, C. (2004). To fission or to fusion: effects of community size on wild chimpanzee (*Pan troglodytes verus*) social organization. *Behavioral Ecology and Sociobiology*, **56**, 207–16.

Matsumoto-Oda, A. & Hayashi, Y. (1997). Nutritional aspects of fruit choice by chimpanzees. *Folia Primatologica*, **70**, 154–62.

Matsumoto-Oda, A. & Kasagula, M.B. (2000). Preliminary study of feeding competition between baboons and chimpanzees in the Mahale Mountains National Park, Tanzania. *African Study Monographs*, **21**, 147–57.

Matsumoto-Oda, A., Hosaka, K., Huffman, M.A., & Kawanaka, K. (1998). Factors affecting party size in chimpanzees of the Mahale Mountains. *International Journal of Primatology*, **19**, 999–1011.

McBeath, N.M. & McGrew, W.C. (1982). Tools used by wild chimpanzees to obtain termites at Mt. Assirik, Senegal: the influence of habitat. *Journal of Human Evolution*, **11**, 65–72.

McGrew, W.C. (1983). Animal foods in the diets of wild chimpanzees (*Pan troglodytes*): why cross-cultural variation? *Journal of Ethology*, **1**, 46–61.

McGrew, W.C., Baldwin, P.J., & Tutin, C.E.G. (1981). Chimpanzees in a hot, dry and open habitat: Mt. Assirik, Senegal, West Africa. *Journal of Human Evolution*, **10**, 227–44.

(1988). Diet of wild chimpanzees (*Pan troglodytes verus*) at Mt. Assirik, Senegal. I. Composition. *American Journal of Primatology*, **16**, 213–26.

McGrew, W.C., Tutin, C.E.G., & Baldwin, P.J. (1979). Chimpanzees, tools, and termites: cross-cultural cmparisons of Senegal, Tanzania, and Rio Muni. *Man*, **14**, 185–214.

Mitani, J.C., Watts, D.P., & Lwanga, J.S. (2002). Ecological and social correlates of chimpanzee party size and composition. In *Behavioural Diversity in Chimpanzees and Bonobos*, ed. C. Boesch, G. Hohmann, & L.F. Marchant, pp. 102–11. Cambridge: Cambridge University Press.

Moore, J. (1996). Savanna chimpanzees, referential models and the last common ancestor. In *Great Ape Societies*, ed. W.C. McGrew, L.F. Marchant, & T. Nishida, pp. 275–92. Cambridge: Cambridge University Press.

Newton-Fisher, N.E. (1999). The diet of chimpanzees in the Budongo Forest Reserve, Uganda. *African Journal of Ecology*, **37**, 344–54.

Newton-Fisher, N.E., Reynolds, V., & Plumptre, A.J. (2000). Food supply and chimpanzee (*Pan troglodytes schweinfurthii*) party size in the Budongo Forest Reserve, Uganda. *International Journal of Primatology*, **21**, 613–28.

Nishida, T. (1976). The bark-eating habits in primates, with special reference to their status in the diet of wild chimpanzees. *Folia Primatologica*, **25**, 277–87.

(1990). *The Chimpanzees of the Mahale Mountains*. Tokyo: University of Tokyo Press.

Nishida, T. & Uehara, S. (1983). Natural diet of chimpanzees (*Pan troglodytes schweinfurthii*): long-term record from the Mahale Mountains, Tanzania. *African Study Monographs*, **3**, 109–30.

Pruetz, J.D. & Knutsen, P. (2003). Scrambling for a common resource: chimpanzees, humans, and *Saba senegalensis*, in southeastern Senegal. *American Journal of Physical Anthropology, Supplement*, **120**, 172.

Pruetz, J.D. & McGrew, W.C. (2001). What does a chimpanzee need? Using natural behavior to guide the care of captive populations. In *Special Topics in Primatology, Volume 2: The Care and Management of Captive Chimpanzees*, ed. L. Brent, pp. 17–37. San Antonio, TX: American Society of Primatologists.

Pruetz, J.D., Marchant, L.F., Arno, J., & McGrew, W.C. (2002). Survey of savanna chimpanzees (*Pan troglodytes verus*) in southeastern Senegal. *American Journal of Primatology*, **58**, 35–43.

Rodman, P.S. (1984). Foraging and social systems of orangutans and chimpanzees. In *Adaptations for Foraging in Nonhuman Primates*, ed. P.S. Rodman & J.G.H. Cant, pp. 134–60. New York, NY: Columbia University Press.

Sakura, O. (1994). Factors affecting party size and composition of chimpanzees (*Pan troglodytes verus*) at Bossou, Guinea. *International Journal of Primatology*, **15**, 167–83.

Stanford, C.B. & Nkurunungi, J.B. (2003). Behavioral ecology of sympatric chimpanzees and gorillas in Bwindi Impenetrable National Park, Uganda: diet. *International Journal of Primatology*, **24**, 901–18.

Sterck, E.H.M., Watts, D.P., & van Schaik, C.P. (1997). The evolution of female social relationships in nonhuman primates. *Behavioral Ecology and Sociobiology*, **41**, 291–309.

Sugiyama, Y. & Koman, J. (1987). A preliminary list of chimpanzees' alimentation at Bossou, Guinea. *Primates*, **28**, 133–47.

(1992). The flora of Bossou: its utilization by chimpanzees and humans. *African Study Monographs*, **13**, 127–69.

Suzuki, A. (1966). On the insect-eating habits among wild chimpanzees living in the savanna woodland of western Tanzania. *Primates*, **7**, 481–7.

Takahata, Y., Hasegawa, T., & Nishida, T. (1984). Chimpanzee predation in the Mahale Mountains from August 1979 to May 1982. *International Journal of Primatology*, **5**, 213–33.

Teaford, M.F. & Ungar, P.S. (2000). Diet and the evolution of the earliest human ancestors. *Proceedings of the National Academy of Sciences*, **97**, 13506–11.

Tutin, C.E.G., Ham, R.M., White, L.J.T., & Harrison, M.J.S. (1997). The primate community of the Lope Reserve, Gabon: diets, responses to fruit scarcity, and effects on biomass. *American Journal of Primatology*, **42**, 1–24.

Tweheyo, M. & Obua, J. (2001). Feeding habits of chimpanzees (*Pan troglodytes*), red-tail monkeys (*Cercopithecus ascanius schmidti*) and blue monkeys (*Cercopithecus mitis stuhlmanii*) on figs in Budongo Forest Reserve, Uganda. *African Journal of Ecology*, **39**, 133–9.

van Schaik, C.P. (1989). The ecology of social relationships amongst female primates. In *Comparative Socioecology: The Behavioural Ecology of Humans and Other Mammals*, ed. V. Standen & R. Foley, pp. 195–218. Oxford: Blackwell Scientific Publishers.

Wrangham, R.W. (1977). Feeding behaviour of chimpanzees in Gombe National Park, Tanzania. In *Primate Ecology: Studies of Feeding and Ranging Behaviour in Lemurs, Monkeys and Apes*, ed. T.H. Clutton-Brock, pp. 503–38. London: Academic Press.

(1979). Sex differences in chimpanzee dispersion. In *The Great Apes*, ed. D.A. Hamburg & E.R. McCown, pp. 481–9. Menlo Park: Benjamin Cummings.

(1980). An ecological model of female-bonded primate groups. *Behaviour*, **75**, 262–99.

(2000). Why are male chimpanzees more gregarious than mothers? A scramble competition hypothesis. In *Primate Males: Causes and Consequences of Variation in Group Composition*, ed. P. Kappeler, pp. 248–58. Cambridge: Cambridge University Press.

Wrangham, R.W., Chapman, C.A., Clark-Arcadi, A.P., & Isabirye-Basuta, G. (1996). Social ecology of Kanyawara chimpanzees: implications for understanding the costs of great ape groups. In *Great Ape Societies*, ed. W.C. McGrew, L.F. Marchant, & T. Nishida, pp. 45–57. Cambridge: Cambridge University Press.

Wrangham, R.W., Conklin, N.L., Chapman, C.A., & Hunt, K.D. (1991). The significance of fibrous foods for Kibale Forest chimpanzees. *Philosophical Transactions of the Royal Society of London*, **334**, 171–8.

Wrangham, R.W., Conklin, N.L., Etot, G. *et al.* (1993). The value of figs to chimpanzees. *International Journal of Primatology*, **14**, 243–56.

Wrangham, R.W., Conklin-Brittain, N.L., & Hunt, K.D. (1998). Dietary response of chimpanzees and cercopithecines to seasonal variation in fruit abundance. I. Antifeedants. *International Journal of Primatology*, **19**, 949–70.

7   Food choice in Taï chimpanzees: are cultural differences present?

CHRISTOPHE BOESCH, ZORO BERTIN GONÉ BI, DEAN
ANDERSON, AND DANIEL STAHL

Introduction

Historically, culture has been defined as a uniquely human attribute. However, in recent years, new observations and different perspectives have led
some researchers to extend the concept to some animal species, e.g., chimpanzees, orangutans, capuchin monkeys, dolphins, and whales (Rendell &
Whitehead, 2001; Whiten *et al.*, 2001; Panger *et al.*, 2002; Boesch, 2003; van
Schaik *et al.*, 2003). Many different definitions of culture can be found in the
literature, but the consensus is based on the notion that socially learned

Feeding Ecology in Apes and Other Primates. Ecological, Physical and Behavioral Aspects, ed.
G. Hohmann, M.M. Robbins, and C. Boesch. Published by Cambridge University Press.
© Cambridge University Press 2006.

behavioral traits are population specific, and therefore cannot be explained in terms of ecological or genetic differences (Boesch, 2003). The increasing evidence of population-specific behavioral differences that could not be explained by either ecological or genetic differences in some animal species lay at the heart of such a proposition. Chimpanzees have been shown to possess the largest number of different cultural behaviors, while macaques, dolphins, and whales show the least (Perry & Manson, 2003; Boesch, in press). In discussions of non-human culture, explanations for differences in cultural behaviors are required to be explained by factors other than the ecological or genetic. This rigorous criterion of determining what culture is has never been used when discussing human cultures, but was considered necessary to apply to other animal species since we know that many differences between animal populations are a direct response to ecological differences observed between populations (Galef, 1990; Tomasello, 1990). When applying this criterion to chimpanzees, over 50 behavior patterns have been proposed as having a cultural origin, and were observed in social, communication, and foraging contexts (Whiten *et al.*, 2001; Boesch, 2003, in press).

Food choice in a given species is directly influenced by the presence in the habitat of different species of fruits and leaves as well as by their availability in the habitat. Experimental studies on food choice in deer and primates revealed that food selection is strongly affected by the content in energy, proteins, and tannins (Janson *et al.*, 1986; Berteaux *et al.*, 1998; Alm *et al.*, 2002; Nakagawa, 2003; Wasserman & Chapman, 2003). Since availability as well as nutrient components may influence choice, comparison of food selection between different food species is difficult to interpret. Therefore, cultural differences have not often been sought in the context of feeding. Despite the obvious importance of cultural difference in eating habits among humans, only a limited number of attempts have been made to identify cultural differences in food choice in non-human primates. A first comparison of local differences in feeding habits between Gombe and Mahale chimpanzees in Tanzania revealed that most differences were explained by differences in availability of the plant species concerned (Nishida *et al.*, 1983).

Similarly, when exploring possible cultural differences in food choice, it should not be forgotten that food selection by an individual will also be affected by the availability of the food in the habitat in terms of ripening and distribution of the different food types. In turn these factors will be strongly influenced by rainfall and temperature. Thus, in examining cultural differences, comparing food choice between widely different forest areas is not practical, and studies that have detailed observations on feeding behavior from different groups within the same habitat remain limited. As an example, two neighboring groups of Bwindi gorillas with overlapping home ranges had some dietary differences that could not be explained by availability (Ganas *et al.*, 2004). In capuchin monkeys, a more systematic study comparing two groups within the same population and groups belonging to different populations proposed 17 different processing techniques of similar food to be cultural (Panger *et al.*, 2002). Detecting cultural differences in food choice thus requires detailed data on food availability of all the different food types in the habitat consumed by the individuals within a group.

At first glance, it might seem paradoxical to propose exploring differences in food choice between neighboring groups within the same habitat, as it is often assumed that food availability and production hardly vary over small distances, and therefore, neighboring groups would experience similar food availability. If it is true that fewer differences should be expected, then at the same time, the ones we do see are more likely to occur because of factors other than ecological ones. For example, in a study of three neighboring groups of capuchin monkeys in Costa Rica the time spent eating plant items was not predicted by the availability of these plants in the habitat (Chapman & Fedigan, 1990). Thus, comparing neighboring groups could make efforts to find cultural differences easier. In this chapter, we attempt such an approach with wild chimpanzees.

Chimpanzees are omnivores who primarily eat fruits but who also consume leaves, flowers, insects, and meat (Goodall, 1986; Nishida, 1990; Anderson *et al.*, in press). Their diet is quite diverse, containing, in most studied populations, more than 100 different plant species, and they have been seen to eat a number of food types on most days. Taï chimpanzees eat 223 species of plants from 51 families, and that includes 301 types of food; and they have been seen to eat an average 36 different species each month of the year (Goné Bi, 2004). Tropical forests in Africa are characterized by one or two rainy seasons that influence fruit and leaf production.

We collected data on the Taï chimpanzees, both on feeding behavior and on fructification patterns in the forest, thereby allowing us to answer questions about food choice. We specifically posed the following questions: do Taï chimpanzees select their food according to abundance only? What affects their food choice other than abundance? Were there differences we observed between neighboring groups that cannot be explained by ecological differences?

Methods

Subject and study communities

The Taï National Park, Côte d'Ivoire, includes about 4540 km^2 of evergreen lowland rainforest (for detailed description of the study site see Boesch &

Boesch-Achermann, 2000). Taï chimpanzees have been studied since 1979 (Boesch & Boesch-Achermann, 2000), and since 1995 three distinct and neighboring communities have been studied on a regular basis (Herbinger *et al.*, 2001). The animals are well-habituated to human observers, which allowed detailed observations to be made of feeding behavior. The data presented in this study are part of a long-term study and were collected from September 1999 to December 2000. They include daily observations of the three chimpanzee communities studied in the project: the North, Middle, and South groups. The North Group has been followed since 1979 by human observers and has been habituated to them since 1984. Since 1989, field assistants have recorded daily observations of their feeding behavior on standardized check-sheets. The South group was followed daily starting in 1995 and the Middle group in 1997.

Data collection

The first set of data we collected were the feeding data of the chimpanzees in the three groups. The project's field assistants recorded feeding data on standardized check-sheets, following target adult individuals of the three groups from nest to nest (mean daily observation time $= 9.7$ hours ± 2.1). In this data set, all feeding events of the target individuals were recorded and the time and the species eaten precisely specified. From September 1999 through December 2000, 110 food species were seen to be eaten by the chimpanzees. Scans were performed every 15 minutes and the activity of the individuals within the field assistants' sight was recorded. When the activity consisted of feeding, the name of the species eaten was noted. Henri Téré and Laurent Aké Assi, botanists from the Swiss Centre and the University of Abidjan, respectively, identified the different plant species eaten by the chimpanzees. As meat and insects made up only a small proportion of the monthly feeding time, we did not include them in the present analysis.

The second set of data collected concerned the food availability within the territory of each of the three different chimpanzee groups. DA conducted the first phenological transects system in the North group (Anderson *et al.*, 2002), while ZGB expanded it to the South and Middle group, organizing the data collection with the help of two field assistants specially trained in this data collection procedure. For each territory, about 740 trees were marked, which included an average of 64 tree species. This sampling is representative of the chimpanzee diet as it includes all plant food species that were observed to be eaten in any 1 month of the 16 months sampled for at least 3% of the feeding time, and 8 of the 11 species seen eaten 1–2% of the feeding time in any given month (n = 35 tree species). Each tree had a number and its position was measured along the transect. The diet of the chimpanzees in the Taï forest contains 97.5% vegetable material, 85% of them being fruits (Goné Bi, 2004). Monthly feeding on each species varies widely and many species are rarely eaten more than 1% of the feeding time.

All of the transects for each territory were conducted during 8 days in the middle of a month by ZGB and two field assistants, Camille Dji and Jonas Tahou. All marked trees in the three territories were visited monthly and the presence of ripe fruits was recorded. Based on four categories, we also recorded the quantity of fruit in the trees, but did not use these data in the present analysis.

Data analysis

To answer the questions posed earlier, we analyzed the data in different ways. First, we extracted a "food abundance index" (A_m) for each species, which quantifies the proportion of trees that carry ripe fruits (Anderson *et al.*, 2002). The calculation followed the formula

$$A_m = \sum_{k=1}^{n} D_{kq} B_{kq} P_{km}$$

where D_{kq} stands for the density of adult trees of species k in quadrate q, where B_{kq} stands for the basal area of species k in quadrate q, and where P_{km} stands for the percent of monitored trees of species k with ripe fruit. For our analysis we only used ripe fruits because Taï chimpanzees almost always wait for fruits to ripen before eating them, contrary to some other primate species.

Second, we calculated a "food preference index" to measure the tendency of the chimpanzee to select a food species specifically rather than at random (Johnson, 1980). Random preference here is defined as eating food in accordance with its abundance in the habitat. Such measures of food preferences have been subject to much discussion since mathematically it is very difficult for an index to represent preference totally independent of availability (Chesson, 1978, 1983; Johnson, 1980; Lechowitz, 1982; Manly *et al.*, 2002). A rare plant species will have a high preference index whenever it is eaten, whereas a more abundant plant must be eaten much more often to produce a similar preference index value. Thus, preference index will be biased in favor of rare species and due to chance might even classify some rare species as highly preferred. To address the question of food preference, we used two indexes. The first is a classic measure of food preference. For each month we ranked the feeding time and the fruit abundance of all species for the months when food was available. For each species we calculated the difference between its feeding time rank and its abundance rank. The Food Preference Index (F_{sp}) for a given species was then calculated as

$$F_{sp} = \frac{\sum(Rank_{feeding\ time} - Rank_{abundance})}{N_{months,\ when\ food\ was\ available}}$$

which is a mean rank score of all months with observations. The food preference index was calculated for each group separately as well as for all three groups combined.

Although this index is also known to bias preference in favor of rare food species (Lechowitz, 1982), that bias is less than the one that would result from using other classical preference indexes, e.g., the Jacob's Index or Chesson's standardized selection ratio.

Food preference indexes are not appropriate for judging absolute level of feeding (Lechowitz, 1982), but at the same time direct observations in the forest suggest that individuals may disproportionately increase feeding time on one species the more it is available. To specifically include this critical aspect, we used a second preference measure called "dynamic food preference." The dynamic food preference takes into account the proportion of trees of the food species that bore ripe fruits. This dynamic food preference describes the slope of the feeding time related to abundance of the food after the formula

$$y = \exp(c + \beta x)$$

where y stands for relative feeding time and x for relative abundance measures. In this formula, c describes the time chimpanzees feed on the food when the availability is (close to) zero, whereas "β" describes the increment in feeding time as abundance increases. The relationship between abundance and feeding time is, therefore, modeled exponentially, not linearly. It is relevant to note that c in this formula is rather similar to the food preference index (F_{sp}) since both measure feeding time compared to a general measure of availability. Indeed, in our data c correlates strongly with our food preference index ($r_s = 0.624$, n = 32, p < 0.0001). Beta greater than 0 indicates that the food is preferred the more trees carry ripe fruits, while a negative β would indicate that the food is neglected as its fruiting increases. The dynamic food preference, measured by β, quantifies how chimpanzee's preference varies as more trees of the species produce ripe fruits (see Figure 7.1). We shall use both measures of preference to analyze chimpanzee food choices in the three groups.

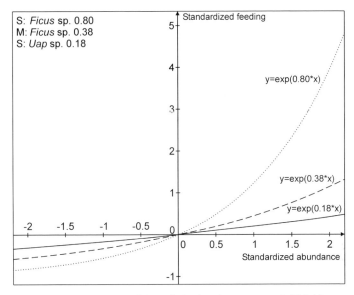

Figure 7.1. Three examples of dynamic food preference curves in Taï chimpanzees. Dotted line = *Ficus* fruits in the South group; dashed line = *Ficus* fruits in the Middle group; solid black line = *Uapaca* fruits in the South group.

Statistical analysis

To determine the relationship between feeding time and several continuous and categorical independent variables we conducted a General Linear Mixed Model (GLMM) analysis (Pinheiro & Bates, 2000). As the dependent variable, we chose relative feeding time (relative time a species was eaten in relation to the daily activity budget). Data were only included in the analysis if the species had fruits during the respective month. The distribution of the data was positively skewed (most relative feeding times were small), and therefore we transformed the data by taking the natural logarithm. In addition, since there were several zero feeding observations, we transformed all data by $Ln(x+1)$ (Zar, 1999). As independent variables we used the density and mean diameter at breast height (DBH) of each tree species, monthly rainfall, chimpanzee group, and relative abundance (A_m of a species in a month divided by the total A_m of all species during the respective month). Plant species was included as a random factor. We used Akaike Information Criteria (AIC) to select the best fitting model. AIC is an objective measure to find the predictor variables, which explains most of the variation of the dependent variable in relation to the number of variables included in the model (Quinn & Keough, 2002). To further analyze the interaction between relative abundance and species, we conducted a linear regression between the Ln-transformed relative feeding and the relative abundance for each species. We checked the assumptions of parametric tests by visual inspections of the residuals (Quinn & Keough, 2002). In addition, to compare the effect of abundance of a species between the groups, we conducted simple linear regression for each group separately. We only conducted regressions for each group if a species was abundant for at least 6 months within each of the group's territory.

Dynamic food preference index β

By using the natural logarithm of the relative feeding times, we model an exponential rather than a linear relationship between relative feeding time and relative abundance:

$$Ln(x + 1) = c + \beta x$$

which is equivalent to:

$$x = e^{(c+\beta x)} - 1$$

We therefore derive the regression coefficient β for each species, which represents the above described "dynamic food preference" index. In order to compare β of different species, we use the standardized β (Quinn & Keough, 2002). The standardized β enables comparisons to be made of models with variables of differing magnitudes and dispersions. In our case, it allows the comparison β of different species with different initial abundance of fruiting trees. Beta is calculated by using standardized (z-transformed) data for calculating the usual regression slope.

Results

Analysis of feeding time

The results of the analysis of different factors that might influence relative feeding time, namely tree species, density and dbh, rainfall, and relative abundance, revealed that the model that best described feeding time was the one that only included relative abundance and tree species. There was a significant interaction between relative abundance and fruit species, which means that the relationship between relative abundance and relative feeding time differed between the species. (GLMM: relative abundance: $F_{1,331} = 0.022$, p = 0.88, random factor species: Wald Z = 2.58, p = 0.01, abundance*species: $F_{31,449}$ = 3.071, p < 0.001). To further analyze the interaction between relative abundance and plant species, we conducted simple linear regressions between relative abundance and relative feeding time for each species separately. Of the 32 species, there were 12 that showed a significant positive relationship between relative abundance and (Ln-transformed) relative feeding time, while in the other 20 species no significant relationships were detected. The standardized regression coefficient beta for these regressions represents our dynamic preference index, which will be discussed in the following section. There was no significant effect of group on relative feeding time if group was included as an additional factor in the model. However, the small sample size of observations per species within each group did not allow testing for possible interactions of abundance of fruits and fruit species.

Food choice as observed in all three groups

Of the 32 species of fruits eaten most frequently by the chimpanzees in Taï, 16 had a rank food preference index above 0, indicating that they were eating these foods more frequently than they were present in the forest (see Figure 7.2). From these species we classified some as "preferred" whenever they were part of the 20% top-ranked species ($F_{sp} > 2.68$). Following this criterion, six species were considered preferred: *Dacryodes aubrevillei, Ficus* spp., *Dialium guinensis, Parinari excelsia, Treculia africana,* and *Chrysophyllum taiensis.* Similarly, some species can be classified as "neglected" when they are part of the 20% bottom ranked species ($F_{sp} < -3.23$). The neglected species are: *Pycnanthus angolensis, Pentadesma butyrea, Tarietta utilis, Nauclea* spp. and *Diospyros manii.*

For all three chimpanzee groups, preferred food species are not the most abundant trees in the forest, nor the largest. They actually are the tree species with the smallest proportion of trees producing fruits. And so, as those producing the least fruit ($r_s = -0.531$, n = 32, p = 0.002), they therefore correspond with the preference index bias, which we discussed earlier in the Methods section. On the other hand, preferred food species are eaten more frequently when few trees of the species are producing fruits ($r_s = 0.641$, n = 32, p < 0.001). In general, preferred foods are eaten over a longer period than are the other food species ($r_s = 0.571$, n = 32, p < 0.001).

As revealed by the linear regression analysis of the 32 species (see '*Analysis of feeding time*' in the Results section), only 12 showed a significant positive dynamic food preference, where feeding time increases exponentially

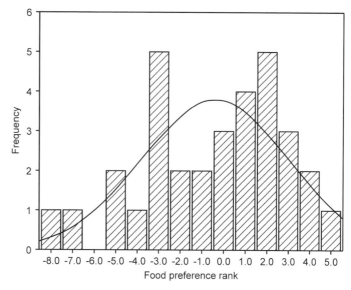

Figure 7.2. Distribution of food preference indexes for 32 fruit species eaten by Taï chimpanzees.

with increasing fruit production (Standardized β, p < 0.05). Surprisingly only three of those were classified as preferred species under F_{sp}, *D. aubrevillei*, *Ficus* spp. and *P. excelsia*, while two were classified as neglected species, *P. angolensis* and *Nauclea* spp. Preferred species are not those eaten more when the proportion of trees bearing ripe fruits increases (Correlation of F_{sp} and β: $r_s = 0.164$, n = 32, p = 0.36).

Group-specific aspect of food choice

In order to address the question of potential cultural differences, we specifically looked at differences we could find in food choice between the three groups. We found three differences between the three groups that might reflect cultural differences in the way chimpanzees select their food.

First, chimpanzees in the North and Middle groups increase their feeding time relatively more for species that have more fruit producing trees (β correlates with abundance; North: r = 0.385, n = 30, p < 0.05, Middle: r = 0.374, n = 28, p = 0.05), whereas the effect is totally absent in the South (r = 0.084, n = 29, p = 0.67) (see Figure 7.3). In other words, the North and Middle chimpanzees eat over a longer period food species that have higher fruit production.

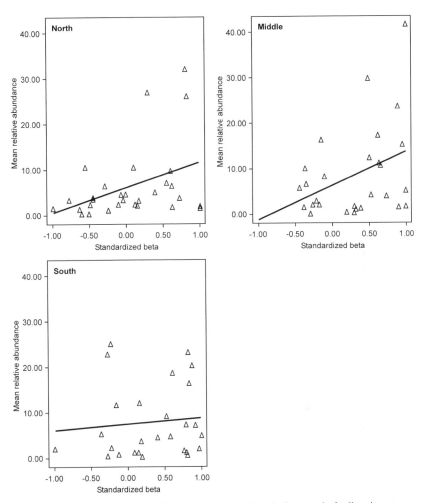

Figure 7.3. Group specific reaction, measured by the increase in feeding time, to an increase in mean relative abundance in ripe fruit trees for the three chimpanzee groups.

Second, if chimpanzees eat more of some species of fruits the more trees carry them, this effect is significantly more observed for less synchronous species in the chimpanzees of the North and the South group (β correlates with c: North: $r_s = -0.513$, n = 30, p < 0.005; South: $r_s = -0.587$, n = 27, p < 0.001), while this effect is totally absent in the Middle group ($r_s = -0.058$, n = 28, p = 0.77) (see Figure 7.4). In other words, the fewer the trees that bear fruits at the same time, the more chimpanzees on the North and South will eat them when production increases. It is only the North and

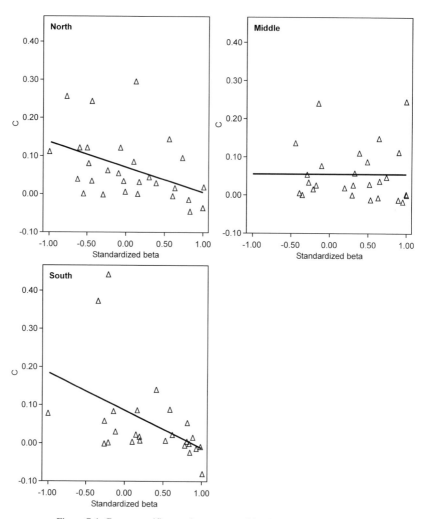

Figure 7.4. Group specific reaction, measured by the increase in feeding time, to an increase of synchrony in ripe fruit production.

South chimpanzees who thereby reveal a genuine preference for rare species.

Third and finally, we compared the beta values for the few species that were observed to be eaten at least 6 months of the year so that we could test statistically for differences between the three communities in response to an increase in the number of trees bearing fruits. As can be seen in Table 7.1, eight tree species fulfilled the conditions. Despite the small sample size, and

Table 7.1. *Different dynamic preference values (standardized β-value) in the three chimpanzee groups for eight frequently consumed plant species that were observed to be eaten at least for 6 months of the year to permit statistical analysis*

Food species	North group	Middle group	South group	p values
Ficus sp.	− 0.24	0.38	0.80	**
Irvingia gabonensis	0.63	0.20	−0.13	**
Sterculia rhinopetala	−0.60	−0.27	0.14	ns
Uapaca sp.	−0.03	−0.18	0.18	ns
Klainedoxa gabonensis	0.62	0.20	0.87	ns
Parinari excelsia	0.55	0.74	0.40	ns
Sacoglottis gabonensis	0.81	0.64	0.72	ns
Pycnanthus angolensis	0.82	0.63	0.92	ns

(General Linear Model analysis with Ln (relative feeding time) as dependent variable and relative abundance and species as independent variables. ** = $p < 0.01$: significant interaction between group and relative abundance).

hence low statistical power, two out of eight tree species show significant differences in dynamic preferences between the three groups: For *Ficus* spp., the North group decreases while the South group strongly increases its feeding time as more trees bear fruits. For *Irvingia gabonensis*, it is exactly the opposite: as the South group decreases the North group increases its consumption as more trees bear fruits. This is especially notable since food abundance indexes are very similar for all three groups for those two species. For *Sterculia oblongata, Klainedoxa gabonensis*, and to a lesser degree for *Uapaca* spp., we see a similar difference between groups; it, however, does not reach statistical significance. Thus, it appears that for some species the three groups react in very different ways when productivity increases.

Group-specific differences due to alternative food availability

When searching for possible cultural differences, we should keep in mind that some group-specific differences could be explained by different interactions between food species. For example, *Irvingia* could be eaten differently (Table 7.1) if another food species comes into abundance in the South group but not in the North group. Such interactions may be an important factor in explaining food choice since chimpanzees in the Taï forest eat a lot of different fruits and seasonality is quite strong. On average, chimpanzees were seen to feed every month on five food species for more than 5% of the feeding time, and on eight species for more than 2% of the feeding time. So interactions between different food species could occur. At the same time, the large number of possible alternative food sources that chimpanzees find in the forest makes it more difficult to prove such an interaction effect.

In an attempt to find an interaction effect, we analyzed, along with other fruit species, eight main food species that are eaten for at least 6 months within each group. For each of the 8 main fruit species, we conducted a multiple regression to determine the effect of the relative feeding time of each of the main food species on the other 7 main fruit species, as well as on the 14 non-main food species that were eaten during at least 2 months for 5% of the feeding time. The relative feeding times of the other fruit species were therefore used as independent variables. We also included the relative abundance of the food species, which was studied as an independent variable. The small sample size did not allow the analysis of complex regression models and only forward model selection was used to find the best model. With all three groups combined, we found three negative relationships ($p < 0.05$) between three main species and other food species. In two cases the effect of abundance of the main species remained significant and therefore, the outcome of the correlation remains ambiguous. In the third case, the correlation was due to one data point (outlier). When considering the three groups separately, we found six negative relationships ($p < 0.05$) in one of the three groups, but here again either a variation in abundance ($n = 3$) or one data point explained the relationship ($n = 3$).

In a second analysis, we looked at the specific cases where we saw a difference between abundance and feeding time for a main food species and which appeared that it could be explained by the presence of alternative food species. The problem was again that in most situations there were simply too many other alternative food species to be able to draw a clear-cut conclusion. For example, with *Ficus* spp., feeding behavior between the North and South chimpanzees differs a lot (Table 7.1). In April, the abundance of figs is high in the North and decreases later, while in the South it is high in May and decreases thereafter. North chimpanzees barely eat figs in April, eating them mainly in May, when availability is low, while the South chimpanzees eat them in accordance with availability. At the same time, both groups eat *Klainedoxa, Parinari, Sacoglottis*, and *Chrysophyllum*. *Klainedoxa* fruits are generally eaten more in the South, where they are more abundant than in the North. Both groups increasingly consume *Sacoglottis* and *Chrysophyllum* fruits beginning in May, and progressively more so in June, and in April both groups eat the flesh of the *Parinari* and in June crack the nuts of the same species. So interactions with other available fruits do not seem to easily explain why figs are eaten so little in May in the North when they are abundant.

Discussion

In the three neighboring groups of Taï chimpanzees, food choice shows some strong similarities in preference for some food species and neglect of others. These food preferences apply to 16 out of 32 species, the remaining ones being eaten in proportion to their availability. Using a dynamic food preference measurement, another 12 species are selected more frequently the more these trees produce fruits. However, the response of the chimpanzees to an increase in the number of trees bearing fruits is group specific and independent of availability, as the standardized β controls for different levels of availability. Some groups show a strong preference for rare or more synchronized tree species in a way not seen in other groups. Such group-specific differences were also observed when chimpanzees ate the fruits of the same species, e.g., *P. angolensis*.

Some strong similarities in food choice between the three groups of Taï chimpanzees could have been predicted because food availability is expected to be quite similar within the same stretch of forest. The three groups discussed here are neighboring ones, with the northern limit of the home range of the North group only 13 km apart from the southern limit of the South group, a distance too small to expect many climatological or pedological differences (Fritsch, 1980; de Rouw, 1991). When using one of the classic food preference indexes (F_{st}), we did not find any group-specific differences between the three groups considered, even for fruit species not eaten in accordance with their availability. Since we found some drastic differences when using our dynamic measure of food preference, it suggests that those classic measures are too static to detect some dynamic responses of the animals when feeding on foods that have varying degrees of production.

Intriguingly, 50% of 32 fruit species were not eaten by Taï chimpanzees in accordance with their availability in the forest, illustrating the limited predictive value of this ecological parameter to understand food choice in Taï chimpanzees. Some studies convincingly confirmed the importance of different nutritional parameters on food choice. For example, tannins influence food choice in fallow deer (Alm *et al.*, 2002) and in capuchin monkeys (Janson *et al.*, 1986), while protein and fiber content were important to two species of colobus monkeys but energy was not (Wasserman & Chapman, 2003). It is still possible that some species are preferred because some nutrients are favored, and therefore, those species are eaten more often than would be expected if one just considered their availability. We hope to be able to test this possibility in the near future.

Beside these important similarities in food choice, we have also observed some clear differences in food choice as groups presented specific responses to variation in productivity of different tree species (Figures 7.3 and 7.4). These differences are not explained by availability or by differences in nutritional content, as they emerged only after we included the varying degree of productivity in the analysis. This may be due to interactions with other fruits that are available in different proportions in each group's territory. However, our attempt to find such an interaction effect did not reveal any clear effect. So at this point we cannot conclude that large interaction effects seem to explain such differences.

Differences in food selection that cannot be explained by differences in the availability of the food concerned are normally proposed to be cultural differences. French people are known to eat snails and frog legs, while Germans neglect them, eating with delight raw pork grease that the French would not touch. While pigs, frogs, and snails are common in both countries, the different habits in terms of eating them are considered cultural. This is just one example of the myriad of cultural differences among humans. In capuchin monkeys of two or three different populations different processing techniques have been observed being used to eat 20 different food types (Panger *et al.*, 2002). For example, some Acacia fruits are eaten directly by the capuchins in Lomas Barbudal, while those of Santa Rosa first rub them before eating them. Similarly, caterpillars were wrapped in leaves in Lomas Barbudal but rubbed in Palo Verde. We have also observed these kinds of differences between the North and South groups of Taï chimpanzees (Boesch, 2003): *Thoracotermes* termites are eaten in the North but not in the South; day nests are made for resting in the North and as a play component in the South; South chimpanzees eat both the pith and adult stem of *Haloplegia* plants, while the North chimpanzees only eat the pith.

The data presented here are not about techniques or about consumption of plants, rather they detail differences in the frequency of consumption of food. Preference for the same plants varies between the three chimpanzee groups within this Taï population, and the reaction to changes in the level of fruiting varies as well. The difference we observed between the three groups – in the feeding time of the same food species and differences in fruiting patterns – is intriguing. It is possible that these differences are cultural and maintained within each group through a social learning process, since no difference in availability or fruiting differences can explain them. Further studies are needed, and if they confirm these group-specific reactions, it would show that it is possible to find cultural differences between neighboring groups within the same population.

Acknowledgments

We thank the Ivorian authorities for supporting this study since its inception in 1979, especially the Ministry of the Environment and Forests as well as the Ministry of Research, the Direction of the Taï National Park, and the Swiss Research Centre in Abidjan. We thank Grégoire Kouhon Nohan, Honora Néné, Jonas Tahou, Camille Dji, François Yro, Camille Bolé, Niçaise Daurid, Bally Louis Bernard, and Sylvain Guy who helped to collect chimpanzee samples and behavioral data. The Max Planck Society financially supported this work.

References

Alm, U., Birgersson, B., & Leimar, O. (2002). The effect of food quality and relative abundance on food choice in fallow deer. *Animal Behaviour*, **64**, 439–45.

Anderson, D., Nordheim, E., & Boesch, C. (2006). Environmental factors influencing the seasonality of estrus in chimpanzees. *Primates*, **47**, 43–50.

Anderson, D., Nordheim, E., Boesch, C., & Moermond, T. (2002). Factors influencing fission-fusion grouping in chimpanzees in the Taï National Park, Côte d'Ivoire. In *Behavioural Diversity in Chimpanzees and Bonobos*, ed. C. Boesch, G. Hohmann, & L.F. Marchant, pp. 90–101. Cambridge: Cambridge University Press.

Berteaux, D., Crête, M., Huot, J., Maltais, J., & Ouellet, J.-P. (1998). Food choice by white-tailed deer in relation to protein and energy content of the diet: a field experiment. *Oecologia*, **115**, 84–92.

Boesch, C. (2003). Is culture a golden barrier between human and chimpanzee? *Evolutionary Anthropology*, **12**, 26–32.

(in press). Culture in evolution: towards an integration of chimpanzee and human culture. In *Examining Culture Scientifically*, ed. M. Brown. Washington, DC: University of Washington Press.

Boesch, C. & Boesch-Achermann, H. (2000). *The Chimpanzees of the Taï Forest: Behavioural Ecology and Evolution*. Oxford: Oxford University Press.

Chapman, C. & Fedigan, L. (1990). Dietary differences between neighboring *Cebus capucinus* groups: local traditions, food availability or responses to food profitability? *Folia Primatologica*, **54**, 177–86.

Chesson, J. (1978). Measuring preference in selective predation. *Ecology*, **59**, 211–15.

(1983). The estimation and analysis of preference and its relationship to foraging models. *Ecology*, **64**, 1297–304.

de Rouw, A. (1991). *Rice, Weeds and Shifting Cultivation in a Tropical Rain Forest: A Study of Vegetation Dynamics*. Unpublished Ph.D. thesis, Wageningen University.

Fritsch, E. (1980). *Etude Pédologique et Représentation Cartographique à 1/15 000ème d'une Zone de 1.600 ha Représentative de la Région Forestière de Sud-Ouest Ivoirien*. Adiopodoumé: Rapport ORSTOM.

Galef, B. (1990). Tradition in animals: field observations and laboratory analyses. In *Interpretation and Explanation in the Study of Animal Behavior*, ed. M. Bekoff & D. Jamieson, pp. 74–95. Boulder, CO: Westview Press.

Ganas, J., Robbins, M., Nkurunungi, J.B., Kaplin, B., & McNeilage, A. (2004). Dietary variability of mountain gorillas in Bwindi Impenetrable National Park, Uganda. *International Journal of Primatology*, **25**, 1043–72.

Goné Bi, Z. (2004). *Distribution Spatio-temporelle des Plantes dont les Fruits sont Consommés par les Chimpanzés dans le Parc National de Taï, Côte d'Ivoire*. Unpublished Ph.D. thesis, Université d'Abidjan.

Goodall, J. (1986). *The Chimpanzees of Gombe: Patterns of Behavior*. Cambridge, MA: The Belknap Press of Harvard University Press.

Herbinger, I., Boesch, C., & Rothe, H. (2001). Territory characteristics among three neighbouring chimpanzee communities in the Taï National Park, Ivory Coast. *International Journal of Primatology*, **32**, 143–67.

Janson, C., Stiles, E., & White, D. (1986). Selection on plant fruiting traits by brown capuchin monkeys: a multivariate approach. In *Frugivores and Seed Dispersal*, ed. A. Estrada & T.H. Fleming, pp. 83–92. Dordrecht: Dr. W. Junk Publishers.

Johnson, D. (1980). The comparison of usage and availability measurements for evaluating resource preference. *Ecology*, **61**, 65–71.

Lechowitz, M. (1982). The sampling characteristics of electivity indices. *Oecologia*, **52**, 22–30.

Manly, B.F.J., McDonald, L.L., Thomas, D.A., McDonald, T.L., & Erickson, W.E. (2002). *Resource Selection by Animals: Statistical Design and Analysis for Field Studies*, 2nd edn. Dordrecht: Kluwer Academic Publishers.

Nakagawa, N. (2003). Difference in food selection between patas monkeys (*Erythrocebus patas*) and tantalus monkeys (*Cercopithecus aethiops tantalus*) in Kala Maloue National Park, Cameroon, in relation to nutrient content. *Primates*, **44**, 3–11.

Nishida, T. (1990). *The Chimpanzees of the Mahale Mountains: Sexual and Life History Strategies*. Tokyo: University of Tokyo Press.

Nishida, T., Wrangham, R.W., Goodall, J., & Uehara, S. (1983). Local differences in plant-feeding habits of chimpanzees between the Mahale Mountains and Gombe National Park, Tanzania. *Journal of Human Evolution*, **12**, 467–80.

Panger, M., Perry, S., Rose, L. *et al.* (2002). Cross-site differences in foraging behavior of white-faced capuchins (*Cebus capucinus*). *American Journal of Physical Anthropology*, **119**, 52–66.

Perry, S. & Manson, J. (2003). Traditions in monkeys. *Evolutionary Anthropology*, **12**, 71–81.

Pinheiro, J.C. & Bates, D.M. (2000). *Mixed-Effects Models in S and S-PLUS. Statistics and Computing Series*. New York: Springer-Verlag.

Quinn, G.P. & Keough, M.J. (2002). *Experimental Design and Data Analysis for Biologists*. Cambridge: Cambridge University Press.

Rendell, L. & Whitehead, H. (2001). Culture in whales and dolphins. *Behavioral and Brain Sciences*, **24**, 309–24.

Tomasello, M. (1990). Cultural transmission in tool use and communicatory signaling of chimpanzees? In *Comparative Developmental Psychology of Language and Intelligence in Primates*, ed. S. Parker & K. Gibson, pp. 274–311. Cambridge: Cambridge University Press.

van Schaik, C.P., Ancrenaz, M., Brogen, G. *et al.* (2003). Orangutan cultures and the evolution of material culture. *Science*, **299**, 102 5.

Wasserman, M. & Chapman, C. (2003). Determinants of colobine monkey abundance: the importance of food energy, protein and fibre content. *Journal of Animal Ecology*, **72**, 650–9.

Whiten, A., Goodall, J., McGrew, W. *et al.* (2001). Charting cultural variations in chimpanzees. *Behaviour*, **138**, 1489–525.

Zar, J.H. (1999). *Biostatistical Analysis*, 4th edn. Upper Saddle River: Prentice Hall.

8 The effects of food size, rarity, and processing complexity on white-faced capuchins' visual attention to foraging conspecifics

SUSAN PERRY AND JUAN CARLOS ORDOÑEZ JIMÉNEZ

Introduction

Capuchin monkeys (*Cebus*) are noteworthy for their strong convergence with chimpanzees and humans with regard to many characteristics. Socially, they

Feeding Ecology in Apes and Other Primates. Ecological, Physical and Behavioral Aspects, ed. G. Hohmann, M. M. Robbins, and C. Boesch. Published by Cambridge University Press.

have complex alliance formation patterns and lethal coalitionary aggression (Gros-Louis *et al.*, 2003; Perry, 2003; Perry *et al.*, 2004;) as well as the capacity to form traditions (Panger *et al.*, 2002; Perry *et al.*, 2003a, 2003b; Fragaszy *et al.*, 2004). Ecologically, they exhibit omnivory (e.g., Freese, 1977; Terborgh, 1983; Fragaszy, 1986; Robinson, 1986; Chapman & Fedigan, 1990; Fragaszy *et al.*, 2004), predation on vertebrates (Fedigan, 1990; Perry & Rose, 1994; Rose, 1997; Rose *et al.*, 2003), and extractive foraging and tool use (Visalberghi, 1990; Westergaard & Suomi, 1993; Ottoni & Mannu, 2001; Panger *et al.*, 2002). In an omnivorous primate relying on a broad diet that requires extensive manual processing, a young capuchin, before it can exhibit full adult competence, has much to learn about how to identify foods, catch and/or extract prey items, and process foods that are protected by mechanical defenses. What are the roles of different forms of learning (e.g., trial and error learning/individual learning vs. socially biased learning (Fragaszy & Visalberghi, 2004)) in acquiring this expertise? The contribution of socially biased learning to foraging techniques has only recently entered discussions of primate feeding ecology (e.g., McGrew, 1992; van Schaik, 2002), but social learning may in fact be a critical adaptation for some primates. Social learning may enable individuals to access foods they would otherwise have difficulty accessing or, at the very least, enable them to acquire such skills earlier in life than they would if they relied exclusively on trial and error learning to acquire all of their foraging skills (e.g., use of termiting tools in chimpanzees (Lonsdorf *et al.*, 2004); foraging on palm hearts, pith, or fruits in orangutans: Russon, 2003; use of tools by orangutans to access *Neesia* seeds (van Schaik & Knott, 2001); use by chimpanzees (Boesch & Boesch, 1990; Matsuzawa, 1994) and capuchins (Ottoni & Mannu, 2001) of hammers and anvils to crack nuts). Foraging patterns established early in life can have a profound impact on lifetime reproductive success (Altmann, 1998), and more attention should be paid to the factors affecting the early development of foraging skills, particularly during the ages when mortality is most pronounced. Some researchers have speculated that the need to solve foraging-related problems has been a critical factor in the evolution of primate intelligence (e.g., Byrne, 1995), and that social learning about foraging may also be a factor promoting enhanced encephalization (Reader, 2003).

In both the field and the lab, capuchins are remarkably tolerant of the close proximity of conspecifics while they are foraging, allowing groupmates to closely scrutinize their foraging activities (Fragaszy & Visalberghi, 2004; Fragaszy *et al.*, 2004). However, a wealth of experimental evidence (reviewed by Visalberghi & Addessi, 2003; Fragaszy & Visalberghi, 2004; and Fragaszy *et al.*, 2004) demonstrates that brown capuchin monkeys (*C. apella*) learn remarkably little about the motor details of the processing techniques they observe or about the fine details of food choice. For example, Visalberghi (1993) presented capuchins with a problem in which they had to push a stick through a tube to obtain a peanut located in the center of the tube. Three capuchins were provided with seemingly ample time to observe a capuchin demonstrator solve the problem correctly, yet they all failed to copy the demonstrator's actions (although they did make contact between the stick and the tube more often after their "lessons" than before them). Although Visalberghi & Fragaszy (1995) found that capuchins ate more of a novel food if their companions were eating the same novel food, Visalberghi & Addessi (2000) showed that the monkeys ate more of a novel food when their groupmates were eating *any* other food, not just the same novel food presented to the subject. Furthermore, when the experimenters dyed the foods different colors and gave the subject an option of colors, the subject did not select the novel food that matched the color of the familiar food being eaten by its groupmates (Visalberghi & Addessi, 2001). Nor, in a similar experiment, did the subject prefer to eat the novel food that matched the odor of the food eaten by the demonstrator; however, offspring of the demonstrator (but not non-offspring) did pay more attention to novel-smelling rather than familiar-smelling foods when held by a demonstrator (Drapier *et al.*, 2003).

If the monkeys are neither learning what to eat nor how to eat it, what are they learning? Visalberghi & Addessi (2000) suggest that the simple mechanism of social facilitation (i.e., an increase in the performance of a particular behavior that is contingent on concurrent performance of the same behavior by others in the vicinity) encourages monkeys to eat while others within their view are eating. If in wild situations they are, on average, most likely to have foraging individuals within view when they are in the same food patch as their more knowledgeable foraging groupmates, then this simple mechanism might, on average, be sufficient to guide an animal towards acquiring the same diet as groupmates. Likewise, given the propensity of capuchins to pound, probe, and scrub every object they handle until they achieve a desirable outcome, social facilitation along with tenacity and a predisposition to experiment with the properties of physical objects may be sufficient to guide capuchins towards the same processing solutions that are exhibited by their groupmates. Perhaps all they learn from observing conspecifics process food is that the objects handled by the model are worthy of exploration. Trial and error learning may lead the observer to acquire similar behavior patterns as the model, even if no sophisticated imitation process is occurring.

In this chapter, we test functional explanations for a common behavior of wild white-faced capuchins (*C. capucinus*) that may be difficult for the foregoing model to accommodate. Individuals sometimes approach foraging conspecifics in order to establish body contact or come within a few centimeters' distance, peering alternately at the food item and at its owner's face. This behavior, which we call "food interest," occasionally elicits mild aggression (nips, pushes) from the recipient. This risk of aggression is not surprising, given that wild capuchins are often hungry and therefore irritable (Janson & Vogel, Chapter 11, this volume), despite the generally high social tolerance level of *Cebus*. If capuchins observing foraging conspecifics gain nothing more than a generalized interest in the objects (fruits, embedded insects) found in the current food patch, there should be no need to observe at close range and risk recipients' aggression. Therefore, despite the large amount of experimental evidence bearing on this issue (Fragaszy *et al.*, 2004), we regard as still unresolved the question of whether capuchins can gain specific knowledge about food choices and food-processing techniques via socially biased learning.

Here we describe the diet of a group of wild white-faced capuchins living at Lomas Barbudal, Costa Rica, in order to illustrate the breadth of the monkeys' dietary options and the degree to which they engage in complex extractive foraging. Having described the diet, we then test predictions from the following three hypotheses accounting for food interest behavior:

1. Capuchins exhibit food interest in order to gain knowledge about what to eat (i.e., which items are toxic or nutritious). Prediction: Monkeys should be disproportionately more interested in observing the consumption of items that are rare in the diet because they probably already have knowledge of the more commonly eaten items.

2. Capuchins exhibit food interest in order to learn how to process foods for consumption. Prediction: Monkeys should be disproportionately more interested in those foods that require more expertise to process (as measured by the minimum number of processing techniques required to obtain food).

3. Capuchins exhibit food interest not in order to acquire information, but in order to obtain food. Prediction: Monkeys should exhibit food interest most often in large food items (i.e., items that are easily shared or confiscated).

These hypotheses are not mutually exclusive, and all three could be valid to some extent. For example, young capuchins might initially exhibit food interest in order to beg, but later when they are more proficient at foraging on their own, their food interest might be geared more towards gaining knowledge rather than gaining an immediate food reward.

Methods

The study site

This study was conducted at Lomas Barbudal Biological Reserve, and the surrounding forested areas, in Guanacaste, Costa Rica. The site is a tropical dry forest at low elevation (10 180 m) and is composed of various forest types, the most predominant being riverine forest, dry deciduous forest, and savanna (see Frankie *et al.*, 1988 for a more detailed description of the habitat). The forest is highly seasonal, with a rainy season starting in mid-May and extending into November. The leaves start to fall from the trees around January so there is virtually no foliage from February to mid-May except in riverine forest. Lomas Barbudal has a species composition quite similar to other dry forest sites where *C. capucinus* feeding behavior has been studied, such as Santa Rosa (Freese, 1977; Chapman & Fedigan, 1990; Rose, 1998) and Palo Verde (Hartshorn & Poveda, 1983; G. Frankie, pers. comm., 1992; Panger, 1997; Chavarría *et al.*, 2001). However, northern Lomas Barbudal and the adjacent lands differ from Santa Rosa and Palo Verde in the extent to which they include riverine forest. Water is a scarce resource at Santa Rosa, where the monkeys must rely on very few permanent waterholes (L. Rose, pers. comm., 1991).

The study subjects

The vast majority of the data presented in this paper comes from observations conducted in 1992–3 of a single *C. capucinus* group, Abby's group. At this time, the group contained 3–4 adult males (10+ years), 5–6 adult females (6+ years), and 11–12 immatures (see Perry, 1995 for further details; during this time period, the oldest immature was believed to be 5–6 years old). The group was well habituated, having been studied for 15 months prior to the inception of this study. Although the bulk of the data presented here come from Abby's group, supplemental observations were made from four other habituated groups between 2001–2004 (see below).

The data sets

We used three different data sets for the behavioral analyses reported here. The primary data set was collected on Abby's group, from May 1,

1992–May 1, 1993, as part of SP's dissertation research. Most plants in the diet had been identified during observations conducted in May–August 1990 and May 1991–April 1992. In this data set, 21 individuals of all age-sex classes were observed during 10-minute focal follows, and their activities and proximities to other group members were recorded during point samples collected at 2.5-minute intervals. Proximity categories included contact within 1 body length (i.e., approximately 40 cm), within 1–5 body lengths, and within 5–10 body lengths. These point samples were used to estimate the amount of foraging time devoted to each food type. If foods were consumed only between point samples, this was noted *ad libitum* as well, and such foods are listed as part of the diet and reported as comprising <1% of the diet. These data were recorded directly onto microcassette recorders by SP, with the help of two field assistants (see Perry, 1995, 1996 for further details on behavioral sampling). Only data collected by SP in combination with a field assistant were used for the analysis of diet composition. The data set used for calculating the diet composition consisted of 576.83 hours (17305 point samples, 5533 of which were of foraging individuals).

During the 1992–3 study, all close-range observations of the foraging activities of non-focal animals were recorded on a continuous basis during the 10-minute focal follows. The categories of food-related social interactions included (a) "food interest" (peering at the food in another's hand or mouth as it was being processed or consumed, typically with a distance of 3 inches (7.5 cm) or less between the food item and the observer's face); (b) food-touching (in which the forager's food was touched by the observing monkey); (c) attempted food theft (in which the observer tugged at the food but did not succeed in taking it away from the owner); (d) tolerated food theft (in which the observing monkey took the food away from the owner without receiving any resistance); and (e) forced theft (in which the observing monkey forcibly stole the food from a resisting food owner). We recorded these events, both when they were directed towards the focal animal by others, and when the focal animal performed the behaviors. Only food interest observations by immature animals were included in the analyses to follow because immatures are conspicuously less talented at obtaining and processing food relative to adults, all of whom are competent foragers (Perry, unpublished data from subsequent field seasons, 2001–2004), and therefore it can be assumed that adults have less need to learn socially than do immatures. The number of observations on which to conduct analogous analyses with adults was insufficient because 92% of all food interest observations were performed by juveniles. The entire May 1992–May 1993 database of continuous observations (859.67 hours of focal observation) was used to calculate "food interest" frequencies.

The second data set was collected between January–May, 2001. The protocol for this phase of the study was identical to the above protocol except that it involved all-day follows of individual monkeys up to 13 hours, such that there was little or no bias towards data collection on monkeys that are lower in the canopy. The study subjects included 31 adult monkeys from two social groups, Abby's and Rambo's group. The total data set yielded 3129 feeding point samples. This data set was used primarily to determine the proportion of time that monkeys spent foraging on plants vs. insects or vertebrates.

The third data set was collected between January 2002–August 2004. Behavioral sampling during this time period consisted of 10-minute follows of immature individuals from five social groups, collected, as in 1992–3, by teams of two field assistants who were subjected to regular reliability tests. Any plant samples unknown to the assistants were identified by SP or JCOJ at the end of the day. This data set was not analyzed quantitatively in this study, but was used to augment the list of foods utilized by the study population since some plants that are relatively common items in the diet are not eaten in all years. If a plant food was not eaten during 1992–3, but was eaten during other study periods, it did not receive a score for "% of diet in the month during which it was eaten most" in Table 8.1.

Data on processing of plant foods (i.e., manipulation of foods prior to ingestion) were taken from all three studies and collected *ad libitum* (see Panger *et al.*, 2002) in most cases. Those foods that require complex processing (i.e., a large number of processing steps) were the subjects of a special protocol in 2001–2004 in which we systematically recorded, on a regular basis, the details of processing techniques employed by all individuals. In Table 8.1., we report only the minimum number of processing steps required to prepare a plant for consumption, even though there are often individuals who employ a more elaborate processing strategy for particular foods.

Data analysis

Proportion of each item in the diet

For the purposes of constructing Table 8.1, we calculated, for each monkey in each month, the number of point samples in which it was seen foraging on each food type, and then divided that by the total number of point samples in which the monkey was feeding. The diet composition for each month was constructed by averaging the diet compositions for all monkeys in the group during that month. In Table 8.1, we report each plant's contribution to the diet in two ways. In one column, we report proportion of foraging time

Table 8.1. *Plants consumed by capuchins at Lomas Barbudal*

Species	Plant type	Processing steps prior to insertion in mouth	Part ingested	Period of use	% of overall diet, averaged across months	% of diet in month it was eaten most
Acacia collinsii	Tree	1	Fruit, ant larvae	Feb–May	0.18	0.93
Acroceras sp.	Grass	0	Fruit	Nov–Feb	0.02	0.17
Adenocalymna inundatum?	Liana	1	Fruit, insects	Dec–Mar		
Alibertia edulis	Shrub	0	Fruit	Nov–Feb		
Allophylus occidentalis	Shrub	0	Fruit	June–Sep, Oct		
Anacardium excelsum	Tree	0	Fruit (fleshy receptacle)	Feb–May	0.16	1.2
Anacardium occidentale	Tree	1	Fruit (fleshy receptacle)	Feb–May	0.42	2.56
Anona spp. (multiple species)	Tree	1	Fruit	July–Aug, Dec–Mar		
Apeiba tibourbu	Tree	1	Seeds	Mar–May	0	0.04
Ardisia revoluta	Shrub	0	Fruit	Feb–May	0.15	5.22
Blepharodon sp.?	Vine		Fruit	Nov–Jan		
Bactris major	Shrub	0	Fruit and stems	Jan–Feb, Jul–Sep		
Bambusa sp.	Bamboo	1	Pith, shoots	Dec		
Bauhinia sp.	Liana	1	Seeds	Mar–May	0.8	8.11
Bixa orellana	Shrub	1	Seeds	Apr–May	3.9	11.46
Bromelia pinguin	Bromeliad	3	Fruit, flower and apical meristems	year round	5.67	17.35
Brosimum alicastrum	Tree	0	Fruit	May–July		
Bursera simaruba	Tree	0	Fruit, sap and pith	Jan–Mar	1.36	9.02
Bursera tomentosa	Tree	0	Fruit	Dec–Jan, May–Jul		

Byrsonima crassifolia	Tree	0	Fruit	May–Jul		
Caesalpinia eriostachys	Tree	0	Flower (nectar)	Feb		
Calliandra bijuga	Shrub	0	Flower	Jan–Feb		
Capparis baduca	Shrub	0	Fruit	Jan–Feb		
Capparis indica	Shrub	0	Fruit	Jun–Jul		
Carex sp.?	Grass	0	Fruit	Jun–Jul		
Carica papaya	Tree	1	Fruit	May	0.65	4.72
Casearia arguta	Tree	1	Fruit	Apr–Jul	0	0.4
Casearia sp.	Tree	0	Fruit	Aug–Sep		
Casearia sylvestris	Tree	0	Fruit	Apr		
Cassia grandis	Tree	2	Fruit	May		
Cayaponia attenuata	Vine	0	Fruit	Mar		
Cecropia peltata	Tree	0	Fruit	Jun–Aug	0.04	0.49
Celtis iguanaea	Liana	0	Fruit	Jun		
Chomelia spinosa	Shrub	0	Fruit	Nov–Jan	3.74	22.4
Cissus sp.	Vine	0	Fruit	Sep–Nov	0.49	3.12
Cissus sp.	Vine	0	Fruit	Sep–Nov		
Clidenia sericea	Shrub	0	Fruit	May–Jun		
Cnidoscolus urens	Herb	0	Fruit	May–Sep	0.01	0.1
Cochlospermum vitifolium	Tree	1	Seeds, pith, roots and flowers	May (seeds) Nov–Mar		
Coccoloba caracasana	Tree	0	Fruit	Apr–May	0.15	0.92
Coccoloba guanacastensis	Tree	0	Fruit	Mar–May	0.03	0.31
Combretum decandrum (farinosum)	Liana	0	Flower (nectar)	Feb–Apr		
Cordia panamensis	Tree	0	Fruit	Aug–Oct		
Couepia pyliandra	Tree	1	Fruit	Apr–May		
Cupania guatemalensis	Shrub	1	Fruit	Mar–Apr	0.14	1.43
Curatella americana	Tree	0	Fruit	Mar–Jun	0.17	1.15
Dalechampia scandens	Herb	0	Fruit	Jan–Feb		

Table 8.1. (cont.)

Species	Plant type	Processing steps prior to insertion in mouth	Part ingested	Period of use	% of overall diet, averaged across months	% of diet in month it was eaten most
Davila ruyosa	Liana	0	Fruit	Apr–May		
Dilodendron costaricense	Tree	0	Fruit	Mar–Apr		
Diospyrus nicaraguensis	Tree	1	Fruit	Jan–Feb, Sep–Oct		
Doleocarpus dentatus	Liana	0	Fruit	Apr–May		
Dorstenia drakena	Herb	0	Fruit	Jun–Aug	0.04	0.45
Erythroxylum havanense	Shrub	0	Fruit	May–Jun	0.06	0.49
Eugenia sp. (monticola?)	Shrub	0	Fruit	Feb–Apr	0.24	1.51
Eugenia panamensis	Tree	0	Fruit	Aug–Oct	0.16	1.25
Ficus maxima	Tree	0	Fruit	irregular fruiting		
Ficus sp.	Tree	0	Fruit	irregular fruiting		
Garcinia edulens	Tree	0	Fruit	Feb–Apr		
Genipa americana	Tree	1	Fruit	Apr–Jun	0.03	0.2
Godmania aesculifolia	Tree		Insects in fruit	Dec–Feb		
Guarea glabra	Tree	0	Fruit	Feb–Apr	0.02	0.16
Guazuma ulmifolia	Tree	0	Fruit	Jan–Jun	0.85	7.89
Guettarda macrosperma	Tree	0	Fruit	Jun–Oct	2.13	19.42
Hamelia patens	Shrub	0	Seeds	July–Sep		
Hemiangium excelsum	Tree	1	Fruit	Aug–Sep		
Solanum sp.	Herb	0	Fruit	Jul–Aug		
Hirtella racemosa	Tree	0	Fruit	Feb–Apr	0.56	3.41
Hylocereus costaricensis	Cactus	0	Fruit	Jun–Jul		
Hymenaea courbaril	Tree	0	Flowers	Mar–Apr	0.01	0.15
Inga vera	Tree	0	Fruit, flowers	Apr–Jun		
Ipomoea sp.	Vine	0	Leaves and stem	Feb–May	0.02	0.18

Ixora floribunda	Shrub	0	Fruit	Apr–May	8.83	39.61
Jacquinia pungens	Shrub	1	Fruit	Sep–Jan		
Krugiodendrum ferreum	Tree	0	Fruit	Jan–Jun	4.38	15.36
Luehea candida	Tree	1	Seeds	Jan–May		
Luehea speciosa	Tree	1	Seeds	Feb		
Maclura tinctoria	Tree	0	Fruit	May–Aug	0.38	2.53
Malvaviscus arborea	Shrub	0	Fruit	Dec–Feb		
Mangifera indica	Tree	1	Fruit	Feb–May	0.09	0.61
Manilkara chicle/zapota	Tree	1	Fruit and flower	Jun(flowers), Nov–May (fruit)	0.01	0.1
Margaritaria nobilis	Shrub	0	Fruit	July–Sep	0.07	0.76
Matelea quirosii	Vine	0	Fruit	Aug–Sep, Dec–Jan	0.04	0.42
Melothria. sp.?	Vine	0	Fruit	Mar	0.01	0.09
Mesechites sp.	Liana	0	Fruit	Jan–Feb		
Miconia argentea	Shrub	0	Fruit	Mar–Jul	0.01	0.15
Monstera adansonii	Hemiepiphyte	0	Fruit	Dec–Jan		
Mouriri myrtiloides	Shrub	0	Fruit	Jul		
Muntingia calabura	Shrub	0	Fruit	Jan–Jun	1.38	8.46
Musa sp.	Banana	1	Flower (nectar and maybe pollen)	Feb–Apr	0.22	1.6
Ochroma pyramidale	Tree	0	Flower (nectar and maybe pollen)	Feb–Apr	0.02	0.24
Ocotea veraguensis	Tree	0	Fruit	Apr–Jun		
Ouratea lucens	Shrub	0	Fruit	Jan–Apr		
Passiflora sp.	Vine	1	Fruit	Jun–Sep	0.57	6.23
Paullinia cururu	Liana	0	Fruit	Aug–Sep	0.02	0.22
Phaseolus. sp.?	Vine	0	Flowers	Oct–Dec		
Phoradendron sp.	Epiphyte	0	Flower and fruit	Sep–Oct		
Piper amargo	Herb	0	Fruit	May–Jun	0.06	0.69
Pisonia aculeata	Tree	0	Fruit	Dec–Jan		
Prockia crucis	Shrub	0	Fruit	May–Aug	0.27	2.01

Table 8.1. (cont.)

Species	Plant type	Processing steps prior to insertion in mouth	Part ingested	Period of use	% of overall diet, averaged across months	% of diet in month it was eaten most
Quercus oleoides	Tree	1	Fruit	Aug–Oct	0.09	0.56
Randia spinosa	Tree	1	Fruit	Dec–Feb	0.84	4.3
Randia subcordata	Tree	1	Fruit	Nov–May		
Rhynchosia sp.?	Vine	1	Fruit	Jan–Mar		
Rourea glabra	Liana	0	Fruit	May		
Rytidostylis sp.?	Vine	0	Fruit	Oct		
Samanea saman	Tree	2	Fruit	Feb–Apr		
Sapranthus palanga	Tree	0	Fruit	Mar–Sept	0.1	1.05
Sciadodendron excelsum	Tree	0	Fruit	May–July	0.15	1.6
Sideroxylon capiri	Tree	0	Flower (nectar and maybe pollen)	Jan–May		
Simaruba glauca	Tree	1	Seeds and fruit (ripe)	Feb–Mar	0.02	0.26
Sloanea terniflora	Tree	3	Fruit	Jan–Jun	6.08	36.85
Spondias mombin	Tree	0	Fruit	July–Oct		
Solanum hazenii	Shrub	0	Fruit	Jun–July		
Spondias purpurea	Tree	0	Fruit	Jan–May	0.06	0.65
Stemmadenia obovata	Shrub	1	Fruit	Jun–July	0.02	0.18
Sterculia apetala	Tree	5	Fruit	Jan–Feb	1.19	7.66
Tabebuia impetiginosa	Tree	0	Insects in fruit	Jan–April		
Tabebuia ochracea	Tree	1	Flower (nectar) and fruit	Jan–Jun	0.17	1.59
Tetracera volubilis	Liana	0	Fruit	Jan–Apr	0.28	1.32
Trichilia americana	Tree	0	Pith, fruit	Feb (pith), Jun–Jul	0.07	0.79

Trichilia martiana	Tree		Fruit	Jun–Jul		
Vigna sp.?	Vine	0	Flowers	Dec–Jan		
Ximenia americana	Shrub	0	Fruit	Feb–Mar		
Xylosma flexuosa	tree	0	fruit	Apr		
Zanthoxylum setulosum	Tree	0	Fruit	Aug–Sep	0.01	0.12
Zuelania guidonia	Tree	2	Fruit	May–Jan	0.19	1.08
Zygia longifolia	Tree	0	Seed	Jan–May	0.19	0.91
(Leguminaceae)	Lianas		Seeds	Feb–May	0.12	0.74
(unidentified vine)	Vine	0	Fruit	Apr	0.04	0.48

(averaged across the 11 months of sampling) the monkeys spent foraging on that plant type. In the last column, we report the proportion of the diet during the month when that food was most important in the diet.

Calculation of food interest index and proportion of items in observed diet

For the analyses correlating diet composition with frequency of "food interest" and testing the association between "food interest" and complexity of processing technique or food size, we did not use the monthly rates calculated in Table 8.1. Because some months (the dry season months) provided better observation conditions and therefore higher data yields than others, our results would have been distorted. Therefore, we used raw frequencies of food interest rates and raw frequencies (point samples) of feeding observations in order to determine whether monkeys showed "food interest" most often in foods that were rare in the diet or in hard-to-process foods. We calculated an expected number of food interest episodes for each food by multiplying the overall proportion of the food in the diet (i.e., the number of point samples in which that food was being handled or eaten, divided by the total number of foraging point samples in the database, n = 5533) by the total number of episodes of "food interest" observed (n = 335). Then we subtracted this expected value from the observed frequency of "food interest" events for each food to obtain a "food interest index" (Table 8.2). Thus a positive number on this "food interest index" means that *more* interest was shown in the food than would be expected if the monkeys exhibited "food interest" in each food type at random, i.e., based solely on the amount of opportunity they had to observe conspecifics handle that food type. A negative "food interest index" indicates less food interest than would be expected by chance. A score of zero on the food interest index would mean that the animals were exhibiting food interest in precise accordance with the expectation that animals pay more attention to foods that are handled more frequently by conspecifics due merely to the greater opportunity they have to watch these items being eaten or processed. This "food interest index" was used as the dependent variable in all of the hypotheses tested. Note that the use of this food interest index (as opposed to raw frequencies or rates of food interest) enables us to control for the problem of variation in handling times among food types. A finding that the monkeys have higher food interest scores for hard-to-process foods that have longer handling times *cannot* be interpreted as meaning that the monkeys show higher rates of food interest in foods *because* they have longer handling times. Handling time has been taken into account. Our measure of "proportion in the diet" is effectively a measure of opportunity to observe foraging on that item, and "food interest" was

Table 8.2. *Characteristics of foods to which monkeys exhibited food interest*

Food type	# Processing steps[a]	Size category[a]	Proportion in diet	Food interest index
Wasp nests (*Polistes* and *Polybia*)	2	2	0.0103	10.55
Bromelia pinguin (common fruit)	3	2	0.0658	7.96
Caterpillars		2	0.0195	7.46
Acacia (fruit and thorns)	1	1	0.0093	5.89
Sterculia apetala	5	2	0.0177	5.07
Fruit 6 (legume)	1	2	0.0016	4.46
Coati (*Nasua narica*)	2	3	0.0029	4.03
(Zapote)	1	3	0.0001	3.97
Ardisia revoluta	0	1	0.0025	3.15
Fruit 7 (cactus)		2	0.0001	2.97
Mouse	2	3	0.0007	2.76
Trichilia sp.	1		0.0011	2.64
Bromelia pinguin (rare fruit)	1	2	0.0001	1.97
Fruit 4		2	0.0001	1.97
Vine	0	1	0.0001	1.97
Bromelia pinguin apical meristem	1	2	0.0002	1.94
Bird nest			0.0005	1.82
Mangifera indica	1	3	0.0007	1.76
Sapranthus palanga	0	3	0.0007	1.76
Squirrel	2	3	0.0007	1.76
Flower 3		1	0.0018	1.39
Cassia grandis	2	3	0.0001	0.97
Diospyrus nicaraguensis	1	1	0.0001	0.97
Garcinia edulens?		1	0.0001	0.97
Piper amarga	0	2	0.0001	0.97
Fruit 1			0.0001	0.97
Fruit 2		2	0.0001	0.97
Fruit 5		1	0.0001	0.97
Pink flower		1	0.0001	0.97
White fluffy flowers from vine		1	0.0001	0.97
Shoots			0.0001	0.97
Leaf stalk 1	0	1	0.0001	0.97
Anacardium occidentalis	1	2	0.0061	0.94
Monstera adansonii	0	2	0.0002	0.94
Passiflora sp.	1	2	0.0033	0.91
Eugenia panamensis	0	1	0.0005	0.82
Genipa americana	1	2	0.0005	0.82
Simaruba glauca		1	0.0005	0.82
Piper sp. 3			0.0005	0.82
Apeiba tibourbu	1	2	0.0007	0.76
Egg	1	2	0.0009	0.70
Spondias purpurea	0	1	0.0011	0.64
Cissus sp.	0	1	0.0018	0.39
Coccoloba guanacastensis	0	1	0.0020	0.33

Table 8.2. (*cont.*)

Food type	# Processing steps[a]	Size category[a]	Proportion in diet	Food interest index
Dorstenia drakena	0	1	0.0002	−0.06
Stemmadenia sp.	1	2	0.0002	−0.06
Ximenia americana	0	1	0.0002	−0.06
Luehea speciosa	1	2	0.0002	−0.06
Cnidoscolus urens	0	1	0.0002	−0.06
Manilkara chicle	1	1	0.0002	−0.06
Doleocarpus dentatus	0	1	0.0002	−0.06
Acroceras sp.	0	2	0.0002	−0.06
Melothria sp.	0	1	0.0002	−0.06
Zanthoxylum setulosum	0	1	0.0002	−0.06
Cecropia peltata	0	1	0.0004	−0.12
Guarea glabra	0		0.0004	−0.12
Matelea quirosii?		1	0.0004	−0.12
Casearia corymbosa	0	1	0.0004	−0.12
Ocotea veraguensis	0	1	0.0004	−0.12
Pith 1			0.0004	−0.12
Musa sp.	1	3	0.0034	−0.15
Mud		1	0.0034	−0.15
Erythroxylum havanense	0	1	0.0005	−0.18
Randia spinosa	1	1	0.0005	−0.18
Piper sp. 2			0.0005	−0.18
Epiphyte			0.0005	−0.18
Margaritaria nobilis	0	1	0.0007	−0.24
Sciadodendron excelsum	0	2	0.0007	−0.24
Fruit 3	0	1	0.0007	−0.24
Zuelania guidonia	2	2	0.0013	−0.42
Casearia arguta	1	1	0.0043	−0.45
Cupania guatemalensis	1	1	0.0016	−0.54
Randia subcordata	1	2	0.0107	−0.57
Curatella americana	0	1	0.0020	−0.67
Prockia crucis	0	1	0.0020	−0.67
Anacardium excelsum	0	1	0.0023	−0.79
Maclura tinctoria	0	1	0.0023	−0.79
Zygia longifolia	0	2	0.0025	−0.85
Tabebuia ochracea	1	2	0.0025	−0.85
Eugenia sp.	0	1	0.0040	−1.33
Tetracera volubilis	0	2	0.0040	−1.33
Guazuma ulmifolia	0	2	0.0164	−1.51
Guettarda macrosperma	0	1	0.0137	−1.60
Bixa orellana	1	1	0.0154	−2.15
Hirtella racemosa	0	1	0.0098	−2.27
Muntingia calabura	0	1	0.0125	−4.18
Bauhinia spp.	1	2	0.0161	−4.39
Bursera simaruba	0	1	0.0262	−7.78

Table 8.2. (*cont.*)

Food type	# Processing steps[a]	Size category[a]	Proportion in diet	Food interest index
Luehea candida	1	2	0.0669	−9.40
Jacquinia pungens	1	1	0.0521	−10.44
Chomelia spinosa	0	1	0.0369	−12.35
Sloanea terniflora	3	1	0.0969	−15.45

Note:
[a]missing values indicate cases in which an accurate measure could not be obtained and this food was not included for statistical analysis.

scored only once per item consumed, no matter how long the monkey watched the processing of that item, thereby biasing the data *against* finding a tendency for monkeys to show more food interest in items that are handled for longer periods of time. The food interest indices, as well as the scores for all of the independent variables (rarity in observed diet, size of food, and complexity of processing) are reported in Table 8.2 for all foods in which monkeys exhibited food interest at least once.

In order to determine whether monkeys selectively paid more attention to foods eaten more rarely by groupmates, we used a Spearman's rank correlation test to compare each food type's proportion in the overall diet with the "food interest index" for that food. In some cases, a food was not actually seen being consumed in the feeding point samples, but was present in the continuous data on "food interest." This is because some foods that are very rarely eaten will not be sampled in point samples if they are not eaten by the focal animal or if they are only eaten between point samples. Obviously, these foods do not merit a score of zero in the rankings of proportions in the diet, so they were arbitrarily given a score of half-of-one point sample.

For the comparison of the food interest index with processing complexity, we created three categories of processing complexity: (a) no processing required before the food is placed in the mouth (though skins and pits can be spit out after mastication); (b) one processing step (e.g., removal of a husk) is required before the food can be ingested; and (c) two or more steps are necessary before ingestion. To give one example of a 2 +-step food, *Sterculia* fruits must be processed in the following way: several fibers must first be gnawed off the thick, woody husk at the seam of the fruit; then the monkey must pry open the husk at the seam, typically using teeth, hands, and feet; then the pulp-covered seeds must be removed carefully to avoid the stinging hairs that cover both the inner husk and the pulp; and the hairs in many cases must then be wiped off the pulpy seed before ingestion. We used a Kruskal–Wallis ANOVA by ranks test to test whether more "food interest" was exhibited towards consumption of foods that are hard to process, though identical results were obtained using a median test. Likewise, we used a Kruskal–Wallis test to determine whether the monkeys selectively exhibited food interest toward foods that were of larger size and therefore more shareable. The categories for this test were (a) 1-bite item; (b) 2–9 bite item; and (c) items yielding 10 or more bites of food.

There were some confounding associations between the three independent variables (rarity, processing complexity, and size). In general, bigger food items also required more processing (Fisher's exact p = 0.0000). To reduce this confound, we tested the association between rarity in the diet and food interest index separately within each broad size and within each processing complexity category.

In all of these analyses, some insect and vertebrate prey were included (see Table 8.2), even though they are not shown in Table 8.1. Vertebrate prey categories included mice, squirrels, coatis, and eggs, all of which were large enough to share and required some complex processing. Insects were typically hard to identify in the field because they were small and too quickly consumed, but we did have separate categories for wasps and caterpillars. The broad category of unidentified insects was dropped from analysis because it included a mixture of rare and common insects, as well as a mixture of processing categories.

Results

Characterization of the white-faced capuchin diet

The diet in the 1992–3 field season consisted of 45.49% plant matter, 47.43% invertebrates, and 0.53% vertebrates (Figure 8.1), though there were fairly extreme fluctuations in the proportions of plant and animal matter in the diet from month to month. The 2001 data set (using only data from all-day follows of adults in January–May) had a somewhat higher proportion of fruit in the diet (73% fruit, 25% insects, and 2% vertebrates), though when only adult data from the same months are considered for the 1992–3 data set, the difference between data sets is not so extreme (62% fruit, 31% insects, 1% vertebrates in 1992–3). Insects were more heavily represented in the rainy

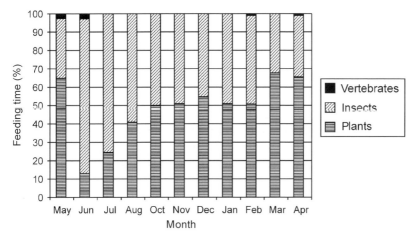

Figure 8.1. Proportions of plants, insects, and vertebrates in the capuchin diet, May 1992–April 1993, Abby's group.

season months, and during the second month of the rainy season, when caterpillars were in abundance, the proportion of insects in the diet was as high as 80%. The plant portion of the diet consisted almost entirely of fruit (see Table 8.1), and although the monkeys certainly ate ripe fruit, they typically began foraging on fruits long before they were ripe, employing their manual dexterity to defeat the mechanical defenses of plants in order to access fruits before their seeds were ready to be dispersed. Plants from 62 families were utilized. Fruits of at least 109 different species were consumed, and for 10 species, just the seeds were eaten. Flowers and/or nectar of 14 plants were consumed. In rare cases, the pith (4 species), stems or shoots (4 species), sap (1 species), roots (1 species), or leaves (1 species) were eaten.

Identification of insects down to the species level was impossible since the insects were small and typically consumed before we could get a good look at them. However, the monkeys frequently consumed caterpillars (even *Automeris*, which is covered with urticating hairs), wasp larvae, ants (adults and larvae, particularly *Pseudomyrmex*), katydids, stinkbugs, grasshoppers, lepidopterids, cicadas, spiders, scorpions, solitary bees, thornbugs, mantids, walking sticks, and roaches. Caterpillars made up 30% of the overall diet in the month of June, and wasp larvae (primarily *Polistes instabilis*) made up 15% of the diet in the month of July. Many of the insect species on which capuchins specialize reside in woody matrices (e.g., ants dwelling in acacia thorns or inside hollow sticks; bees that live deep inside small treeholes; egg cases hidden under bark). Therefore, capuchins spend much of their day stripping bark from trees, breaking apart branches, and unrolling leaves to look for hidden insect prey. Such behavior is typical not only of the Lomas capuchins, but apparently of the entire genus (see for example Terborgh, 1983, p. 99, who describes capuchins as "destructive foragers" and Fragaszy *et al.*, 2004, chapter 2).

Vertebrates made up a small but very highly valued part of the diet (see Rose *et al.*, 2003 for a complete list of vertebrates consumed by the Lomas capuchins). Vertebrates are typically difficult to process because they first must be killed and then the skin and hair or feathers removed before the flesh and guts are exposed for consumption. Monkeys foraging on vertebrate prey receive more begging attempts than foragers on any other food category, and because these prey items are often extremely large, the food is often shared (see, for example, Perry & Rose, 1994).

A broad overview of the plants that make up the majority of the diet over the course of a year are shown in Table 8.3. In this table, we show only those foods that make up at least 5% of the foraging time for a particular month, for the period May 1, 1992–May 1, 1993. The month of September is omitted from consideration because the observers were not present during enough days that month to produce accurate estimates. Each number in Table 8.3 represents the percentage of the total feeding time (for both plants and animals) for a particular plant in a particular month. These values are averaged across all group members. Nine of these 14 species require handling before ingestion, and three of these require three or more steps, so extractive foraging is often necessary.

During the end of the rainy season and the beginning of the dry season, when there is still foliage on the trees and some water in the treeholes, the monkeys spend most of their fruit foraging time on shrubs growing in deciduous forest. As terrestrial bromeliads come into their peak fruiting season in January–February, they spend several hours per day foraging on these, and also begin to forage extensively on *Luehea candida* and *Sterculia apetala* fruits, all of which require sophisticated processing techniques. During the late dry season, the monkeys spend almost all of their time near rivers, streams, and springs, foraging primarily in large trees such as *Sloanea terniflora* (which bears fruits demanding elaborate processing techniques), though they make occasional forays into drier parts of the forest to forage on *Luehea candida, Bursera simaruba, Guazuma ulmifolia, Bixa orellana*, and *Bauhinia*. Once the rains begin, insects (especially caterpillars and wasps) become the most important item in the diet, though they are supplemented by easy-to-process fruits from shrubs and small trees such as *Casearia arguta, Muntingia calabura*, and *Guettarda macrosperma*, and fruits from the *Passiflora* vines. Thus, the capuchins at Lomas Barbudal rely quite extensively

Table 8.3. *Most common plant foods in the capuchin diet, 1992–3[a] and the number of steps necessary to extract food from each species*

Species name	# Processing steps	May 92	Jun 92	Jul 92	Aug 92	Oct 92	Nov 92	Dec 92	Jan 93	Feb 93	March 93	Apr 93
Jacquinia pungens	1	0	0	0	1.1	39.6	28.3	19.1	8.3	0.8	0	0
Chomelia spinosa	0	0	0	0	0	0	15.6	22.4	3.2	0	0	0
Luehea candida	1	15.4	1.4	0	0	0	0	0.2	8.9	10.1	10.1	2.2
Sterculia apetala	5	0	0	0	0	0	0	0	5.3	7.7	0.2	0
Sloanea terniflora	3	13.7	0.8	0	0	0	0	0	0	0.2	15.3	36.8
Bursera simaruba	0	0	0	0	0	0	0	0	0.2	0.4	9	5.4
Guazuma ulmifolia	0	0.7	0	0	0	0	0	0	0.3	0.2	7.9	0.2
Bixa orellana	1	11.5	0	0	0	0	0	0.1	0	0.1	0	0
Casearia arguta	1	1.8	4.7	0	0	0	0	0	0	0	0	0.3
Muntingia calabura	0	1	1.9	8.5	0	0	0	0	0	0.8	2.2	0.9
Guettarda macrosperma	0	0	0	4.1	19.4	0	0	0	0	0	0	0
Bromelia pinguin	3	5.2	2.1	0.5	3.4	6.6	1	9	17.3	14.7	0.7	0.5
Passiflora	1	0	0	0	6.2	0	0	0	0	0	0	0
Bauhinia	1	0.1	0	0	0	0	0	0	0	0	8.1	0.5

Note:
[a]Numbers represent the percentage of the total diet; only foods that exceeded 5% of the diet during at least one month are shown.

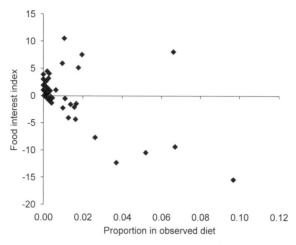

Figure 8.2. Proportion of observed diet vs. food interest index for all foods.

on hard-to-process foods for 7 months out of the year. They utilize both riverine forest and secondary forest extensively. Even recently burned areas are frequented – although they have lower fruit productivity, burned areas are often productive places to hunt for insect prey in rotting wood.

Effects of food rarity, processing complexity, and size on monkeys' visual attention to groupmates' foraging activities

Table 8.2 shows the raw data used to determine the effects of food rarity, processing complexity, and food size on the magnitude of the food interest index. Monkeys paid more attention to foods that were rarer in their diet (Spearman's rank correlation test, $r_s = -0.42$, n = 92, p < 0.01; Figure 8.2). They also attended more to foods that were more difficult to process (Kruskal–Wallis ANOVA by ranks test: H (2, n = 73) = 9.068, p = 0.0107) (Figure 8.3). Size of the food also had an effect on the amount of visual attention (Kruskal–Wallis ANOVA by ranks test H (2, n = 83) = 11.92, p = 0.003) (Figure 8.4).

Recall that bigger food items also required more processing in general (see Methods). To evade the possible confound that bigger fruit may require more processing than smaller fruit and that more processing may stimulate more interest, the same analysis was restricted according to processing difficulty. Restricting the analysis of effects of rarity on food interest index to foods that required no processing, the significance level increased ($r_s = -0.72$,

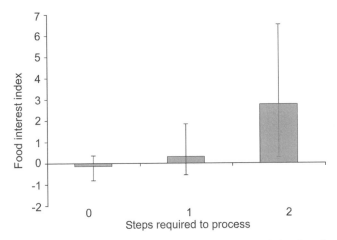

Figure 8.3. Effects of processing complexity on food interest index (bars denote interquartile range).

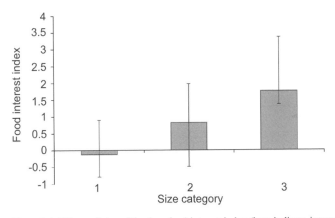

Figure 8.4. Effects of size of food on food interest index (bars indicate interquartile range).

$n = 37$, $p < 0.001$; Figure 8.5). The same was true when the analysis was restricted to foods that were bite-sized ($r = -0.70$, $n = 46$, $p < 0.001$; Figure 8.6). The effect of rarity in the diet on food interest frequency was non-significant when the analysis was restricted either to foods requiring complex processing or to foods that were larger than bite-sized.

Restricting the analysis of the effects of food processing complexity on food interest rates to small (bite-sized) foods, we found no statistically

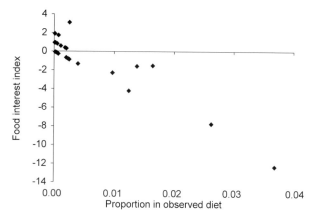

Figure 8.5. Effects of food rarity on food interest index for foods requiring no processing.

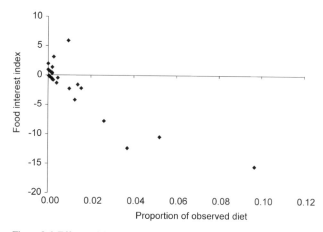

Figure 8.6. Effects of food rarity on food interest index for 1-bite foods.

significant effect. However, when we looked within the category of larger foods, we found again that food interest indices were higher for food types requiring greater processing complexity (Kruskal–Wallis H (2, n = 33) = 8.23, p = 0.016). The median of the food interest index was −0.06 for 0-step foods, 0.73 for 1-step foods, and 3.39 for foods requiring two or more steps to process.

Restricting the analysis of the effects of food item size on food interest index to foods requiring no processing, we found no statistically significant effects (Mann–Whitney U, U = 99, n = 32 small foods, n = 7 larger foods, n.s.). However, within the category of foods requiring one or more processing steps, there was still a marginally significant effect of food item size (Kruskal–Wallis ANOVA by ranks test H (2, n = 35) = 5.76, p = 0.056). Using a median test, the p-value dropped just below p = 0.05 (x^2 = 6.38, df = 2, p = 0.041). The median value of the food interest index was –0.45 for 1-bite foods, 0.76 for medium-sized foods, and 1.76 for foods requiring 10 or more bites to consume.

Discussion

Comparison of the diet of the Lomas Barbudal capuchins with that of white-faced capuchins at other sites

Broadly speaking, all other previous studies of white-faced capuchins have reported results similar to ours. White-faced capuchins everywhere are omnivores, focusing most of their foraging time on insects and fruits, though occasionally sampling seeds, vertebrates, flowers, pith, and on rare occasion saps or leaves and stems. Thirty-five of the 42 plant species on Freese's (1977) capuchin food list from Santa Rosa and 54 of the 62 plants on Rose's (1998) food list from Santa Rosa were also on the list of plant foods consumed by the Lomas Barbudal capuchins. Twenty-three of the 32 foods on the capuchin food list for Palo Verde (Panger, 1997) were also eaten by Lomas monkeys. We compiled a far longer plant food list than have other sites (130 species if data from all years are included), but this is probably just because we conducted a longer study. Approximately 65 plant foods were used by a single social group during an 11-month period. *Bromelia pinguin, Jacquinia pungens, Sterculia apetala, Bixa orellana,* and *Passiflora* were among the most important staples of the capuchin diet at Lomas, particularly during times when other foods were scarce; however, they were virtually absent in the diet of the Santa Rosa monkeys.

The Santa Rosa capuchins (Rose, 1998) had a diet composed of 73% plants and 25% insects in the dry season (which is almost exactly what we found in our 2001 field season), and 48% plants and 51% insects in the rainy season, which is roughly comparable to what we found at Lomas. It appears that capuchins in rainier habitats such as Barro Colorado Island also have similar diets (Watts, 1977; Phillips, 1995).

As noted by Paul Rozin (1976), omnivory is beneficial to animals because it gives them greater flexibility in changing their environments. The capuchin's omnivorous habits no doubt account for the fact that the genus *Cebus* is one of the most widely distributed New World primates (Fragaszy *et al.*, 1990). However, being a flexible generalist has its costs. How is a primate to learn which foods are good to eat and which are dangerous? Multiple mechanisms are used by omnivores to solve this problem, some of which involve paying attention to asocial cues such as taste and smell, which can alert foragers to potential dangers. Social cues can also be useful in a variety of circumstances, but just as asocial cues are not infallible, social cues are only useful if the model has better knowledge of the potential food's properties than the observer does. The question of when to attend to asocial cues and when to attend to social cues is a thorny issue and one that is critical to understanding why primates adopt the diets they choose (Rozin, 1976; Dewar, 2003a, 2003b).

Using social cues to learn about food: what might capuchins be learning from watching their groupmates?

All three of the hypotheses that we tested were supported in this study. First, we found that capuchins pay more attention to foods that are rare in the diet, which might indicate that they are trying to learn the relevant characteristics of the food (e.g., how to distinguish it from similar items, or what parts are eaten by conspecifics). This result was not statistically significant when the sample was restricted only to those food types that are large and/or require processing before ingestion. We suggest that effects of rarity on food interest vanish when small, easily ingested foods are removed from the sample because processing complexity and large size also have effects on food interest rates, to the point that the effect of rarity is swamped by these other effects. We cannot in this current study test whether watching conspecifics eat a relatively novel food has an effect on the immature monkey's tendency to sample the same food type. Research currently underway will permit such analyses.

One might expect that a combination of social cues and asocial cues would be involved in monkeys' decisions about what to eat and whom to watch eating (Dewar, 2003a, 2003b). For example, there is suggestive evidence from Whitehead's (1986) study of wild howler monkeys that infants typically observe adults' foraging activities before consuming leaves, but they are less attentive to conspecifics when the group is foraging on fruit, apparently using trial and error learning to guide their feeding preferences for fruit. This makes sense, as fruits rarely contain toxins, but leaves commonly do. In light of these results, it may be profitable for laboratory researchers to continue to search for testing paradigms in which capuchins learn something about which foods to eat by watching conspecifics. Perhaps by presenting monkeys with novel foods of types in which asocial cues are likely to be unreliable, they would be able to elicit a stronger and/or more specific effect of social learning on food choice than has been found in previous experiments (e.g., Visalberghi & Addessi, 2001).

It is also worth noting that capuchins exhibiting food interest are close enough to attend to many factors (smell, shape, and in some cases even texture) of the novel foods being eaten by groupmates. Thus far, most of the laboratory experiments on the role of observation have focused primarily on color or smell of the food (e.g., Visalberghi & Addessi, 2001; Drapier *et al.*, 2003) rather than the shape or texture. Perhaps the monkeys in these Visalberghi and Addessi experiments realize that the novel food(s) at their disposal are not identical to either of the colored familiar foods available to demonstrators, and therefore do not attend to color. Although it might be difficult to train a demonstrator to eat a novel food enthusiastically, it would be interesting to perform an experiment in which the demonstrator eats only one of two novel foods (perhaps one could be adulterated with pepper only for the demonstrator) and the subject has access to the same two novel foods (neither adulterated with pepper).

Second, we found that capuchins pay more attention to others' foraging efforts when they are manipulating foods that are hard to process. This result cannot be accounted for by the fact that hard-to-process foods have longer handling time than easy-to-process foods, because we used the proportion of feeding time that the food was being handled and/or ingested as the measure of opportunity to observe the food eaten, and we used frequencies of "food interest" (only scored once per foraging bout per dyad) rather than durations of "food interest" time. Nor can this result be accounted for (entirely) by the fact that the monkeys are begging for food that is large enough to be worth sharing. We did not, in this study, demonstrate that the monkeys actually learned something about the fine-grained details of food processing techniques by focusing their observations on harder-to-process foods. However, Panger *et al.* (2002) did find that wild capuchins who share a specific quirky foraging technique are closer associates (i.e., have more opportunity to observe one another) than dyads that do not share the same technique. Preliminary results from Lomas Barbudal accord with this pattern. For example, a pilot study of the role of social influence on the acquisition of *Sloanea* processing techniques in Abby's group reveals that monkeys who spend more time in association are more likely to share the same *Sloanea* processing technique (logistic regression, n $= 406$ dyads, $\times^2 = 4.25$, df $= 1$, p $= 0.04$; Perry, unpublished data). Though not subject to the same controls as experimental studies, these results are difficult to explain without invoking some role of social learning in determining the fine details of motor movements during food processing (i.e., some process more specific than stimulus enhancement or social facilitation).

So what could account for the differences in our findings and the laboratory researchers' findings? One unlikely possibility is that *C. apella* (the species that has been the subject of virtually all captive research) has substantially different problem-solving abilities from *C. capucinus*. It is somewhat more likely that the socioecological conditions favoring reliance on social learning differ between the two conditions.

It is also possible, of course, that the laboratory researchers have not yet identified tasks that fully reveal the monkeys' capacities to learn socially. Perhaps if the motor actions required of them during testing were elements more commonly performed spontaneously in nature and included auditory cues to supplement the visual input (such as pounding or scrubbing the food item itself, both of which produce easily recognized sounds, as opposed to inserting sticks in holes, turning wheels or pins, or lifting latches), it would be easier for them to attend to and copy the motor pattern. Another possibility is that the luxury of captive settings makes animals more lax about what they eat (since they are never fed anything particularly toxic) and/or less attentive to details of processing food (since they are less hungry than wild animals and therefore less motivated to learn these tasks). It is also possible that the monkeys were given insufficient time to learn from models in captivity (though they were given as many as 75 "lessons" for the stick in the tube task; Fragaszy *et al.*, 2004). Clearly it is not easy for the monkeys to learn about processing techniques: we find in the wild that capuchins take several years to settle into one technique for processing of *Sloanea*, for example (Perry, unpublished data). Much individual practice and, most likely, many observations of a model performing a behavior are necessary before skills as complex as *Sloanea* foraging are perfected.

The basic social learning model proposed by Visalberghi, Fragaszy, and their colleagues (see, for example, Fragaszy *et al.*, 2004, chapter 13) is that capuchins learn socially simply by coordinating their behaviors with one another. By maintaining tolerance, close spatial proximity, and a general coordination of activities, monkeys' attention is channeled such that the monkeys tend to receive the same sorts of inputs for their trial and error learning. Our results indicate that wild capuchins are interested in foods not simply because groupmates have them (although seeing another monkey holding an object clearly does make it more attractive), but that they also attend somewhat to the properties of the object being held. In this study, they paid more attention to items that were rare, difficult to process, and large (i.e., perhaps more likely to be shared). It has yet to be demonstrated what exactly is learned by focusing more on items with these characteristics, but developmental studies are under way at Lomas to address that question. So our results are consistent with the Fragaszy & Visalberghi (2004) view of socially biased learning in most respects, but the novel finding is that our results imply a more active seeking of particular types of socially acquired information than would have been expected from Fragaszy & Visalberghi's (2004) previous results.

We hope that laboratory researchers will continue to study the mechanisms of social learning in capuchins. Although we field researchers cannot hope to achieve the same level of experimental control that lab researchers can when investigating social learning mechanisms, we can provide critical information about the ecological context in which these mechanisms have evolved, i.e., the adaptive challenges that shape the cognitive mechanisms. Information about the circumstances in which capuchins exhibit visual attention to conspecifics may prove useful to lab researchers in designing more ecologically valid experiments regarding social influences on foraging skills.

Acknowledgments

We thank Gordon Frankie, M. Panger, K. MacKinnon, L. Rose, Ulises Chavarría, and I. Salas for assistance in identifying plant specimens. Assistance in data collection in 1992–3 was provided by J. Manson and J. Gros-Louis. We used supplemental observations by H. Gilkenson, J. Anderson, and A. Fuentes in 2001–2004 to augment the food list. D. Stahl and J. Manson provided statistical assistance. This study was funded primarily by the National Science Foundation, the National Geographic Society, the Leakey Foundation, the Max Planck Institute for Evolutionary Anthropology, the Wenner–Gren Foundation, the University of Michigan Rackham Graduate School, and Sigma Xi. J. Manson, E. Addessi, E. Visalberghi, M. Robbins, D. Fragaszy, G. Hohmann, and one anonymous reviewer provided helpful comments on the manuscript.

References

Altmann, S. A. (1998). *Foraging for Survival: Yearling Baboons in Africa*. Chicago, IL: University of Chicago Press.

Boesch, C. & Boesch, H. (1990). Tool use and tool making in wild chimpanzees. *Folia Primatologica*, **54**, 86–99.

Byrne, R. (1995). *The Thinking Ape: The Evolutionary Origins of Intelligence*. Oxford: Oxford University Press.

Chapman, C. A. & Fedigan, L. M. (1990). Dietary differences between neighboring *Cebus capucinus* groups: local traditions, food availability or responses to food profitability? *Folia Primatologica*, **54**, 177–86.

Chavarría, U., González, J., & Zamora, N. (2001). *Árboles Communes del Parque Nacional Palo Verde, Costa Rica / Common Trees of Palo Verde National Park*. Heredia: INBio.

Dewar, G. C. (2003a). *Innovation and Social Transmission in Animals: A Cost-Benefit Model of the Predictive Function of Social and Nonsocial Cues*. Unpublished Ph.D. thesis, University of Michigan-Ann Arbor.

(2003b). The cue reliability approach to social transmission: designing tests for adaptive traditions. In *The Biology of Traditions: Models and Evidence*, ed. D. M. Fragaszy & S. Perry, pp. 127–58. Cambridge: Cambridge University Press.

Drapier, M., Addessi, E., & Visalberghi, E. (2003). Response of *Cebus apella* to foods flavored with familiar or novel odor. *International Journal of Primatology*, **24**, 295–315.

Fedigan, L. M. (1990). Vertebrate predation in *Cebus capucinus*: meat eating in a neotropical monkey. *Folia Primatologica*, **54**, 196–205.

Fragaszy, D. M. (1986). Time budgets and foraging behavior in wedge-capped capuchins (*Cebus olivaceus*): age and sex differences. In *Current Perspectives in Primate Social Dynamics*, ed. D. M. Taub & F. A. King, pp. 159–74. New York, NY: Van Nostrand Reinhold.

Fragaszy, D. M. & Visalberghi, E. (2004). Socially biased learning in monkeys. *Learning and Behavior*, **32**, 24–35.

Fragaszy, D. M., Visalberghi, E., & Fedigan, L. M. (2004). *The Complete Capuchin: The Biology of the Genus Cebus*. Cambridge: Cambridge University Press.

Fragaszy, D. M., Visalberghi, E., & Robinson, J. G. (1990). Variability and adaptability in the genus *Cebus*. *Folia Primatologica*, **54**, 114–18.

Frankie, G. W., Vinston, S. B., Newstrom, L. E., & Barthell, J. F. (1988). Nest site and habitat preferences of *Centris* bees in the Costa Rican dry forest. *Biotropica*, **20**, 301–10.

Freese, C. (1977). Food habits of white-faced capuchins *Cebus capucinus* (Primates: Cebidae) in Santa Rosa National Park, Costa Rica. *Brenesia*, **10/11**, 43–56.

Gros-Louis, J., Perry, S., & Manson, J. H. (2003). Violent coalitionary attacks and intraspecific killing in wild capuchin monkeys (*Cebus capucinus*). *Primates*, **44**, 341–6.

Hartshorn, G. S. & Poveda, L. J. (1983). Checklist of trees. In *Costa Rican Natural History*, ed. D. H. Janzen, pp. 158–83. Chicago, IL: University of Chicago Press.

Lonsdorf, E. V., Eberly, L. E., & Pusey, A. E. (2004). Sex differences in learning in chimpanzees. *Nature*, **428**, 715–16.

Matsuzawa, T. (1994). Field experiments on use of stone tools by chimpanzees in the wild. In *Chimpanzee Cultures*, ed. R. W. Wrangham, W. C. McGrew, F. B. M. de Waal, & P. G. Heltne, pp. 351–70. Cambridge, MA: Harvard University Press.

McGrew, W. C. (1992). *Chimpanzee Material Culture: Implications for Human Evolution*. Cambridge: Cambridge University Press.

Ottoni, E. B. & Mannu, M. (2001). Semifree-ranging tufted capuchin monkeys (*Cebus apella*) spontaneously use tools to crack open nuts. *International Journal of Primatology*, **22**, 347–58.

Panger, M. A. (1997). *Hand Preference and Object-Use in Free-Ranging White-Faced Capuchin Monkeys (Cebus capucinus) in Costa Rica.* Unpublished Ph.D. thesis, University of California-Berkeley.

Panger, M. A., Perry, S., Rose, L. M. *et al.* (2002). Cross-site differences in the foraging behavior of white faced capuchin monkeys (*Cebus capucinus*). *American Journal of Physical Anthropology*, **119**, 52–66.

Perry, S. (1995). *Social Relationships in Wild White-Faced Capuchin Monkeys*, Cebus capucinus. Unpublished Ph.D. thesis, University of Michigan-Ann Arbor.

   (1996). Female-female relationships in wild white-faced capuchin monkeys, *Cebus capucinus. American Journal of Primatology*, **40**, 167–82.

   (2003). Coalitionary aggression in white-faced capuchins, *Cebus capucinus*. In *Animal Social Complexity: Intelligence, Culture and Individualized Societies*, ed. F. B. M. de Waal & P. L. Tyack, pp. 111–14. Cambridge, MA: Harvard University Press.

Perry, S. & Rose, L. M. (1994). Begging and transfer of coati meat by white-faced capuchin monkeys, *Cebus capucinus. Primates*, **35**, 409–15.

Perry, S., Baker, M., Fedigan, L. M., *et al.* (2003a). Social conventions in wild white-faced capuchin monkeys: evidence for traditions in a neotropical primate. *Current Anthropology*, **44**, 241–68.

Perry, S., Barrett, H. C., & Manson, J. H. (2004). White-faced capuchin monkeys show triadic awareness in their choice of allies. *Animal Behaviour*, **67**, 165–70.

Perry, S., Panger, M. A., Rose, L. M., *et al.* (2003b). Traditions in wild white-faced capuchin monkeys. In *The Biology of Traditions: Models and Evidence*, ed. D. M. Fragaszy & S. Perry, pp. 391–425. Cambridge: Cambridge University Press.

Phillips, K. A. (1995). Resource patch size and flexible foraging in white-faced capuchins (*Cebus capucinus*). *International Journal of Primatology*, **16**, 509–19.

Reader, S. M. (2003). Relative brain size and the distribution of innovation and social learning across the nonhuman primates. In *The Biology of Traditions: Models and Evidence*, ed. D. M. Fragaszy & S. Perry, pp. 56–93. Cambridge: Cambridge University Press.

Robinson, J. (1986). Seasonal variation in use of time and space by the wedge-capped capuchin monkey *Cebus olivaceus*: implications for foraging theory. *Smithsonian Contributions to Zoology*, **431**, 1–60.

Rose, L. M. (1997). Vertebrate predation and food-sharing in *Cebus* and *Pan*. *International Journal of Primatology*, **18**, 727–65.

   (1998). *Behavioral Ecology of White-Faced Capuchins* (Cebus capucinus) in Costa Rica. Unpublished Ph.D. thesis, Washington University-St. Louis.

Rose, L. M., Perry, S., Panger, M. A., *et al.* (2003). Interspecific interactions between *Cebus capucinus* and other species: data from three Costa Rican sites. *International Journal of Primatology*, **24**, 759–96.

Rozin, P. (1976). The selection of foods by rats, humans, and other animals. In *Advances in the Study of Behavior, Volume 6*, ed. J. S. Rosenblatt, R. A. Hinde, E. Shaw, & C. Beer, pp. 21–76. New York: Academic Press.

Russon, A. E. (2003). Developmental perspectives on great ape traditions. In *The Biology of Traditions: Models and Evidence*, ed. D. M. Fragaszy & S. Perry, pp. 329–64. Cambridge: Cambridge University Press.

Terborgh, J. (1983). *Five New World Primates: A Study of Comparative Ecology.* Princeton: Princeton University Press.

van Schaik, C. P. (2002). Fragility of traditions: the disturbance hypothesis for the loss of local traditions in orangutans. *International Journal of Primatology*, **23**, 527–38.

van Schaik, C. P. & Knott, C. D. (2001). Geographic variation in tool use on *Neesia* fruits in orangutans. *American Journal of Physical Anthropology*, **114**, 331–42.

Visalberghi, E. (1990). Tool use in *Cebus*. *Folia Primatologica*, **54**, 146–54.

(1993). Tool use in a South American monkey species: an overview of the characteristics and limits of tool use in *Cebus apella*. In *The Use of Tools by Human and Non-Human Primates*, ed. A. Berthelet & J. Chavaillon, pp. 118–31. Oxford: Clarendon Press.

Visalberghi, E. & Addessi, E. (2000). Response to changes in food palatability in tufted capuchin monkeys, *Cebus apella. Animal Behaviour*, **59**, 231–8.

(2001). Acceptance of novel foods in capuchin monkeys: do specific social facilitation and visual stimulus enhancement play a role? *Animal Behaviour*, **62**, 567–76.

(2003). Food for thought: social learning and feeding behavior in capuchin monkeys. Insights from the laboratory. In *The Biology of Traditions: Models and Evidence*, ed. D. M. Fragaszy & S. Perry, pp. 187–212. Cambridge: Cambridge University Press.

Visalberghi, E. & Fragaszy, D. M. (1995). The behaviour of capuchin monkeys, *Cebus apella*, with novel food: the role of social context. *Animal Behaviour*, **49**, 1089–95.

Watts, D. (1977). *Activity Patterns and Resource Use of White-Faced Cebus Monkeys* (Cebus capucinus) *on Barro Colorado Island, Panama Canal Zone.* Unpublished M.A. thesis, University of Chicago.

Westergaard, G. C. & Suomi, S. J. (1993). Use of a tool-set by capuchin monkeys (*Cebus apella*). *Primates*, **34**, 459–62.

Whitehead, J. M. (1986). Development of feeding selectivity in mantled howling monkeys (*Alouatta palliata*). In *Primate Ontogeny, Cognition and Social Behavior*, ed. J. G. Else & P. C. Lee, pp. 105–17. Cambridge: Cambridge University Press.

Part II

Testing theories:

New directions in ape feeding ecology

Introduction

RICHARD W. WRANGHAM

The six chapters in this second section share several features beyond using dietary data to test ideas. Empirically, their data come from long-term field-work reflecting enormous research efforts and challenging field conditions, they are very careful in their methods, and they achieve tight, clear results. From a theoretical standpoint, the material presented here assumes that individuals strive to maximize fitness, it builds importantly on prior concepts of behavioral ecology, and the approaches it describes are novel.

But beyond those similarities, the diversity of material is rich, and there is little overlap in the theoretical backgrounds for these investigations. Chapter 9 (Altmann) addresses the causes of particular feeding patterns, while the remaining five explore their consequences, which range from social relationships (Chapter 10, Koenig & Borries), aggression (Chapter 11, Janson & Vogel), population density (Chapter 12, Marshall & Leighton), and grouping and reproduction (Chapter 13, Wich *et al.*), to juvenile ranging behavior and the pattern of food transfers (Chapter 14, Marlowe). Strikingly for a book about apes, only three of the chapters address apes (and only two of those are non-human apes).

If primate ecology were a mature field, these investigations might fit together like pieces of a jigsaw to create a larger picture out of their separate details. But primatology is too young for such a result. Although its methods have been systematized since the 1970s, its youth is reflected in its theoretical achievements still being tentative and diverse. As this section's chapters tend to show, the current phase of primatology is more concerned with creating viable ideas than with consolidating old ones.

By summarizing his study of juvenile baboons (Altmann, 1998), Altmann's aim is to show that field data can be used to rigorously test the adaptive significance of particular food choices. This is a daunting notion for most primatologists. Their subjects often have long lists of food items, each of which potentially has several relevant primary nutrients, one or more indices

Feeding Ecology in Apes and Other Primates. Ecological, Physical and Behavioral Aspects, ed. G. Hohmann, M.M. Robbins, and C. Boesch. Published by Cambridge University Press. © Cambridge University Press 2006.

of fiber concentration, and a number of secondary compounds. There are many obstacles to getting high-quality data on feeding or adequate sample sizes of individuals. But Altmann encourages the faint-hearted by showing how a strong research design based on testing the relationship between food choices and performance variables (such as energy gain) can lead to strong conclusions. In his own study, even a small sample of individual yearlings (11) and mothers (6) produced remarkably clean results. Altmann showed that the diets for none of the yearlings was optimal, and remarkably, not only that variance in energy intake was the principal factor influencing fitness, but also that the fitness of females as adults was highly correlated with what they ate as yearlings. His study was conducted under near-ideal conditions, observing fully habituated, group-living monkeys on flat open ground in a simple habitat. Its success suggests that replication under more difficult conditions is a worthwhile risk.

Koenig and Borries also use data on monkeys to test basic ideas. Noting that even the strongest of socioecological principles are still vulnerable to empirical and theoretical refinement, they ask three questions that lie at the heart of the theory. How well does the distribution of food predict the pattern of within-group contest competition? Do rates of aggression or dominance rank better predict energy gain? And, how well do rates of aggression predict the strength and linearity of the dominance hierarchy?

In each case, their langur data yield counter-intuitive results. For example conventional wisdom predicts that in groups with more aggression, there will be a greater difference in feeding success between high-ranking and low-ranking individuals. But Koenig and Borries found that in a comparison of two groups, a low rate of aggression was associated with a high skew in physical condition. The implication is that the effect of aggression is specific to context, depending, for example, on the micro-distribution of the food items. Their conclusions resonate with ape data. Among female chimpanzees there is normally such a low rate of overt aggression that dominance hierarchies are difficult to describe, yet current evidence suggests intense rank-related skew in their reproductive success (Pusey *et al.*, 1997).

Koenig and Borries' chapter suggests that an important step in clarifying the causal links among food distribution, feeding behavior, and social conse-quences is making our ecological and behavioral variables more theoretically relevant. When assessing the impact of food distribution, for example, we may need to abandon purely phenological measures such as "clumping" in favor of indices that take feeding behavior into account, such as "contestability." Again, their conclusions make sense for ape studies. No female apes form alliances to compete with each other for food, yet their fruit diets are often markedly clumped. Koenig and Borries point the way to solving this conundrum.

In a third illustration of the value of monkey studies, Janson and Vogel advocate a shift in the way that primatologists assess the fitness gain from behaviors such as feeding. They note that traditionally, biologists have indexed fitness gain by measures that are independent of the internal state of the animal, such as "energy ingested per minute." But since a hungry animal benefits more than a satiated individual from a given amount of energy, a better measure of success should take into account the degree of hunger. Janson and Vogel's review of the literature on satiation and gastric emptying is a fascinating introduction to an area that will be unfamiliar to many, and is clearly potentially significant for a range of studies. They use their review to show how "state-dependent fitness functions" such as hunger can be assessed. This allows them to test the hypothesis that in fruiting trees, it is the hungrier capuchins that are aggressive more often.

Their results were striking. First, by controlling for a wide range of possible confounds, including a remarkable indexing of "alternative foods available," they conclude that the main factor affecting the rate of aggression was the size of the tree crown. Smaller crowns had more aggression, which makes sense. But surprisingly, although Janson and Vogel found hunger to be a significant influence also, it was responsible for only 1–2.6% of the variance in frequencies of aggression. Given that Janson and Vogel found that their indices of hunger were robust and varied widely, this seems small. But the aim of this chapter is more theoretical than empirical. The authors suggest various ways their approach can be improved, which opens the possibility that in future hunger may be found to have stronger effects.

Their chapter raises questions of particular interest with respect to apes. Janson and Vogel indexed hunger by estimating the fractional fullness of the forager's gut. Great apes spend about 50% of their daytime chewing, and for an undisturbed individual most of the time not chewing is spent immobile, apparently digesting the prior meal. So my sense of a chimpanzee's day is that their digestive system operates at close to maximal capacity all day. Might this mean that hunger levels vary little over the day in great apes, compared with a species that spends less time eating, e.g. capuchins and humans? If so, what effect does this have on foraging strategy and aggression? Janson and Vogel's focus on the dynamics of gastric emptying raises numerous possibilities for comparative studies across and within taxa.

In this section's first ape chapter, Marshall and Leighton probe the familiar notion that food density affects population density. Their aim is to find out whether the foods that contribute most to this relationship are preferred or fallback. Data come from the western Bornean park of Gunung Palung, where a combination of strong seasonality in plant production and strongly differentiated habitat types provide rich variation in both food abundance and gibbon densities. Like most questions about ape feeding ecology, the simplicity of the problem masks substantial complexity in research design. Marshall and Leighton describe how their enormous data sets were contracted to a comparison among seven habitat types, and they provide a recipe for operationalizing foods as preferred or fallback.

Their result is amazingly clean and somewhat surprising. In Gunung Palung, population densities were correlated far more strongly with the density of figs (a fallback food) than with preferred food items, suggesting that fig density controls population density. Subsequent between-population comparisons have strongly supported this result (Marshall, pers. comm.). The apparent importance of figs as a fallback food raises the question of whether this is a widespread rule in primates, and how it happens (e.g. by affecting juvenile survival). More immediately, it has very important consequences for conservation management strategies.

Like Marshall and Leighton, Wich *et al.* also take advantage of the intense seasonality of fruit production in South-East Asian forests. Their focus is on grouping, ranging and mating in orangutans, and once again they produce remarkable results. For example, orangutans spend much time alone, but they do occasionally associate. In previous studies, associations have sometimes been found to be larger during periods of fruit availability and sometimes not. Wich *et al.* show that in Ketambe, Sumatra, orangutan groups did not respond to food availability. This result is strong because of the care that the authors took to standardize their methods. Among chimpanzees the size of parties generally varies with food abundance, so the orangutan result implies that the economics of group formation varies among apes. Wich *et al.* conclude that predator pressure has very little effect on the grouping of orangutans. This suggests that orangutans may be an exceptional species in which to assess the impact of feeding pressures uncontaminated by the trade-off with predation.

Overall, Wich *et al.* found no evidence that the large seasonal changes in fruit production at Ketambe affected orangutan grouping or ranging. Even mating and reproduction were unaffected. This result challenges much previous work, and like Marshall and Leighton, Wich *et al.* explain it by focusing on figs. In their case, they point to the high fig density at Ketambe as possibly releasing orangutans from the typical pressures of an intermittently low food supply. These two chapters remind us that specific habitat variables may importantly influence socioecological relationships among apes, as among other primates.

The feeding strategies of humans are clearly very different from those of other apes. For example, no primates can abandon a large pile of food without risking its loss to competitors, whereas humans can relax in camp knowing that their gathered foods are unlikely to be stolen. Yet as Marlowe notes in his chapter on foraging among the Hadza hunter-gatherers of northern Tanzania, the differences are still not well characterized. The traditional focus has been on diet choice, but Marlowe addresses the way in which foods are obtained. He draws attention to the distinction between foraging from a central place, and provisioning group members at a sleeping-site. Marlowe argues that central place provisioning allows juveniles to reduce their foraging effort, and also allows group members to exchange more food than they otherwise could. Marlowe's analysis challenges researchers to develop a comparative social ecology that builds on these important empirical advances.

These chapters should be an inspiration to future researchers. They show that elegant and important results are abundantly available to theoretically inspired students.

References

Altmann, S. A. (1998). *Foraging for Survival: Yearling Baboons in Africa*. Chicago: University of Chicago Press.

Pusey, A., Williams, J., & Goodall, J. (1997). The influence of dominance rank on the reproductive success of female chimpanzees. *Science*, **277**, 828–31.

9 *Primate foraging adaptations: two research strategies*

STUART A. ALTMANN

Introduction

In the course of evolution, probably every organ system of animals has been altered by the exigencies of obtaining food. An outcome of our rapidly

Feeding Ecology in Apes and Other Primates. Ecological, Physical and Behavioral Aspects, ed. G. Hohmann, M. M. Robbins, and C. Boesch. Published by Cambridge University Press. © Cambridge University Press 2006.

expanding knowledge of primates in the wild is that numerous potential adaptations for foraging have been described. These proposed adaptations are at virtually every level of biological organization, including social and individual behavior, mental states, physiology, anatomy, and morphology (Rodman & Cant, 1984; Whiten & Widdowson, 1992; Miller, 2002). However, an adaptation proposed is not an adaptation confirmed.

Gould & Vrba (1982) pointed out the presence of two distinct adaptation concepts in the literature, one historical, emphasizing traits' origins and their past histories of selection, the other nonhistorical, emphasizing current functions of traits and their contributions to fitness. My discussion is limited to the latter.

From the standpoint of the current functions of phenotypic traits and their impact on selection, a trait variant is better adapted, relative to competing variants in other individuals of the same species, to the extent that it directly contributes to fitness. Competing variants of traits are those that potentially can become relatively more common at the expense of others. Trait variants are more likely to be competing the closer the subjects are in other respects, for example, in order of increasing proximity: members of the same species, same deme or local population, same group, cohort, season of birth, sex, and with mothers of nearly the same dominance rank.

Thus, in studies of living organisms, one can confirm that a trait is adaptive and measure its degree of adaptiveness by determining its effect on fitness. However, just identifying which traits are adaptive tells us little about the functional processes by which traits affect fitness or their relative contributions to it. For studying the adaptive significance of traits in extant species, several methods are available (Endler, 1986; Rose & Lauder, 1996; Altmann, 2005). The choice among them depends on what aspects of adaptation one wishes to study.

In what follows, I focus on two closely related strategies for studying foraging adaptations in extant, wild primates or other animals. Both are based on quantitative relationships between phenotypic traits, their proximate effects ("performances," "functions"), and biological fitness. This focus reflects an important distinction, that between phenotypic selection and the genetic response to selection.

> Natural selection acts on phenotypes, regardless of their genetic basis, and produces immediate phenotypic effects within a generation that can be measured without recourse to principles of heredity or evolution. In contrast, evolutionary response to selection, the genetic change that occurs from one generation to the next, does depend on genetic variation. *(Lande & Arnold, 1983)*

Of these two strategies for the study of adaptations in wild primates or other animals, one is based on a priori design specifications for optimal phenotypes and has been applied to primates in their natural habitats. The other is based on multivariate selection theory, which deals with the effects of selection acting simultaneously on multiple characters, and has been applied to various other animals. Emphasis here will be placed on what each strategy can reveal about adaptations, the types of data that each requires, and how one can get from one to the other. For details of field techniques, logistics, sampling methods, assumptions, data analysis, and so forth, the reader should turn to the primary literature.

Measuring adaptiveness in extant species

First strategy

This strategy utilizes methods developed by Russell Lande and Stevan Arnold for studying adaptations by measuring the impact of traits and their proximate effects on biological fitness (Arnold, 1983, 1988; Lande & Arnold, 1983). To make this approach concrete, consider a study of a fictitious primate.

> The study is carried out on a local population of arboreal monkeys for which long-term birth and death records are maintained. We suspect that the monkeys' fitness is limited primarily by their intakes of proteins and energy. Although all have access to and eat the same foods, some individuals eat more nuts than others, others more flower nectaries, and still others, more insects. Each day, we record the intake of each food that they consume. We also measure any other traits that are suspected of being correlated with intakes of nuts, flowers, and insects, our three prime candidates for limiting the monkeys' fitness. We collect and preserve samples of each food and have them analyzed for nutrients and any suspected toxins.
>
> Suppose that we want to measure the impact on biological fitness of individual differences in a given diet component, such as the amount of nut-meat consumed, independent of the quantities of other foods in the diet. How can we do this?

Selection gradients
To measure the potential impact of any given phenotypic trait on relative fitness, regress relative fitness w on it (Lande & Arnold, 1983). However, because fitness-affecting traits may be correlated, use partial regression β_{wz_i}

(ordinary, not standardized) to measure the direct impact of the ith trait, z_i, on relative fitness w, with indirect effects from correlated traits thus held constant. Repeat for each of the other traits that may affect relative fitness. Then, to document any correlations among these traits, calculate their covariances (unstandardized correlations).

Relative fitness w of an individual is defined with respect to the mean fitness in the population: $w = W/W^-$, where W is the absolute fitness of an individual and W^- is the mean absolute fitness in the population. The fitness of individuals can be estimated in several ways, particularly by using aspects of reproductive success, e.g., the number of surviving offspring. Individual fitness values can be determined at long-term study sites of populations for which birth and death data are consistently recorded. For data sets of shorter duration, it can be estimated from various components of fitness (Howard, 1979).

The partial regression of relative fitness on a given character is its *selection gradient*. It measures the change in relative fitness expected if that character were changed by a unit amount but none of the other characters varied. It can be thought of as an indication of the sensitivity of fitness to changes or differences in the character.

> So, from the birth and death records of our exemplar monkeys, we calculate the relative fitness of each subject. The partial regression of relative fitness on, e.g., nut consumption is the latter's selection gradient. It indicates the change in the monkeys' fitness per unit increase in nut consumption, with all other foods in the regression held constant at their mean value, and it thus provides a measure of the potential adaptiveness of nut consumption.

Contributions of traits to fitness

Of course, to understand the realized adaptiveness of traits, we want to know the *magnitude* of their influences on fitness, not just the latter's *sensitivity* to them. For this purpose, we use each trait's selection gradient to calculate that trait's average contribution to mean fitness.

> Suppose that, as suspected, the fitness gradients reveal three traits, nut-eating, flower-eating, and insect-eating, to which fitness is particularly sensitive. We would like to know how much each of these phenotypic traits contributes to mean fitness. Each such contribution is the product of that trait's average value and its selection gradient.

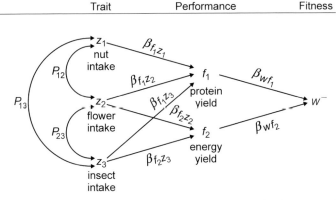

Figure 9.1. Path diagram representation of statistical relationships (beta terms, single-headed arrows) between any phenotypic trait z_i (diet), each of its proximate effects f_j (performance), and the latter's effects on fitness, w^-. The covariances among the traits are represented by double-headed arrows. For example, P_{13} is the covariance between nut consumption and insect consumption. For simplicity, arrows indicating residual influences on performance and fitness are not shown.

$$
\begin{aligned}
w^- &= \text{average fitness} \\
&= \beta_{wz_1 \cdot \bar{z}_1} = \text{average contribution of nut-eating} \\
&+ \beta_{wz_2 \cdot \bar{z}_2} = \text{average contribution of flower-eating} \qquad (9.1)\\
&+ \beta_{wz_3 \cdot \bar{z}_3} = \text{average contribution of insect-eating} \\
&+ \dots \quad = \text{contribution from other elements.}
\end{aligned}
$$

A remarkable result, due to Russell Lande, is that for a set of characters that affect fitness, their selection gradients, each calculated with effects of all correlated traits that directly affect fitness partialled out, include all the information about phenotypic selection (but not inheritance) that is needed to predict the directional response to selection.

Partitioning selection gradients

Although selection gradients evaluate causal links from traits to fitness, they do not tell us anything about the intervening functional effects of traits that augment fitness. Selection requires a mechanism. Arnold (1983) used Sewall Wright's method of path analysis to provide a convenient means of partitioning selection gradients (Figure 9.1). He showed that for any trait z_i (such as nut consumption, z_1 in Figure 9.1) that affects only one fitness-related performance variable f_j (e.g., protein yield, f_1), a selection gradient β_{wz_i} can be partitioned into two parts, a *performance gradient* $\beta_{f_j z_i}$, representing the effect of the trait on that aspect of performance, and a *fitness gradient* β_{wf_j}, representing the effect of performance on fitness. That is,

$$\beta_{wz_i} = \beta_{f_j z_i} \bullet \beta_{wf_j},$$

$$\begin{bmatrix} \text{selection} \\ \text{gradient} \end{bmatrix} = \begin{bmatrix} \text{performance} \\ \text{gradient} \end{bmatrix} \bullet \begin{bmatrix} \text{fitness} \\ \text{gradient} \end{bmatrix} \qquad (9.2)$$

where β_{wf_j} is the partial regression of relative fitness w on the jth performance variable, and $\beta_{f_j z_i}$ is the partial regression of that performance variable on the ith trait variable.

A trait may affect more than one performance variable, resulting in branching paths. For example, the second trait z_2 in Figure 9.1 (flower-eating) affects two performance variables, f_1 and f_2 (protein and energy). In that case, the total path connecting character z_2 with relative fitness is the sum of the two paths, one through performance variable f_1 and one through performance variable f_2, as shown in Figure 9.1. The corresponding relationship in partial regression coefficients is $\beta_{wz_2} = \beta_{f_1 z_2} \beta_{wf_1} + \beta_{f_2 z_2} \beta_{wf_2}$. Thus, the total selection gradient can be partitioned into additive parts, corresponding to branching paths of influence on fitness, as well as factored along paths. These elementary results can readily be expanded for analysis of selection in situations considerably more complicated than that of the fictitious primate depicted in Figure 9.1.

Contributions of traits to performance

We can use each trait's performance gradient to calculate that trait's average contribution to the mean value of a given proximate effect. To illustrate, consider our exemplar primate.

Suppose that, as we suspected, the fitness gradients reveal two performance variables, protein intake and energy intake, to which fitness is particularly sensitive. We would like to know how much, on average, each phenotypic trait contributes to the animals' protein intake (and similarly, to their energy intake). The protein intake of the average animal in the local population can be expressed as the sum of the contributions made by each forage-related trait. Each such contribution is the product of that trait's average value and its performance gradient for protein yield:

$$\begin{aligned} \bar{f}_1 &= \textit{average protein yield} \\ &= \beta_{f_1 z_1 \cdot \bar{z}_1} = \textit{average contribution of nut-eating} \\ &+ \beta_{f_1 z_2 \cdot \bar{z}_2} = \textit{average contribution of nectary-eating} \qquad (9.3) \\ &+ \beta_{f_1 z_3 \cdot \bar{z}_3} = \textit{average contribution of insect-eating} \\ &+ \dots \quad\;\; = \textit{contribution from other elements.} \end{aligned}$$

Contributions of performance variables to fitness

We can proceed similarly for the second causal link, evaluating the average contribution made by each performance variable to mean fitness.

> For example, the mean fitness w^- of the monkeys can be partitioned into additive components:
>
> $$w^- = \beta_{wf_1} \bullet f_1 + \beta_{wf_2} \bullet \bar{f}_2 + contribution\ from\ other\ elements,$$
>
> (9.4)
>
> where the first two terms in the summation on the right are the contributions to mean fitness made by the average monkey's intake of protein and energy, respectively.

Such contributions of performance variables to fitness are excellent indicators of their relative adaptiveness.

In sum, the first strategy enables us to study the effects of correlated traits on fitness, by measuring both the sensitivity of relative fitness to variability in individual phenotypic traits, holding other traits constant, and the independent contribution that each trait makes to fitness. In so doing, it measures the adaptiveness of traits. In addition, the first strategy enables us to partition that sensitivity into two causal links, those from traits to performance variables, and those from performance variables to fitness. It enables us to estimate the mean contribution of each phenotypic trait to the average value of each performance variable, and—perhaps the best measure of a performance variable's adaptiveness—each performance variable's contribution to fitness.

Second strategy

An optimality model of foraging behavior specifies how an individual of a given species should behave, under prevailing circumstances, in order to optimize (maximize or minimize, as appropriate) a performance variable. For example, what selection of foods would maximize energy intake? What hunting strategy would minimize hunting time? That performance variable, the "currency," is selected because it is expected to be a major contributor to mean fitness. The objective variable that is to be optimized is written as a function (the "objective function") of its causative traits.

> Consider a model for maximizing mean daily energy intake E applied to our paradigm monkeys, with their diet of nuts, flower nectaries, and insects. Suppose that the energy obtained from foods is 15 kilojoules per gram of nuts, 3 kJ per gram of nectaries, and 4 kJ

per gram of insects, and let c_i represent respectively the amounts (grams consumed per day) of the Ith food. The objective then is to find those values of c_1, c_2, *and* c_3 *that would maximize E, where* $E = 15c_1 + 3c_2 + 4c_3$.

Of course, there are no benefits without costs, "no such thing as a free lunch." The second component of an optimality model consists of *costs* (*constraints, limiting factors*), such as nutrient requirements at the lower end, toxins or other hazards at the upper, that represent the animals' limitations and that keep the currency from going to zero or infinity. They too are written as functions.

Some costs may vary continuously with the objective variable. If such costs can be expressed in the same units as the objective variable, then the objective can be to optimize the difference between them, the net benefit, or "trade-off."

> Suppose for simplicity that for our exemplar monkeys, flower nectar is their only source of energy and that as they begin to exhaust their local supply of flowers with filled nectaries, they range ever farther away from their home range center. As they do so, they encounter adjacent groups with increasing frequency, resulting in progressively more energy-consuming chases. They eventually reach a "point of diminishing returns" beyond which the extra energy gained from additional nectar is less than the loss from being chased. Going just that far to feed on nectaries is the optimal solution, unless doing so would not already have put them beyond some other upward-limiting constraint.

Other costs can each be approximated by a step function, a discrete boundary beyond which the animal cannot remain indefinitely without seriously impairing some vital function, perhaps fatally, but within which further increases have no significant effect. For example, nutrient intakes above the minimum required to prevent deficiency symptoms are claimed by nutritional scientists to have no further beneficial effect. The same may be true of many other constraints. So, for example, if our monkeys' gut sizes and food passage times limit them to 500 g of food per day, their consumption constraint, in grams/day, is $c_1 + c_2 + c_3 \le 500$.

In short, such models consist of an equation for the objective function (a putative fitness-enhancing performance variable that is to be maximized or minimized, written as a function of contributing phenotypic traits), and various other functions representing constraints on the objective function (attributes of the organism or its relationship with the environment). For example, in an optimal diet model whose objective is maximizing daily energy intake, the simultaneous solution to the equations of the model would indicate an amount of each available food, which, if consumed by the subjects, would maximize mean daily energy intake and give them an otherwise adequate diet without taking too much time, exceeding the animals' consumption capacity, and so forth. A simultaneous solution to these equations is required. The animals' actual diets and their effects can then be compared with their optimal diets. In 1978, Steve Wagner and I published a method for providing closed-form solutions to such sets of equations if linear, not knowing that an iterative procedure, developed by Dantzig, was already well established.

Data requirements

Note that the first and second strategies include the same three empirical components. First, they require quantitative samples, taken in a local population of a species, of individual variants in a set of phenotypic traits. Typically, these are traits known or thought to affect a particular vital activity, such as getting food or obtaining mates. Next, they require quantitative samples of proximate effects (performances) of these variants, whether known a priori to be functional or otherwise. Third, they require estimates of the biological fitness of each subject.

In addition, the second strategy makes use of an optimality model that relates phenotypic traits to their fitness-enhancing proximate effects, and so, requires specific information, as follows. Quantitative data – presumably obtained in the performance samples described above – are required for each individual's success on that performance variable (or those variables) that are the basis of the model's objective function. Beyond that, information is required to establish the model's constraints: quantitative data on various traits of the animals and of their relationships with the environment, as needed to establish upper and lower constraints on the optimization.

An example

The optimal diet model that I applied to the foraging of yearling baboons illustrates the feasibility of such models (Altmann, 1998, chapter 8). On the assumption that energy is the primary fitness-limiting component of the baboons' diets, I took optimal diets to be those that maximize the yearlings' daily energy intake while simultaneously keeping them above their minima for nutrients and below their maxima for various constraints such as toxin tolerances, time limits, gut processing capacities, and so forth. In addition to the linear objective function, the model had 72 constraint functions, each in the form of a linear equation! Yet, models much larger than this can now quickly be solved by readily available computer programs. The model was adjusted and recalculated for data taken at 10-week intervals to take into account seasonal changes in available foods and age-related changes in the yearling's requirements and tolerances.

Obtaining data on the diets of 11 yearlings occupied a year, during which I recorded food intakes in 18 460 feeding bouts during 333 hours of in-sight sample time. For practical reasons, data analysis was limited to 52 core foods on which the yearlings spent the most time feeding and that collectively accounted for 93% of their feeding time. Data on the subsequent survivorship and reproduction of these subjects were obtained as part of routine long-term demographic monitoring in the Amboseli baboon project, and continued for the rest of the subjects' lives, which in the extreme case, female Dotty, took 27.7 years (Bronikowski *et al.*, 2002). However, well before that, strong patterns became apparent, and one need not wait until the last subject has died before examining available data. Eight components of fitness were evaluated as of a cut-off date that was, on average, 14.4 years after the yearlings were born. (By that date, all but two were dead.) These data on each individual's dietary intake, its costs and benefits, and the subject's fitness were then used to test the optimal diet model.

Confirmation

Models are elaborate hypotheses, and are often regarded as attempts to describe some small aspect of the world. From this descriptive perspective and given an assumption that natural selection tends to eliminate suboptimal traits, a model of optimal foraging or diet would be considered confirmed to the extent that the observed foraging behavior or food consumption matches the model's specifications. A goodness-of-fit test usually would be appropriate and would tell us whether the traits of individuals or their population mean were significantly different from specified optimal trait values. Yet, if so, we would not be able to distinguish the shortcomings of the model from the shortcomings of the animals. For a host of reasons (reviewed by Maynard Smith, 1978; Emlen, 1987; Rose & Lauder, 1996; Altmann, 1998), organisms may not perform at or near their optima.

An alternative strategy, and the one that I adopted (Altmann, 1998, chapter 1), is to consider optimality models as normative, not descriptive. In a normative strategy, deviations of traits from values specified by an optimality model of adaptive traits are regarded not as tests of the model but as indications of potential differences in fitness. The model itself would be tested by showing that those individuals whose traits are closer to the putatitive optimum have higher fitness as a consequence and, conversely, that those that deviate sufficiently from the optimum have predictable functional impairments, such as nutrient deficiencies or toxicity effects, which lead to a reduction in fitness. Confirming an optimality model in this way is, at the same time, a confirmation of adaptive differences in the specified traits and in the functional mechanisms by which these traits affect fitness.

Not surprisingly, the yearlings' dietary intakes were not optimal. Every baboon at every age at which I sampled it took in suboptimal quantities of virtually all macronutrients (water, minerals, proteins, fiber, other carbohydrates, energy and lipids), relative to quantities specified by their age-specific diet for maximizing energy. Indeed, there was but one exception out of 227 yearling-age-macronutrient comparisons: female Eno, like all the others, took in suboptimal quantities of all macronutrients, except that, at 30–40 weeks of age, she took in the optimal quantity of lipids, but just barely (Altmann, 1998 table 7.8).

All of these macronutrient shortfalls of the yearlings deviated less than 8% from the shortfall in their total dietary mass, but that total was, on average, just 52% of the mass of an energy-maximizing diet. That is, on average, the yearlings ate close to a balanced diet, in the sense that the macronutrients in their diets were in ratios moderately close to what would be needed to obtain an energy-maximizing diet, they just didn't eat enough of them. None of their diets, at any age in the 30–70 week age-interval in which they were sampled, were within two standard deviations of that optimal mass.

In short, many more of the yearlings' nutrient shortfalls were attributable to inadequate food consumption than to poor choices of foods – to the quantity of their diet, rather than its quality. I do not know the source of these food deficiencies, but suspect that in Amboseli the sparse distribution of foods, not their quality, abundance, or seasonal unavailability, is the primary factor limiting the baboons' intakes. For energy, this may have affected the lactation capacity of the yearlings' mothers: 45% of the yearlings' mean energy shortfall resulted from the discrepancy (1.31 MJ/day in milk energy) between the amount of nursing that they did and the amount prescribed by the energy-maximizing diet.

Clearly, the energy-maximizing diet model for Amboseli's yearlings would be rejected outright by anyone following the descriptive strategy. Neither the yearlings nor their mean were anywhere near optimal. However, with a normative strategy, the optimal diet model was confirmed to a remarkable degree (Altman, 1991, 1998). Several characteristics of the yearlings' diets and other traits, many of which are intercorrelated, provide good predictions of components of fitness. I here focus on two, *energy shortfall* (deviation of energy intake from optimal energy intake, as a percentage of the latter) and *protein surplus* (deviation of protein intake from amount in an energy-maximizing diet, as a percentage of the latter). *Reproductive success* – the

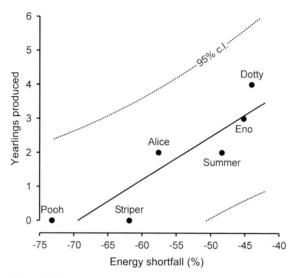

Figure 9.2. Reproductive success of females, on the y-axis (numbers of surviving yearlings they each produced in their lifetime) as a function of the females' energy shortfalls when they were yearlings, x-axis (percent deviation of their actual intakes from their respective optimal intakes). Regression: $n = 6$, adjusted $R^2 = 0.76$, $p \leq 0.05$, $y = 8.8 + 0.127x$. From Altmann (1998, Fig. 8.9).

number of yearlings that each female produced by the cutoff date (mean age 14.4 years, two females surviving) – was taken as the fitness value for the six females in my study. (No paternities for the male subjects were known at the time.) A linear regression of fitness on energy shortfall indicated that energy shortfall accounted for 76% of the variance in female fitness ($p \leq 0.05$). Those females who took in more energy, and so had smaller energy shortfalls, produced more surviving infants (Figure 9.2). On average, each eight percentage point difference in energy shortfall during a female's childhood translated into an additional surviving yearling over her lifetime.

I then calculated a multivariate linear regression of female fitness (reproductive success, as defined above) on a combination of energy shortfall and protein surplus (Figure 9.3). The resulting linear equation is:

$$\text{fitness} = 33.22 + 30.53(\text{energy shortfall}) - 21.57(\text{protein surplus}) \tag{9.5}$$

This linear combination of energy shortfall and protein surplus accounted for 94% of the variance among females in their fitness. In short, the fitness of these females *as adults* was highly predictable from what they ate when they were yearlings, just 30–70 weeks of age.

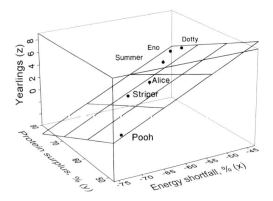

Figure 9.3. Reproductive success, on the *z*-axis – that is, numbers of yearlings produced by the females – is predicted as a linear combination (the tilted plane) of their energy shortfalls as yearlings, on the *x*-axis (percent deviation of their actual energy intakes from their respective optimal intakes) and their protein surpluses, on the *y*-axis (energy intakes above requirements), each calculated with the effect of the other predictor variable held constant at its mean value. Named dots show individual observed values; deviations are vertical lines (very short) from observed to predicted values. Regression equation is equation (9.5) in text, $n = 6$, adjusted $R^2 = 0.94$, $p \leq 0.01$. From Altmann (1998, Fig. 8.13).

Considering how highly correlated these two performance variables are (product moment correlation 0.95), adding protein as a second predictor variable might have seemed an unlikely way to improve the prediction of fitness. However, the coefficients of a multiple regression are, in fact, the beta values of partial regression, and each indicates the rate of change in the dependent variable (fitness) per unit in that predictor variable, with the others held constant at their mean value. Thus, the protein surplus coefficient in Equation 9.5 indicates the rate of change in fitness per unit change in daily protein intake in excess of requirement, independent of the effects of energy on fitness; and conversely, for the energy shortfall.

The resemblance of Equations 9.4 and 9.5 is no coincidence: the coefficients of a multivariate linear regression are the partial regression coefficients of the variables. The equations differ in two respects. In Equation 9.4, the dependent variable is relative fitness, whereas I used absolute fitness – a simple difference in units. Second, Equation 9.4 regresses fitness values on values of performance variables. In contrast, Equation 9.5 regresses fitness values on *deviations* in values of performance variables from values in an optimality model. It provides an answer to the central question: Do those individuals whose diets come closer to the putative optimum have higher fitness? Confirming an optimality model in this way is a confirmation of adaptive differences in specified traits and in our identification of the functional mechanisms by which these traits affect fitness. At the same time, the model provides testable hypotheses for relationships between traits and their proximate effects, not just the null hypothesis that some effects occur.

In the process of applying standard methods for linear optimization to calculate the constituents of an optimal diet, numerous related questions are also answered (Altmann, 1984, 1998 chapter 5). Here are examples. What attributes of the animals or of their environment limit the amount of energy in the diet? How much could each attribute of the organisms or environment that does not limit energy intake change before it became a limiting factor? How sensitive is the energy content of the diet to the values of each of the limiting factors? How much could the energy density of a food change, or its estimated value be in error, without changing the composition of the optimal diet? Is the optimal diet unique? That is, is there more than one combination of foods that would maximize the objective function while satisfying all constraints?

The second strategy, like the first, is concerned with the effects of traits on performance and of each of these on fitness. Nothing in the second strategy would preclude making use of gradients of selection, performance, and fitness, described in the first strategy, to clarify these causal links. On the contrary, the two methods can fruitfully be combined (Arnold, 1988).

Nonlinearities

In my model of energy-maximizing diets for yearling baboons, the various equations – for the objective function, for lower-bound constraints (nutrients) and for upper-bound constraints (toxin tolerances, gut capacity, and so forth)— were all linear. In that study, 94% of the yearlings' individual differences in fitness were accounted for by a linear combination of their deviations from the amounts of protein and energy in an energy-maximizing diet. However, in some situations, some of these relationships may be appreciably nonlinear, and their treatment should be considered.

Some examples

1. In optimal foraging studies, the most common examples of nonlinearity are ones in which the objective function is a cost–benefit ratio or any other ratio of two random variables, such as energy obtained per minute of foraging time. Michael Altmann (in Altmann, 1998, Appendices 7 and 8) has provided a method for maximizing or minimizing objective functions of this type. For methods of

optimizing other types of nonlinear objective functions, an appreciable literature on nonlinear optimization is available.

2. Nutrient requirements, toxin tolerances, and probably many other constraints that are approximated by lines (in two dimensions, e.g., just two foods) or by hyperplanes (multiple dimensions) are actually probability distributions, e.g., the probability that any given dietary intake of ascorbic acid would result in scurvy. By drawing the lines at, say, the mean plus two standard deviations of tolerances for toxins (and conversely for nutrients), one would reduce the probability of advocating a diet that is high in energy but debilitating or even lethal.

3. Many nutrients interact with each other and with toxins in plants or other forage-related hazards. Some of these interactions result in nonlinear constraints. For example, because of the genetic variability of malarias, the chance and severity of malaria infection probably increases exponentially with time spent foraging in malaria-infested areas. However, many other such interactions may be nonlinear. For example, oxalic acid in some foods reacts with calcium ions to form insoluble calcium oxalate, thereby rendering that much calcium biologically unavailable. However, this reaction just requires representing intakes of available calcium as mols of calcium consumed minus mols of oxalate consumed.

4. The relationships described herein between phenotypic traits, their proximate effects, and fitness are based on directional selection. For stabilizing and other forms of nonlinear selection, see Arnold (2003).

Practicability

Knowledge of mechanisms, and of requirements and limits for food components, is the great advantage that studies of foraging behavior have over studies of many other forms of behavior, in that we can say, at least to a first approximation, what a well-adapted primate should eat. If it takes in too little iron, it will become anemic, too much and it may suffer from siderosis. If it takes in too little vitamin D, rickets (infants) or osteomalacia (adults) results; too much, and demineralization of bone and mineralization of soft tissue result. If it eats too many seeds of certain legumes, it may suffer the toxic effects of trypsin inhibitor, too few and it may not get enough protein. For most nutrients and a few toxins in foods, quantitative upper and lower limits are moderately well established (Altmann, 2005). Yet, who could do the same for, say, play behavior? What kinds and what amounts are better? Where is the research that could be used to advise a maturing male primate of when and how hard to hit or slash at his opponent in a fight and when to back off, or to tell a primate mother which adult females she should allow to hold her infant and for how long?

Whether the two types of study that I have described are practicable for a set of traits or their proximate effects depends on whether the requisite data can be obtained, namely, quantitative data on individual differences in traits, in proximate trait effects, and in components of fitness. To secure all three requires a combination of short- and long-term research plans.

Two major practical problems occur in implementing either of these research strategies for studying primate foraging adaptations. First, sampling large numbers of subjects is difficult. My sample size, 11 infants – or, for reproductive success, just six females – might not have revealed statistically significant differences were it not for the subjects' great variability in survivorship and reproductive success and strong individual differences in their dietary intakes. During a year of field work, I sampled the foraging of the entire cohort of infants in the main study group that were between 30 and 70 weeks of age at any time during my study (nine subjects); in a pilot study the previous year, I had sampled two others in the same age range. To enlarge the sample appreciably would have required additional observers and groups. In addition, the amount of sampling that I did per individual was less than ideal. Data that I obtained in a year of sampling the foraging of 11 yearling baboons and the chemical characteristics of their foods were sufficient to differentiate many but not all pairs of individual intakes on the basis of seven macronutrients during each of four, 10-week age classes.

The other major practical problem is the time required to obtain good estimates of fitness. However, basic demographic data – dates of births, deaths, emigrations of individuals from and immigrations into groups, and sightings of solitary individuals – are routinely obtained in almost all long-term studies. They provide information that is needed for a wide variety of projects. The advantage of studying infants, as I did, is that one can thereby capitalize on the high mortality that is characteristic of many mammals. As for reproductive success, my cutoff date for evaluating the fitness of my subjects was at an age that was half the life span of our oldest animal of known age.

Even with very long-lived primates such as chimpanzees, the study of adaptations in the wild is feasible. Of more than 40 chimp study sites in Africa, four have resulted in studies of chimp communities extending more than 15 years, and two – Kasakela at Gombe (Wilson & Wrangham, 2003) and the Boussou group in Guinea (Sugiyama, 2004) – have been studied for

43 years and 26 years, respectively, through 2003. Compare these numbers with some key demographic values. Mean age at first parturition averages 10.9 years to 14.6 years in chimps, depending on location. Survival to first parturition ranges from 22%–58% (Sugiyama, 2004). Thus, by a happy coincidence, these apes, with one of the longest life expectancies among primates, are also the subject of some of the most sustained programs of field research. With appropriate combinations of short- and long-term planning, the adaptive value of a wide variety of traits could be confirmed, particularly by taking advantage of high infant mortality.

Other options

What can one do to study the adaptiveness of forage-related traits if data on lifetime fitness values of the subjects are not (yet) obtainable? Several opportunities are available (Altmann, 2005), including both of the strategies described above. The first strategy is based on Arnold's (1983) separation of fitness into two parts: a performance gradient representing the effect of the trait on some aspect of performance and a fitness gradient representing the effect of performance on fitness. "The point of this distinction," he wrote, "is that even when effects on fitness cannot be measured, it will often be possible to measure the effects on performance." We can take advantage of the ability of performance gradients to isolate the effect that each trait has on a given performance from effects of correlated traits, and we can quantitatively evaluate the contribution made by each trait to each performance variable. If we assume that, through their impact on vital processes, each of these performance variables affects fitness, they are indicative of the adaptiveness of the traits, even though in the absence of fitness data, we would be unable to test that assumption.

Similarly, if we have an optimality model (second strategy), we would already have hypothesized how to combine trait variables into quantitative predictions of each individual's level of performance on a major fitness-enhancing performance variable, and thus to predict its fitness relative to other members of its local population. Far more optimality models have been applied to humans than to any other species of primates, albeit without fitness correlates (Winterhalder & Smith, 1981, 2000; Smith, 1983; Smith *et al.*, 2001, and references therein).

Coda

In long-term studies, estimates of fitness components, such as survivorship, become possible first, then eventually, lifetime fitness. As these estimates become available, we can evaluate the impacts both of traits on functions and of functions on fitness with increasing accuracy. The benefits that we then reap go far beyond being able to say, yes, these traits are demonstrably adaptive. We obtain a far richer understanding of the mechanisms and processes by which they affect natural selection and so, in turn, are shaped by it.

I end with Arnold's (1983) ending. It is not enough to complain that adaptation is often invoked without critical evidence (Williams, 1966; Lewontin, 1979; Rowell, 1979; Baldwin & Baldwin, 1979; Gould & Lewontin, 1979). We also need an analytical approach that emphasizes what can be accomplished. The strategy outlined here is a step in the right direction.

Acknowledgments

I am grateful to Stevan J. Arnold for discussions of strategies for relating foods, functions, and fitness, and to Jeanne Altmann for providing helpful comments on an earlier version of this article. My research on the foraging behavior of yearling baboons was supported by research grants MH 19617 from the National Institute of Mental Health and 15007 from the National Institute of Child Health and Development, and by the Abbott Fund of the University of Chicago. Finally, I am grateful to the Max Planck Institute for Evolutionary Anthropology (Leipzig) for the invitation to participate in this publication and the Institute's conference where it was born.

References

Altmann, S.A. (1984). What is the dual of the energy-maximization problem? *The American Naturalist*, **123**, 433–41.
  (1991). Diets of yearling female baboons (*Papio cynocephalus*) predict lifetime fitness. *Proceedings of the National Academy of Sciences of the United States of America*, **88**, 420–3.
  (1998). *Foraging for Survival: Yearling Baboons in Africa*. Chicago, IL: University of Chicago Press
  (2005). Adaptation. In *Encyclopedia of Anthropology*, ed. J. Birxh. Thousand Oaks, CA: Sage Publications.
Altmann, S.A. & Wagner, S.S. (1978). A general model of optimal diet. In *Recent Advances in Primatology*, ed. D. J. Chivers & J. Herbert, pp. 407–14. London: Academic Press.

Arnold, S.J. (1983). Morphology, performance and fitness. *American Zoologist*, **23**, 347–61.

Arnold, S.J. (1988). Behavior, energy and fitness. *American Zoologist*, **28**, 815–27.

(2003). Performance surfaces and adaptive landscapes. *Integrative and Comparative Biology*, **43**, 367–75.

Baldwin, J.D. & Baldwin, J.I. (1979). The phylogenetic and ontogenetic variables that shape behavior and social organization. In *Primate Ecology and Human Origins: Ecological Influences on Social Organization*, ed. I.S.I.S. Bernstein & E.O. Smith, pp. 89–116. New York, NY: Garland STMP Press.

Bronikowski, A.M., Alberts, S.C., Altmann, J. *et al.* (2002). The aging baboon: comparative demography in a nonhuman primate. *Proceedings of the National Academy of Sciences of the United States of America*, **99**, 9591–5.

Emlen, J.M. (1987). Evolutionary ecology and the optimality assumption. In *The Latest on the Best*, ed. J. Dupré, pp. 163–77. Cambridge, MA: MIT Press.

Endler, J.A. (1986). *Natural Selection in the Wild*. Princeton, NJ: Princeton University Press.

Gould, S.J. & Lewontin, R.C. (1979). The spandrels of San Marco and the panglossian paradigm: a critique of the adaptationist programme. *Proceedings of the Royal Society of London, Series B*, **205**, 581–98.

Gould, S.J. & Vrba, E.S. (1982). Exaption: a missing term in the science of form. *Paleobiology*, **8**, 4–15.

Howard, R.D. (1979). Estimating reproductive success in natural populations. *The American Naturalist*, **114**, 221–31.

Lande, R. & Arnold, S.J. (1983). The measurement of selection on correlated characters. *Evolution*, **37**, 1210–26.

Lewontin, R.C. (1979). Sociobiology as an adaptationist program. *Behavioral Science*, **24**, 1–10.

Maynard Smith, J. (1978). Optimization theory in evolution. *Annual Review of Ecology and Systematics*, **9**, 31–56.

Miller, L.E. (2002). *Eat or Be Eaten: Predator Sensitive Foraging Among Primates*. Cambridge: Cambridge University Press.

Rodman, P.S. & Cant, J.G.H. (1984). *Adaptations for Foraging in Nonhuman Primates*. New York, NY: Columbia University Press.

Rose, M.R. & Lauder, G.V. (1996). *Adaptation*. New York: Academic Press.

Rowell, T.E. (1979). How would we know if social organization were not adaptive? In *Primate Ecology and Human Origins: Ecological Influences on Social Organization*, ed. I.S.I.S. Bernstein & E.O. Smith, pp. 1–22. New York, NY: Garland STMP Press.

Smith, E.A. (1983). Anthropological applications of optimal foraging theory: a critical review. *Current Anthropology*, **24**, 625–51.

Smith, E.A., Borgerhoff Mulder, M., & Hill, K. (2001). Controversies in the evolutionary social sciences: a guide for the perplexed. *Trends in Ecology and Evolution*, **16**, 128–35.

Sugiyama, Y. (2004). Demographic parameters and life history of chimpanzees at Boussou, Guinea. *American Journal of Physical Anthropology*, **124**, 154–65.

Whiten, A. & Widdowson, E.M. (1992). *Foraging Strategies and Natural Diet of Monkeys, Apes, and Humans.* Oxford: Clarendon Press.

Williams, G.S. (1966). *Adaptation and Natural Selection.* Princeton, NJ: Princeton University Press.

Wilson, M.L. & Wrangham, R.W. (2003). Intergroup relations in chimpanzees. *Annual Review of Anthropology,* **32**, 363–92.

Winterhalder, B. & Smith, E.A. (1981). *Hunter-Gatherer Foraging Strategies.* Chicago, IL: University of Chicago Press.

    (2000). Analyzing adaptive strategies: human behavioral ecology at twenty-five. *Evolutionary Anthropology,* **9**, 51–72.

10 The predictive power of socioecological models: a reconsideration of resource characteristics, agonism, and dominance hierarchies

ANDREAS KOENIG AND CAROLA BORRIES

Introduction

Beginning with Wrangham's work (1979, 1980), testing predictions regarding patterns of female agonistic behavior, social structure, and dispersal as a

Feeding Ecology in Apes and Other Primates. Ecological, Physical and Behavioral Aspects, ed. G. Hohmann, M.M. Robbins, and C. Boesch. Published by Cambridge University Press.

function of resource distribution and feeding competition (overviews in Koenig, 2002; Isbell & Young, 2002) have become a major focus of socio-ecological studies. Although other factors such as predation pressure and risk of infanticide (van Schaik, 1989; Sterck *et al.*, 1997) may be of additional or even greater importance and demark crucial differences between competing models (e.g., Isbell & Young, 2002), for simplicity these factors will be ignored in what follows.

Generally, resource characteristics (i.e., abundance, distribution, quality, size) have been suggested as determining the foraging, feeding, and agonistic behavior of females (van Schaik, 1989; Isbell, 1991) as well as the energy intake per unit time or patch (Janson, 1985, 1988a; Janson & van Schaik, 1988). Measurement of individual energy gain, i.e., overall energy intake minus energy expenditure, in relation to dominance rank and group size can then be used to determine the competitive regime (e.g., Janson & van Schaik, 1988; van Schaik & van Noordwijk, 1988; van Schaik, 1989; Koenig, 2002). Based on the competitive regime, the occurrence or strength of within-group and between-group contest, social consequences in terms of dominance characteristics, affiliation, and dispersal can be predicted (Wrangham, 1980; van Schaik, 1989; Isbell, 1991; Isbell & van Vuren, 1996; Sterck *et al.*, 1997; Koenig, 2002; Isbell & Young, 2002).

This model has been successful regarding several sets of predictions. For instance, resource distribution appeared to predict competitive regimes (e.g., Whitten, 1983; Barton & Whiten, 1993; Saito, 1996), resource distribution has been linked to social structure and dispersal (e.g., Mitchell *et al.*, 1991; Barton *et al.*, 1996; Koenig *et al.*, 1998; Boinski *et al.*, 2002; Schülke, 2003), and aggression has been the major factor predicting feeding success (e.g., Janson, 1985; Vogel, 2004; Janson & Vogel, this volume, Chapter 11). However, several problems have been identified as well. Resource distribution was not always a good predictor for contest competition (Boinski, 1999; Koenig, 2000, 2002). In addition, rates of aggression across species did not always match the suspected competitive regimes (Sterck & Steenbeek, 1997; Pruetz & Isbell, 2000). Finally, competitive regimes were not necessarily good predictors of social structure (e.g., Borries, 1993; Cords, 2000; Koenig & Borries, 2001; reviews in Koenig, 2002; Isbell & Young, 2002).

The purpose of this overview is therefore to return to some of the original predictions and scrutinize their predictive power and potential pitfalls, suggesting possible perspectives for future work. To that end, we pose the following three questions: (1) Do resource characteristics (particularly clumpedness) predict the mode of competition (particularly within-group contest competition)? (2) Do rates of agonism instead of dominance rank predict overall variation in energy gain? (3) Do rates of agonism predict social structure in terms of linearity or strength of hierarchies?

Resource characteristics and feeding competition

Traditional approach – botanical indices

Originally, it was suggested that resources of high quality, occurring in well-defined, clumped patches that are smaller than group size should elicit within-group contest competition (van Schaik, 1989). Such patches can be monopolized by a single individual or a subset of a group and consequently may lead to differential energy gain based on dominance rank (Janson & van Schaik, 1988; van Schaik & van Noordwijk, 1988; van Schaik, 1989).

In order to capture these resource characteristics, botanical measures of clumpedness via, e.g., mean-to-variance indices or similar botanical measurements (e.g., Krebs, 1999) appeared to be initially successful. Feeding time, energy intake, or energy gain could indeed be linked to spatially clumped resources (e.g., Barton & Whiten, 1993; Saito, 1996; Koenig *et al.*, 1998). In addition, spatially clumped or monopolizable resources quite often induce more conflicts over food (e.g., Phillips, 1995a; Pruetz & Isbell, 2000; Wittig & Boesch, 2003). At the same time, however, exceptions have been observed (Boinski, 1999; Cords, 2000; Boinski *et al.*, 2002) and the whole concept of patchiness or spatial clumpedness criticized (Isbell *et al.*, 1998; summarized in Isbell & Young, 2002). Thus, it seems necessary to ask whether resource characteristics indeed can be used to predict modes of competition. If they cannot, which additional factors need to be incorporated?

Resources, group spread, and energy gain in Hanuman langurs

In an attempt to examine the relevance of botanical measures of clumpedness, we re-examined data from a long-term study on Hanuman langurs (*Semnopithecus entellus*) at Ramnagar in southern Nepal (details on study site, population, and life history characteristics in Koenig *et al.*, 1997; Borries, 2000; Borries *et al.*, 2001). Data were derived from botanical measures of more than 2500 trees and climbers, their clumpedness was calculated via mean-to-variance ratios, and nutrient content of food and non-food was determined (Chalise, 1995; Koenig *et al.*, 1997; Koenig *et al.*, 1998). Results are combined with location and activity data for a medium-size group derived from scan sampling of all group members every 30 minutes (18–19 group

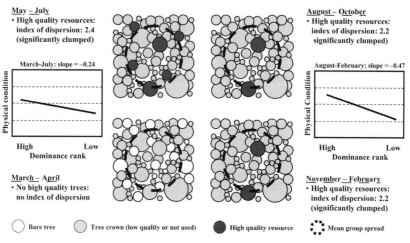

Figure 10.1. Resource characteristics and energy gain of the females in one medium-sized group of Hanuman langurs over seasons. The central four graphs depict an average 0.4 ha-forest area with the actual density of trees and corresponding crown sizes (two-dimensional) as measured via 2542 trees and climbers on 31 botanical plots (Koenig *et al.*, 1997). The botanical distribution of major food plant species of superior nutritional quality are given for each of the four "seasons" (cf. Koenig *et al.*, 1998; Koenig, 2000). All other resources are either bare, not used, or of low nutritional quality (see legend within figure). Superimposed (bold broken line) is the average group spread of a medium-sized group (cf. Koenig *et al.*, 1998). The schemes depicting physical condition in relation to dominance ranks summarize the results for March–July and August–February and are based on average slopes of nonparametric regression (see Table 4, P troop, in Koenig, 2000; see text for an explanation of why these periods were summarized).

members). During a scan the location of group members was plotted on a 16 × 16-m grid map and a minimum convex polygon was calculated (782 scans with at least 75% (n ≥ 12) of independently moving individuals in a given scan; Beise, 1996; Koenig *et al.*, 1998). In addition, we estimated female physical condition on a seven-point scale (from 1 = meager to 7 = fat) every month, judging the degree of visibility of shoulder blades, spinal column, ribs, hips, and tail bones (details in Koenig *et al.*, 1997; Koenig, 2000).

Figure 10.1 details the situation in a 0.4-ha area in the forest across seasons. During May–July the langurs fed mostly on fruits and young leaves of two particular food species with superior nutritional quality (containing significantly more sugar and/or protein than the other food plants; Koenig *et al.*, 1998; Danish *et al.*, this volume, Chapter 18). The index of dispersion for these two tree species showed a significantly clumped pattern (indices: 2.2 each species individually; 2.4 both species combined; Koenig *et al.*, 1998).

During the seasons August–October (flowers and leaves) and November–February (fruits and leaves) the langurs relied heavily on one climber species of superior quality, which was spatially clumped (index: 2.2). Finally, during March–April when the forest did not provide much food, the langurs fed on a wide variety of different items, none of which was of high nutritional quality (for details see Koenig *et al.*, 1998).

If traditional indices of dispersion are used, one would predict stronger within-group contest for the seasons May–July, August–October, and November–February (clumped distribution) as compared with March–April (dispersed, i.e., no high-quality resources). Hence, one would predict a stronger skew in energy gain for the period from May to February. In order to test this prediction, we compared the monthly regression slopes between dominance rank and physical condition for the "clumped" versus the "dispersed" months, but the result was nonsignificant (Mann–Whitney U test: $U = 10.0$, $n_1 = 11$, $n_2 = 3$, $p = 0.18$, one-tailed; observations from January 1994 to March 1995, no data for August 1994; based on values provided in Table 4, troop PA, in Koenig, 2000).

Even though the small sample for the "dispersed" months might have influenced the result, we suggest another explanation. Using an average group spread overlaid on a patch of the forest (Figure 10.1, bold broken line), we checked how many resources would fall within the spread of an average group. It became apparent that the number of high-quality resources within the average group spread varied in the course of a year. During August–February the group spread covered one to two high-quality resources, during March–April none, and during May–July three to six resources. Thus, during May–July high-quality resources effectively became more evenly distributed. This could allow lower ranking individuals to increase their energy intake, reducing the skew in energy gain. Consequently, one would expect weak or no contest during the months from March–July and stronger contest during August–February. Indeed, compared to the situation in March–July, the average slope of the regression lines during August–February was almost twice as steep, indicating stronger contest (Figure 10.1; Mann–Whitney U test: $U = 10.0$, $n_1 = 8$, $n_2 = 6$, $P = 0.04$, one-tailed; no data for August 1994, based on values provided in Table 4, troop PA, in Koenig, 2000).

The logic of resource characteristics and contest competition

The results derived from Hanuman langurs seem to indicate that resource dispersion based on botanical scores alone may not necessarily predict the mode of competition accurately (see also Phillips, 1995b; Saito, 1996; Barton *et al.*, 1996; Isbell *et al.*, 1998; Vogel, 2004). It appears that a more accurate prediction requires additional information such as resource size, resource quality, and particularly group spread. But why does it appear to be so difficult to botanically determine "contestable resources"?

The answer to this question lies in the resource characteristics supposedly predicting contest, or more precisely, in the relativity of these characteristics. The first characteristic, high quality, is a characteristic of the food item. It means that there should be a high gain rate from a particular resource compared to other items. Gain rate might be a function of nutrient content but also of handling time and of toughness of items (see also discussion of patch depletion time in Isbell *et al.*, 1998). In essence, high quality refers to the nutrient content and potentially to other aspects of preferred items relative to other items. In addition, individuals may differ in terms of their satiation level (Janson & Vogel, Chapter 11, this volume), or in their manipulative abilities or gut capacities, etc. (depending on age, sex, or species). Thus, the quality and importance of a food item varies as a function of the individual and the species.

If indeed some items are of superior quality relative to the surrounding resources, contestability of such resources depends on their size in relation to group size and clumpedness of resources (Janson, 1988b; van Schaik, 1989). Resource size clearly is a relative term. Since feeding group size varies from one to several hundred individuals across non-human primate species (Clutton-Brock & Harvey, 1977), the exact same resource may or may not be contestable for individuals of another species or of another group, even within the same population (Figure 10.2a). Therefore, contestability of resources may vary within a given population depending on the size of the group observed. The difference may be even larger when comparing across species. Holding group size constant, species and individuals of different ages will differ in body mass and the willingness and ability to defend areas of different sizes (Vogel & Janson, unpublished data). So far, individual spacing in resources has remained largely unexplored (e.g., Pruetz & Isbell, 2000). However, within the primate order body mass varies by a factor of 10 000 (Smith & Jungers, 1997). Thus, the same resource, which may elicit contest in a large-bodied primate, may have enough space and food to satiate all group members of a small-bodied primate.

Similarly, clumpedness of resources is not necessarily a fixed variable. Clumpedness, and hence the potential for contest, refers to the fact that resources of similar quality are clumped in space, and the next – the alternative – resource is outside the reach of an individual because group spread might be constrained, for instance due to predation (Janson, 1988b). A skew in energy gain is related to the fact that lower ranking individuals gain less

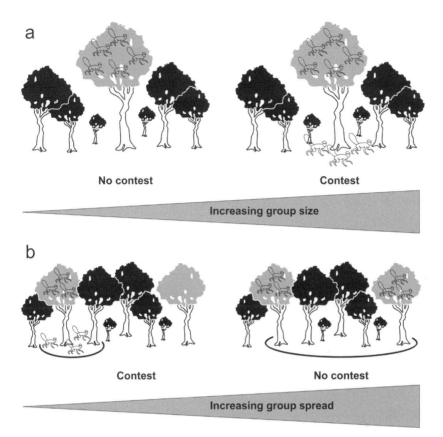

Figure 10.2. The effect of (a) increasing group size (holding group spread constant) and (b) increasing group spread (holding group size constant) on the likelihood of within-group contest competition. Trees in lighter gray indicate high-quality resources, bold line indicates group spread.

or none of the food from such a high-quality patch. Skew in energy gain is thus contingent on the nearby availability of alternative resources of similar quality. If excluded from high-quality patches, lower ranking individuals have the option of feeding on lower quality food and would need to increase feeding time to meet a similar energy intake. Alternatively, lower ranking individuals can opt to feed far from the rest of the group, which may increase predation rate (e.g., van Schaik & van Noordwijk, 1988). This indicates that group spread is an important variable mediating the potential for contest (Figure 10.2b). The idea that changes in group spread mediate strength or occurrence of contest has been mentioned occasionally (e.g., Janson, 1988b; Phillips, 1995b; Cords, 2000; Pruetz & Isbell, 2000), but systematic approaches are rare (but see Robinson, 1981; van Noordwijk & van Schaik, 1987; Janson, 1990; Barton, 1993; Vogel, 2004).

In sum, it is not just resource characteristics that are important. The potential for contest of an individual resource depends on an interplay of the quality, abundance, and size of resources, and at least three other factors: (1) the primate species – with differing body size, gut size, and morphology, manipulative abilities, individual space, etc.; (2) the group size – which may vary within as well as across populations and species; and (3) the group spread – which may vary across seasons and within as well as across populations and species.

Integrating botanical and observational measures

Given the effects of species, group size, and group spread on feeding competition, it seems clear that botanical measures, nutrient quality of food, and ingestion rates are just the first steps in predicting competition. Reliable predictions would require further information on resource use (i.e., how many individuals feed or can feed together) in relation to group size as well as information on simultaneous use of resources of similar quality and overall group spread.

Such information can be obtained via scan sampling that monitors simultaneous use of resources, the size of the resource, and the amount of food it provides (see Koenig *et al.*, 1998). If intake rates are an additional focus of a study, the method of "focal tree sampling" (Vogel, 2004; Vogel & Janson, unpublished data) is even more appropriate. As a further plus, this method can also be applied if group spread is large or groups fission. Under such circumstances a scan sampling technique is less feasible if not impossible. Thus, in addition to data on botanical characteristics it is important to broaden the approach by including factors relevant to the study animals.

Rates of agonism and energy gain

Common assumptions about agonism and energy gain

Socio-ecological models attempt to measure and predict modes of feeding competition and rates of agonism. It has been suggested that the slope of the regression of energy gain on dominance rank measures strength of contest, i.e., a negative slope indicates contest competition and the stronger the slope the stronger the contest (Janson & van Schaik, 1988; van Schaik &

van Noordwijk, 1988). Thus, contest competition as measured by the skew in energy gain represents "despotic societies" *sensu* Vehrencamp (1983). In addition, the suggestion has been made that displacements over food and spontaneous aggression are rare if within-group contest is absent and common if within-group contest is present (van Schaik, 1989; Isbell, 1991).

Over the years these two separate predictions have been slightly modified. Particularly noteworthy are the ideas that "intensity of competition" or "steepness of hierarchies" can be described by frequency of aggression/ agonism or fighting success (e.g., Hemelrijk, 2002; Boinski *et al.*, 2002). If true, rates of agonism would be easy measures of the strength of contest competition because they are not based on extensive and difficult measures of energy intake or gain. Thus, it seems necessary to ask: do rates of agonism or aggression predict overall skew in energy gain? If true, a higher rate of agonistic behavior should coincide with a steeper slope of regression of energy gain on dominance rank.

Agonism and energy gain in Hanuman langurs

In order to test this prediction we compared data for two groups of Hanuman langurs from the same population (see references above). Rates of agonistic behavior (acts per hour) were derived from focal observations for a small group (2–3 adult females) and a large group (14–15 adult females) with a generally similar age structure living in habitats of similar quality. We distinguished between aggression, submission, and displacements (see Dolhinow, 1978; Borries, 1989). Contextual information was used to distinguish food and non-food related interactions. Energy gain was estimated from monthly scores of the physical condition of all adult females based on a seven-point scale (see above).

Agonistic behavior (both food and non-food related) was 2 to 3 times higher in the large group than in the small group (Table 10.1). This was true for all types of agonistic behavior (aggression, submission, and displacement). Given an increase in agonism in the larger group and assuming that agonism predicts strength of contest competition, a stronger skew in energy gain in the larger versus the smaller group was expected. This was, however, not the case. Regression slopes between dominance rank and physical condition for the females of the two groups during the 7 months for which data were available for both groups (for some months no data were available for the small group; see details in Koenig, 2000) indicated a stronger skew for the small group, contrary to the prediction (Wilcoxon signed ranks test (one-tailed) with T = 0.0, n = 7, p < 0.01; because a two-tailed test reveals

Table 10.1. *Rates of agonism among adult females in two groups of Hanuman langurs*

	Small group (2–3 females)			Large group (14–15 females)		
	Food	Non-food	Sum	Food	Non-food	Sum
Aggression	0.06	0.04	0.11	0.17	0.16	0.34
Submission	0.01	0.01	0.02	0.04	0.08	0.12
Displacements	0.19	0.03	0.22	0.41	0.19	0.60
Total per context	0.27	0.09	–	0.62	0.44	–
Overall rate		0.35			1.06	

Rates are based on 30-minute focal continuous recording and reported as acts per focal hour (small group: 94.0 hours; large group: 273.5 hours; study period: 15 months).

($p < 0.02$) that the two groups indeed differ). This result might have occurred because of the small number of females in the small group. However, the result held when the data were tested using three rank classes instead of individual data points (see Koenig, 2000).

The logic of agonism, rank, and contest

In this particular case it appeared that agonistic rates did not correctly predict skew in overall energy gain. Even though this is only a single case and a two-group comparison, the data come from two neighboring groups of the same population, using identical methods for collecting the data. Thus, the results seem to suggest that using overall agonistic or aggression rates as indicators of the strength of within-group contest competition is based on a false assumption. Agonistic behavior was predicted by van Schaik (1989) to relate to the absence (rare) and presence (common) of within-group contest competition. But this does not necessarily mean that the overall skew in energy gain is a direct function of the frequency of agonism. Instead, frequency of agonism (given and/or received) or frequency of winning should predict energy intake in relation to particular resources (holding resource quality, resource size, and group size constant; see, e.g., Janson, 1985, 1988a; Vogel, 2004). In other words, if food is worth contesting for (and all else being equal), frequency of aggression should predict feeding success, i.e., individuals acting more often aggressively may have a higher intake or decreasing energy gain may be coupled with increasing aggression received (Janson, 1985, 1988a). However, this general idea is usually true only for some resources and not for others, and agonism is a good predictor of energy intake only for certain resources (Janson, 1985, 1988a; Vogel, unpublished data). Looking at a scale of total energy gain, agonism given or received may correspond to the skew in energy gain, but it does not have to fit (Robinson, 1981; Janson, 1985). In order for concepts concerning the strength of within-group contest to apply, the skew in energy gain must be linked to dominance rank and not to agonistic rates (see van Schaik, 1989).

Rates of agonism and social structure

Predictions for linearity and strength of hierarchies

A third set of hypotheses of the socioecological models relates feeding competition and agonism to social structure. In the traditional version, the argument was that, among other things, dominance hierarchies of females differ in the degree of linearity (van Schaik, 1989; Sterck *et al.*, 1997). Despotic societies should be characterized by frequent food-related displacements and spontaneous aggression. This should lead to linear (and matrilineal) hierarchies. In contrast, egalitarian societies have been linked to few agonistic interactions and nonlinear and individualistic hierarchies (van Schaik, 1989).

More recently it has been argued that hierarchies of female primates do not truly differ in linearity but in strength (Isbell & Young, 2002). Strength is supposedly expressed via the time it takes to detect hierarchies. In other words, strong hierarchies are those where frequent agonism leads to short detection times of (linear) hierarchies. In contrast, weak hierarchies are those with few interactions, leading to long observation times in order to detect hierarchies (Isbell & Young, 2002). Given these two competing ideas, it seems reasonable to ask whether rate of agonism can accurately predict social structure in terms of linearity or strength of hierarchies.

What's wrong with linearity?

Van Schaik's (1989) original prediction, summarized above, cannot currently be tested because the available data are insufficient. Such testing would require a large dataset, including rates of agonism (from focal data and for different contexts) and dominance matrices with all relationships known (see below) across different taxonomic groups. Some researchers do indeed provide rates of agonistic behavior in different contexts or even separated into aggressive or submissive behaviors. Often, however, hierarchies are insufficient to conduct statistical tests, i.e., there are too many empty cells (see below) or groups are too small for statistical testing (n < 6 females; see Appleby, 1983). In other cases, dominance hierarchies are provided that do allow for testing. In such cases, however, data often come from provisioned populations or are collected via *ad libitum* sampling, which does not permit calculating rates of agonistic behavior (Martin & Bateson, 1993). In any case, even if appropriate data were available, the index used for evaluating linearity is not necessarily an appropriate tool for two reasons. The first problem concerns the index's insensitivity to particular relationships. The second problem is that in many cases relationships of some individuals are unknown, making comparisons across groups, populations, and species impossible.

Linearity of hierarchies relates to transitivity of relationships, e.g., if in a group of three individuals A dominates B, B dominates C, and A also dominates C, perfect transitivity exists. Via Landau's linearity index h (Landau, 1951) or Kendall's K (Kendall, 1962), and appropriate statistics (Appleby, 1983), linearity can be assessed. The linearity index, h, for instance, measures relationship transitivity resulting in a maximum value of 1 for perfectly transitive relationships and a linear hierarchy. However, unknown relationships in dominance matrices reduce the index and significance of linearity is underestimated. To counter this problem de Vries (1995) suggested a new improved procedure (index h') implemented in the program MatMan™ (Noldus Information Technology, 2003). While this innovation has resulted in a much improved test procedure, serious problems remain.

The index in its current form (h or h') is sensitive to intransitive relationships (more entries below the diagonal; Table 10.2a) and to tied relationships (the same number of interactions larger than zero above and below the diagonal; effects not shown here). Any intransitive and tied relationship plus its strength (the rank distance between the actors) reduces the linearity index and affects the p-value of the statistics. The index is, however, insensitive to inconsistent relationships, where entries below the diagonal are smaller than those above (Table 10.2b). Note that in the examples shown in Tables 10.2a and 10.2b the directional consistency indices (van Hooff & Wensing, 1987) remain the same. Therefore, a hierarchy can be constructed where all cells below the diagonal have entries of 1 or more (completely inconsistent relationships) but the linearity index is still 1.0, indicating perfect (and significant) linearity, while the directional consistency has dropped markedly (Table 10.2c). Thus, in its current form the linearity index measures the transitivity of relationships based on a particular definition of intransitive relationships (see above). This definition is perfectly fine. Unfortunately, however, it matches neither the common assumption that transitive

Table 10.2. *The effect of intransitive and inconsistent relationships on linearity and directional consistency measures using the program MatMan*[TM]

(a) Intransitive relationships ($DCI = 0.94$, $h' = 0.89$, $p = 0.05$)

	Recipient					
	A	B	C	D	E	F
A		4	2	4	4	4
B			4	4	4	4
C	4			4	4	4
D					4	4
E						4
F						

(b) Inconsistent relationships ($DCI = 0.94$, $h' = 1.00$, $p = 0.02$)

	Recipient					
	A	B	C	D	E	F
A		4	4	4	4	4
B			4	4	4	4
C	2			4	4	4
D					4	4
E						4
F						

(c) Completely inconsistent relationships ($DCI = 0.33$, $h' = 1.00$, $p = 0.02$)

	Recipient					
	A	B	C	D	E	F
A		4	4	4	4	4
B	2		4	4	4	4
C	2	2		4	4	4
D	2	2	2		4	4
E	2	2	2	2		4
F	2	2	2	2	2	

relationships in a linear hierarchy are also entirely unidirectional nor the observations of non-human primates. In hierarchies of female primates, entries below the diagonal can either be very rare (e.g., baboons, Hausfater *et al.*, 1982) or common, despite significant linearity (e.g., Hanuman langurs, Hrdy & Hrdy, 1976; Borries *et al.*, 1991; Koenig, 2000). Consequently, it is important to realize that linearity and directionality measure two different traits of a matrix: linearity measures a hierarchical trait whereas directionality

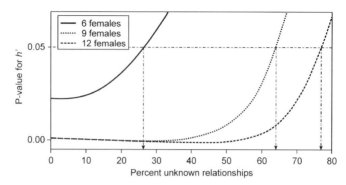

Figure 10.3. Significance level (P value) for the linearity index h' in relation to the percentage of unknown relationships for three group sizes (6, 9, and 12 adult females). Simulation is based on one-zero relationships starting with a matrix of all known relationships and gradually increasing the number of unknown relationships. Plotted are least-square fits to the actual p-values. Broken lines indicate the percentage when the p value reaches a 0.05 level beyond which hierarchies are no longer significantly linear.

judges relationships. With regard to the former (linearity), so far it seems that female hierarchies do not vary much (but see below) and directional consistency should receive more attention in future studies (Isbell & Young, 2002).

In addition, the problem remains that unknown relationships confound the index of linearity. The new procedure calculating the index h' takes these unknown relationships into account, increases the value for h', and improves the linearity testing (de Vries, 1995). However, the new index cannot make unknown relationships disappear (for an attempt to counter this problem see Singh *et al.*, 2003).

In order to demonstrate the effect of unknown relationships and its consequences on linearity assessments, we ran a simulation with the MatMan™ program. Starting with a perfectly linear hierarchy (one interaction per dyad) in groups of 6, 9, and 12 females, we introduced unknown relationships by removing entries and repeatedly calculating the linearity index and p-values up to the point where the p-value was above 0.05 (not significantly linear). In addition, for each of these cut-off points we calculated the percentage of unknown relationships MatMan™ would allow but still give a significantly linear hierarchy (Figure 10.3). As it turns out, in a group of 6 females, 4 out of 15 relationships (27%) could be unknown, but the linearity would roughly meet the alpha level of 5% ($h' = 0.91$, p = 0.056). In a group of 9 females, 23 of 36 relationships could be unknown (64%) ($h' = 0.62$, p = 0.047), and in a group of 12 females, even 51 of 66 (77%) ($h' = 0.44$, p = 0.043). Thus, even if only a minor fraction of the relationships is known, the linearity could still be significant. More importantly, even with a significant linearity the indices are very different. The linearity index is strongly affected by group size and by unknown relationships, i.e., h' declines almost linearly with increasing number of unknown relationships. Thus, the values of linearity indices are not comparable unless all or almost all relationships in a hierarchy are known.

In sum, the linearity index catches the effect of intransitive and tied relationships, but much of the differences across primate dominance hierarchies are not captured because those seem to relate to differences in directional consistencies or inconsistent relationships. Moreover, relevant traits such as intransitive and tied relationships cannot be compared since many of the available matrices contain numerous unknown relationships, not allowing comparisons of corresponding indices.

Is strength of hierarchies any better?

In contrast to ideas concerning the rate of agonism and linearity, it has been suggested that three other factors might be more variable or easier to measure than linearity, i.e., directional consistency, stability, and strength of hierarchies (Isbell & Young, 2002). In the following, we will focus on the strength of hierarchies because testing it is straightforward. Isbell & Young (2002) suggested that the rate of agonism should determine the time to detect hierarchies, which supposedly is an indicator of the strength of hierarchies and a factor varying across primate societies. They predict that high rates of agonism should lead to short detection times of hierarchies. Based on such interaction frequencies one can estimate how long it will take to obtain a certain amount of data.

In the following test, we assume a model population of two groups differing in size (seven and 15 females); these groups could be different or the same species, in a different or the same habitat. Since unknown relationships would affect the results (see above), we assume that it is necessary to collect five interactions for each of the dyads. Stipulating to the condition that all relationships be known requires departing slightly from the original assumption by Isbell & Young (2002). They suggested that instead of information for all dyads only information for the cells immediately above the central diagonal is necessary. The total number of interactions that must then be collected would be the product of the number of relationships and the number of interactions per dyad. Given seven females and 21 dyads this would equal 105 interactions and 15 females with 105 dyads would equal 525 interactions.

Table 10.3. *Time needed to record five interactions per dyad for virtual groups of primate females varying in interaction frequency and group size*

(a) Interaction rates (per hour) stable for 7 and 15 females

	7 females (21 dyads) (105 interactions)	15 females (105 dyads) (525 interactions)
Low interaction frequency (0.2 per female focal hour)	525 h	2625 h
High interaction frequency (0.6 per female focal hour)	175 h	875 h

(b) Interaction rates (per hour) increasing by 50% in groups of 15 females versus 7 females

	7 females (21 dyads) (105 interactions)	15 females (105 dyads) (525 interactions)	
Low interaction frequency (0.2 per female focal hour)	525 h	1750 h	Low interaction frequency (0.3 per female focal hour)
High interaction frequency (0.6 per female focal hour)	175 h	583 h	High interaction frequency (0.9 per female focal hour)

Setting the interaction rates at 0.2 per hour (rare) and 0.6 per hour (frequent) one can calculate how long it would take to collect the required total number of interactions using a focal animal sampling technique (Altmann, 1974). Based on these calculations it is of course clear that (all other things being equal) with lower interaction frequencies it takes longer to acquire the necessary data (compare column entries in Table 10.3a). However, group size apparently overrides this effect. For a group with 15 females and high interaction frequency it takes much longer to acquire the necessary data than for a group of 7 females with low interaction frequency. It might be argued, however, that interaction frequencies change with group size, i.e., interaction frequencies increase with the number of interactants available (see also above). We therefore repeated the calculation including a 50% increase of interaction rates from seven to 15 females, i.e., in the group with 15 females rates were 0.3 per hour (rare) and 0.9 per hour (frequent; Table 10.3b). Still, under this condition larger groups have very long "detection" times. The larger group with a high interaction rate is indistinguishable from a small group with a low interaction rate. Thus, even though the suggested link between interaction rates and detection of hierarchies seems straightforward, when differences in group size exist, no direct relationship remains. Detection time does not depend simply on a single factor, such as interaction rates, but on a number of factors, including group size, observational technique, and visibility.

Conclusions

In this overview we revisited three sets of predictions of the socioecological model. The first set relating resource characteristics to feeding competition appears to be strong. Resource characteristics have often turned out to be good predictors of feeding competition. The exceptions, however, have revealed that this view is a bit too simplistic. A better match could be achieved if information that is relevant for the primates themselves – describing the resources used by the study animals – was included (Janson, 1985, 1988a, 1988b; Isbell *et al.*, 1998; Vogel, 2004; Vogel & Janson, unpublished data; Janson & Vogel, Chapter 11, this volume). Thus, botanical measures are useful and have some predictive power, but future studies should go further and at the very least include the number of animals per resource, group size, resource size, group spread, alternative resource availability, and satiation level.

Secondly, using rates of aggression or agonism as predictors for the strength of contest competition is based on a false assumption. Apparently, aggression is one way of obtaining or safeguarding resources; hence, it is one mechanism of how energy is obtained. As such, rates of agonism should be a strong predictor of feeding success in particular resources (depending on resource characteristics, group size and spread, etc.). This should, however, not be confused with the overall (ultimate) consequences of feeding competition, the overall skew in energy gain, which should depend on dominance rank. And the skew in this relation depicts the strength of contest. Both may be related to rates of agonism, but need not be. Therefore, obtaining the relevant energetic data in order to investigate the strength of contest is still necessary. Relevant tests of the socioecological model rely on such data becoming available.

Finally, the predictions of the social consequences of feeding competition are still a part of the model that so far has not been tested properly. It may appear that rates of agonism do not necessarily tell us anything about linearity or other hierarchy characteristics. However, the available tools and data are currently not sufficient to allow us to reach final conclusions.

As it stands, it appears to be necessary to focus more on directional consistency and stability of hierarchies instead of linearity as a way of describing dominance relationships and hierarchies (Isbell & Young, 2002). In addition, it might be useful to consider other measures such as the number of inconsistent relationships and the number and strength of intransitive relationships. Beyond these issues concerning analytical methods, it is of prime importance to gather data sets that allow for an overall testing of the existing models. Furthermore, investigating the underlying causes of differences in directional consistency and stability of relationships appears to be necessary. The current models assume that strong contest competition is coupled with high consistency and stability of dominance relationships. This idea is perhaps too simple for two reasons. It neglects individual differences in willingness to challenge existing relationships leading to unstable hierarchies. In addition, it fails to factor in probabilities for coalitions occuring, which can either stabilize or destabilize hierarchies depending on the form the coalitions take. Mathematical modeling of the conditions of feeding competition and the hierarchical consequences could help to refine the existing verbal models.

Acknowledgments

We would like to thank G. Hohmann, M. Robbins, C. Boesch and their helpers for a great conference in Leipzig and the invitation to contribute to this book. We gratefully acknowledge very constructive comments on the manuscript by D. Doran-Sheehy, A. Green, L. Isbell, C. Janson, J. Kamilar, J. Ostner, O. Schuelke, E. Vogel, and two anonymous reviewers. Results on Hanuman langurs are based on fieldwork conducted between 1991 and 1997. We thank Dr M. K. Giri, Dr P. Shresta, and Prof. S. C. Singh (Natural History Museum, Katmandu), the Research Division of the Tribhuvan University (Katmandu), and the Ministry of Education, Culture, and Social Welfare (His Majesty's Government, Katmandu) for cooperation and permission to conduct the study; H. Acharya, U. Apelt, R. Armbrecht, J. Beise, H. K. Cchetri, M. K. Chalise, D. B. Ghale, S. S. Lama, K. Launhardt, E. Mittra, J. Nikolei, J. Ostner, D. Podzuweit, Y. B. Rana, O. Schuelke, K. B. Thapa, S. Thurnheer, K. Wesche, P. Winkler, and T. Ziegler for contributions to the database; and the Deutsche Forschungsgemeinschaft (Vo124/19–1+2, Wi966/4–3; Institute of Anthropology, Göttingen; C.B.) and the Alexander von Humboldt-Stiftung (V-3-FLF-1014527; A.K.) for financial support. The study on Hanuman langurs was non-invasive and complied with the laws of Nepal and Germany.

References

Altmann, J. (1974). Observational study of behavior: sampling methods. *Behaviour*, **49**, 227–67.

Appleby, M. C. (1983). The probability of linearity in hierarchies. *Animal Behaviour*, **31**, 600–8.

Barton, R. A. (1993). Sociospatial mechanisms of feeding competition in female olive baboons, *Papio anubis*. *Animal Behaviour*, **46**, 791–802.

Barton, R. A. & Whiten, A. (1993). Feeding competition among female olive baboons, *Papio anubis*. *Animal Behaviour*, **46**, 777–89.

Barton, R. A., Byrne, R. W., & Whiten, A. (1996). Ecology, feeding competition and social structure in baboons. *Behavioral Ecology and Sociobiology*, **38**, 321–9.

Beise, J. (1996). *Zur Raeumlichen Organisation einer Langurengruppe (Presbytis entellus DUFRESNE 1797) in Ramnagar, Suednepal*. Unpublished Diploma thesis, University of Göttingen.

Boinski, S. (1999). The social organization of squirrel monkeys: implications for ecological models of social evolution. *Evolutionary Anthropology*, **8**, 101–12.

Boinski, S., Sughrue, K., Selvaggi, L. *et al*. (2002). An expanded test of the ecological model of primate social evolution: competitive regimes and female bonding in three species of squirrel monkeys (*Samiri oerstedii, S. boliviensis*, and *S. sciureus*). *Behaviour*, **139**, 227–61.

Borries, C. (1989). *Konkurrenz unter freilebenden Langurenweibchen (Presbytis entellus)*. Unpublished Ph.D. thesis, University of Göttingen.

(1993). Ecology of female social relationships: Hanuman langurs (*Presbytis entellus*) and the van Schaik model. *Folia Primatologica*, **61**, 21–30.

(2000). Male dispersal and mating season influxes in Hanuman langurs living in multimale groups. In *Primate Males – Causes and Consequences of Variation in Group Composition*, ed. P. M. Kappeler, pp. 146–58. Cambridge: Cambridge University Press.

Borries, C., Koenig, A., & Winkler, P. (2001). Variation of life history traits and mating patterns in female langur monkeys (*Semnopithecus entellus*). *Behavioral Ecology and Sociobiology*, **50**, 391–402.

Borries, C., Sommer, V., & Srivastava, A. (1991). Dominance, age, and reproductive success in free-ranging female Hanuman langurs (*Presbytis entellus*). *International Journal of Primatology*, **12**, 231–57.

Chalise, M. K. (1995). *Comparative Study of Feeding Ecology and Behaviour of Male and Female Langurs (Presbytis entellus)*. Unpublished Ph.D. thesis, Tribhuvan University-Katmandu.

Clutton-Brock, T. H. & Harvey, P. H. (1977). Primate ecology and social organization. *Journal of Zoology, London*, **183**, 1–39.

Cords, M. (2000). Agonistic and affiliative relationships in a blue monkey group. In *Old World Monkeys*, ed. P. F. Whitehead & C. J. Jolly, pp. 453–79. Cambridge: Cambridge University Press.

de Vries, H. (1995). An improved test of linearity in dominance hierarchies containing unknown or tied relationships. *Animal Behaviour*, **50**, 1375–89.

Dolhinow, P. (1978). A behavior repertoire for the Indian langur monkey (*Presbytis entellus*). *Primates*, **19**, 449–72.

Hausfater, G., Altmann, J., & Altmann, S. A. (1982). Long-term consistency of dominance relations among female baboons (*Papio cynocephalus*). *Science*, **217**, 752–5.

Hemelrijk, C. K. (2002). Self-organization and natural selection in the evolution of complex despotic societies. *Biological Bulletin*, **202**, 283–8.

Hrdy, S. B. & Hrdy, D. B. (1976). Hierarchical relations among female Hanuman langurs (Primates: Colobinae, *Presbytis entellus*). *Science*, **193**, 913–15.

Isbell, L. A. (1991). Contest and scramble competition: patterns of female aggression and ranging behavior among primates. *Behavioral Ecology*, **2**, 143–55.

Isbell, L. A. & van Vuren, D. (1996). Differential costs of locational and social dispersal and their consequences for female group-living primates. *Behaviour*, **133**, 1–36.

Isbell, L. A. & Young, T. P. (2002). Ecological models of female social relationships in primates: similarities, disparities, and some directions for future clarity. *Behaviour*, **139**, 177–202.

Isbell, L. A., Pruetz, J. D., & Young, T. P. (1998). Movements of vervets (*Cercopithecus aethiops*) and patas monkeys (*Erythrocebus patas*) as estimators of food resource size, density, and distribution. *Behavioral Ecology and Sociobiology*, **42**, 123–33.

Janson, C. H. (1985). Aggressive competition and individual food consumption in wild brown capuchin monkeys (*Cebus apella*). *Behavioral Ecology and Sociobiology*, **18**, 125–38.

 (1988a). Food competition in brown capuchin monkeys (*Cebus apella*): quantitative effects of group size and tree productivity. *Behaviour*, **105**, 53–76.

 (1988b). Intra-specific food competition and primate social structure: a synthesis. *Behaviour*, **105**, 1–17.

 (1990). Social correlates of individual spatial choice in foraging groups of brown capuchin monkeys, *Cebus apella*. *Animal Behaviour*, **40**, 910–21.

Janson, C. H. & van Schaik, C. P. (1988). Recognizing the many faces of primate food competition: methods. *Behaviour*, **105**, 165–86.

Kendall, M. G. (1962). *Rank Correlation Methods*, 3rd edn. London: Charles Griffin.

Koenig, A. (2000). Competitive regimes in forest-dwelling Hanuman langur females (*Semnopithecus entellus*). *Behavioral Ecology and Sociobiology*, **48**, 93–109.

 (2002). Competition for resources and its behavioral consequences among female primates. *International Journal of Primatology*, **23**, 759–83.

Koenig, A. & Borries, C. (2001). Socioecology of Hanuman langurs: the story of their success. *Evolutionary Anthropology*, **10**, 122–37.

Koenig, A., Beise, J., Chalise, M. K., & Ganzhorn, J. U. (1998). When females should contest for food – testing hypotheses about resource density, distribution, size, and quality with Hanuman langurs (*Presbytis entellus*). *Behavioral Ecology and Sociobiology*, **42**, 225–37.

Koenig, A., Borries, C., Chalise, M. K., & Winkler, P. (1997). Ecology, nutrition, and timing of reproductive events in an Asian primate, the Hanuman langur (*Presbytis entellus*). *Journal of Zoology, London*, **243**, 215–35.

Krebs, C. J. (1999). *Ecological Methodology*, 2nd edn. Menlo Park: Benjamin/ Cummings.

Landau, H. G. (1951). On dominance relations and the structure of animal societies. I. Effect of inherent characteristics. *Bulletin of Mathematical Biophysics*, **13**, 1–19.

Martin, P. & Bateson, P. (1993). *Measuring Behaviour: An Introductory Guide*, 2nd edn. Cambridge: Cambridge University Press.

Mitchell, C. L., Boinski, S., & van Schaik, C. P. (1991). Competitive regimes and female bonding in two species of squirrel monkeys (*Saimiri oerstedi* and *S. sciureus*). *Behavioral Ecology and Sociobiology*, **28**, 55–60.

Noldus Information Technology (2003). *MatMan*^*TM*. *Software for Matrix Manipulation and Analysis*. Wageningen, the Netherlands: Noldus Information Technology.

Phillips, K. A. (1995a). Foraging-related agonism in capuchin monkeys (*Cebus capucinus*). *Folia Primatologica*, **65**, 159–62.

(1995b). Resource patch size and flexible foraging in white-faced capuchins (*Cebus capucinus*). *International Journal of Primatology*, **16**, 509–19.

Pruetz, J. D. & Isbell, L. A. (2000). Correlations of food distribution and patch size with agonistic interactions in female vervets (*Chlorocebus aethiops*) and patas monkeys (*Erythrocebus patas*) living in simple habitats. *Behavioral Ecology and Sociobiology*, **49**, 38–47.

Robinson, J. G. (1981). Spatial structure in foraging groups of wedge-capped capuchin monkeys, *Cebus nigrivittatus*. *Animal Behaviour*, **29**, 1036–56.

Saito, C. (1996). Dominance and feeding success in female Japanese macaques, *Macaca fuscata*: effects of food patch size and inter-patch distance. *Animal Behaviour*, **51**, 967–80.

Schülke, O. (2003). To breed or not to breed – food competition and other factors involved in female breeding decisions in the pair-living nocturnal fork-marked lemur (*Phaner furcifer*). *Behavioral Ecology and Sociobiology*, **55**, 11–21.

Singh, M., Singh, M., Sharma, A. K., & Krishna, B. A. (2003). Methodological considerations in measurement of dominance in primates. *Current Science*, **84**, 709–13.

Smith, R. J. & Jungers, W. L. (1997). Body mass in comparative primatology. *Journal of Human Evolution*, **32**, 523–59.

Sterck, E. H. M. & Steenbeek, R. (1997). Female dominance relationships and food competition in the sympatric Thomas langur and long-tailed macaque. *Behaviour*, **134**, 749–74.

Sterck, E. H. M., Watts, D. P., & van Schaik, C. P. (1997). The evolution of female social relationships in nonhuman primates. *Behavioral Ecology and Sociobiology*, **41**, 291–309.

van Hooff, J. A. R. A. M. & Wensing, J. A. B. (1987). Dominance and its behavioral measures in a captive wolf pack. In *Man and Wolf: Advances, Issues, and Problems in Captive Wolf Research*, ed. H. Frank, pp. 219–52. Dordrecht, the Netherlands: Dr. W. Junk Publishers.

van Noordwijk, M. & van Schaik, C. P. (1987). Competition among female long-tailed macaques, *Macaca fascicularis*. *Animal Behaviour*, **35**, 577–89.

van Schaik, C. P. (1989). The ecology of social relationships amongst female primates. In *Comparative Socioecology: The Behavioural Ecology of Humans and Other Mammals*, ed. V. Standen & R. A. Foley, pp. 195–218. Oxford: Blackwell Scientific Publications.

van Schaik, C. P. & van Noordwijk, M. A. (1988). Scramble and contest in feeding competition among female long-tailed macaques (*Macaca fascicularis*). *Behaviour*, **105**, 77–98.

Vehrencamp, S. (1983). A model for the evolution of despotic versus egalitarian societies. *Animal Behaviour*, **31**, 667–82.

Vogel, E. R. (2004). *The Ecological Basis of Aggression in White-Faced Capuchin Monkeys, Cebus capucinus, in a Costa Rican Dry Forest*. Unpublished Ph.D. thesis, State University of New York-Stony Brook.

Whitten, P. L. (1983). Diet and dominance among female vervet monkeys (*Cercopithecus aethiops*). *American Journal of Primatology*, **5**, 139–59.

Wittig, R. M. & Boesch, C. (2003). Food competition and linear dominance hierarchy among female chimpanzees of the Taï National Park. *International Journal of Primatology*, **24**, 847–67.

Wrangham, R. W. (1979). On the evolution of ape social systems. *Social Science Information*, **18**, 335–68.

(1980). An ecological model of female-bonded primate groups. *Behaviour*, **75**, 262–300.

11 Hunger and aggression in capuchin monkeys

CHARLES JANSON AND ERIN VOGEL

Introduction

Optimal foraging theory has been applied sparingly to primates (e.g., Terborgh, 1983; Post, 1984; Robinson, 1986; Nakagawa, 1990; Grether *et al.*, 1992; Barton & Whiten, 1994; Altmann, 1998). Many assumptions of foraging theory have been challenged as not being applicable to primates (Post, 1984), with some justification. However, many of these original objections (failure to consider patchy food sources, inability to incorporate social food competition) have been addressed in recent years. One of the relatively

Feeding Ecology in Apes and Other Primates. Ecological, Physical and Behavioral Aspects, ed. G. Hohmann, M. M. Robbins, and C. Boesch. Published by Cambridge University Press. © Cambridge University Press 2006.

285

recent innovations of foraging theory that has not yet been used in interpreting primate foraging behavior is state-dependent fitness functions (e.g., Houston & McNamara, 1999; Clark & Mangel, 2000). This approach to optimality is particularly important in explaining time-varying behaviors that may not correlate in any obvious way with external environmental variables. In this chapter, we argue that incorporating state-dependent fitness into analyses of social food competition may greatly increase the explanatory power of socio-ecological analyses. In particular, much of the moment-to-moment variation in food choices and rates of aggression in primates may turn out to be explained by parallel changes in the state of the interacting animals. The state that we focus on in this chapter is hunger level and how it may affect the rate of the animals' aggression in food trees.

A new look at optimality

State-dependent fitness functions differ from conventional optimality theory in a fundamental yet simple way (Stephens & Krebs, 1987). In conventional theory, the fitness value of a given increment in some proximal currency (say, energy gain) is the same regardless of the state of the animal. Thus, to predict the fitness maximizing behavior, one need only sum or average the increments over some fixed interval to derive the long-term average rate of fitness gain; in foraging theory, this is called the long-term rate of energy gain (Stephens & Krebs, 1987). In state-dependent theory, fitness is evaluated once at the end of some time period, and is typically based on the "state" of the animal at that time. Thus, the state replaces the proximal currency as a predictor of fitness (Clark & Mangel, 2000). State variables include fat reserves, body size, hunger level, etc. The time period may be a day, a breeding season, a winter, a year, or any other relevant period. This period is in turn broken down into a sequence of intervals. During each interval, an animal's behavioral decisions (where to forage, which food to choose, whether or not to fight over food, etc.) can change the state variable. These state changes are accumulated across intervals, and the animal's fitness at the end of the set of intervals is evaluated depending on its state. Usually fitness is a nonlinear function of the state variables so that increases in a state variable are highly beneficial when the state is low, but of negligible, or even negative value when the state is high. A common example is the level of fat reserves in a small bird at the end of a winter day (Houston & McNamara, 1999). How much and when the bird consumed food during the day is important only to the extent that the individual items contribute to the fat reserves at the end of the day. These fat reserves, in turn, determine the

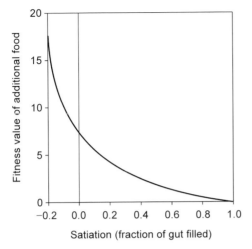

Figure 11.1. An illustration of the expected change in fitness from a unit of additional food ingested, which varies inversely with the satiation level of the animal. Negative satiation values refer to energy deficits that would require food just to bring the animal back to an energy-neutral state.

chance of the bird's surviving the night, which is the actual measure of fitness in this case. In general, an animal's optimal choice depends on both its current state and the amount of time it takes to reach the end of the accumulation period. Evaluating the optimal strategy is a bit complex, but the technique has been presented in recent reviews of state-dependent fitness theory (Houston & McNamara, 1999; Clark & Mangel, 2000).

We feel that it is time to expand primate foraging and competition studies to explicitly incorporate dynamic state variables. An obvious variable to focus on in primates is hunger, which will often vary dramatically during the course of a day and from individual to individual within a group, depending on their short- and long-term foraging success. Over longer time periods, hunger will be related to an individual's fat reserves or body condition, which is known to correlate with reproductive success in several primate species (Koenig, 2000; Richard *et al.*, 2000). We first explain how incorporating hunger can affect theoretical predictions about the frequency and intensity of aggression. Second, we present a brief review of the physiological determinants of hunger. We then discuss one possible way to estimate a correlate of hunger, as the fractional fullness of the forager's gut. Simply put, when the forager is satiated, it has low hunger and additional food should have zero marginal fitness value to the forager (Figure 11.1). We present a simple input-output model that should allow the fractional satiation to be estimated from detailed field data on feeding histories. Finally, we present an empirical analysis of rates of agonism at food trees, in which average fractional satiation was estimated for white-faced capuchin monkeys (*Cebus capucinus*). We test whether, in this case, fractional satiation helps to explain variation in rates of agonism across different feeding trees.

Hunger, resource value, and aggression in cohesive primate groups

Despite the widespread use of socioecological models to relate forms of food competition to the distribution of resources (e.g., Janson, 1988b; van Schaik, 1989; Sterck *et al.*, 1997; Koenig, 2002), there has until recently been no formal theory relating the frequency and intensity of aggression to characteristics of the resource being contested. Several recent papers have attempted to unite social foraging theory with game theory and optimal foraging models to produce a theory of socially dependent food competition (Dubois & Giraldeau, 2003; Dubois *et al.*, 2003). These theories are based on Hawk–Dove models and were devised mainly for bird flocks. They do not incorporate several of the important biological characteristics of primate groups including: (1) long-term relationships leading to prior knowledge of Resource-Holding Power (RHP); (2) the use of patchy renewable resources leading to prior knowledge of the availability of alternative food sources; and (3) graded contests with the ability of any one contestant to break off the contest at any time. And none of these theoretical treatments incorporates hunger as a state variable.

This is not the place to provide a complete discussion or model of food-related aggression in primate groups. The main issue is what is the expected effect of hunger on the frequency or intensity of aggression? Taking hunger into account is likely to have two effects on agonistic interactions. First, hungrier animals are expected to fight more often over a given food resource. This pattern should hold because the fitness value of a given amount of food is likely to differ substantially depending on whether the animal is well-fed or starving (Caraco, 1979). In particular, hungry animals should gain more fitness value from the resource than should satiated ones (Figure 11.1). In conventional game theory models of aggression (Maynard Smith, 1974), higher mean fitness values for gaining access to a resource lead to a higher chance that any contestant will fight (play "Hawk"). Second, increased hunger is expected to bring about more escalated contests. This result occurs because in models of protracted contests ("wars of attrition," e.g., Maynard Smith & Parker, 1976), an individual's willingness to escalate a contest is

Table 11.1. *List of potential explanatory variables used in the multiple regression model to predict increased incidence of aggression per feeding bout*

Variable	Hypothesis	Expected sign of effect
Feeding bout length in minutes	Opportunity for agonism	+
Average no. of feeding adult females in focal tree	Opportunity for agonism	+
Average no. of feeding adult males in focal tree	Opportunity for agonism	+
Average no. of feeding nonadults in focal tree	Opportunity for agonism	+
Crown volume	Crowding	−
Maximum no. of feeding animals in focal tree	Crowding	+
Crown energetic value	Resource value	+ (small values)
	Opportunity costs	− (large values)
No. alternate fruiting trees of same species as focal tree within 40 m of focal tree	Opportunity costs	−
No. alternate fruiting trees of different species from focal tree within 40 m of focal tree	Opportunity costs	−
Minutes lost feeding = time until displaced individual feeds again	Opportunity costs	+
Distance moved by displaced individual before it feeds again	Opportunity costs	+
Alternate resource score, visual fruit count, K = 20	Opportunity costs	−
Fractional satiation at time of entry into the focal tree	Inverse of hunger	−

proportional to its ratio of V/C, where V is the fitness value for gaining a resource and C is the fitness cost of the contest per unit time. In this case, contests are expected to be longer and more escalated the higher V is, and V should be higher when animals are hungrier.

Hypotheses

A large number of social and ecological factors is likely to affect the amount of agonism observed in a feeding tree (Janson, 1987). We measured 13 potential explanatory variables. These can be grouped according to broad hypotheses (Table 11.1). First, the simplest null hypothesis is that agonism occurs at a constant rate per unit time per individual. If so, then the number of agonistic interactions should correlate positively with feeding bout length and with the average numbers of individuals feeding in the tree crown. Second, if agonism occurs with increased crowding, then the number of interactions should increase with the density of individuals (maximum number of feeding animals in the tree crown) and decrease with increasing crown volume.

A more detailed analysis would include social attributes of the potential contestants in a feeding tree, but that is beyond the scope of this paper.

If agonism is based on cost–benefit considerations, one would expect agonism to be more frequent the greater the value of gaining access to the resource (e.g., Maynard Smith, 1974). The value of access to a resource can be partitioned into at least three components. One is the nutritional content of the resource, which we assess here by estimates of total energetic value of fruit pulp in the tree crown (see Methods). All other things being equal, we expect that agonism should increase with increasing nutritional value of the resource, at least when resource value is small.

A second aspect affecting the fitness value of access to a resource is the opportunity cost – the nutrition that the interacting individuals could use at that moment if they did *not* gain access to the contested site. The nearest food to a contested feeding site is likely to be within the same tree crown, if it is large enough or has enough fruit. Once a tree crown contains enough food to satiate more than one monkey, we would expect increasing nutritional value to lead to reduced agonism, as potential contestants move to other feeding locations rather than contest access to occupied sites. Thus, depending on the actual amounts of food in the feeding trees visited, the effect of crown energetic value on agonism could be positive, negative, or both (positive for small values and negative for large values of crown energetic value). To test for all these possible patterns, we included both linear and quadratic (squared) terms for crown energetic value in the analysis; if crown energetic value has a "hump-shaped" effect on agonism, the linear term should be positive and the quadratic term should be negative.

Monkeys may avoid agonism in a given feeding tree by using other feeding locations outside the crown of that tree, so these should also be assessed as part of the opportunity cost. To assess this possibility, ERV measured in detail the presence and productivity of other fruiting trees in the local neighborhood of each focal tree observed. The variables used here are: (1) the number of fruiting trees of the same species as the focal and within 40 m of it (see Methods for justification of the 40 m criterion); (2) the number of fruiting trees of species different from the focal and within 40 m of it; (3) an "alternative resource score" based on visual fruit counts and $K = 20$ (see Methods); (4) the minutes lost feeding by the displaced individual when agonism occurred; and (5) the distance the displaced individual traveled from the location of the agonism before feeding again.

The final component affecting the fitness value of access to a resource (whether in the focal tree or outside of it) is the mapping of nutritional gain onto fitness, as all adaptive scenarios for agonism must ultimately be based on fitness gains and losses, rather than energy or other proximate currencies of fitness. In this paper, we focus on hunger (see above). Hunger is a subjective sensation that we cannot assess in free-ranging monkeys. Instead we estimated the inverse of hunger, namely satiation. Our estimate of satiation is an integrated measure over time of food intake and digestion prior to the entry of potential contestants into a feeding tree (see next section for details).

Hunger: physiological correlates and field sampling

The physiological determinants of hunger are relatively well understood in humans, although different researchers disagree about the relative importance of different feedback signals that control appetite (as reviewed in de Graaf *et al.*, 2004, the most recent review available at the time of this writing). We will assume that the physiology of hunger is similar between human and nonhuman primates. Researchers on eating behavior in humans distinguish "satiation" and "satiety" as, respectively, the subjective feelings that control the termination of a given meal vs. the feelings that control initiation of the next meal. In practice, however, a given physiological signal may affect both aspects, so the distinction between the two may be more semantic than real (de Graaf *et al.*, 2004). Hunger, or appetite, is considered to be the inverse of satiety, as it relates to desire to initiate the next meal. Subjective hunger is usually assessed in humans on some psychological scale, sometimes numerical, sometimes verbal. A more objective measure of hunger is the size of the next meal consumed, and this is usually highly correlated with subjective hunger scores within studies (de Graaf *et al.*, 2004).

There are at least 17 biological markers of satiation or satiety, but currently only eight of these have plausible causal mechanisms, a strong relationship to subjectively rated appetite, and at least moderately consistent effects across studies (de Graaf *et al.*, 2004). These eight include brain-imaging correlates of satiety, degree of stomach distension, and levels of six blood-borne chemicals, one of which is a direct consequence of eating and digestion (glucose) and the rest of which are hormones or neuropeptides produced by various parts of the digestive system (de Graaf *et al.*, 2004). Meal termination is often signaled by stomach distension, but feelings of hunger appear to respond more readily to nutrient intake – drinking a large amount of water may make you feel "full," yet leave you still eager to eat a meal (Näslund *et al.*, 2001). Some blood-borne chemicals (CKK and GLP-1) affect satiety at least in part by controlling the rate of emptying of the stomach, so that the flow of nutrients into the intestines is relatively constant (Schwartz *et al.*, 2000; de Graaf *et al.*, 2004). The higher the concentrations of these gut-produced chemicals, the slower the gut empties, and the longer the sensation of having a full stomach. These and other chemicals can affect satiety more directly, either by binding to receptors in the central nervous system or to peripheral receptors linked to sympathetic nerve input to the central nervous system (Schwartz *et al.*, 2000). These gut-secreted hormones and neuropeptides are produced in response to the presence of relevant nutrients in different parts of the gut (carbohydrates in the stomach for ghrelin, fats and proteins in the duodenum for CKK, all three major nutrient classes in the duodenum for protein YY, and carbohydrates and fats in the ileum or lower gut for GLP-1; de Graaf *et al.*, 2004). Experimentally reducing the rate of absorption of nutrients into the blood maintains high concentrations of the nutrients in the gut for longer, keeps production of these gut hormones high, and increases feelings of satiety (French & Read, 1994; de Graaf *et al.*, 2004). The remaining hormone, leptin, is produced by fat cells and is thought to affect long-term energy balance by moderating sensitivity to the chemicals that affect short-term appetite (de Graaf *et al.*, 2004); insulin may have a similar effect (Schwartz *et al.*, 2000). Leptin and insulin may be important in increasing the perception of average hunger in individuals with below-average energetic intake (e.g., subordinates in a capuchin group: Janson, 1985). Evidence from humans suggests that long-term reductions in food intake both increase baseline levels of hunger and reduce sensitivity to short-term feedback mechanisms for hunger, thus allowing individuals to eat larger meals before feeling satiated (de Graaf *et al.*, 2004).

What does the above synopsis mean for practical measures of estimating hunger in wild primates? We will argue that knowing the pattern of food ingestion (how much is eaten and when) is often sufficient to calculate short-term changes in hunger. Clearly, hunger develops as a function of time after the previous meal ends. For most fruit-eating primates, gut distension alone is unlikely to be the major factor affecting hunger, as it is uncommon for fruit trees to produce enough ripe fruit to satiate all individuals in a group (e.g., see Janson, 1988a). After a meal, whether or not the stomach is full, hormonal feedback controls stomach-emptying rate. At first there is a lag period when no food is released into the intestines (French *et al.*, 1993), then a brief period when the release of nutrients is rapid (while the intestine has low levels of nutrients), followed by a prolonged time of relatively constant release of nutrients until the stomach is empty (Näslund *et al.*, 2001). Thus, as time proceeds without feeding, stomach emptying must be associated with lower nutrient concentration in the stomach and this should lead to higher concentrations of ghrelin, which is highly correlated with hunger (de Graaf *et al.*, 2004). In addition, how long the stomach takes to empty following a particular set of meals should be easy to calculate by assuming that energy leaves the stomach at a constant rate. Shortly after the stomach empties, nutrient concentrations in the intestine should start to decline, with associated decreases in intestine-derived hormones and increasing levels of hunger. How quickly hunger increases after the stomach empties should be a function of the nutrient concentration in the intestine, which should in turn be affected by how quickly the nutrients are digested and absorbed into the blood. Fruit sugars, the main component of energy intake for fruit-eating primates, are very soluble and quickly absorbed, so that it is likely that hunger increases rapidly after the stomach empties. Thus, for diets dominated by nutrients that are easily digested and absorbed (such as simple sugars in fruits), hunger should increase during stomach emptying until the stomach is empty, with little residual effect of nutrients remaining in the intestine, as this effect will at best be brief. In a future comparative analysis, we will consider the expected patterns of hunger with respect to digestion for other primate diets (Vogel & Janson, unpublished data).

The scenario outlined above suggests a simple calculation for estimating hunger levels in wild fruit-eating primates. As no previous study on primates has attempted to estimate hunger levels explicitly, we are free to invent our own method. Based on the above logic, we use a formula that estimates amount of food remaining in the stomach, the inverse of hunger. We call the result "fractional satiation," S. S decreases linearly with time since ingestion, and increases with the amount ingested. If the maximum time to digest a full stomach's worth of food is P, then the fractional rate of stomach emptying is $1/P$. If a full meal were ingested, the fraction of the meal remaining at time i after ingestion would approximately equal $(P-i)/P$, disregarding either the lag phase or the succeeding brief burst of rapid food passage out of the stomach (Figure 11.2). If we apply the same time discounting to any amount of food ingested, F, the amount of food left after time i is $F * (P-i)/P$. Because we are interested in S at a given time, which we shall call time 0, we need to sum up the fractional amounts remaining from food ingested at all times i prior to 0, for $i < P$. Thus, $S = \sum[F_i^* (P-i)/P]$ where F_i is the amount of food ingested at time i before now, and P is the time needed to empty the stomach. When $I = 0$, the food is 100% present in the stomach. When $i = P$, the food is 0% present in the gut. When $0 < i < P$, the food is considered partially present, as a part of it has been removed at a fractional rate $1/P$ (Figure 11.2). An alternative assumption roughly consistent with the synopsis of appetite presented above is that the food is removed from the stomach at an absolute rate of M/P, where M is the maximum amount of fruit that the stomach can hold; the corresponding formula would be $S = \sum[F_i - (Mi/P)]$ for all $i < F_iP/M$. This formula may be slightly more realistic but requires estimating an extra parameter M, so we will not use it in this analysis. We hope that future research on gut emptying parameters

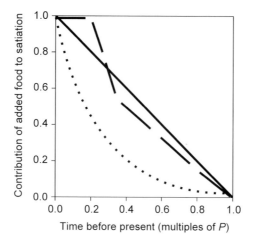

Figure 11.2. Different possible ways of discounting the contributions of ingested food to gut occupancy as a function of time before the present. For fruit-eating primates such as capuchins, we take P to be the time to empty the stomach (see text). One plausible model is that of exponential absorption (dotted line), which would be expected if nutrient uptake were proportional to nutrient concentration. Studies of human digestion suggest that a more accurate pattern is that given by the dashed line – with a lag time, brief rapid loss phase, and longer period of constant release of nutrients (see text for details). As an approximation of this more realistic model, we use the simplest form of linear discounting (solid line). Other shapes may be appropriate for species that ferment foliage or digest more complex foods that require longer residence in the intestines for full nutrient release and absorption. Researchers should use whichever function makes the most sense for a given analysis.

may indicate a better approximation than the one we have chosen. Given the necessary information, it is easy to use different curves to weight food ingested at any time prior to a specific observation point (Figure 11.2). For the linear decrease model used here, the maximum value of S equals $P/2$.

Determining P for capuchins is not easy because few studies of digestive function have been conducted on primates, and nearly all have only measured aspects of total gut (ororectal) transit time (but see Campbell *et al.*, 2004). The only published values for transit time for capuchins is 3.5 hours (both *C. capucinus* and *Cebus apella*; Milton, 1984); this number represents the time to first defecation of the ingested markers and is from captive monkeys fed a mixed diet of domestic fruit, bread soaked in milk, and monkey chow. The only direct measurements of gastric emptying time for nonhuman primates are for lemurs (Campbell *et al.*, 2004). For the two species of fruit-eating lemurs they studied, the mean gastric emptying time was almost exactly half of the transit time, using small beads (a better approximation to chewed up fruit pulp than the large beads). Using the same ratio and Milton's

(1984) data on capuchins, the gastric emptying time for capuchins should be about 2 hours. In the absence of more precise data, we take $P = 120$ minutes.

Methods

Study site and species

Cebus capucinus were studied by ERV in Lomas Barbudal Biological Reserve, IDA Property, and Finca El Pelón de la Bajura. Lomas Barbudal Biological Reserve and the surrounding area, hereafter referred to as Lomas, is a 2400-ha reserve located in northwest Costa Rica, in Guanacaste Province (10°30'N and 85°22'W). The forest of Lomas is generally classified as a tropical deciduous forest. Over 200 tree species have been identified in the reserve (Frankie, unpublished data; Vogel, unpublished data), 120 of which are used for food by the monkeys.

Cebus capucinus are primarily arboreal monkeys living in mixed-sex, cohesive groups of 15–30 individuals, with a male to female ratio close to 1:2 (Oppenheimer, 1968; Fedigan *et al.*, 1985). Males are larger than females, weighing an average of 3.2 kg and 2.3 kg, respectively (Glander *et al.*, 1991). Females are philopatric, although some rare occurrences of female transfer have been documented (Oppenheimer, 1968; Manson *et al.*, 1999). Although the genetic work is not complete, it is generally thought from long-term studies that most females within the groups are close kin (Perry, 1995; Rose, 1998). Males typically emigrate prior to or at sexual maturity and may transfer to different groups throughout their lifetimes.

Cebus capucinus show frequent agonism in food trees, and outcomes are generally predictable based on dominance rank (Perry, 1996; Vogel, 2004). Dominance hierarchies were calculated for each study group based on subtle (i.e., cowers, avoids, fear grins) and active (i.e., chases, bites, lunges) aggressive outcomes, plus the direction of dyadic displacements during continuous focal follows (Vogel, 2004). Dominance hierarchies were calculated using MatMan software (Noldus Information Technology, 2003). In all three groups, hierarchies were significantly linear for the entire group, although linearity indexes were typically well below 1.0 (0.30–0.55; Vogel, 2004).

All data reported here were collected by ERV from one study group of *C. capucinus* (Group QQ) from January 2002 through July 2002, for a total of 1950 contact hours. Habituation of QQ's Group was started in August 2001. The group was habituated to observers by the middle of January 2002. Data collection began once all observers were tested and agreed on the identification of all group members 3 or more years old (estimated from their size), and none of the group members showed signs of fear or avoidance of ERV or her field assistants. On average, we spent 25 consecutive days per month with the study group. During the study period, QQ's Group was composed of 4 adult males, 9 adult females, 1 subadult male and female, 7 juveniles, and 4–5 infants. The home range of QQ's group during the study period extended over 276 hectares. With rare exceptions, the group foraged and rested as a cohesive unit (length and width less than 100 m, Vogel, unpublished data). Most of the data collection took place during the dry season, although it did carry over a few months into the rainy season.

Observation techniques

Individuals within a social group were identified using differences in size, facial fur patterns, and other distinctive features (i.e., spots, scars, freckles, broken tails). Because there were two assistants collecting behavioral data during the field season, periodic interobserver reliability checks were conducted to assure agreement in identification of group members, behavioral activity categories, and identification of fruit trees. Throughout the study period, one assistant collected all of the ecological data described below.

In addition to the Focal Tree Method described below, systematic, instantaneous scan samples at 5-min intervals were recorded, at which time the general activity (i.e., feeding, foraging, resting/sleeping, grooming, traveling, other social) of all group members within view of the observers was recorded. To ensure that there was no overlap, the observers communicated via walkie-talkies and identified individuals that were located between the two observers. For any individuals that were feeding or foraging, we recorded the type of food (as well as species in the case of plant foods). Because feeding in fruit trees was usually relatively conspicuous, and two observers at a time could move through the study group to look for feeding individuals, we took the number of animals observed feeding in a given scan as a complete count for the group. Our values are likely to be underestimates, but the bias is probably not large.

For baseline social and feeding data, 10-minute continuous focal animal samples were recorded (Altmann, 1974), in which all social behaviors and identities of all individuals that interacted with the focal animal were recorded into a Psion handheld computer. Focal samples were recorded for all adult males and females. A focal session was discarded if the focal animal was out of sight for more than one minute of the 10-minute focal sample. These samples were taken when tree follows were not being conducted. All dyadic and polyadic interactions involving the focal animal and all individuals involved in the interaction were recorded. A random order of focal subjects was used – focal samples on all adult group members were completed before beginning another round of samples. In cases where an animal was missing from the group, we proceeded to the next round without sampling that animal.

Focal tree follow method

All data on frequency of aggression analyzed in this study were obtained using the Focal Tree Method (Vogel, 2004). Specifically, one observer moved to the front of the group and stood under a feeding tree that the group was likely to visit (particularly ones used in the recent past). This observer recorded the time and identification of first arrival, each successive arrival, and time and identification of each departing monkey from the tree. The time between the first arrival to the tree to the last departure is the *feeding bout length*. If there were gaps in feeding during the feeding bout, the amount of time in which the monkeys were not feeding was subtracted from the total feeding bout length. At the time of an aggressive interaction, this observer recorded the individuals involved and the proximity of all visible individuals in the tree to the interacting animals. This observer also recorded data on who won the interaction and whether the aggressor(s) started feeding at the site of the interaction, then continued recording data on all arrivals and departures. A second observer systematically recorded at 2.5-minute intervals the number, identification, activity, and relative position of all monkeys in the tree. Thus, at the time of an aggressive interaction, the number of all monkeys and the identity of all adult and older juvenile monkeys within the feeding tree were known. Data were taken until the last group member left the tree. All focal trees were marked with a unique ID for further ecological processing.

For all focal trees, we recorded the fruiting crown volume and fruit abundance (Vogel, 2004). Fruit abundance of the focal tree was measured using three different methods (Vogel, 2004), but we report the one that was used consistently across all focal trees and was the most direct estimate of fruit mass. This measure was called "visual fruit count." For small-crowned trees, this was obtained by a direct count of all visible ripe fruits on the tree. For large-crowned trees, a different procedure was used. Within 2 days of marking the focal tree, we measured crown height and diameter; crown volume was calculated from the latter two measurements using the equation for the volume of an ellipsoid (as in Janson, 1988a). Counts of fruit numbers were made in five 1-m^3 areas of the crown, selected for clear visibility and to ensure that sampling of the entire crown was relatively even. Visual fruit count was then obtained by multiplying the average of these five fruit counts by the tree crown volume in m^3.

Slight variations of the visual fruit count method were used depending on whether the unripe and ripe fruit were easily distinguished. For 14 of the 22 species with more than one recorded feeding visit in this study, the ripe and unripe fruits were easy to distinguish by color or dehiscence, and only ripe fruits were counted. For an additional two species, fruits in many stages of maturation were present in the tree at one time; in this case, all fruits large enough to be ripe were counted, but the number of actual ripe fruit may be overestimated. For two more species with very large fruits (*Mangifera indica* and *Genipa americana*), the monkeys systematically sniffed every fruit in the tree, rejecting the large majority at each visit. In this case, we counted as ripe fruits in the focal tree only those fruits actually consumed by the monkeys, on the basis that the remainder had been rejected at that visit. To estimate availability in trees of these same species where the monkeys were not observed to feed, we counted all fruits large enough to be ripe and so probably overestimated the numbers of ripe fruits in these cases. Finally, for the remaining four species with relatively synchronously ripening fruits with little obvious signals of ripeness, we counted all fruits as ripe. Despite these minor variations in methods, visual fruit count was highly correlated (on a log-log scale) with other robust but less direct measures of fruit production, such as diameter at breast height (DBH) (Chapman *et al.*, 1994). Because it was too difficult to gather detailed fruit availability data while taking behavioral observations, the focal tree was marked, and the following day one observer, the same person throughout the study period, recorded the focal resource tree data.

In addition to food availability in each focal tree, we recorded all alternative resources used by the monkeys at that time of year and which were within the average group spread around the focal tree (cf. Janson, 1987). Although group spread varied from day to day and hour to hour within a day, a single value for average group spread was used in calculating alternative resource availability. Group spread was periodically determined by observers standing at opposite edges of the spread, who then paced the group diameter. In this study, the average group diameter was about 80 m. Thus, all potential feeding trees within a radius of 40 m of the focal tree were recorded, yielding a sample area of almost exactly 0.5 ha. It did not matter if we had observed the monkeys use any given tree within this plot as long as it held ripe fruit known to be consumed by the capuchins. For each alternative resource, the following variables were measured: visual fruit count, fruiting crown volume, and the distance of the alternative resource to the focal tree (Vogel, 2004). The same person who recorded fruiting estimates of the focal tree also did so for the alternative resource trees.

All fruit species eaten by the capuchins during the study period were collected from trees in which the monkeys had fed, but not necessarily the focal trees. Fruits were weighed whole, divided into components (seed, husk, and pulp), each component weighed again, and then dried until they maintained a constant weight (Hladik, 1977). We tried to obtain specimens of equivalent ripeness to those ingested by the monkeys. The dried samples were then placed in a sealed bag containing 1/8″ silica gel beads and sent to Prof. Dr. J. Ganzhorn's laboratory in Hamburg, Germany for nutritional analysis. The composition of the food (% soluble carbohydrates, % protein, and % fat) was analyzed as in Ganzhorn (1988). The total kJ/g of pulp dry mass was calculated as (0.1674[sum of % pulp dry mass that was soluble carbohydrates plus protein] + 0.3766[% pulp dry mass that was fats]; as in Janson, 1985). These conversions are standards taken from the analysis of human foods as described in a variety of nutrition references (e.g., Merrill & Watt, 1973), and converted from kcal/g to kJ/g.

A combined measurement of the nutritional value of all alternative resources within the group spread is also necessary to gain an understanding of the value of a particular resource to an animal. This comparison is needed because the value of a resource is based on the nutritional gain *relative* to all other potential resources. For this reason, we calculated an *Alternative Resource Score* (ARS) using the following equation:

$$\sum \frac{(\#fruits/tree)(gpulp/fruit)(energy/gpulp)}{D + K}$$

where D is the distance (m) of a tree from the focal tree and K is a constant used to weight those resources closer to the focal tree more heavily. The number of fruits per tree was estimated as described above. Thus, each individual focal tree had an alternate resource score, which is a sum of energy availability in each alternative tree, discounted by the distance of the alternative resource from the focal tree plus a scaling constant K in meters. For this analysis, we chose K equal to 20 because previous analyses indicated that the trees within 20 m of the focal tree may be more important to the capuchin monkeys than more distant ones (Vogel, 2004).

Amounts and timing of food ingested for calculation of fractional satiation

We estimated food intake using the systematic instantaneous scan samples at 5-min intervals throughout the day. These samples provided data on how

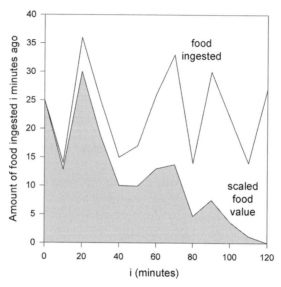

Figure 11.3. Illustration of the calculation of S in a hypothetical example. The upper solid line represents the actual food intake at time i before the present. The upper boundary of the lower, shaded area represents the linearly discounted food intake at each time i. The shading indicates that the linearly discounted food intakes need to be summed up from i = 0 to P (120 min in this case).

many animals in the group were ingesting food. Each point sample of a feeding monkey counts in the formula as though that individual fed for the entire interval between samples. Thus, for any desired time point, we knew the prior 2-hour history of feeding by the "average" group individual, and used this to calculate S. For instance, three individuals feeding 35 minutes ago for a group of N capuchins (with $P = 120$ min) would increase current *average* S by $[3/N]*5*(120-35)/120$. We then summed these S increments for all times less than P minutes prior to the "present." This measure of S is an index of gut filling for the "average" individual in the group. A graphical example of the procedure applied to hypothetical data is given in Figure 11.3. In this analysis, we calculated fractional satiation for the average group member at the time of entry of the first individuals to the focal tree. Ideally, we would have measured S precisely for each of the participants in the agonistic encounter, based on their personal feeding histories. However, as we could not predict which individuals would interact in a given tree, having such precise data would have required obtaining complete focal-animal data for every single group member simultaneously, a feat we could not accomplish.

Statistical considerations

Except for the fractional satiation measure, we used logarithms of the dependent and independent variables to reduce the effect of extreme measurements and consequent heteroskedasticity of residuals. None of the pair-wise correlations between any of the 13 independent variables was greater than 0.68, so we did not consider colinearity to be a major problem in interpreting the results below. In any case, we do not emphasize the importance of any one variable within a set belonging to one of the broader hypotheses; rather, we use the results for any of the variables merely to indicate the importance of the hypothesis of which it forms a part. Because we predicted a unique direction of the effect of each variable on the frequency of agonism, we report one-tailed tests in evaluating statistical significance.

Results

Satiation measures

The fractional satiation values at the time of entry to the focal tree ranged from 0.19 to 25.4 minutes, out of a possible maximum value of 60 if all group members had been feeding continuously for the 2 hours prior to entering a given feeding tree (Figure 11.4). Expressed as a fraction of the maximum possible, the scores ranged from 0.0003 to 0.423. If these reflect satiation, the average group member varied from very hungry to nearly half-full as they entered distinct feeding trees.

Results of the multiple regression

We used the 13 variables in Table 11.1 to explain variation in the number of agonistic interactions for each visit to a given feeding tree. Elsewhere, we will examine the effect of these variables on the level of escalation of agonistic interactions (cf. Vogel, 2004). A total of 306 visits to over 176 individual trees of 24 species were included in the analysis; the original data are too extensive to include here, but are available upon request from ERV. We first present the results including all the independent variables. This complete analysis limits the sample size of focal tree visits to 181, and only includes visits in which at least one agonistic interaction was observed, since Minutes Lost Feeding and Distance Moved after displacement were defined only when agonism was observed. The regression model including all 13 variables and the quadratic term for crown energetic value was highly significant

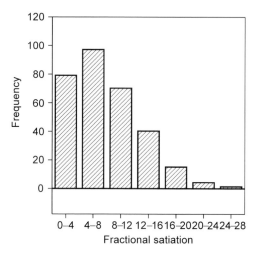

Figure 11.4. Distribution of observed values for white-faced capuchins (*Cebus capucinus*) of the QQ group of average individual fractional satiation S (in feeding minutes) at the time of entry into 306 focal trees, using a linear discounting function (see Figure 11.3).

(Table 11.2). Four of the independent variables had slopes that differed significantly from zero, and all were in the direction predicted in Table 11.1. The number of agonistic acts per visit to a focal tree increased with increasing feeding bout length and average numbers of females and of males, and decreased with increasing crown volume. Fractional satiation had a marginal effect in the expected direction (one-tailed p = 0.06).

Neither Minutes Lost Feeding nor Distance Moved after displacement had slopes that differed significantly from zero in this analysis. In addition, their inclusion in the analysis severely restricted sample size and possibly biased the results, as trees with no agonism were necessarily excluded. Therefore, we recalculated the analysis with these two variables removed, thus increasing the sample to 298 focal trees, including all those in which no agonism occurred. The revised multiple regression model was also highly significant (Table 11.3). Five of the independent variables have slopes that differ significantly from zero at a one-tailed rejection level of 0.05, and all of them are consistent with predictions in Table 11.1.

Given the redundancy in measuring different possible influences on aggression, we make no claim that the particular parameters that emerged as statistically significant from the multiple regressions are uniquely better predictors than other parameters that were excluded. We merely take the presence of a particular parameter in the final model as indicative that the

Table 11.2. *Results of the complete multiple regression model to explain variation in number of agonistic interactions/visit to a feeding tree by white-faced capuchin monkeys* (Cebus capucinus)

Variable	Expected sign	Slope (beta)	Std Error (beta)	F Ratio	% of r^2	Prob > F (1-tailed)
Feeding bout length	+	0.1865	0.0406	21.09	43.62	<0.0001
Average no. of adult females feeding in the focal tree	+	0.2261	0.1079	4.39	9.08	0.0188
Average no. of adult males feeding in the focal tree	+	0.2481	0.1023	5.88	12.17	0.0082
Average no. of nonadults feeding in the focal tree	+	0.01142	0.09778	0.014	0.03	0.4536
Crown volume	–	–0.08492	0.02602	10.65	22.04	0.0007
Maximum no. of individuals feeding in the focal tree	+	0.06872	0.1225	0.31	0.65	0.2878
Crown energetic value	+	0.03341	0.03262	1.049	2.17	0.1537
Square of Crown energetic value	–	–0.00355	0.00535	0.44	0.91	0.2542
No. of alternate trees of same species as focal tree	–	–0.0117	0.03095	0.14	0.29	0.3534
No. of alternate trees of different species than focal tree	–	0.00778	0.03542	0.048	0.10	0.5868
Minutes lost feeding	+	0.00661	0.04516	0.021	0.04	0.4419
Distance moved by loser after displacement	+	–0.01026	0.03524	0.085	0.18	0.6144
Alternate resource score, visual fruit counts, K = 20	–	–0.01326	0.01003	1.75	3.62	0.0940
Fractional satiation at time of entry to the focal tree	–	–0.00391	0.00249	2.46	5.10	0.0592

The r^2 for this model is 38.1% (adjusted $r^2 = 32.9\%$) and it is highly significant (F[14,166] = 7.31, p < 0.0001). Although data on 306 tree crown visits were included in the analysis, only 181 crowns had complete data for all the variables included here. The column "expected slope" duplicates the predictions from Table 11.1 The column "% of r^2" represents the percentage contribution of each variable to the total explained variance attributable to the additive effects of single variables.

Table 11.3. *Results of the reduced multiple regression model to explain variation in number of agonistic interactions/visit to a feeding tree by white-faced capuchin monkeys* (Cebus capucinus)

Variable	Expected sign	Slope (beta)	Std Error (beta)	F Ratio	% of r^2	Prob > F (1-tailed)
Feeding bout length	+	0.2070	0.0432	22.99	28.04	<0.0001
Average no. of adult females feeding in the focal tree	+	0.3277	0.1066	9.46	11.54	0.0012
Average no. of adult males feeding in the focal tree	+	0.1709	0.1121	2.33	2.84	0.0642
Average no. of nonadults feeding in the focal tree	–	−0.0074	0.1043	0.01	0.01	0.5283
Crown volume	–	−0.1595	0.0289	30.56	37.28	<0.0001
Maximum no. of individuals feeding in the focal tree	+	0.2521	0.1292	3.81	4.65	0.0260
Crown energetic value	+	0.0104	0.0386	0.07	0.09	0.3935
Square of Crown energetic value	–	0.0007	0.0063	0.01	0.02	0.5463
No. of alternate trees of same species as focal tree	–	−0.0281	0.0362	0.60	0.73	0.2192
No. of alternate trees of different species than focal tree	–	−0.0065	0.0427	0.02	0.03	0.4400
Alternate resource score, visual fruit counts, K = 20	–	−0.0075	0.0117	0.41	0.50	0.2620
Fractional satiation at time of entry to the focal tree	–	−0.0088	0.0026	11.71	14.28	0.0004

Minutes lost feeding and distance moved by loser after displacement are omitted from the analysis in Table 11.2, allowing inclusion of data for trees with no agonism (total $N = 298$). The r^2 for this model is 36.6% (adjusted $r^2 = 33.9\%$) and it is highly significant (F[12,285] = 13.68, p < 0.0001).

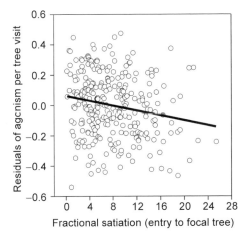

Figure 11.5. Effect of increasing satiation on the number of agonistic interactions/visit to focal trees. The residuals plotted are the values for number of agonistic acts, holding all the variables in Table 11.3 constant except for fractional satiation. Fractional satiation explains only a small fraction of the total variation in number of agonistic interactions/tree visit, but is important relative to most other ecological and social variables tested (Table 11.3).

biological hypothesis it represents (Table 11.1) has a significant impact on aggression. Among the five significant variables found in Table 11.3 were ones representing the hypotheses that total number of agonistic interactions increased with sampling (longer feeding bout length, more adult females in the tree crown), increased with greater crowding (greater maximum number of animals in the crown, smaller crown volumes), and decreased with satiation (greater fractional satiation at time of entry to the tree). The largest explanatory effect on increased agonism came from decreasing crown volume. The second-largest contribution to explained variance came from increased feeding bout length. Hunger, as reflected in fractional satiation at the time of entry to the tree, was the variable with the third largest individual contribution to the explained variance in the number of agonistic interactions (Figure 11.5). The effect of fractional satiation was highly significantly different from zero (one-tailed $p = 0.0004$) and accounted for over 14% of the total explained variance in the model (Table 11.3). Fourth and fifth in amount of variance explained were, respectively, the average number of adult females and the maximum number of individuals feeding in the focal tree. Both these variables may indicate effects of increased crowding.

The significant effect of fractional satiation on agonism was not an artifact of comparing different food species in the same analysis. We restricted the

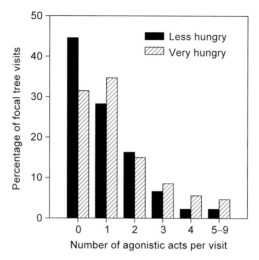

Figure 11.6. During feeding tree visits when the capuchins arrive less hungry (S = 10–25, black bars, N = 92, mean number of agonistic acts/visit = 1.03), they engage in less agonism than when they arrive very hungry (S = 0–10, hatched bars, N = 213, mean number of agonistic acts/visit = 1.42). The difference in agonism between the two conditions is statistically significant (Kruskal–Wallis nonparametric ANOVA, z = −2.13, P = 0.033).

analysis to variation in agonism among focal trees of a single species. In the two species (*Sloanea terniflora* and *Anacardium occidentalis*) with sample sizes of more than 40 focal trees, increased frequency of agonism was significantly related to decreasing fractional satiation. Of the remaining four species for which the multiple regression of Table 11.3 could be run at all, three of them showed an effect of fractional satiation on agonism in the expected direction, albeit none were significantly different from zero.

Satiation and agonism in focal trees

Fractional satiation was essentially uncorrelated with other independent variables in Table 11.1 – the largest pair-wise correlation was 0.16. Therefore, it is justifiable to directly contrast the frequency of agonism in trees where the capuchins arrived very hungry (S = 0–10, mean S = 5.1 minutes) versus less hungry (S = 10–25, mean S = 13.9 minutes). The frequency of agonistic interactions in tree visits when the animals arrived less hungry was 28% less than when the animals were very hungry (Figure 11.6). Extrapolating this significant trend would lead to the expectation that completely satiated animals (S = 60 minutes) would show almost no food-related agonism at all.

Discussion

The explanatory power of hunger on agonism

Taking an animal's current state into account can help to resolve some of the variation in its behavior over time (Houston & McNamara, 1999). Increased hunger, as reflected in reduced values of our empirical measure of fractional satiation, was related to increased agonism per visit to feeding trees, holding all the other variables constant (Table 11.3). Fractional satiation explained only a small percentage (1–2.6%) of the total variation in frequency of agonism. However, even the variable with the largest effect in the model explained less than 18% of the variation in frequency of agonism. The contribution of fractional satiation to explaining variation in frequency of agonism was the third largest of all variables in Table 11.3 and was highly significant statistically. This analysis is based on the study of a single social group of capuchins, so it must be regarded with caution until it can be replicated with other groups and other species.

The low overall fraction of the variation in numbers of agonistic interactions explained by the full regression model (36.6–38.1%; Tables 11.2, 11.3) is likely due to several factors. First, even though the number of agonistic interactions is a simple count variable, it may be underestimated in some cases. Subtle approach–avoids are likely to be missed and should ideally be included when they center on contestable resources. Second, many of the independent variables are at best estimates of the parameters perceived by the capuchins. These include "hunger," energy available in the tree crown, and alternative resource availability, for all of which practical considerations in the field forced a tradeoff between speed of measurement and precision of the values estimated. Even the number of other monkeys in the tree crown, a parameter easy to count, assumes that each other monkey was a potential source or target of aggression, thus ignoring the specifics of dominance rank, kinship, alliances, etc.

Testing for the effects of hunger on behavior could be strengthened in several ways. First, one could use more refined measures based on the food intake history of the actual participants, rather than the group average value that we used. Such complete data on individuals would require multiple simultaneous focal-animal follows, an investment in sampling that is not often possible except in small study groups (e.g., Pochron *et al.*, 2003). Second, a far better understanding of the physiology of hunger in different species of primates would help to determine how food intake at different times prior to a particular moment actually affects hunger (the curves illustrated in Figure 11.2). Future studies should try a variety of curve shapes and a variety of P values based on the digestive parameters of their study species, when known. Several candidate values for P include the time it takes to empty the stomach (e.g., Campbell *et al.*, 2004), the mean transit time of food through the entire gut (Milton, 1984), and the time it takes to metabolize the energy content of a stomach-full of food (after which, presumably, blood glucose levels would drop and possibly signal increased hunger: de Graaf *et al.*, 2004). Further work on these questions, probably best carried out in captive animals, would be useful. In any case, the fact that a simple index reflecting a group's average feeding history can be used to calculate a useful measure of hunger means that future studies can test the importance of this state variable in greater detail.

Fitness consequences of hunger-mediated variation in agonism

Even though our measure of hunger explains only a small fraction of variation in rates of agonism, the fitness consequences may be important. If each agonistic interaction has the potential to escalate into serious injury-causing aggression, the reduced agonism initiated when animals were more satiated could represent a marked increase in fitness. The results in Figure 11.6 suggest that capuchins would engage in almost no agonism in feeding trees if they arrived at them fully satiated.

Conversely, incorporating hunger into resource value means that a very hungry subordinate individual should be willing to approach a dominant individual in a food source, even if the subordinate would normally avoid the dominant's vicinity. If such approaches at least occasionally result in the dominant allowing the subordinate to feed in the resource (whether or not the dominant leaves), the subordinate gains access to food that it otherwise would not have. Even if such events are rare, the large fitness consequences of obtaining food when very hungry (Figure 11.1) suggest that subordinates should pursue such opportunities.

Behavioral consequences of state-dependent behavior

Understanding time-varying fitness values for a given "objective" resource can have important consequences for several aspects of primate socioecology. First, the example given here suggests that increased hunger leads to increased willingness to challenge possession of a food source. Perhaps one reason that capuchin monkeys show high rates of agonism is that their guts empty relatively quickly (compared with many primates; Milton, 1984), and so are more likely to be hungry. Conversely, the low rates of agonism seen in many folivores may be due not to superabundant food or low metabolic rates but to their slow gut passage rates, which make the marginal value of added food relatively low. Second, the tradeoffs that animals may make between foraging and other activities, such as avoidance of predators, will depend on their fitness valuation of added food. Thus, it should not be surprising to see that as individuals become hungrier, they forage farther from group mates, because as the fitness return for added food increases, they are willing to risk a higher chance of predation to increase food intake. Likewise, peripheral animals in a social group may be kept at the periphery not because they fear aggression from central dominant animals (cf. Barta *et al.*, 1997), but because the greater hunger of subordinate individuals makes taking a higher risk of predation acceptable (Janson, 1990). Third, if much of the social behavior of primates occurs in the context of a biological market, and if access to food is one of the commodities in this market, then time-varying levels of hunger in group members could dictate parallel changes in patterns of social exchange. For instance, animals of low status should be willing to groom "control" animals far more than they receive grooming from them because a given increase in food intake is worth much more to the low-status animal than to the "control" animal. When food is distributed so that differences in overall food intake among individuals of different status are minimal, the asymmetry in grooming relationships should be reduced or eliminated. Once researchers become sensitive to the likely variation in fitness values over time due to changes in an animal's state (hunger, age, etc.), the number of applications of state-dependent behavioral analysis will quickly increase.

References

Altmann, J. (1974). Observational study of behavior – sampling methods. *Behaviour*, **49**, 227–67.

Altmann, S. A. (1998). *Foraging for Survival: Yearling Baboons in Africa*. Chicago, IL: University of Chicago Press.

Barta, Z., Flynn, R., & Giraldeau, L. A. (1997). Geometry for a selfish foraging group: a genetic algorithm approach. *Proceedings of the Royal Society of London, Series B, Biological Sciences*, **264**, 1233–8.

Barton, R. A. & Whiten, A. (1994). Reducing complex diets to simple rules: food selection by olive baboons. *Behavioral Ecology and Sociobiology*, **35**, 283–93.

Campbell, J. L., Williams, C. V., & Eisemann, J. H. (2004). Characterizing gastrointestinal transit time in four lemur species using barium-impregnated polyethylene spheres (BIPS). *American Journal of Primatology*, **64**, 309–21.

Caraco, T. (1979). Time budgeting and group size: a theory. *Ecology*, **60**, 611–17.

Chapman, C. A., Wrangham, R. W., & Chapman, L. J. (1994). Indices of habitat-wide fruit abundance in tropical forests. *Biotropica*, **6**, 160–71.

Clark, C. W. & Mangel, M. (2000). *Dynamic State Variable Models in Ecology: Methods and Applications*. Oxford: Oxford University Press.

de Graaf, C., Blom, W. A. M., Smeets, P. A. M., Stafleu, A., & Hendriks, H. F. J. (2004). Biomarkers of satiation and satiety. *American Journal of Clinical Nutrition*, **79**, 946–61.

Dubois, F. & Giraldeau, L. A. (2003). The forager's dilemma: food sharing and food defense as risk-sensitive foraging options. *The American Naturalist*, **162**, 768–79.

Dubois, F., Giraldeau, L. A., & Grant, J. W. A. (2003). Resource defense in a group-foraging context. *Behavioral Ecology*, **14**, 2–9.

Fedigan, L. M., Fedigan, L., & Chapman, C. A. (1985). A census of *Alouatta palliata* and *Cebus capucinus* in Santa Rosa National Park, Costa Rica. *Brenesia*, **23**, 309–22.

French, S. J. & Read, N. W. (1994). Effect of guar gum on hunger and satiety after meals of differing fat content: relationship with gastric emptying. *American Journal of Clinical Nutrition*, **59**, 87–92.

French, S. J., Murray, B., Rumsey, R. D. E., Sepple, C. P., & Read, N. W. (1993). Is cholecystokinin a satiety hormone? Correlations of plasma cholecystokinin with hunger, satiety and gastric emptying in normal volunteers. *Appetite*, **21**, 95–104.

Ganzhorn, J. U. (1988). Food partitioning among Malagasy primates. *Oecologia*, **75**, 436–50.

Glander, K. E., Fedigan, L. M., Fedigan, L., & Chapman, C. (1991). Field methods for capture and measurement of three monkey species in Costa Rica. *Folia Primatologica*, **57**, 70–82.

Grether, G. F., Palombit, R. A., & Rodman, P. S. (1992). Gibbon foraging decisions and the marginal value model. *International Journal of Primatology*, **13**, 1–17.

Hladik, C. M. (1977). Field methods for processing food samples. In *Primate Ecology: Studies of Feeding and Ranging Behaviour in Lemurs, Monkeys, and Apes*, ed. T. H. Clutton-Brock, pp. 595–601. New York, NY: Academic Press.

Houston, A. I. & McNamara, J. M. (1999). *Models of Adaptive Behaviour: An Approach Based on State*. Cambridge: Cambridge University Press.

Janson, C. H. (1985). Aggressive competition and individual food intake in wild brown capuchin monkeys. *Behavioral Ecology and Sociobiology*, **18**, 125–38.

(1987). Ecological correlates of aggression in brown capuchin monkeys. *International Journal of Primatology*, **8**, 431.

(1988a). Food competition in brown capuchin monkeys (*Cebus apella*): quantitative effects of group size and tree productivity. *Behaviour*, **105**, 53–76.

(1988b). Intra-specific food competition and primate social structure: a synthesis. *Behaviour*, **105**, 1–17.

(1990). Ecological consequences of individual spatial choice in foraging brown capuchin monkeys (*Cebus apella*). *Animal Behaviour*, **38**, 922–34.

Koenig, A. (2000). Competitive regimes in forest-dwelling Hanuman langur females (*Semnopithecus entellus*). *Behavioral Ecology and Sociobiology*, **48**, 93–109.

(2002). Competition for resources and its behavioral consequences among female primates. *International Journal of Primatology*, **23**, 759–83.

Manson, J. H., Rose, L. M., Perry, S., & Gros-Louis, J. (1999). Dynamics of female-female relationships in wild *Cebus capucinus*: data from two Costa Rican sites. *International Journal of Primatology*, **20**, 679–706.

Maynard Smith, J. (1974). The theory of games and the evolution of animal conflicts. *Journal of Theoretical Biology*, **47**, 209–21.

Maynard Smith, J. & Parker, G. A. (1976). The logic of asymmetric contests. *Animal Behaviour*, **24**, 159–75.

Merrill, A. L. & Watt, B. K. (1973). Energy value of foods. basis and derivation. In *Agriculture Handbook 74*. Washington, DC: United States Department of Agriculture.

Milton, K. (1984). The role of food-processing factors in primate food choice. In *Adaptations for Foraging in Nonhuman Primates: Contributions to an Organismal Biology of Prosimians, Monkeys, and Apes*, ed. P. S. Rodman & J. G. H. Cant, pp. 249–79. New York, NY: Columbia University Press.

Nakagawa, N. (1990). Choice of food patches by Japanese monkeys (*Macaca fuscata*). *American Journal of Primatology*, **21**, 17–29.

Näslund, E., Hellström, P. M., & Kral, J. G. (2001). The gut and food intake: an update for surgeons. *Journal of Gastrointestinal Surgery*, **5**, 556–67.

Noldus Information Technology (2003). *MatMan. Software for Matrix Manipulation and Analysis*. Wageningen, the Netherlands: Noldus Information Technology.

Oppenheimer, J. R. (1968). *Behavior and Ecology of the White-faced Capuchin Monkey* (Cebus capucinus) *on Barro Colorado Island*. Unpublished Ph.D. thesis, University of Illinois-Urbana.

Perry, S. (1995). *Social Relationships in White-faced Capuchin Monkeys, Cebus capucinus*. Unpublished Ph.D. thesis, University of Michigan-Ann Arbor.

  (1996). Female-female social relationships in wild white-faced capuchin monkeys, *Cebus capucinus*. *American Journal of Primatology*, **40**, 167–82.

Pochron, S. T., Fitzgerald, J., Gilbert, C. C., Lawrence, D., Grgas, M., Rakotonirina, G., Ratsimbazafy, R., Rakotosoa, R., & Wright, P. C. (2003). Patterns of female dominance in *Propithecus diadema edwardsi* of Ranomafana National Park, Madagascar. *American Journal of Primatology*, **61**, 173–85.

Post, D. G. (1984). Is optimization the optimal approach to primate foraging? In *Adaptations for Foraging in Nonhuman Primates: Contributions to an Organismal Biology of Prosimians, Monkeys, and Apes*, ed. P. S. Rodman & J. G. H. Cant, pp. 280–303. New York, NY: Columbia University Press.

Richard, A. F., Dewar, R. E., Schwartz, M., & Ratsirarson, J. (2000). Mass change, environmental variability and female fertility in wild *Propithecus verreauxi*. *Journal of Human Evolution*, **39**, 381–91.

Robinson, J. G. (1986). Seasonal variation in use of time and space by the wedge-capped capuchin monkey, *Cebus olivaceus*: implications for foraging theory. *Smithsonian Contributions to Zoology*, **431**, 1–60.

Rose, L. M. (1998). *Behavioral Ecology of White-faced* Capuchins (Cebus capucinus) in Costa Rica. Unpublished Ph.D. thesis, Washington University-Saint Louis.

Schwartz, M. W., Woods, S. C., Porte, Jr., D., Seeley, R. J., & Baskin, D. G. (2000). Central nervous system control of food intake. *Nature*, **404**, 661–71.

Stephens, D. W. & Krebs, J. R. (1987). *Foraging Theory*. Princeton, NJ: Princeton University Press.

Sterck, E. H. M., Watts, D. P., & van Schaik, C. P. (1997). The evolution of female social relationships in nonhuman primates. *Behavioral Ecology and Sociobiology*, **41**, 291–309.

Terborgh, J. W. (1983). *Five New World Primates: A Study in Comparative Ecology*. Princeton, NJ: Princeton University Press.

van Schaik, C. P. (1989). The ecology of social relationships amongst female primates. In *Comparative Socioecology of Mammals and Humans*, ed. V. Standen & R. Foley, pp. 195–218. Oxford: Blackwell Publishers.

Vogel, E. (2004). *The Ecological Basis of Aggression in White-faced Capuchin Monkeys,* Cebus capucinus, *in a Costa Rican Dry Forest*. Unpublished Ph.D. thesis, State University of New York-Stony Brook.

12 How does food availability limit the population density of white-bearded gibbons?

ANDREW J. MARSHALL AND MARK LEIGHTON

Feeding Ecology in Apes and Other Primates. Ecological, Physical and Behavioral Aspects, ed.
G. Hohmann, M.M. Robbins, and C. Boesch. Published by Cambridge University Press.
© Cambridge University Press 2006.

Introduction

For decades primatologists have assumed that food availability is the primary determinant of primate population density (Terborgh & van Schaik, 1987; Davies, 1994). Observations of temporal changes in population density related to food availability have supported this assumption. For example, Dittus (1979) documented a 15% decrease in the size of a population of Toque macaques (*Macaca sinica*) during a period of reduced food availability. Similarly, populations of vervet monkeys (*Cercopithecus aethiops*) (Struhsaker, 1973; Lee & Hauser, 1998) and of yellow baboons (*Papio cynocephalus*) (Hausfater, 1975) also diminished in size when their resource base was reduced. Altmann *et al.* (1985) reported a 34-fold decrease in the population of yellow baboons in Amboseli over a 15-year period and attributed this decline to drastic environmental change that reduced the availability of high-quality food. Finally, the population density of mangabeys (*Cercocebus albigena*) at Kibale increased over a 20-year period, most likely because the forest was regenerating after logging and the density of mangabey food trees had substantially increased (Olupot *et al.*, 1994).

Comparisons of the same or closely related species at different locations have also provided some insight into the role of food availability in limiting primate populations. However, such studies typically involve populations separated by substantial distances (e.g., McKey, 1978; Oates *et al.*, 1990; Ganzhorn, 1992; Gupta & Chivers, 1999), thereby defying interpretation because of the many potentially confounding factors such as human disturbance (Struhsaker, 1999), biogeographic history (Gupta & Chivers, 1999), and differences in research methodology (Chapman *et al.*, 1999). More recent studies, notably those of Chapman and colleagues (e.g., Chapman & Chapman, 1999), have examined variation in primate densities on a more refined spatial scale and shown that variation in density between sites is correlated with the abundance of food resources.

These results, and many others, suggest that food availability is an important force in limiting primate biomass. This robust conclusion prompts the question of whether there exist specific classes of foods whose abundance sets the carrying capacity for tropical forest primates. A key distinction may be between preferred and fallback foods. Preferred foods are those that are eaten more often than would be predicted based on their availability at any given time (i.e., overselected, *sensu* Leighton, 1993; Manly *et al.*, 2002). Fallback foods are classified as foods whose use is negatively correlated with the availability of preferred food items across time (Conklin-Brittain *et al.*, 1998). Basic foraging theory (Stephens & Krebs, 1986) suggests that preferred foods are those that can be efficiently harvested (i.e., yield high energy returns/foraging effort compared with other food items). Empirical study has confirmed that this suggestion is valid, at least for species that have been investigated (e.g., orangutans: Leighton, 1993). Gibbons exhibit a strong preference for fleshy fruits (McConkey *et al.*, 2002; Leighton, unpublished data) whose sugar-rich pulp provides relatively high rates of energy return. Fallback foods are utilized when these energy-rich fleshy fruits are scarce. These low-preference "fallback foods" may be available at other times but are ignored during periods of high preferred fruit availability (Leighton & Leighton, 1983). As the spatial and temporal availability of preferred and fallback foods may vary independently, their effects on primate populations may be different.

Primate dietary intake is subject to wide seasonal variation because of fluctuations in the phenology of food resources (Terborgh, 1986; Oates, 1987; Janson & Chapman, 1999). During periods of high food availability, surplus energy (i.e., above the amount required for physiological mainten-ance) becomes available, enabling growth and reproduction (Charnov & Berrigan, 1993). The greater the abundance of these items in a habitat, the greater the maximum energy availability during productive periods. Since an increase in net energy availability during these periods should enable higher reproductive rates, areas with a higher availability of preferred foods should maintain a higher primate density than areas that are relatively depauperate of these resources. During times of low fruit availability, most primates tend to rely more heavily on less preferred food resources to fulfill the caloric demands of physiological maintenance (Leighton & Leighton, 1983; Terborgh, 1986). In most organisms that are limited by resources, classic ecological theory predicts that periods of food shortage will set the population size. This is especially true for species whose populations grow at relatively slow rates (e.g., primates) since they are unable to closely track temporal fluctuations in food availability (Wiens, 1977). For such species, food may be superabundant most of the time, and populations can go for many months (or years) without experiencing any resource limitation (Cant, 1980). However, occasional periods of food scarcity may cause an increase in mortality levels (Foster, 1974; Wiens, 1977), resulting in bottlenecks that ultimately limit population size (Milton, 1982; Davies, 1994).

Plausible arguments can be made in support of either preferred foods or fallback foods serving as the limiting factor on primate populations, yet to date no studies have specifically addressed the relative importance of these two classes of foods. Even studies claiming to have demonstrated the effects of preferred food availability have found correlations between important foods (those comprising a substantial percentage of the diet) and biomass, rather than between preferred foods and biomass (e.g., Caldecott, 1980;

Mather, 1992). Though interesting and suggestive, these analyses are inconclusive because they neither explicitly differentiate between preference and importance nor consider the effects of temporal variation in food availability. In addition, since other possible food variables, (e.g., patch size or the availability of fallback foods) were not measured, these results may be confounded by other factors. Thus, the issue of whether different classes of foods have different effects on primate populations remains untested.

Patterns of food availability in Bornean forests

Compared with most other rainforests, Malesian tropical forests exhibit patterns of food availability that are both more temporally variable and less predictable (Leighton & Leighton, 1983; van Schaik, 1986; Whitmore, 1990; Wich & van Schaik, 2000). This heightened variability is largely due to the phenomenon of mast fruiting, which involves the gregarious fruiting of many individual plants, presumably as an adaptation to avoid seed predation (Janzen, 1974; Ashton *et al.*, 1988; Curran & Leighton, 2000). During masts in Malesian forests virtually all taxa in the dominant family Dipterocarpaceae, along with many other common taxa, fruit in synchrony after several years of reproductive inactivity (Medway, 1972; Appanah, 1981; Ashton *et al.*, 1988). This pattern results in extreme temporal fluctuations in food availability for vertebrates (Leighton & Leighton, 1983; van Schaik & van Noordwijk, 1985; Knott, 1998).

In order to generate testable predictions derived from our hypotheses, we assigned each month in the study period to one of four classes based on food availability: masts, high fruit periods, low fruit periods, and crunches. Below we provide a brief description of each category. Figure 12.1 provides schematic diagrams of food availability and dietary composition during each category. As indicated above, mast periods are times during which a large set of the woody plants in the forest (predominantly trees, although there are masting liana species) produce fruit. In addition to this high density of large fruit patches, it is probable that masting taxa produce fruits that are of higher quality (i.e., higher rates of net energy return) than most fruits produced outside the mast, presumably because plants that fruit during masts are subject to heavy intra- and interspecific competition for vertebrate dispersers. Although this concept has yet to be explicitly examined, it is a logical deduction considering the competitive milieu of plants, and both our unpublished data and published results (Leighton, 1993; van Schaik & Knott, 2001) support this claim. This suggests that during masts both the amount

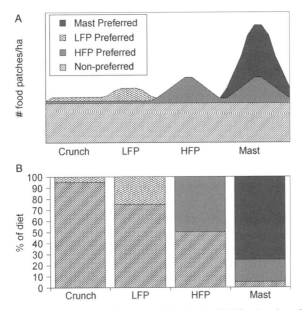

Figure 12.1. Schematic diagrams of (A) food availability (number of food patches/ha) and (B) diet composition during crunches, periods of low (LFP), high (HFP) food availability, and masts. X-axes represent categories of food availability, with total food availability increasing from left to right.

and quality of food resources may be maximal, resulting in diets that are comprised mostly of preferred foods.

High fruit periods (HFP) are non-mast periods during which a substantial number of food taxa fruit together. These peaks occur on a roughly annual basis, although their arrival seems less predictable in forests on Indonesian islands than in most African or Neotropical rainforests, presumably due to less predictable patterns of rainfall at the Indonesian sites (Whitmore, 1990; Janson & Chapman, 1999). We found that most gibbon fruit taxa that fruit during HFP also fruit during mast periods and are thus likely to be subjected to intense competition for dispersal agents. Therefore, although food availability during HFP is not as high as during masts, the selection of fruits with high net rates of energy return for frugivores results in the availability of a substantial number of preferred foods during HFP. Consequently, diets during HFP comprise a substantial proportion of high-quality, preferred items.

Low fruit periods (LFP) are periods during which both the availability and relative quality of foods are low. As plants require some period of reproductive inactivity following fruiting so as to recoup sufficient nutrient status in order to fruit again, and since the majority of food trees fruit during masts or HFP, LFP are a direct consequence of masts and HFP. Plant taxa whose reproductive strategy is to fruit during periods of low overall food availability experience relatively little competition for dispersers, and therefore can afford to produce fruits that provide meager energetic returns. Therefore, the quality of fruits produced during LFP is expected to be substantially lower than those produced during times of higher food availability, and the diets of consumers comprise a lower percentage of preferred foods. This claim is also supported by available data (Leighton, 1993, and unpublished data). Vertebrate consumers must make up the balance of their diets during LFP with foods that provide low rates of net energy return and are of low preference rank.

Crunch periods refer to rare times of extreme food shortage. Crunch periods differ from LFP in degree; food availability is drastically reduced and there are virtually no plants fruiting that provide preferred, high-quality foods. During these periods reproduction is probably impossible for most frugivorous primates and the lack of food may temporarily cause elevated mortality and potentially limit population size.

Phenological patterns of plant taxa

The overall temporal pattern of food availability in Bornean forests is the result of the summation of the phenological patterns of numerous individuals from hundreds of different plant taxa. In order to characterize the phenological patterns of gibbon food taxa, we defined four general classes of phenology. These categories classify plant taxa according to their availability relative to overall patterns of food availability in the forest and were defined to highlight the temporal variation that is presumably most relevant to gibbons. These categories are: (1) *masting taxa*, defined as taxa that fruit predominantly during mast periods; (2) *HFP fruiters*, defined as taxa that fruit mainly during the regular forest-wide peaks in food availability and may or may not also fruit during mast periods; (3) *LFP fruiters*, defined as taxa that fruit mainly during times of low overall food availability and may or may not fruit at other times; and (4) *crunch fruiters*, defined as taxa that are available predominantly during periods of extreme food shortage. Figure 12.2 provides examples of each of the four categories from the phenology data used for this analysis.

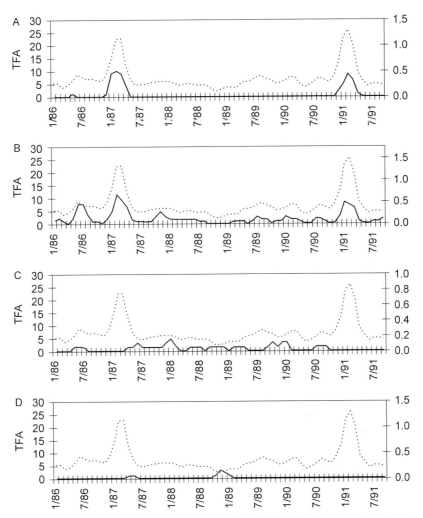

Figure 12.2. Phenology of sample taxon in each of the four classes used to categorize gibbon foods in Bornean rainforests between January 1986 and September 1991 at Gunung Palung. The dotted line indicates Total Food Availability (TFA; patches/ha), and the black line indicates the number of patches/ha of a sample masting taxon (A: *Neoscortechinia kingii*), HFP taxon (B: *Rourea minor*), LFP taxon (C: *Porterandia sessiliflora*), and crunch taxon (D: *Gnetum sp. 1*). See text for additional details.

Specific hypotheses to be tested

The primary goal of this chapter is to examine how food availability limits gibbon population density across seven habitat types at our site in Gunung Palung National Park (GPNP), West Kalimantan, Indonesia. In particular, we assess whether preferred or fallback foods set the habitat-specific carrying capacity (i.e., maximum population size that can be sustained indefinitely) for gibbons. Gibbon densities at GPNP have remained stable in each habitat over the last 20 years and no hunting has occurred within the study site, suggesting that these populations are at carrying capacity (Mitani, 1990; Marshall, 2004). Therefore correlations between the availability of certain types of food and gibbon density would support the hypothesis that a certain type of food limits gibbon populations. We therefore tested the following hypotheses.

H1: Gibbon density is limited by the total availability of food
If food resources are the principal limitations on gibbon population density, then it is plausible to hypothesize that variation in population density between habitats is related to the total availability of foods. Although this hypothesis seems oversimplified and does not incorporate temporal variation in food availability or food quality, it is essentially the logic upon which many previous attempts to explain variation in population density are based (e.g., Mather, 1992; Chapman & Chapman, 1999). In these studies, the densities of food plants or their basal area serve as the measure of food availability. *H1* predicts that variation in the density of gibbons across the seven habitats at GPNP is positively correlated with measures of the overall abundance of food in each habitat.

H2: Gibbon density is limited by the average availability of food
A second hypothesis is that average food availability in a habitat may serve as the key limitation on gibbon population density. This hypothesis implies that short-term fluctuations in food availability have little effect on gibbon populations, and that it is mean food availability over time that sets the carrying capacity of a habitat. *H2* predicts that variation in the density of gibbons across the seven habitats at GPNP is positively correlated with measures of the mean food availability/month in each habitat.

H3: Gibbon density is limited by the availability of preferred foods
H3 predicts that the density of gibbons is positively correlated with the availability of preferred taxa. We test three versions of this hypothesis: that gibbon density is correlated with the availability of preferred mast taxa, preferred HFP taxa, or the overall density of preferred foods.

*H4: Gibbon density is limited by the availability of fallback
food resources*

H4 predicts that gibbon density is positively correlated with the availability
of fallback food resources. We tested this hypothesis in two ways. First we
examined the correlation between gibbon density and the availability of
foods during LFP and crunch periods. Our second test involved the identification of specific food resources as fallback foods. We identified fallback
foods using the definition provided above and examined correlations between the habitat-specific density of specific fallback food items and gibbon
density.

Materials and methods

Study site

We conducted research at the Cabang Panti Research Station (CPRS) and
surrounding areas in GPNP ($1°13'$S, $110°7'$E). GPNP comprises a small
coastal mountain range surrounded by seasonally flooded swamp, peat, and
mangrove forests (Cannon & Leighton, 2004). We gathered data in seven
distinct habitat types within the CPRS trail system: peat swamp, freshwater
swamp, alluvial bench, lowland sandstone, lowland granite, upland granite,
and montane forests. General descriptions and detailed data on the plant
composition of each habitat are provided in Webb (1998), Cannon &
Leighton (2004), and Marshall (2004). These habitats are contiguous, thus
minimizing differences in rainfall, latitude, seasonality, gamma diversity, and
predation pressure that could confound results (Cannon & Leighton, 2004).
Since GPNP is located near the Bornean coast, its elevational gradient is
compressed (the Massenerhebung effect, see Whitmore, 1984), allowing for
ease of data collection in several different upland habitats.

Study subjects

Our study species was the Bornean white-bearded gibbon (*Hylobates albibarbis*). The taxonomic status of the gibbons living between the Kapuas and
Barito Rivers in southwest Borneo has been the source of considerable debate
(Groves, 1984, 2001; Marshall & Sugardjito, 1986). Here we use the species
designation *H. albibarbis*, but remain open to the possibility that this taxon
could be more appropriately designated as a subspecies of either *H. muelleri*
or *H. agilis*. Gibbons typically specialize on ripe, non-fig fruit and augment

their diet with figs, flowers, and young leaves during periods of low preferred fruit availability (Leighton, 1987).

Field methodology and data analysis

Habitat-specific gibbon density

We measured gibbon density by establishing a pair of replicate census routes in each of the seven habitats found at CPRS (n = 14 total census routes). Census routes averaged 3.42 km in length (range 2.90–3.80 km) and followed existing trails through the forest. AJM and his assistants walked a total of 409 censuses (1374 km) between September 2000 and June 2002. All walks were conducted at the same speed and data collection followed a standard protocol. An average of 58 censuses (range 38 to 87) were walked in each habitat. We followed standard methodologies for the analysis of line transect data using Distance 4.0 Release 2 (Thomas *et al.*, 2002), calculating detection functions separately in each habitat and controlling for size bias in sampling (Buckland *et al.*, 2001).

Diet composition

We used all the independent feeding observations gathered opportunistically and on censuses between April 1985–December 1991 and August 2000– August 2002 to compile a list of food taxa utilized by gibbons. Following the logic described below, we used the genus as the taxonomic level of analysis. Our resulting food list contained 91 genera.

Habitat-specific density of food trees and lianas

AJM randomly placed 10 plots in each habitat to assess the density of key tree and liana taxa in each of the seven forest types. All fig roots and liana stems within 10 m on either side of the transect midline and whose diameters at breast height (DBH, 137 cm above the ground) were greater than 4.5 cm were included, resulting in 5 ha of plots (0.5 ha/plot × 10 plots/habitat) for these forms in each habitat. The same plot size was used for trees with DBH greater than 34.5 cm. Five meters on either side of the transect midline, all trees with boles greater than 14.5 cm were included, resulting in 2.5 ha plots/habitat for smaller trees (0.25 ha/plot × 10 plots/habitat). We used these data to calculate the mean density and total basal area of each food taxon in each habitat.

Identification of trees and liana

In the field, AJM and his three field assistants identified family, genus, and (when possible) species of each stem in the density plots. We used genera as the taxonomic unit for food items in all analyses for several reasons. First, after a substantial training period and validation through double-blind tests involving the identification of trees identified at our site by recognized experts (e.g., P. Ashton, P. Stevens, M. van Balgooy) we were able to reliably identify the genus of individuals in the field without collecting leaf and bark samples. This made our plot work highly efficient and allowed us to sample a far larger area than would have been possible if we had collected the voucher specimens that would have been required to identify stems to the species level. Second, since some of the largest and most diverse plant families and genera have not been taxonomically revised recently, and systematic sampling in highly diverse Bornean forests such as those found in GPNP has been limited, many taxa at CPRS have not yet been assigned formal species names. Third, our data show that most fruits within the genera eaten by gibbons at our site are similar in the aspects of phenology and chemistry relevant to these analyses (Leighton, 1993, and unpublished data). Therefore, the lumping of two or more species under a single taxonomic designation probably did not obscure important differences relevant to gibbons. We retained more fine-grained taxonomic classification for genera in which lumping all species into one genus would have introduced bias (e.g., genera containing species that exhibited different growth forms or whose fruits were of different dispersal syndromes and/or of substantially different nutritional value). *Ficus* stems were identified to subgenus, following taxonomy in Laman & Weiblen (1998). Nomenclature followed that of the *Tree Flora of Malaya* series (Whitmore, 1972, 1973; Ng, 1978, 1989).

Temporal availability of food

We used data from 126 phenological plots that were monitored monthly between January 1986 and September 1991 (n = 69 months) to assess temporal variation in food availability. Phenology plots were 0.10 ha in size and were placed using a stratified random design across all seven habitat types (Cannon & Leighton, 2004). In these plots all trees larger than 14.5 cm dbh, all lianas larger than 3.5 cm dbh, and all hemiepiphytic figs whose roots reached the ground were measured and tagged. The phenological phase of each tagged stem in these phenology plots was recorded each month as one of six mutually exclusive categories: reproductively inactive, or containing flower buds (i.e., developing flowers were visible, but no flowers were at anthesis); mature flowers (i.e., at least one flower on the tree was at anthesis); immature fruit (i.e., fruits in which the seed was undeveloped); mature fruit (i.e., full-sized fruits that were unripe but had seeds that were fully developed and hardened); or ripe fruit (i.e., at least one fruit on the tree was ripe, usually signaled by a change in color or softness).

These data were used to calculate total food availability/ha and total number of fruiting taxa in each habitat in each month. Since the vast majority of gibbon feeding observations (93%) were of ripe fruit eating, we counted only food taxa stems with ripe fruit crops in a particular month as available food.

Classification of periods of food availability

In order to determine whether fruiting peaks were synchronized between habitats, we calculated the correlation between food availability in each habitat and the food availability in all other habitats during the same month. We conducted food availability analysis separately for any habitat whose phenological patterns were negatively correlated or weakly positively correlated ($r < 0.4$), with those of other habitats. We used objective, operational definitions based on the number of fruiting stems/ha and the diversity of fruiting species to identify each month as a mast, HFP, LFP, or crunch based on examination of the patterns of fruit availability over the 69 months sampled (see Marshall, 2004 for details).

Temporal availability of each food taxon

We utilized observations collected in the 126 phenology plots that were monitored for 69 months to assess the temporal availability of each food taxon over time. We developed decision rules that assigned each taxon to one of the four categories of food availability based on their phenology relative to our operationally defined seasons (e.g., masting species were defined as those for which >75% of all fruiting events were in mast months, LFP fruiters were those for which >50% of fruiting occurred in LFP; see Marshall, 2004 for details).

Preference ranking of each food taxon

For each habitat we compiled a list of all taxa fed upon by gibbons in each of the four food availability categories (i.e., mast, HFP, LFP, crunch) and arranged them in order of decreasing habitat-specific preference. Preference values were calculated by dividing the use of each item (number of independent feeding observations) by its availability (number of patch months/ha). We plotted the cumulative importance (% in the diet during that season) as each new food taxon was added to the list (in order of decreasing preference). For mast periods, all food taxa required to comprise 75% of the diet were categorized as "preferred"; the final 25% were characterized as non-preferred. During HFP the taxa required to explain 50% of the diet were categorized as preferred taxa and the taxa required to explain the remaining 50% of the diet were classified as non-preferred foods. For LFP and crunch periods, the top 25% of foods were classified as preferred and the remainder were characterized as non-preferred. These cut-off rules reflect the fact that during mast periods the majority of available foods are of high objective preference (high net rates of energy return) whereas during LFP most foods are of low preference. This methodology incorporated habitat- and season-specific information on preference and importance to assign each food taxon into one of two classes.

Statistical analyses

We used ranged major axis regression (RMA) for all analyses because our independent variable (food availability) was subject to measurement error (Legendre & Legendre, 1998). We performed all RMA analyses using the program Model II (Legendre, 2001) and conducted non-parametric Monte Carlo randomization tests of 10 000 iterations to test for significance of slope and correlation estimates (Sokal & Rohlf, 1981; Manly, 1997). All regression analyses were performed on appropriately transformed variables to reduce non-normality and heteroskedasticity.

Results

Habitat-specific gibbon density

Gibbon densities at CPRS varied by more than an order of magnitude, with point density estimates ranging from 0.44 individuals/km^2 in montane forest to 10.27 individuals/km^2 in lowland sandstone habitats (Table 12.1).

Table 12.1. *Gibbon population density in seven habitat types*

Habitat	# censuses	Total Distance (km)	# Gibbon obs.	Density (gibbons/km^2)	SE
Peat swamp	87	290.3	50	7.28	1.25
Freshwater swamp	87	140.4	27	5.90	1.34
Alluvial bench	108	148.9	24	7.10	1.70
Lowland sandstone	101	129.3	34	10.27	2.12
Lowland granite	61	176.6	37	6.23	1.33
Upland granite	142	229.7	30	4.17	1.17
Montane	166	258.6	4	0.44	0.25

Data include the total number of censuses, distances sampled, number of independent gibbon observations made on censuses, and the point estimate (gibbons/km^2) and standard error (SE) of gibbon density in each habitat.

Periods of food availability in each habitat

Examination of interhabitat correlations suggested that fruiting patterns in freshwater swamp, alluvial bench, lowland sandstone, lowland granite, and upland granite habitats were well synchronized ($r > 0.7$, $p < 0.0001$) and that for purposes of classification these habitats could be lumped as mast, HFP, LFP, or crunch months. Montane and peat swamp forests were less well synchronized with the overall phenological patterns of the forest ($r < 0.4$, $p > 0.001$), and were therefore analyzed separately. Figure 12.3 shows temporal variation of food availability in each habitat and the classification of each month.

Hypothesis tests

As a test of *H1* (gibbon density is limited by total food availability), we calculated two measures of the total food availability in each habitat – the total stems of food taxa/ha, and the total basal area (TBA) of food taxa/ha. As TBA/ha incorporated the size of trees, it had the benefit of weighing larger stems (and their larger associated fruit crops) more heavily. However, TBA calculations obscured the effects of lianas. Therefore, we also used the measure of stems/ha in order to incorporate the effects of lianas. There was no relationship between habitat-specific density and either food stems/ha ($r^2 = 0.06$, $p = 0.61$) or TBA of food trees ($r^2 = 0.18$, $p = 0.34$) in a habitat. Taken together, these tests showed that total food availability is uncorrelated with density, and that total abundance of food does not limit gibbon population density. In our test of *H2* (gibbon density is limited by average food availability), mean food availability/month in a habitat was similarly unrelated to density ($n = 7$ habitats, RMA $r^2 = 0.0001$, $p = 0.98$, Kendall's Tau 0.05, $p = 0.88$), indicating that the average food availability in a habitat does not serve as the key limitation of gibbon population density.

To test *H3* and *H4* (gibbon density is limited by preferred and fallback foods, respectively), we used the data from the phenology plots to calculate the mean food availability/month in each habitat during each of the four categories of food availability. The mean food availability, regardless of category, did not predict gibbon density ($n = 7$ habitats, $r^2 < 0.15$, $p > 0.30$ for all tests). We also analyzed the relationship between gibbon density and the stem density and TBA/ha of preferred and non-preferred foods for each of the four seasons. Of the eight classes of foods formed by these combinations, only the preferred HFP foods were significant predictors of gibbon density (TBA/ha: $r^2 = 0.71$, $p = 0.01$; stems/ha $r^2 = 0.72$, $p = 0.02$).

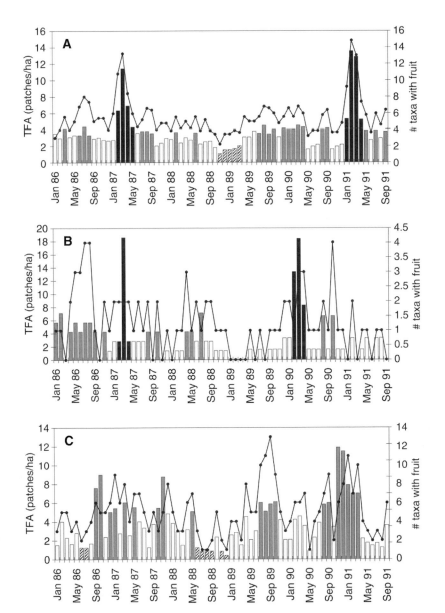

Figure 12.3. Phenological characterizations of gibbon food availability in lowland (A), montane (B), peat swamp (C) forest habitats. Line indicates the number of taxa with fruit in each month. Bars indicate mast (black), HFP (gray), LFP (white), and crunch (hatched).

Although the relationship between two measures of preferred HFP foods and gibbon density were significant, they were clearly strongly influenced by a single point, representing the montane forest. Using the logic that if gibbon densities are constrained by the availability of HFP preferred foods then the correlation between density and the availability of these food resources should remain if one habitat is removed, we removed the montane habitat and recalculated the correlations. Following the removal of the highly influential montane habitat datum from the analysis, both the strength and significance of both correlations dropped sharply (n = 6 habitats, $r^2 < 0.47$, p > 0.14 for both). This suggested that the availability of HFP preferred foods may not serve as the most important limiting factor on gibbon population density.

In order to conduct our final test of *H4*, we examined the relationship between feeding observations on various food taxa and overall food availability. Our analysis showed that the percentage of gibbon feeding observations on figs was significantly negatively correlated to overall food availability (n = 21 3-month periods, $r^2 = 0.36$, p = 0.003, Marshall, 2004). This indicated that figs served as fallback foods for gibbons at GPNP. This conclusion is supported by substantial evidence from a range of tropical sites that figs serve as important fallback foods for other vertebrate frugivores during periods of resource scarcity (e.g., Leighton & Leighton, 1983; Terborgh, 1986; Wrangham *et al.*, 1993; O'Brien *et al.*, 1998).

Gibbon densities were highly correlated with the density of figs (n = 7 habitats, $r^2 = 0.93$, p = 0.0005, Figure 12.4). In order to establish that this relationship was not driven by a single, overly influential outlier, we removed the montane habitat and reran this analysis. In contrast to the results for HFP preferred foods, figs remained a strong and significant predictor of gibbon density when the montane habitat was removed (n = 6 habitats, $r^2 = 0.90$, p = 0.004). These analyses suggest that fig density is the key constraint on gibbon density, and that the density of preferred HFP foods is merely correlated with fig density, with little independent explanatory power.

Discussion

In this chapter we tested hypotheses about the role of various types of food resources in limiting gibbon population density. We systematically classified food resources based on their phenology, preference, and importance for gibbons and examined how variation in different classes of food was related to differences in population density across seven different habitat types. We found that gibbon population density was highly correlated with the

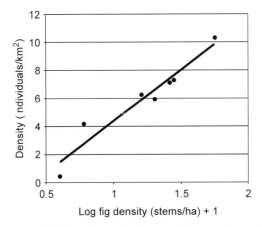

Figure 12.4. The relationship between gibbon density and the density of figs in seven habitats ($r^2 = 0.93$, p = 0.0005). Fig density was logged to reduce non-normality.

abundance of figs. This result suggests that gibbon populations are limited by the availability of their most important fallback food, figs. Thus, gibbons conform to the classic view that primate populations are primarily limited by food availability during periods of overall low resource availability.

Gibbons exhibit relatively risk-averse life history strategies (i.e., long interbirth intervals, high infant and juvenile survivorship), suggesting that there has been strong selection in this species on traits that promote survivorship. Figs may provide female gibbons with the energy required to maintain body condition during food-poor times when the intense obligate investment required to insure offspring survival is metabolically costly. In habitats with high densities of figs, gibbon females may be able to maintain better health and physical condition during LFP and crunches than they could in habitats with low fig densities. This would allow them to recover faster from costly periods of pregnancy and lactation and thus reproduce more frequently. Under this scenario, fig densities limit gibbon populations through their effects on female condition during periods of low food availability, which in turn affect birth rates.

The results provided in this chapter are based on small sample sizes (n = 7), and therefore preclude the use of multivariate models required to simultaneously test the effects of different types of food resources. In addition, these small sample sizes reduce the power of statistical tests to reject our null hypotheses. Several refinements are possible that would allow us to more fully address this question. First, subdividing the habitats into smaller sections and comparing gibbon density to the food availability within each smaller section would increase our sample size and might provide sufficient data to conduct multivariate tests. Second, long-term demographic monitoring of groups of both species across the range of habitats found at GPNP would uncover temporal variation in birth and death rates that would shed light on the factors constraining population density. Third, formal demographic modeling would provide an opportunity to examine the actions of various regulatory mechanisms and their theoretical implications in more depth. Finally, the predictive power of the models proposed here could be tested using data from other sites.

Broader applications and implications

Although there is broad general interest in identifying the factors that limit primate population density, attempts to do so have to date met with limited success. This is primarily due to the fact that the theoretical framework from which we examine these questions is underdeveloped. Few general hypotheses have been proposed about the role of various ecological variables as limiting factors on primate populations, and formal models that explicitly consider the effects of temporal and spatial variation in resource availability have yet to be developed for primates. In this chapter we suggest a conceptual approach that explicitly considers the importance of different classes of foods and stresses the importance of classifying food resources relative to overall resource availability. Although temporal variation in food availability is unusually pronounced at our site, and despite the fact that the importance of temporal fluctuations in resource ability is strongly related to the dietary adaptations of a species, this approach could be profitably applied at a wide range of sites. This methodology requires that one have long-term data on both primate population density and food availability. Fortunately, these data are collected regularly by primate fieldworkers, suggesting that data are available from a wide range of sites that could be used to test the general applicability of the hypotheses we propose here. Ideally, studies would examine the correlates of population density of the same species at several sites that are close in proximity but variable in quality (e.g., Chapman & Chapman, 1999).

Studies that advance our understanding of the ultimate constraints on primate populations have the potential for providing important information to conservation managers. For example, if the key food resources that regulate primate populations can be identified, then these tree and liana taxa can be spared during selective logging operations (Wasserman & Chapman, 2003). In addition, techniques that rapidly assess the quality of habitat for primates can be developed and used to either identify areas that deserve special conservation attention or to determine which of several areas are the most deserving of limited conservation funds. Finally, an understanding of the key factors constraining primate populations would suggest valuable ways to artificially manipulate carrying capacity in forests that have been degraded or are being specifically managed to maximize primate population size.

Acknowledgments

Permission to conduct research at Gunung Palung National Park was kindly granted by the Indonesian Institute of Science (LIPI), the Directorate General for Nature Conservation (PHKA), and the Gunung Palung National Park Office (UTNGP). AJM gratefully acknowledges the support of the J. William Fulbright Foundation, the Louis Leakey Foundation, a Frederick Sheldon Traveling Fellowship, and the Department of Anthropology at Harvard University for funding fieldwork, and the Cora Du Bois Charitable Trust, the Department of Anthropology at Harvard University, and an Elliot Dissertation Completion Fellowship for financial support while writing his dissertation. Special thanks to John Harting for assistance with GIS mapping, and to three reviewers for helpful comments on this manuscript. In addition, AJM thanks T. Blondal, R. McClellan, N. Paliama, J. Whittle, and C. Yeager for friendship and logistical support throughout the course of his fieldwork in Indonesia. We appreciate the assistance and support of the many students, researchers, and field assistants who worked at Cabang Panti Research Station over the past two decades, particularly M. Ali A. K., Busran A. D., Morni, Hanjoyo, and Edward Tang, without whose help this work would not have been possible. Many thanks to Tim Laman, who kindly provided the photograph for this chapter.

References:

Altmann, J., Hausfater, G., & Altmann, S. A. (1985). Demography of Amboseli baboons, 1963–1983. *American Journal of Primatology*, **8**, 113–25.
Appanah, S. (1981). Pollination in Malaysian primary forests. *Malaysian Forester*, **44**, 37–42.
Ashton, P. S., Givnish, T. J., & Appanah, S. (1988). Staggered flowering in the *Dipterocarpaceae*: new insights into floral induction and the evolution of mast fruiting in the aseasonal tropics. *The American Naturalist*, **132**, 44–66.
Buckland, S. T., Anderson, D. R., Burnham, K. P. *et al.* (2001). *Introduction to Distance Sampling: Estimating Abundance of Biological Populations*. Oxford: Oxford University Press.

Caldecott, J. O. (1980). Habitat quality and populations of two sympatric gibbons (*Hylobatidae*) on a mountain in Malaya. *Folia Primatologica*, **33**, 291–309.

Cannon, C. H. & Leighton, M. (2004). Tree species distributions across five habitats in a Bornean rain forest. *Journal of Vegetation Science*, **15**, 257–66.

Cant, J. G. H. (1980). What limits primates? *Primates*, **21**, 538–44.

Chapman, C. A. & Chapman, L. J. (1999). Implications of small scale variation in ecological conditions for the diet and density of red colobus monkeys. *Primates*, **40**, 215–31.

Chapman, C. A., Gautier-Hion, A., Oates, J. F., & Onderdonk, D. A. (1999). African primate communities: determinants of structure and threats to survival. In *Primate Communities*, ed. J. G. Fleagle, C. H. Janson, & K. E. Reed, pp. 1–37. Cambridge: Cambridge University Press.

Charnov, E. L. & Berrigan, D. (1993). Why do female primates have such long lifespans and so few babies? *Evolutionary Anthropology*, **1**, 191–4.

Conklin-Brittain, N. L., Wrangham, R. W., & Hunt, K. D. (1998). Dietary response of chimpanzees and cercopithecines to seasonal variation in fruit abundance. II. Macronutrients. *International Journal of Primatology*, **19**, 971–98.

Curran, L. M. & Leighton, M. (2000). Vertebrate responses to spatiotemporal variation in seed production of mast-fruiting *Dipterocarpaceae*. *Ecological Monographs*, **70**, 101–28.

Davies, A. G. (1994). Colobine populations. In *Colobine Monkeys: Their Ecology, Behaviour and Evolution*, ed. A. G. Davies & J. F. Oates, pp. 285–310. Cambridge: Cambridge University Press.

Dittus, W. P. J. (1979). The evolution of behaviors regulating density and age-specific sex ratios in a primate population. *Behaviour*, **69**, 265–302.

Foster, R. B. (1974). Seasonality of fruit production and seed fall in a tropical forest ecosystem in Panama. Unpublished Ph.D. thesis, Duke University. Cited in J. G. H. Cant (1980), What limits primates?, *Primates*, **21**, 538–44.

Ganzhorn, J. U. (1992). Leaf chemistry and biomass of folivorous primates in tropical forests: test of a hypothesis. *Oecologia*, **91**, 540–7.

Groves, C. P. (1984). A new look at the taxonomy and phylogeny of the gibbons. In *The Lesser Apes: Evolutionary and Behavioural Biology*, ed. H. Preuschoft, D. J. Chivers, W. Y. Brockelman, & N. Creel, pp. 542–61. Edinburgh: Edinburgh University Press.

(2001). *Primate Taxonomy*. Washington, DC: Smithsonian Institution Press.

Gupta, A. K. & Chivers, D. J. (1999). Biomass and use of resources in south and south-east Asian primate communities. In *Primate Communities*, ed. J. G. Fleagle, C. H. Janson, & K. E. Reed, pp. 38–54. Cambridge: Cambridge University Press.

Hausfater, G. (1975). Dominance and reproduction in baboons (*Papio cynocephalus*): a quantitative analysis. In *Contributions to Primatology, Volume 7*, ed. A. H. Schultz, D. Stark, & F. S. Szalay, pp. 1–150. Basel: Karger.

Janson, C. H. & Chapman, C. A. (1999). Resources and primate community structure. In *Primate Communities*, ed. J. G. Fleagle, C. H. Janson, & K. E. Reed, pp. 237–67. Cambridge: Cambridge University Press.

Janzen, D. H. (1974). Tropical blackwater rivers, animals, and mast fruiting by the *Dipterocarpaceae*. *Biotropica*, **6**, 69–103.

Knott, C. D. (1998). Changes in orangutan caloric intake, energy balance, and ketones in response to fluctuating fruit availability. *International Journal of Primatology*, **19**, 1061–79.

Laman, T. G. & Weiblen, G. D. (1998). Figs of Gunung Palung National Park (West Kalimantan, Indonesia). *Tropical Biodiversity*, **5**, 245–97.

Lee, P. C. & Hauser, M. D. (1998). Long-term consequences of changes in territory quality on feeding and reproductive strategies of vervet monkeys. *Journal of Animal Ecology*, **67**, 347–58.

Legendre, P. (2001). *Model II regression - User's Guide*. Montreal: Département de sciences biologiques, Université de Montréal.

Legendre, P. & Legendre, L. (1998). *Numerical Ecology*, 2nd edn. Amsterdam: Elsevier.

Leighton, D. R. (1987). Gibbons: territoriality and monogamy. In *Primate Societies*, ed. B. B. Smuts, D. L. Cheney, R. M. Seyfarth, R. W. Wrangham, & T. T. Struhsaker, pp. 135–45. Chicago, IL: University of Chicago Press.

Leighton, M. (1993). Modeling dietary selectivity by Bornean orangutans: evidence for integration of multiple criteria in fruit selection. *International Journal of Primatology*, **14**, 257–313.

Leighton, M. & Leighton, D. (1983). Vertebrate responses to fruiting seasonality within a Bornean rain forest. In *Tropical Rain Forests: Ecology and Management*, ed. S. L. Sutton, T. C. Whitmore, & A. C. Chadwick, pp. 181–96. Boston: Blackwell Scientific Publications.

Manly, B. F. J. (1997). *Randomization, Bootstrap and Monte Carlo Methods in Biology*, 2nd edn. London: Chapman and Hall.

Manly, B. F. J., McDonald, L. L., Thomas, D. L., McDonald, T. L., & Erikson, W. P. (2002). *Resource Selection by Animals: Statistical Design and Analysis for Field Studies*, 2nd edn. London: Kluwer Academic Publishers.

Marshall, A. J. (2004). *Population Ecology of Gibbons and Leaf Monkeys across a Gradient of Bornean Forest Types*. Unpublished Ph.D. thesis, Harvard University.

Marshall, J. T. & Sugardjito, J. (1986). Gibbon systematics. In *Comparative Primate Biology, Volume 1: Systematics, Evolution, and Anatomy*, ed. D. R. Swindler & J. Erwin, pp. 137–85. New York, NY: Alan R. Liss.

Mather, R. J. (1992). *A Field Study of Hybrid Gibbons in Central Kalimantan, Indonesia*. Unpublished Ph.D. thesis, University of Cambridge. Cited in D. J. Chivers (2001). The swinging singing apes: fighting for food and family in far-east forests, In *The Apes: Challenges for the 21st Century, Conference Proceedings of the Brookfield Zoo*, pp. 1–28, Chicago, IL: Chicago Zoological Society.

McConkey, K. R., Aldy, F., Ario, A., & Chivers, D. J. (2002). Selection of fruit by gibbons (*Hylobates muelleri x agilis*) in the rain forests of Central Borneo. *International Journal of Primatology*, **23**, 123–45.

McKey, D. B. (1978). Soils, vegetation, and seed-eating by black colobus monkeys. In *The Ecology of Arboreal Folivores*, ed. G. G. Montgomery, pp. 423–37. Washington, DC: Smithsonian Institution Press.

Medway, L. (1972). Phenology of a tropical rain forest in Malaya. *Biological Journal of the Linnean Society*, **4**, 117–46.

Milton, K. (1982). Dietary quality and demographic regulation in a howler monkey population. In *The Ecology of a Tropical Forest*, ed. E. G. Leigh, A. S. Rand, & D. M. Windsor, pp. 273–89. Washington, DC: Smithsonian Institution Press.

Mitani, J. C. (1990). Demography of agile gibbons (*Hylobates agilis*). *International Journal of Primatology*, **11**, 411–24.

Ng, F. S. P. (1978). *Tree Flora of Malaya*, Volume 3. London: Longman.

   (1989). *Tree Flora of Malaya*, Volume 4. London: Longman.

Oates, J. F. (1987). Food distribution and foraging behavior. In *Primate Societies*, ed. B. B. Smuts, D. L. Cheney, R. M. Seyfarth, R. W. Wrangham, & T. T. Struhsaker, pp. 197–209. Chicago, IL: University of Chicago Press.

Oates, J. F., Whitesides, G. H., Davies, A. G. *et al.* (1990). Determinants of variation in tropical forest primate biomass: new evidence from West Africa. *Ecology*, **71**, 328–43.

O'Brien, T. G., Kinnaird, M. F., Dierenfeld, E. S. *et al* (1998). What's so special about figs? *Nature*, **392**, 668.

Olupot, W., Chapman, C. A., Brown, C. H., & Waser, P. M. (1994). Mangabey (*Cercocebus albigena*) population density, group size, and ranging: a twenty-year comparison. *American Journal of Primatology*, **32**, 197–205.

Sokal, R. R. & Rohlf, F. J. (1981). *Biometry*. New York, NY: W H. Freeman.

Stephens, D. W. & Krebs, J. R. (1986). *Foraging Theory*. Princeton, NJ: Princeton University Press.

Struhsaker, T. T. (1973). A recensus of vervet monkeys in the Masai-Amboseli Game Reserve, Kenya. *Ecology*, **54**, 930–2.

   (1999). Primate communities in Africa: the consequences of long-term evolution or the artifact of recent hunting? In *Primate Communities*, ed. J. G. Fleagle, C. H. Janson, & K. E. Reed, pp. 289–94. Cambridge: Cambridge University Press.

Terborgh, J. (1986). Keystone plant resources in the tropical forest. In *Conservation Biology: The Science of Scarcity and Diversity*, ed. M. Soulé, pp. 330–44. Sunderland: Sinauer.

Terborgh, J. W. & van Schaik, C. P. (1987). Convergence vs. nonconvergence in primate communities. In *Organization of Communities, Past and Present*, ed. J. H. R. Gee & P. S. Giller, pp. 205–26. Oxford: Blackwell Scientific.

Thomas, L., Laake, J. L., Strindberg, S. *et al.* (2002). *Distance 4.0 Release 2*. Research Unit for Wildlife Population Assessment, University of St. Andrews.

van Schaik, C. P. (1986). Phenological changes in a Sumatran rain forest. *Journal of Tropical Ecology*, **2**, 327–47.

van Schaik, C. P. & Knott, C. D. (2001). Geographic variation in tool use on *Neesia* fruits in orangutans. *American Journal of Physical Anthropology*, **114**, 331–42.

van Schaik, C. P. & van Noordwijk, M. A. (1985). Interannual variability in fruit abundance and the reproductive seasonality in Sumatran long-tailed macaques (*Macaca fascicularis*). *Journal of Zoology, London (A)*, **206**, 533–49.

Wasserman, M. D. & Chapman, C. A. (2003). Determinants of colobine monkey abundance: the importance of food energy, protein, and fibre content. *Journal of Animal Ecology*, **72**, 650–9.

Webb, C. O. (1998). *Seedling Ecology and Tree Diversity in a Bornean Rain Forest*. Unpublished Ph.D. thesis, Dartmouth College.

Whitmore, T. C. (1972). *Tree Flora of Malaya, Volume 1*. London: Longman.
  (1973). *Tree Flora of Malaya, Volume 2*. London: Longman.
  (1984). *Tropical Rain Forests of the Far East*, 2nd edn. Oxford: Clarendon Press.
  (1990). *An Introduction to Tropical Rain Forests*. Oxford: Clarendon Press.
Wich, S. A. & van Schaik, C. P. (2000). The impact of El Niño on mast fruiting in
  Sumatra and elsewhere in Malesia. *Journal of Tropical Ecology*, **16**, 563–77.
Wiens, J. A. (1977). On competition and variable environments. *American Scientist*,
  65, 590–7.
Wrangham, R. W., Conklin-Brittain, N. L., Etot, G. *et al.* (1993). The value of figs to
  chimpanzees. *International Journal of Primatology*, **14**, 243–56.

13 Influence of fruit availability on Sumatran orangutan sociality and reproduction

SERGE A. WICH, MARTINE L. GEURTS, TATANG MITRA SETIA, AND SRI SUCI UTAMI-ATMOKO

Introduction

Many primates face fluctuations in inter- and intra-annual food availability and might respond to these shifts by changing their diet (Stanford & Nkurunungi, 2003), ranging pattern (Olupot *et al.*, 1997; Buij *et al.*, 2002), party size (Chapman *et al.*, 1995; Mitani *et al.*, 2002), copulation rates (Knott, 1998a) and timing of conceptions (van Schaik & van Noordwijk, 1985;

Feeding Ecology in Apes and Other Primates. Ecological, Physical and Behavioral Aspects, ed. G. Hohmann, M. M. Robbins, and C. Boesch. Published by Cambridge University Press. © Cambridge University Press 2006.

Koenig *et al.*, 1997; Di Bitetti & Janson, 2000), and by increasing search efforts (Chapman, 1988), and minimizing energy expenditure (Doran, 1997). The purpose of this chapter is to examine whether fluctuations in fruit availability influence orangutan party size, copulation rates, and the timing of conceptions in a long-term study of Sumatran orangutans (*Pongo abelii*).

Studies of fruit availability and orangutan sociality show mixed results. Some studies found an increase in sociality in large fruit trees (Utami *et al.*, 1997) or during periods of high fruit availability (Sugardjito *et al.*, 1987; Knott, 1998a), whereas others did not (van Schaik, 1999; Utami, 2000). Although real differences in sociality between orangutan populations may exist (van Schaik, 1999), interpreting these results remains difficult because different measures of sociality have been applied. For example, some studies used sociality measures composed of lumped data for all age-sex classes (Sugardjito *et al.*, 1987); others examined sociality for age-sex classes separately (van Schaik, 1999; Wich *et al.*, 1999; Utami, 2000). Since factors driving sociality, such as food availability or reproductive partners (Anderson *et al.*, 2002; Mitani *et al.*, 2002), might not be of similar importance for each age-sex class, lumping age-sex classes may confound real patterns or create spurious ones. Although in theory fruit availability measures could have influenced the results of the different studies, the various different fruit availability measures used in Ketambe (e.g., all fruit species or only orangutan fruit species) correlate strongly and therefore were not expected to have resulted in the differences between the Ketambe studies (Wich, unpublished data).

An important related question is what orangutans actually gain from being social. Based on a lack of influence of fruit availability on female orangutan sociality (at Suaq Balimbing, Sumatra), van Schaik (1999) suggests that the benefits that female orangutans gain from being social are not always available. Therefore, van Schaik argues that these benefits cannot be ecological, such as a reduction in predation risk. He argues that in situations where there are consistent benefits of being social, such as a reduction in predation risk, food availability should positively influence sociality (van Schaik, 1999). In situations where benefits from being social only arise in certain contexts, such as social learning during specific feeding events, e.g., using tools to extract honey from treeholes, a relationship between food availability and sociality is not expected (van Schaik, 1999). Thus, other social benefits such as social learning opportunities might currently be factors that influence orangutan sociality even if they evolved as a by-product of sociality itself. Since van Schaik's (1999) analyses concerned one orangutan population, the next step is to conduct similar analyses for another orangutan population. Therefore, the first aim here is to examine the relationship between fruit availability and sociality for another Sumatran orangutan population (Ketambe), using the same definition of sociality employed by van Schaik (1999).

In addition to the possible effects of fruit availability on orangutan sociality, fruit availability might also influence orangutan reproduction (Knott, 1998a, 1998b, 1999, 2001). The hypothesis that fruit availability influences conception stems from the fact that female energy status strongly correlates with ovarian function (humans: Ellison, 1990; Ellison *et al.*, 1993). Thus, to reach a regular ovulatory cycling, females need to reach a certain physical condition threshold (Bercovitch, 1987; Koenig *et al.*, 1997), and that depends on the availability of high-quality food (Knott, 1998b, 2001; Takahashi, 2002). In support of this hypothesis, hormonal analyses indicate that only in periods of high food availability do normal ovulatory cycles occur in Hanuman langurs (Koenig *et al.*, 1997; Ziegler *et al.*, 2000) and orangutans (Knott, 2001).

Knott shows that in Borneo, orangutan (*Pongo pygmaeus wurmbii*) copulations and conceptions occur primarily during periods of high fruit availability (Knott, 1998a). She argues that this is a result of the fact that during periods of extremely low fruit availability orangutans suffer a negative energy balance, as indicated by the presence of ketones in urine (Knott, 1998b), and that as a consequence, females have reduced levels of estrogens, making conception difficult (Knott, 1999, 2001).

It is unknown, however, how general this pattern is and especially whether there are differences between the islands of Sumatra and Borneo. Since it has been suggested that Sumatran forests, due to their younger and more volcanic soils, are more productive than Bornean forests (MacKinnon *et al.*, 1996), differences in behavior between orangutans on the two islands would be expected. Unfortunately, there are no comparisons to support this notion. However, there are some interesting differences between Sumatran and Bornean orangutans that do support the suggestion that Sumatran forests are more productive than Bornean forests (van Schaik, 1999; Delgado & van Schaik, 2000). First, in similar habitats, orangutan density is higher on Sumatra than on Borneo (Delgado & van Schaik, 2000). Second, adult female mean party sizes are larger on Sumatra than Borneo (van Schaik, 1999, table 1). Third, the percentage of fruit in the diet seems slightly higher on Sumatra than Borneo (Table 13.1). Fourth, the percentage of cambium in the diet is lower on Sumatra than Borneo (Table 13.1).

So the second aim of this chapter, then, considers whether fruit availability influences copulations and/or conception rates for the Ketambe orangutans. In orangutans, two types of copulations occur: cooperative and forced (Rijksen, 1978; Fox, 2002). During a cooperative copulation, the female is cooperative, whereas during a forced mating the female resists (for details,

Table 13.1. *Sumatran and Bornean orangutans compared*

Site	Mean adult female party size	% Fruit in diet	% Cambium in diet
Mentoko[a]	1.2	53.8	14.2
Tanjung Putting[b]	1.2	60.9	11.4
Ulu Segama[c]	–	62.0	10.5
Gunung Palung[d]	1.0	66.8	9.3
Ketambe[e]	1.5	67.5	2.6
Suaq[f]	1.9	68.0	1.0

Notes:
[a, b, c]Adult female mean party sizes from van Schaik (1999), except for Ketambe (this study). Average percentages of fruit in diet: Table 3.3 in Rodman (1988).
[d]estimated from Knott (1998).
[e]Wich unpubl. data.
[f]estimated from Fox *et al.* in press.
Sites in italics are on Sumatra; the other sites are on Borneo. All percentages are based on total minutes feeding, except those from Ulu Segama, which are percentages of feeding observations from fig 19 in MacKinnon (1974).

see Fox, 2002). If, as found on Borneo, the probability of conception is higher during periods of high fruit availability, then the expectation is that cooperative copulations will increase during these periods.

Methods

Study area and population

The Ketambe study area (3°41′N, 97°39′E) consists mainly of primary rain-forest (Rijksen, 1978; van Schaik & Mirmanto, 1985) and is located in the Gunung Leuser National Park, Leuser Ecosystem, Sumatra, Indonesia. The forest in the study area ranges from alluvial forest at approximately 350 m a.s.l. to highland forest about 1000 m a.s.l. The orangutan population has been studied more or less continuously since 1971 and is the longest studied population of wild Sumatran orangutans. For the analyses in this chapter we use data from 11 adult females, seven flanged males (adult males with flanges and other secondary sexual characteristics such as long calls), and nine unflanged males (adult males without flanges and lacking other secondary sexual characteristics such as long calls). From 1984–2001 data were collected under the supervision of several researchers (S. Djodjosudarmo, T. Mitra Setia, S. S. Utami, A. H. Lubis, and S. A. Wich). Methods used by these researchers were similar and interobserver reliability was checked between most researchers and showed good correspondence for the variables used here. A total of 18 472 hours of focal data were used for the analyses presented here. Orangutans were generally followed from dawn until dusk (i.e., from their morning to evening nest).

Sociality and reproductive behavior

Associations were defined the same as in van Schaik's (1999) study. An individual was classified as being in association with another independent (this excluded dependent offspring, i.e., offspring always in association with the mother) individual when the distance separating them was less than 50 m. This distance seems to be the distance within which orangutans react to each other (van Schaik, pers. comm., 1999). For each individual that was followed for a minimum of 3 hours per day a determination was made of how many minutes it spent in association with another individual (van Schaik, 1999). The association minutes per day were added and divided by the total number of focal minutes on that day. This yields the mean number of individuals associated with the focal and is expressed as the mean party size per day:

$$PS_i = 1 + (\Sigma t_{ij}/T_i)$$

where T_i is the total number of focal minutes of the focal that day, and Σt_{ij} is the total time the focal spent in association with every other independent individual j, for all j's. The percentage of time spent in association per day was calculated separately for the following age-sex classes: reproductive females, nonreproductive females, flanged males, and unflanged males. The term "age-sex classes" was used in this study because unflanged adult males are normally younger than flanged adult males (Wich *et al.*, 2004a). Adolescent individuals were included as a separate class of dependent individuals with whom the other age-sex classes could associate. Unfortunately, there were not sufficient data from adolescents to include them in the analyses. Reproductive females were defined as those females that were sexually active (i.e., copulating). Non-reproductive females were females that were not sexually active. Sexually active females could be both nulliparous and parous. Nulliparous females were considered sexually active from the first copulation until conception. Parous females were considered sexually active from the resumption of sexual activity until conception. Because of the length of the study, eight of the 11 females occurred in both classes and two of the unflanged males occurred in the flanged class. The only analyses where the same individuals were used simultaneously in two classes were those of mean party size and day journey length (DJL) differences between the age-sex classes. As a result, the data points in these tests cannot be considered independent. However, comparing mean party size and DJL between the different age-classes for the same individuals yielded similar results (Wich, unpublished data) and for the sake of conciseness, overall results are presented here.

We also recorded all copulations. For the analyses, we only included copulations when females were focal individuals. We analyzed forced and cooperative copulations separately since forced copulations depend on male initiative only and cooperative copulations depend on initiative from both male and female.

For behavioral measures such as mean party size, forced, and cooperative copulations, we calculated monthly values for each individual. Monthly means were calculated by taking the mean of the monthly individual means. These means were used in the statistical analyses.

DJL was a proxy for energetic costs and was measured by making a digital photo of the follow map after which the orangutan travel route was carefully tracked on a computer monitor using a mouse. With a specially developed application (by M. Terlou, Utrecht University) in the KS400 image-processing package (Zeiss vision), the DJL was automatically calculated. SAW measured all DJLs and subsequent measures of the same maps showed that measurement error was less than 2%.

Fruit availability

Phenological records for Ketambe are a mix of fruit-trail data and tree plot-based data (for details see van Schaik, 1986; Wich & van Schaik, 2000). For the fruit-trail data, it was not possible to separate orangutan and non-orangutan fruits. Therefore, overall fruit availability data for fruit-trails and plots were used for the analyses here. Because measures of orangutan food availability correlate very strongly with overall food availability (Wich, unpublished data), using overall fruit availability did not influence the results presented here. Between March 1984 and January 1988, monthly counts of ripe fruits on the ground were made on a 4.6-km trail. Whenever ripe fruits were found, the patch from which the fruits came was recorded and counted as one ripe fruit source. From these data the monthly number of ripe fruit sources/km trail were calculated and used in the analyses. Ripe fruit trail data showed a very good correspondence (Spearman $r = 0.7$, $p = 0.02$, $n = 17$) with data based on monitoring trees in plots. Since September 1986 about 420 trees (including fig trees) above a diameter at breast height (dbh) of 39 cm in 17 vegetation plots have been monitored on a monthly basis for the presence/absence of ripe fruits. Details of species composition of these vegetation plots have been published elsewhere by Palombit (1992). Because different researchers used slightly different methods, we combined the various data sets by standardizing within each data set (March 1984–January 1988 (used until August 1986), September 1986–August 1988, September 1988–March 1993, April 1993–May 2001). We standardized the data by computing the z scores for each data point within a data set. We defined periods of high fruit availability as z-score >1, and low fruit availability as z-score <-1. A year in which there were one or more months with a z-score >1.96 was considered a mast-fruiting year. Mast fruiting is a South-East Asian phenomenon characterized by community-wide fluctuations in fruit abundance in which long periods of low production are punctuated by occasional bursts of fruiting activity (Medway, 1972; van Schaik, 1986; Ashton *et al.*, 1988; Wich & van Schaik, 2000).

Reproduction

Determinations were made for each conception event as to whether it occurred in a period of high or low fruit availability or in a year during which mast-fruiting occurred (see above). Since the duration of gestation is approximately 8 months (Markham, 1990) conception was determined by counting back 8 months from the date of birth.

Results

Sociality and fruit availability

The mean orangutan party size in Ketambe was 1.52 (sd $= 0.5$, n $= 240$). The mean party sizes for the different age-sex classes are presented in Figure 13.1. Mean party sizes differed significantly between the age-sex classes (ANOVA $F_{3, 236} = 4.9$, p $= 0.002$, n $= 240$). Unflanged males had significantly larger mean party sizes than other age-sex classes (Least Significant Difference post-hoc tests p < 0.05 between the unflanged males and every other age-sex class after correcting for the number of tests following a Bonferroni method; Hochberg, 1988).

No significant correlation between monthly mean party size and fruit availability was found for any of the age-sex classes (non-reproductive females: r $= 0.12$, NS, n $= 61$; reproductive females: r $= -0.03$, NS, n $= 63$; flanged males: r $= -0.06$, NS, n $= 73$; unflanged males: r $= -0.05$, NS,

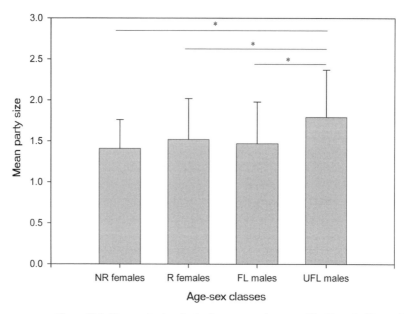

Figure 13.1. Mean party sizes for the four age-sex classes used in this study. The vertical bars indicate standard deviations and the horizontal bars indicate which age-sex classes differ significantly at the $p < 0.05$ level (see text). NR = nonreproductive females, R = reproductive females, FL = flanged males, and UFL = unflanged males.

$n = 43$). Since party size for flanged and unflanged males might be influenced by the number of reproductive females, partial correlations were conducted while controlling for the number of reproductive females. Again, no significant correlations were found between fruit availability and party size for males (flanged males: $r = -0.44$, NS, $n = 70$; unflanged males: $r = -0.03$, NS, $n = 40$).

The mean party sizes of flanged and unflanged males were not significantly related to the number of reproductive females in the population ($r = 0.15$, NS, $n = 73$; $r = 0.12$, NS, $n = 43$). If the same correlations were corrected for fruit availability by using residuals, the results remained similar (flanged males: $r = 0.14$, NS, $n = 70$; unflanged males: $r = 0.11$, NS, $n = 40$).

Because there was no influence of fruit availability on sociality, there might be costs of associating (van Schaik, 1999). A possible reflection of these costs would be an increase in DJL. Since both mean party size and DJL could be influenced by fruit availability, partial correlations were conducted, using residuals to correct for fruit availability (Figure 13.2). For nonreproductive and reproductive females there was no such effect. Flanged males, however, showed a significant increase in DJL, whereas unflanged males

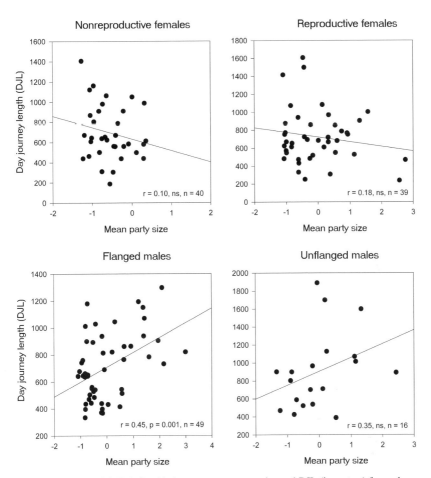

Figure 13.2. Relationship between mean party size and DJL (in meters) for each age-sex class. Mean party size was corrected for fruit availability and residuals are presented here.

did not. For flanged males this increase in DJL might be caused by being in consort with females. Re-analyzing the data while excluding consortships (associations between males and females that last for more than one day) indeed fails to show a significant increase of flanged male DJL with an increase in party size (Figure 13.3). DJLs differed between the age-sex classes (ANOVA $F_{3, \, 236} = 4.4$, p < 0.001, n = 240). Unflanged males had significantly larger DJLs (865.0 m, sd = 361.5) than any other age-sex class (nonreproductive females: 675.4 m, sd = 282.1; reproductive females: 722.1 m, sd = 293.3; flanged males 589.6 m, sd = 206.6, Least Significant

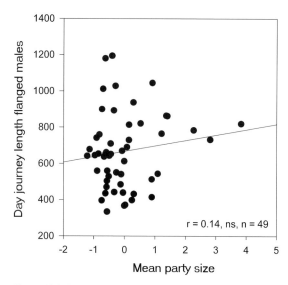

Figure 13.3. Relationship between mean party size and DJL for flanged males, excluding consortships. Mean party size was corrected for fruit availability and residuals are presented here.

Difference post-hoc tests p < 0.05 between the unflanged males and every other age-sex class after correcting for the number of tests following a Bonferroni method; Hochberg, 1988).

Reproduction and fruit availability

There were no significant correlations between the mean number of cooperative or forced copulations/hour and monthly fruit availability (Figure 13.4). Combining forced and cooperative copulations yielded a similar lack of correlation with fruit availability ($r = 0.04$, NS, n = 66). Inspecting the figure for cooperative copulations gives the impression that there are more co-operative copulations during periods when fruit availability is > 0 (i.e., higher than the mean) than during periods when fruit availability is < 0 (i.e., smaller than the mean). However, there is no significant difference between these two periods for the number of cooperative copulations (Mann–Whitney U test: Mean Rank$_{fruit < 0}$ = 31.0, Mean Rank$_{fruit > 0}$ = 36.0, U = 461.5, NS, n = 66)

Conceptions did not occur more often during months with high fruit availability (Binomial test: test proportion = 0.17 (43/249 months, i.e., there

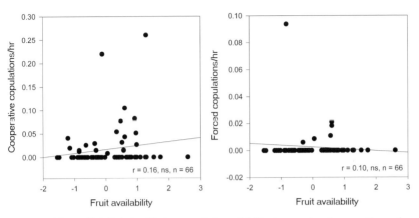

Figure 13.4. Relationship between fruit availability (z-scores) and cooperative and forced copulations.

were 43 months with $z > 1$ in a total of 249 observation months, Figure 13.5), observed proportion = 0.22 (4/18 conceptions), $p = 0.41$). Conceptions were also not more likely to occur during mast-fruiting years (Binomial test: test proportion: 0.23 (7/31), observed proportion: 0.33 (6/18), $p = 0.56$). Conceptions also did not occur more often during months in which fruit availability was >0 (Binomial test: test proportion = 0.43 (108/249), observed proportion = 0.5 (9/18), $p = 0.44$). In addition, conceptions did not occur less during periods of low fruit availability (Binomial test: test proportion: 0.15 (38/249), observed proportion: 0.17 (3/18), $p = 0.56$). There was also no correlation between the number of conceptions per month and the average fruit availability score for that month ($r = 0.37$, NS, $n = 12$).

Discussion

Fruit availability and sociality

The results suggest that the average mean party size at Ketambe (1.5) is lower then the previous estimates of 2.0 (Sugardjito *et al.*, 1987). This is probably due to the different measures of sociality used (see below). The results indicate that unflanged males occur in slightly larger parties than flanged males and females. Even though the differences are small, they indicate that unflanged males are mostly with another individual whereas flanged males and females are not.

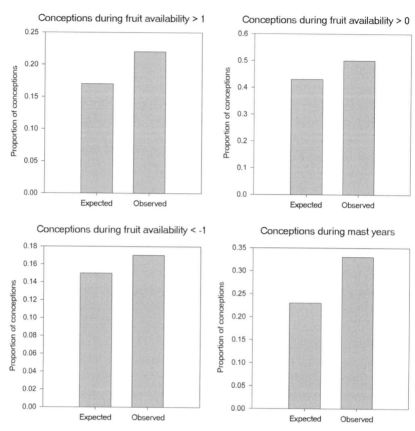

Figure 13.5. Relationship between monthly ripe fruit availability (z-scores), mast-fruiting years, and conceptions.

The analyses of the effect of fruit availability on mean party size were conducted separately on relevant age-sex classes because it has been suggested that the factors driving females and males to associate with other individuals might be different (i.e., fruit for females, fruit and receptive females for males). The results indicate that for none of the age-sex classes was there an increase in mean party size with increasing fruit availability, or alternatively, a reduction in party size during times with decreasing fruit availability. These results are similar to an earlier study at Ketambe that used time spent with neighbors as a measure of sociality for flanged and unflanged males (Utami, 2000). Utami's study also indicated that fruit availability is not correlated to the time flanged and unflanged males spent with neighbors. Similarly, female orangutans at Suaq Balimbing (Sumatra) showed no increase in mean party size with an increase in fruit availability (van Schaik, 1999).

The results presented here from the same orangutan population, however, differ from the results of Sugardjito *et al.* (1987). That study indicated that an increase in fruit availability correlated positively with several measures of orangutan sociality. However, the sociality measures in Sugardjito *et al.* (1987) are different from the mean party size used in this study. For example, as a measure of sociality they calculated the percentage of days an individual was in association regardless of the duration of the grouping event. Thus, in their analyses an ephemeral grouping event of two individuals was weighted similarly as a full day association between two individuals. In addition, they calculated mean group (or travel band) size, but again these calculations were not based on actual duration but on the sizes of the different groups during a day. Thus, the most important difference between the various measures they used and our mean party size measure seems to be in duration. Therefore, their study found a higher mean party size than this study. In addition, the Sugardjito *et al.* (1987) study used a much smaller sample size in both number of individuals (six) and the number of follow days. Finally, their study did not conduct separate analyses on the different age-sex classes.

The lack of a positive relationship between mean party size and fruit availability seems to indicate that the benefits of grouping for orangutans at Ketambe are not always available since in that case a positive relationship is expected (van Schaik, 1999). Thus, one of the most likely general benefits of associating for primates, a reduction in predation risk, is therefore probably not very important in Sumatran orangutans. This is probably because they are almost exclusively arboreal and their main predator, the tiger (*Panthera tigris*), is terrestrial (Sugardjito *et al.*, 1987; van Schaik, 1999). Nevertheless, the clouded leopard (*Neofelis nebula*) has been reported to prey upon rehabilitant infants and might be a threat for wild infants (Rijksen, 1978). However, infants in the wild are always near their mothers and might therefore have sufficient protection against this predator. In any case, mortality is very low in orangutans and therefore predation cannot be high (Wich *et al.*, 2004a).

Hence, it is likely that Ketambe orangutans, as in Suaq (van Schaik, 1999), associate for rare benefits or for no benefit at all (i.e., passively). An obvious rare associating benefit is mating. Another important rare associating benefit might be the transmission of social and foraging skills, suggested to be important in orangutans (van Schaik & Knott, 2001; van Schaik *et al.*, 2003) and other primates (see recent reviews in Fragaszy & Perry, 2003; Reader & Laland, 2003). However, if these social and foraging skills are only rare benefits, a general analysis such as the one conducted in this study might swamp their effects. A detailed investigation of costs of associating in orangutans in areas where, for example, there is often tool-use in the context of feeding, might shed light on this issue. It is important to note that even though social benefits might drive certain associations at this stage in orangutan evolution, social benefits themselves might have only evolved when in the past sociality was driven by other factors such as predation or food availability.

For orangutans and other fission-fusion species, it has been suggested that receptive females attract males (Rodman & Mitani, 1987; van Schaik & van Hooff, 1996; Mitani *et al.*, 2002). Support for this hypothesis, however, has been mixed. Te Boekhorst *et al.* (1990) suggest that the number of nonresident males that temporarily visit an area is more correlated to food availability than the number of reproductive females. Utami (2000) shows that in some years, but not others, the number of reproductive females correlates positively with the number of flanged males in Ketambe. Utami (2000) found no such correlation for unflanged males. If males are indeed attracted by receptive females, it might be expected that mean party size for males will increase with an increase in the number of receptive females. Our results indicate that for both flanged and unflanged males the number of reproductive females does not correlate to mean party size. Several factors could explain the lack of such correlation.

First, males might simply be unable to determine when females are reproductive and associate with females as often as possible regardless of their reproductive state. There are no visual cues (at least not to human observers) to determine when orangutan females are reproductive, let alone when they are ovulating. Nevertheless, males might have other means such as olfactory cues to determine when ovulation occurs; they might also use offspring age as a cue to female receptivity. It appears, at least, that both flanged and unflanged males spent more time associating with reproductive females than with any other age-sex class (Table 13.2) and that therefore they perceive some cue of reproductive state. Another interesting aspect of Table 13.2 is the high association of nonreproductive females with adolescents. After a female reproduces, the previous offspring often follows the mother around, which explains the high association rate.

Second, males might know exactly when to associate with females, but this effect is swamped by the generality of the analyses conducted here because of the small number of days per month during which females are actually receptive. Contrary to this argument is the finding that flanged males do sometimes form lengthy consortships with females. However, these consortships might be mainly between the dominant male and reproductive

Table 13.2. *Percentage of time spent in association with different age-sex classes*

	Reproductive females	Nonreproductive females	Flanged males	Unflanged males	Adolescents
Reproductive females	14.2	15.2	30.5	34.5	5.6
Nonreproductive females	15.9	21.6	12.9	17.4	32.2
Flanged males	47.0	32.2	3.9	11.8	5.1
Unflanged males	38.8	25.8	10.6	16.6	8.2

The percentage in association was only calculated for focal minutes during which associations occurred.

females and as a result have little influence on the overall analyses presented here.

Third, although flanged and unflanged males were analyzed separately we did not analyze the data separately for each male. If, for instance, certain males associate with reproductive females at the peak of their receptivity but not in other periods, and other males show the opposite pattern, this would cancel out the two relationships in the overall analyses we conducted here. Fine-grained analyses are needed to resolve this and will be conducted in the future.

Van Schaik (1999) suggested that the increase in DJL, which he found for adult males in response to a larger mean party size, was an effect of associating with females. Our analyses support this suggestion for flanged males, but not for unflanged males. This difference might be explained by the different mating strategies that flanged and unflanged males use. Unflanged males often force females to copulate whereas flanged males usually have cooperative copulations with females (Rijksen, 1978; Utami, 2000; Fox, 2002). In addition, flanged males can attract females by means of their long call whereas unflanged males do not give long calls and hence cannot do this (van Schaik & van Hooff, 1996). It could therefore be that unflanged males have to actively search for females, supported by the fact that DJLs differ between the age-sex classes. If unflanged males indeed always actively search for females and have a long DJL as a result, this might explain the lack of relationship between mean party size and DJL for unflanged males. Flanged males on Sumatra often engage in consortships with females whereas unflanged males are either not part of such relationships or much less often than flanged males (Schürmann & van Hooff, 1986; Utami, 2000; Fox, 2002). Because reproductive females have longer DJLs than flanged males (Least Significant Difference post-hoc tests $p < 0.05$), flanged males must increase their DJL in order to stay with a reproductive female. Support for this explanation comes from the fact that excluding consortships in the analyses resulted in the disappearance of the significant correlation between DJL and mean party size for flanged males.

An important question is why DJL does not increase with increasing mean party size. The first explanation is that costs are not reflected in an increase in DJL but rather in a decrease in food intake or an increase in travel time (Galdikas, 1988; Mitani, 1989). Travel time might not always be correlated to DJL, due to differences in travel speed, and therefore might show different results. The second explanation could be the fact that there is a high density of large strangling fig trees in Ketambe compared to other areas (Rijksen, 1978; Wich *et al.*, 2004b) and these figs are less seasonal in their fruiting patterns than other fruiting trees in Ketambe (van Schaik, 1986). Large numbers of orangutans commonly aggregate in these large fig trees (Utami *et al.*, 1997). Because of the huge number of fruits they contain, a large number of orangutans can feed in one tree for up to 20 days (Utami *et al.*, 1997; Wich, unpublished data). Because these trees are in fruit pretty much year round, orangutans visit them with an almost equal frequency through the year (Sugardjito *et al.*, 1987), and thus they can maintain high mean party sizes throughout the year without their time budget and/or diet bearing much, if any, costs. In support of this, preliminary analyses from Ketambe indicate that orangutans of all four age-sex classes maintain a very high monthly percentage of fruit in their diet throughout the year and never feed much on bark (see below for difference with Borneo). This might explain why no ketones (a sign of burning body fats) have been found in the urine of orangutans at Ketambe during periods of otherwise low, non-fig fruit availability (Wich *et al.*, in press), whereas they have been found in orangutans at Gunung Palung (GP) (Knott, 1998a).

Thus, it seems that orangutans in Ketambe are able to maintain relatively high mean party size without bearing much cost. As a result, orangutans in Ketambe can reap certain social benefits of grouping while avoiding the costs. For males such benefits might be monitoring of reproductive status of females and copulations outside of consortships. For females and offspring these benefits might be developing and maintaining knowledge of social relationships. Females might also join associations to provide opportunities for their offspring to interact with other orangutans. These benefits can probably be obtained whenever associations with the proper age-sex class occur and therefore also in and around large fig trees. It is therefore likely that the exceptionally high density of large strangling figs explains the lack of relationships between mean party size and DJL.

Fruit availability and reproduction

Knott (1999, 2001) has suggested that variation in fruit availability, such as mast fruiting, has influenced orangutan reproduction. For large frugivores like orangutans, survival may depend on how they cope with the lean periods on both the inter- and intra-annual level. Knott (1998a) has shown that the orangutans at Gunung Palung (GP), Borneo, suffer negative energy balances during periods of prolonged low fruit availability after mast fruiting. She suggests that this negatively influences reproduction due to lower estrogen levels during these periods, resulting in fewer copulations and conceptions (Knott, 1999). It is even conceivable that during prolonged periods of low fruit availability some individuals actually starve or at least have to compromise their immune systems, which then results in a higher susceptibility to diseases. Our results, however, do not show a similar pattern for the orangutans at Ketambe. We found no increase in conceptions during mast years or periods of high fruit availability. In addition, fruit availability did not correlate to the number of copulations.

The question is, how, if at all, fruit availability can produce an island difference in orangutan reproduction patterns. Because the frequency of mast-fruiting does not seem to differ between Sumatra and Borneo (Wich & van Schaik, 2000) there can be three nonmutually exclusive explanations for the lack of influence of fruit availability on reproduction in Sumatra. First, during periods of low fruit availability the fruit availability in Sumatra might simply be higher than on Borneo. Since orangutan fruit trees in both Ketambe and GP are monitored monthly to assess the presence/absence of fruits, a direct comparison is possible. The average monthly percentage of orangutan food trees fruiting in GP is 6.1 (sd = 2.8, range = 2.5–12.5) (from fig. 13.1 in Knott, 1998a), whereas in Ketambe this is 9.6 (sd = 9.6, range = 6.3–14.3, Wich, unpublished data for 1997–2001 period). This difference is significant (Wilcoxon matched pairs-test: $T^+ = 9$, $p = 0.03$) and indicates that on average there is more fruit available to orangutans in Ketambe than GP. This comparison also shows that the number of orangutan food trees fruiting per month can reach a minimum of 2.5% in GP, whereas it is never lower than 6.3% in Ketambe, which is higher than the mean fruit availability in GP.

Second, the nature of the fallback resources might differ between Sumatra and Borneo. During periods of reduced fruit availability Bornean orangutans seem to feed to a large extent on cambium and leaves (Knott, 1998a, 2005), whereas in Ketambe orangutans feed largely on figs (Rijksen, 1978; Sugardjito *et al.*, 1987). As a result, the monthly percentage of fruit in the diet of Ketambe orangutans is never lower than 50%, whereas in GP it might be as low as 20% (Knott, 1998b). Cambium can constitute 37% of the diet in GP (Knott, 1998b); it never constitutes more than 5.3% in Ketambe (Wich *et al.*, in press). At Suaq Balimbing (Sumatra), vegetable matter is more commonly eaten during months of low fruit availability, but there is always a lot of fruit and many insects in the diet, suggesting high diet quality (Fox *et al.*, 2004). Since energy intake from fruits is much higher than from cambium (Knott, 1998a) it is possible that the orangutans in Ketambe do not suffer a prolonged negative energy balance due to the fact that they maintain a high percentage of fruits (figs) in their diet during periods of low fruit availability (Wich *et al.*, in press). A positive side effect of feeding on figs might be reduced travel (i.e., energy) costs since orangutans can spend days around the same fig trees (Rijksen, 1978; Utami *et al.*, 1997).

Third, soils on Sumatra are considered to contain more nutrients than those on Borneo due to their more recent and more pronounced volcanic origins (MacKinnon *et al.*, 1996). Therefore, it could be that fruits in Sumatra are richer in nutrients. A study to examine this is currently under way (Knott & Wich). The first two explanations are probably also related to soil fertility.

An important note remains to be made about the above analyses. They rest on the assumption that fruit availability accurately mirrors what is available for orangutans. This might not be realistic since the concept of fruit availability does not take into account the nutrient content of fruits (Knott, 2005).

The next step in determining the effects of food availability on diet, ranging, reproduction, sociality, and time budgets within and between populations rests on a common food availability currency as suggested by Knott (2005). A monthly index (or indices) of the nutrients available in fruits and other items is an example of such a currency, and with which we could make better comparisons between orangutan populations and ascertain how food availability determines sociality and reproduction within and between sites.

Acknowledgments

We gratefully acknowledge the cooperation and support of the Indonesian Institute of Science (LIPI), the Indonesian Nature Conservation Service (PHKA), Universitas National (UNAS), and the Leuser Development

Programme (LDP). The Netherlands Foundation for the Advancement of Tropical Research (WOTRO) and the Netherlands Organisation for Scientific Research (NWO) provided long-term funding. The L. S. B. Leakey Foundation, the Dobberke Foundation, and Lucie Burgers Foundation for Comparative Behaviour Research generously provided funding to SAW to compile the Ketambe long-term data. We are extremely thankful to Djojosudharmo Suharto for permitting us to use his data, and acknowledge Maarten Terlou for writing the application to analyze DJL data. Finally we thank all the students and field assistants who helped in collecting data. Simon Husson, Cheryl Knott, and three anonymous reviewers made valuable suggestions on earlier versions of this chapter.

References

Anderson, D. P., Nordheim, E. V., Boesch, C., & Moermond, T. C. (2002). Factors influencing fission-fusion grouping in chimpanzees in the Taï National Park, Côte d'Ivoire. In *Behavioural Diversity in Chimpanzees and Bonobos*, ed. C. Boesch, G. Hohmann, & L. F. Marchant, pp. 90–101. Cambridge: Cambridge University Press.

Ashton, P. S., Givnish, T. J., & Appanah, S. (1988). Staggered flowering in the dipterocarpaceae: new insights into floral induction and the evolution of mast flowering. *American Naturalist*, **132**, 44–66.

Bercovitch, F. B. (1987). Female weight and reproductive condition in a population of olive baboons (*Papio anubis*). *American Journal of Primatology*, **12**, 189–95.

Buij, R., Wich, S. A., Lubis, A. H., & Sterck, E. H. M. (2002). Seasonal movements in the Sumatran orangutan (*Pongo pygmaeus abelii*) and consequences for conservation. *Biological Conservation*, **107**, 83–7.

Chapman, C. A. (1988). Patch depletion by the spider monkeys of Santa Rosa National Park, Costa Rica. *Behaviour*, **105**, 99–116.

Chapman, C. A., Wrangham, R. W., & Chapman, L. J. (1995). Ecological constraints on group size: an analysis of spider monkey and chimpanzee subgroups. *Behavioral Ecology and Sociobiology*, **36**, 59–70.

Delgado, R. & van Schaik, C. P. (2000). The behavioural ecology and conservation of the orangutan (*Pongo pygmaeus*): a tale of two islands. *Evolutionary Anthropology*, **9**, 201–18.

Di Bitetti, M. S. & Janson, C. H. (2000). When will the stork arrive? Patterns of birth seasonality in neotropical primates. *American Journal of Primatology*, **50**, 109–30.

Doran, D. (1997). Influence of seasonality on activity patterns, feeding behaviour, ranging and grouping patterns of Taï chimpanzees. *International Journal of Primatology* **18**, 183–206.

Ellison, P. T. (1990). Human ovarian function and reproductive ecology: new hypotheses. *American Anthropologist*, **92**, 933–52.

Ellison, P. T., Panter-Brick, C., Lipson, S. F., & O'Rourke, M. T. (1993). The ecological context of human ovarian function. *Human Reproduction*, **8**, 2248–58.

Fox, E. A. (2002). Female tactics to reduce sexual harassment in the Sumatran orangutan (*Pongo pygmaeus abelii*). *Behavioral Ecology and Sociobiology*, **52**, 93–101.

Fox, E. A., van Schaik, C. P., Sitompul, A., & Wright, D. N. (2004). Intra-and interpopulational differences in orangutan (*Pongo pygmaeus*) activity and diet: implications for the invention of tool use. *American Journal of Physical Anthropology*, **125**, 162–74.

Fragaszy, D. M. & Perry, S. (2003). *The Biology of Traditions*. Cambridge: Cambridge University Press.

Galdikas, B. M. F. (1988). Orangutan diet, range, and activity at Tanjung Putting, Central Borneo. *International Journal of Primatology*, **9**, 1–35.

Hochberg, Y. (1988). A sharper Bonferroni procedure for multiple tests of significance. *Biometrika*, **75**, 800–2.

Knott, C. D. (1998a). Social system dynamics, ranging patterns, and male and female strategies in wild Bornean orangutans (*Pongo pygmaeus*). *American Journal of Physical Anthropology*, Supplement, **26**, 140.

(1998b). Changes in orangutan diet, caloric intake and ketones in response to fluctuating fruit availability. *International Journal of Primatology*, **19**, 1061–79.

(1999). Orangutan behavior and ecology. In *The Nonhuman Primates*, ed. P. Dolhinow & A. Fuentes, pp. 50–7. Mountain View: Mayfield Publishing Company.

(2001). Female reproductive ecology of the apes: implications for human evolution. In *Reproductive Ecology and Human Evolution*, ed. P. T. Ellison, pp. 429–63. New York, NY: Walter de Gruyter.

(2005). Energetic responses to food availability in the great apes: implications for hominin evolution. In *Primate Seasonality: Implications for Human Evolution*, ed. D. Brockman & C. P. van Schaik, 351–57. Cambridge: Cambridge University Press.

Koenig, A., Borries, C., Chalise, M. K., & Winkler, P. (1997). Ecology, nutrition, and the timing of reproductive events in an Asian primate, the Hanuman langur (*Presbytis entellus*). *Journal of Zoology, London*, **213**, 409–22.

MacKinnon, J. (1974). The ecology and behaviour of wild orangutans (*Pongo pygmaeus*). *Animal Behaviour*, **22**, 3–74. Cited in Rodman, P. S. (1988), Diversity and consistency in ecology and behavior. In *Orangutan Biology*, ed. J. H. Schwartz, pp. 31–51. Oxford: Oxford University Press.

MacKinnon, K., Hatta, G., Halim, H., & Mangalik, A. (1996). *The Ecology of Kalimantan*. Singapore: Periplus Editions.

Markham, R. J. (1990). Breeding orangutans at Perth Zoo: twenty years of appropriate husbandry. *Zoo Biology*, **9**, 171–82.

Medway, L. (1972). Phenology of a tropical rainforest in Malaya. *Biological Journal of the Linnean Society*, **4**, 117–46.

Mitani, J. C. (1989). Orangutan activity budgets: monthly variations and the effects of body size, parturition and sociality. *American Journal of Primatology*, **18**, 87–100.

Mitani, J. C., Watts, D. P., & Lwanga, J. S. (2002). Ecological and social correlates of chimpanzee party size and composition. In *Behavioural Diversity in Chimpanzees and Bonobos*, ed. C. Boesch, G. Hohmann, & L. F. Marchant, pp. 102–11. Cambridge: Cambridge University Press.

Olupot, W., Chapman, C. A., Waser, P. M., & Isabirye-Basuta, G. (1997). Mangabey (*Cercocebus albigena*) ranging patterns in relation to fruit availability and the risk of parasite infection in Kibale National Park, Uganda. *American Journal of Primatology*, **43**, 65–78.

Palombit, R. A. (1992). *Social and Ecological Variation in Hylobatid Social Systems*. Unpublished Ph.D. thesis, University California-Davis.

Reader, S. M. & Laland, K. N. (2003). *Animal Innovation*. Oxford: Oxford University Press.

Rijksen, H. D. (1978). *A Field Study on Sumatran Orangutans* (Pongo pygmaeus abelii, *Lesson 1827): Ecology, Behaviour and Conservation*. Wageningen, the Netherlands: Veenman.

Rodman, P. S. (1988). Diversity and consistency in ecology and behavior. In *Orangutan Biology*, ed. J. H. Schwartz, pp. 31–51. Oxford: Oxford University Press.

Rodman, P. S. & Mitani, J. C. (1987). Orangutans: sexual dimorphism in a solitary species. In *Primate Societies*, ed. B. B. Smuts, D. L. Cheney, R. M. Seyfarth, R. W. Wrangham, & T. T. Struhsaker, pp. 146–54. Chicago, IL: University of Chicago Press.

Schürmann, C. L. & van Hooff, J. A. R. A. M. (1986). Reproductive strategies of the orangutan: new data and a reconsideration of existing sociosexual models. *International Journal of Primatology*, **7**, 265–87.

Stanford, C. B. & Nkurunungi, J. B. (2003). Behavioral ecology of sympatric chimpanzees and gorillas in Bwindi Impenetrable National Park, Uganda: diet. *International Journal of Primatology*, **24**, 901–18.

Sugardjito, J., Te Boekhorst, I. J. A., & van Hooff, J. A. R. A. M. (1987). Ecological contraints on grouping of wild orangutans (*Pongo pygmaeus*) in the Gunung Leuser National Park, Sumatera, Indonesia. *International Journal of Primatology*, **8**, 17–41.

Takahashi, H. (2002). Female reproductive parameters and fruit availability: factors determining the onset of estrus in Japanese macaques. *American Journal of Primatology*, **51**, 141–53.

Te Boekhorst, I. J. A., Schürmann, C. L., & Sugardjito, J. (1990). Residential status and seasonal movements of wild orangutans in the Gunung Leuser Reserve (Sumatra, Indonesia). *Animal Behaviour*, **39**, 1098–109.

Utami, S. S. (2000). *Bimaturism in Orangutan Males: Reproductive and Ecological Strategies*. Unpublished Ph.D. thesis, Utrecht University.

Utami, S. S., Wich, S. A., Sterck, E. H. M. & Hooff, J. A. R. A. M. (1997). Food competition between wild orangutans in large fig trees. *International Journal of Primatology*, **18**, 909–27.

van Schaik, C. P. (1986). Phenological changes in a Sumatran rainforest. *Journal of Tropical Ecology*, **2**, 327–47.

(1999). The socioecology of fission-fusion sociality in orangutans. *Primates*, **40**, 69–86.

van Schaik, C. P. & Knott, C. D. (2001). Geographic variation in tool use on *Neesia* fruits in orangutans. *American Journal of Physical Anthropology*, **114**, 331–42.

van Schaik, C. P. & Mirmanto, E. (1985). Spatial variation in the structure and litterfall of a Sumatran rain forest. *Biotropica*, **17**, 196–205.

van Schaik, C. P. & van Hooff, J. A. R. A. M. (1996). Towards an understanding of the orangutan's social system. In *Great Ape Societies*, ed. W. C. McGrew, L. F. Marchant, & T. Nishida, pp. 3–15. Cambridge: Cambridge University Press.

van Schaik, C. P. & van Noordwijk, M. A. (1985). Interannual variability in fruit abundance and the reproductive seasonality in Sumatran long-tailed macaques (*Macaca fascicularis*). *Journal of Zoology, London*, **206**, 533–49.

van Schaik, C. P., Ancrenaz, M., Borgen, G. *et al.* (2003). Orangutan cultures and the evolution of material culture. *Science*, **299**, 102–5.

Wich, S. A. & van Schaik, C. P. (2000). The impact of El Niño on mast fruiting in Sumatra and elsewhere in Malesia. *Journal of Tropical Ecology*, **16**, 563–77.

Wich, S. A., Buij, R., & van Schaik, C. P. (2004b). Determinants of orangutan density in the dryland forests of the Leuser Ecosystem. *Primates*, **45**, 177–82.

Wich, S. A., Sterck, E. H. M., & Utami, S. S. (1999). Are orangutan females as solitary as chimpanzee females? *Folia Primatologica*, **70**, 23–8.

Wich, S. A., Utami-Atmoko, S. S, Mitra Setia, T. *et al.* (2004a). Life history of wild Sumatran orangutans (*Pongo abelii*). *Journal of Human Evolution*, **47**, 385–98.

Wich, S. A., Utami-Atmoko, S. S., Mitra Setia, T., Djoyosudharmo, S., & Geurts, M. L. (in press). Dietary and energetic responses of *Pongo abelii* to fruit availability fluctuations. *International Journal of Primatology*.

Ziegler, T., Hodges, K., Winkler, P., & Heistermann, M. (2000). Hormonal correlates of reproductive seasonality in wild female Hanuman langurs (*Presbytis entellus*). *American Journal of Primatology*, **51**, 119–34.

14 Central place provisioning: the Hadza as an example

FRANK W. MARLOWE

Introduction

Central place foraging (CPF) has long been recognized as an important feature of human foragers that distinguishes them from other primates (Washburn & DeVore, 1961; Isaac, 1978; Lovejoy, 1981). However, there have been only a few attempts to compare human CPF with the ranging and

Feeding Ecology in Apes and Other Primates. Ecological, Physical and Behavioral Aspects, ed. G. Hohmann, M.M. Robbins, and C. Boesch. Published by Cambridge University Press. © Cambridge University Press 2006.

sleeping patterns of other species (Potts, 1987; Sept, 1992, 1998; Fruth & Hohmann, 1994). Most of the literature instead focuses on the signals that might reveal when hominids took up CPF (Binford, 1980; Bunn *et al.*, 1988; Potts, 1994; Rose & Marshall, 1996; Bird & Bliege Bird, 1997; O'Connell, 1997). While there has been considerable discussion of how CPF would have altered life, less attention has been paid to why our ancestors adopted it. Here, I propose a classification scheme of ranging, grouping, and feeding patterns that uses the same set of criteria for all species to illuminate why central places are used. I argue that one reason humans use central places is the benefit gained by leaving young, weaned children behind while foraging. I also argue that one important consequence is the extensive food sharing common among human foragers.

When one tries to explain why animals practice CPF, the first problem encountered is one of terminology. According to Orians & Pearson (1979), "Many foragers . . . do not consume their prey where they are captured but return with them to some fixed central place where they are eaten, stored, or fed to dependent offspring (Central Place Foraging)." This definition implies that CPF consists of taking food back to a central place, and indeed that is how most would define it, but some have begun to call those who return to a central place to sleep – without taking food back – central place foragers as well (Chapman *et al.*, 1989; Whiten, 1992). The term "refuging" once conveyed something similar (Hamilton & Watt, 1970), but a literature search shows that refuging has come to be used exclusively for escaping predators by fleeing to a burrow (Blumstein, 1998), any nearby hiding place (Cooper, 1997), or even one's own shell (Scarratt & Godin, 1992), not for animals that return to a central place even when no predators are present. Because it is important to distinguish between those who return to a central place without food and those who return with food, we need two distinct terms for them.

To avoid confusion, I propose we recognize three patterns of ranging. (1) Feed-as-you-go (and sleep where you are): individuals search for food, eat it where encountered, and sleep wherever they end up, e.g., chimpanzees (*Pan troglodytes*) (Goodall, 1986; Boesch & Boesch-Achermann, 2000). (2) Central Place Foraging (CPF): individuals search for food, eat it where encountered, and then return to a central place to sleep, e.g., hamadryas (*Papio hamadryas*) (Kummer, 1995). (3) Central Place Provisioning (CPP): individuals search for food, may or may not eat some where encountered, but then also take some food back to a central place, usually to share with others, especially the young, or to store it for later consumption, e.g., meerkats (*Suricata suricatta*) (Clutton-Brock *et al.*, 2002), and human foragers (Kelly, 1995). In this scheme, CPF occurs in several species of primates but CPP is rare. On the other hand, CPP is common in many birds, carnivores, rodents, and insects. Because many feed-as-you-go species are not indifferent to where

Table 14.1. *Proposed classification scheme of patterns of ranging (columns)
and patterns of grouping (rows)*

	Feed-as-you-go (and sleep where you are)	Central place foraging	Central place provisioning
Solitary or monogamous Pair	Orangutan *Pongo pygmaeus* Potto *Arctocebus calabarensis* Gibbon *Hylobates lar*	Lesser galago *Galago senegalensis* Greater galago *Galago crassicaudatus* Titi monkey *Callicebus torquatus*	Cheetah *Acinonyx jubatus* Beaver *Castor canadensis* Palestine sunbird *Nectarinia osea*
Cohesive group	Black and white colobus *Colobus guereza* Gorilla *Gorilla gorilla* Tufted capuchin *Cebus appella*	Olive baboon *Papio anubis* Pig-tailed macaque *Macaca nemestrina* Yellow baboon *Papio cynocephalus*	If there are species in this cell it is probably those that store food, as there is no other reason for cohesive groups to take foods back
Fission-fusion	Chimpanzee *Pan troglodytes* Bonobo *Pan paniscus* Bottlenose dolphin *Tursiops truncatus*	Hamadryas baboon *Papio hamadryas* Spider monkey *Ateles geoffroyi* Fruit bat *Artibeus jamaicensis*	Meerkat *Suricata suricatta* African wild dog *Pictus lycaon* Human *Homo sapiens*

they sleep, and occasionally reuse a favored sleeping site (Sept, 1998), CPF
could be operationalized as using the same sleeping place for at least 3 or 4
consecutive days. Table 14.1 has discreet cells but of course reality is more
continuous and the classification is meant only to highlight important ecological differences. Given the rate of nest reuse by chimpanzees (Wrangham,
1992), the 3 or 4 day requirement would distinguish them from CPF species
like hamadryas. Although most optimal foraging models of CPF deal with
taking food back to the central place, others address issues such as travel costs
and location of the central place in relation to food patches, issues that apply
equally to those who do not take food back.

Explaining why one of these three patterns of ranging occurs requires
distinguishing three different grouping patterns as well: (1) solitary (or
monogamous pair); (2) cohesive groups; and (3) fission-fusion groups. Each
cell shown in Table 14.1 is likely explained by varying levels of, and
best responses to predation, resource distribution, and feeding competition
(Terborgh & Janson, 1986; Garber, 1987). For example, hamadryas baboons

break into smaller groups to forage in the day and return to sleep together in very large troops on cliffs that afford protection from leopards at night (Kummer, 1995), a pattern of fission-fusion CPF. This is because there are a limited number of appropriate sleeping sites, or because groups need to be so large for safety at night that if those groups stayed together while foraging there would be too much feeding competition (Swedell, 2002). On the other hand, some savanna baboons (*Papio anubis*), who sleep in central places like rock kopjes or trees, stay together as they forage, a pattern of cohesive CPF, perhaps because they can sleep in more protected sites and thus in smaller groups, small enough that foraging together does not result in too much feeding competition, and yet provides greater safety while foraging. Notice there are no examples in the cell for cohesive CPP since there is little reason to take food back to a central place when the whole group is foraging together. Only with completely cohesive groups could we ignore grouping patterns, such as smaller foraging parties, when describing ranging.

When sleeping in a central place is a response to the threat of predation, diet may be largely irrelevant (Boinski *et al.*, 2000). Without such a constraint we should, in general, expect animals to be feed-as-you-go (and sleep where you are) foragers since this saves them the cost of traveling back to the central place each day. On the other hand, diet is an important factor in CPP because foragers must acquire food they can use to feed others or to store, unless they return to a central place merely to eat in safety, as some squirrels do (Lima *et al.*, 1985). Many carnivores, such as cheetahs (*Acinonyx jubatus*) (Durant *et al.*, 2004), wild dogs (*Lycaon pictus*), and wolves (*Canis lupus*) (Asa, 1997) are central place provisioners – at least when they have young in a den. One cause of CPP, therefore, is the need to feed dependent young who are kept in a safe place. This requires that adults be able to acquire more food than they need for their own consumption (surplus food), which is often the case with carnivores that hunt large prey, but is also true for birds that catch arthropods (Markman *et al.*, 2004) and beavers that take logs to their lodge (Gallant *et al.*, 2004). Human acquisition of foods that are only edible after processing may also lead to CPP. Some species park their young in safe places to which they return even though they do not take food back, e.g., dik dik (*Madoqua kirkii*), thus leaving young in a safe place results in CPF but not necessarily CPP, at least among mammals since mothers can nurse their infants rather than take food back.

Here, I present data on the grouping, ranging, and feeding patterns of one group of hunter-gatherers, the Hadza of Tanzania, to illustrate CPP, and to show how they fall into the cell labeled fission-fusion CPP in Table 14.1. I also test two predictions – one a cause and one a consequence of CPP. Because it is difficult for a woman to carry a nursling and a weanling when she goes foraging, one reason for CPP among the Hadza may be the benefit gained by leaving weanlings in camp while foraging. If so, weanlings should spend more time in camp than older and younger children, and below I test that prediction. Glynn Isaac (1978) suggested that taking food back to a central place rather than eating it on the spot should lead to tolerated scrounging and food transfers (sharing). I therefore predict that more food transfers occur in camp than out of camp, and below I test that prediction as well.

Methods

Subject population

The Hadza, hunter-gatherers in northern Tanzania, number about 1000. They live in a savanna–woodland habitat of 4000 km^2 around Lake Eyasi. Elevation varies from 900–2000 m. Average annual temperature varies little across the year (mean $\sim$ 28 °C) but considerably between day and night (min $= 17$ °C, max 33 °C from mid-July 15 to early September 2004 in the Mangola area). There is a dry season from June through November; the rainy season, within which almost all of the 300–600 mm of rain falls, is from December through May.

Hadza women dig tubers, collect baobab, and gather berries. They take their nursing infants with them foraging, carrying them in a kaross on their backs. A woman will keep her infant on her back even while digging tubers, occasionally swinging it to the front to nurse. Most children are completely weaned by 2.5 years of age, about the time women begin to leave them in camp (Marlowe, 2005). Older children, especially girls, often go foraging with the women (Hawkes *et al.*, 1995).

Men collect baobab and honey and use bows and arrows to hunt birds and mammals. They also scavenge carcasses from carnivores. Men usually hunt alone. With bows and poisoned arrows, one man can kill an animal as large as a giraffe with a good shot, so the advantages of two hunters looking for game are probably outweighed by the increased likelihood of being detected before getting close enough to get a good shot. During the late dry season, men hunt at night by ambushing animals when they come to drink at waterholes; because of the danger from lions and leopards who use the same strategy, night-time ambush hunting is done in pairs.

Most berries come into season a little before the rains begin and continue throughout the first half of the rainy season, when honey becomes plentiful. Tubers are available throughout the year (Vincent, 1985) and appear to be fallback foods, if we consider fallback foods to be those eaten when preferred foods are not available. Although seasonality is quite noticeable, by switching to different foods, the Hadza are able to acquire a fairly consistent amount of food throughout the year and say they do not view any one season as more difficult than any other.

Some foods are always processed, some never processed, and others sometimes processed, sometimes not. For example, baobab is eaten fresh when eaten out of camp. When it is taken back to camp, the pulp and seeds are pounded into a meal with a hammer stone. A little water is added to this meal to make a paste, which is an important weaning food. Meat is always roasted or boiled, though marrow is eaten raw. Game the size of gazelle or smaller is often carried back to camp whole while larger game is butchered at the kill site and pieces taken back. Upon learning that parts of a large carcass remain at the kill site, others will go and cut off their own pieces to carry back. Berries are usually eaten fresh, but toward the end of berry season, they are dried to preserve them for several days. All but one species of tuber is roasted for about 5 minutes before being eaten.

The Hadza live in camps with a fluctuating number of residents. Over the years 1995–2004, my censuses of 53 camps revealed a mean camp population of 30.4 (6–139, sd = 28.4, median = 21). Camps tend to be larger in the dry season, since there are a limited number of permanent waterholes, and that constrains where camps can be located (Woodburn, 1964; Mabulla, 2003). During the rainy season, camps can be almost anywhere and people choose to live in smaller groups. A camp may be referred to by a man's name but it is usually his wife and her sisters or her parents that form the core group that changes least. About 60–70% of those couples in which the mothers of both husband and wife are still living, reside in a camp where the wife's mother lives (Woodburn, 1968; Blurton Jones *et al.*, 2005).

In addition to the frequent visiting and moving between camps, people will move the location of their camp every month or two in response to a variety of factors besides access to water. For example, when berries come into season, they move to live near or in the berry bushes. Camps are moved when women have to walk too far to reach tuber patches. And they may move just because someone has died and they feel it is bad to stay in that spot.

Data collection

Data used here come mostly from eight camps studied in 1995, 1996, 2003, and 2004. All food brought into camp was weighed. Energy estimates per kilogram (kJ) of different food types were taken from others' analyses where available (Vincent, 1985; Blurton Jones *et al.*, 1989; Hawkes *et al.*, 1991) and otherwise from values for similar species (Ulene, 1995).

Forays out from camp by men and women were used to count the number of adults in foraging parties. Forays were defined as any trip out from camp in search of food, and the type of foray was determined by whatever food accounted for most of what was acquired. When men walked about with bow and arrow but acquired nothing, the foray was scored as encounter hunting.

Focal individual observations (follows) of women were conducted in four randomized time blocks of 3 hours each across all daylight hours. However, when a follow included a foray, the observer continued the follow until the woman was back in camp and throughout any eating events that immediately followed. Follows of men were conducted during forays, which sometimes take all day, and so were conducted only once per day. Global positioning system (GPS) units were used to record tracks of foragers in 2003 and 2004 to show the path traveled, the distance covered during forays, and the speed and time spent moving.

Instantaneous observations (scans) were conducted hourly throughout the day to record those who were in camp and what they were doing. These scans were used here to show what percent of the group remained in camp across the hours of the day and to test if weanlings were in camp more than others. Weanlings were defined as those 2–4.9 years old, which includes those being weaned or recently weaned but still unable to walk fast. Adult men and women were those 18 years old or older.

All instances of food transfers during follows were recorded. Food transfers were defined as any instance of a person giving or receiving food from anyone else, whether the food was eaten right away or later. For example, a woman handing a piece of roasted tuber to a child, was scored as one instance of food transfer. A man bringing prey into camp that was butchered and divided among five other men, was scored as 5 instances of food transfer. The frequency of food transfers in camp per hour of observation in camp was compared to the frequency of food transfers out of camp per hour of observation out of camp.

Results

The Hadza diet consists of at least 879 species. Excluding birds, the number falls to 138, and most of these are mammals. In terms of frequency and

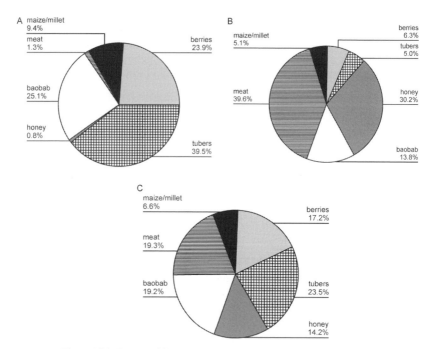

Figure 14.1. Percent of foods by category (daily kilojoules) brought into camp by Hadza: (a) women (n = 59); (b) men (n = 51); and (c) both sexes, all ages (n = 182) in five camps.

kilojoules (kJ), the diet is dominated by far fewer species. Figure 14.1 shows the foods brought back to camp, sorted into the six main categories, as measured by kilojoules per day (daily kJ) across five different camps. The maize and millet shown were received from agro-pastoralist neighbors in exchange for meat or honey, and from a missionary who visited one large camp. Figures 14.1a and 14.1b show a pronounced sexual division of labor, but one that is typical of many human foragers (Marlowe, 2005).

Although not yet analyzed, I estimate about one-third of the total food consumed may be eaten out of camp. Berries would amount to a much larger percentage of the total diet consumed than Figure 14.1c implies because most berries are eaten while out foraging. I estimate that about 2/3 of berries, 1/5 of baobab, 1/4 of tubers, 1/2 of honey, and 1/10 of meat that is acquired is consumed out of camp. When berries are combined with baobab, it is clear that fruit comprises the largest fraction of the diet (36.4%) in Figure 14.1, and this would be even greater if out-of-camp consumption were included.

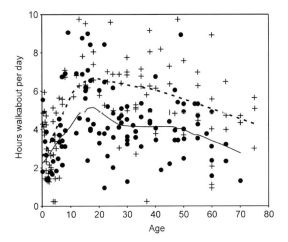

Figure 14.2. Hours per day (in scans) that females (n = 110) and males (n = 108) were out of camp by age (n = 1055 hourly scans). The + symbol and dashed line indicate males, solid dots and solid line indicate females. The regression lines are Lowess smoothed for each sex separately.

Ranging patterns: fission-fusion, day range, and party size

Figure 14.2 shows the amount of time people were gone from camp by age and sex. The mean time spent foraging per day was 4.1 hours for women and 6.1 hours for men. Women walked 5.5 km (0.25–13.50) per foray (n = 110 forays) at an average moving speed of 3.5 km/hour. Men walked an average of 8.3 km (1.57–27.20) per foray (n = 57 forays) at an average walking speed of 3.6 km/hour. Men traveled significantly further than women (t = 4.01, p < 0.0005, (df = 74.8, unequal variances).

The mean size of foraging parties for men was 1.7 ± 2.09 (median = 1, n = 199 forays). For women, mean party size was 5.316 ± 3.93 (median = 4, n = 95 forays). For mixed-sex parties mean party size was 4.06 ± 3.07 (median = 3, n = 33 forays). Party size varies with the type of foray. For women foraging mainly for tubers, the mean party size was 3.8 adult women plus infants and some older children (1–12, n = 55 forays). During berry season there were more mixed-sex parties and mean party size on berry forays was 7.7 (1–24, n = 40 forays). When baobab was the main food brought back to camp, mean party size was 4.6 (1–14, n = 11 forays). Men most often forage for honey alone, but sometimes go in parties of two or three, and sometimes go with their wives. Average party size for honey foraging was 2.0 (1–8, n = 28 forays). When encounter hunting or daytime ambush hunting, men had a mean party size of 1.5 (1–16, n = 124 forays).

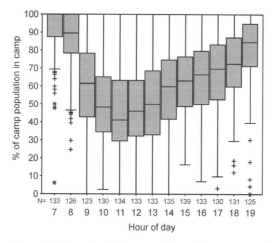

Figure 14.3. Percent of all camp residents who were present in camp by hour of the day across eight different camps in different seasons (n = 1705 scans). The + symbol indicates outliers and the * symbol indicates extremes. Boxes show the interquartile range and midlines show the medians.

Most prime-age adults go foraging every day for some period of time, but occasionally they may stay in camp all day. There are usually some camp residents who are in camp at any given hour of the day, especially in larger camps. Across eight camps, the percent of the camp population present in camp ranged from 0 to 100% (1705 scans). The hour when the fewest people were present was 11:00 when on average 47% (median = 41%) of the camp population was in camp (Figure 14.3). These often included very young children, the very old or disabled, the temporarily injured, and the ill.

Leaving weanlings in camp

Figure 14.4 shows the average number of hours that children through the age of 8 years old spent in camp per day in six camps in 1995/96 (1055 scans). Age was coded into three categories: (1) infants under 2 years old (n = 22); (2) weanlings 2–4.9 years old (n = 24); and (3) children 5–8 years old (n = 22). Both infants and older children were gone from camp more than weanlings. Infants were out of camp 3.5 hours per day compared with 2.3 hours for weanlings (t = 2.897, p = .006, df = 44, equal variances; n_1 = 22; n_2 = 24). Older children were gone from camp 4.2 hours per day, which was also more than weanlings (t = −4.127, p < 0.0005, df = 44, equal variances; n_1 = 24; n_2 = 22). Weanlings are gone from camp least and when they are

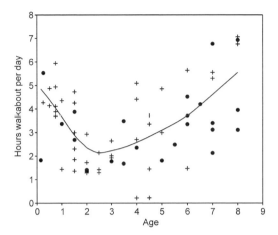

Figure 14.4. Hours per day children spent out of camp by their ages, across six different camps (n = 66 children). The + symbol indicates males and solid dots indicate females. Regression line is Lowess smoothed for both sexes.

gone they are usually with older children in playgroups just beyond the edge of camp.

Food transfers

Table 14.2 shows the frequency of sharing by type of food in two different camps, one in the dry season and one in the wet season. There were 131.6 hours of observation in camp and 187.7 hours of observation out of camp. For all five main categories of foods, there was more sharing in camp than out of camp per hour of observation. The bias toward in-camp sharing was strongest for meat, which is not surprising given that men usually hunt alone and thus have little opportunity to share out of camp. When meat arrives in camp, especially larger game, it is usually shared quite evenly across households (Hawkes *et al.*, 2001). The lowest bias toward sharing in camp is for berries. This too makes sense, given that foragers usually eat berries on the spot. In fact, during berry season, there are days when the Hadza do not take food back to camp but instead live in the berry bushes and are basically feed-as-you-go foragers.

Discussion

In contrast to mothers in many other species who park their infants in a safe place, Hadza mothers take their infants with them but leave their weanlings in

Table 14.2. *Tests of frequencies of food transfers by food category in camp and out of camp for two different camps*

Location of food transfers	Tubers	Berries	Baobab	Meat	Honey
In camp 131.6 hours observed (0.41 of total)					
Expected N	77.5	43.5	24.6	30.3	25.0
Observed N	127	62	51	70	50
Out of camp 187.7 hours observed (0.59 of total)					
Expected N	111.5	62.5	35.4	43.7	36.0
Observed N	62	44	9	4	11
Total transfers 319.3 hours observed	189	106	60	74	61
In camp vs. out of camp test	$\text{Chi}^2 =$ 53.615 $p < 0.003$	$\text{Chi}^2 =$ 13.405 $p < 0.003$	$\text{Chi}^2 =$ 48.020 $p < 0.003$	$\text{Chi}^2 =$ 87.869 $p < 0.003$	$\text{Chi}^2 =$ 42.322 $p < 0.003$

Expected frequencies are derived by assuming the total number of food transfers for each food category would be divided into transfers in camp vs. transfers out of camp in proportion to the number of hours of observation in camp and out of camp. Chi-square tests are Bonferroni corrected for five tests of five food categories.

camp while foraging. In many prosimians, mothers must use their mouths to carry their young and when they go foraging they leave them in secluded nests (Anderson, 1998; Kappeler, 1998; Ross, 2001). The same is true of most carnivores. On the other hand, in monkeys and apes, mothers may use their hands to carry their infants briefly until the young are able to ride clinging to them (Ross, 2001). Among human foragers, women have shorter interbirth intervals, mean = 3.3 years (Marlowe, 2005), than chimpanzees, mean = 5.58 years (Boesch & Boesch-Achermann, 2000). These shorter intervals mean women may have a new nursing infant when their previous child is still just toddling, and traveling with both would be very difficult. Human mothers with carrying devices can tie infants to their backs while foraging but weaned children are too big to carry and too young and small to keep up. Women can solve this problem by leaving weanlings behind in a central place if others will stay behind to look after them. Figure 14.3 shows that many individuals are in camp while others are out foraging and Figure 14.4 shows that weanlings are in camp more than infants and older children. Leaving weanlings in camp appears to be one important reason for having a central place among the Hadza.

Leaving some members in camp necessitates taking food back, at least for those weanlings and their alloparents, who are often older siblings or grandparents. Provisioning depends on the ability to acquire surplus food, thus CPP depends on a certain diet like sizeable game, or an extractive technology such as the digging sticks women use to acquire large amounts of underground tubers, or the rocks used to pound baobab seed and pulp (Hawkes *et al.*, 1997; Kaplan *et al.*, 2000). Hadza women with young children, and possibly men as well, may bring food back to camp with the goal of provisioning the members of their households and alloparents, but because they live in groups composed of several families, they are likely to encounter others (Marlowe, 2004). Due to the fissioning into smaller foraging parties, surplus food arrives in camp asynchronously. This means that some have food while others do not, which increases the opportunity for scrounging. The more people there are demanding shares, the costlier it will be to say no (Blurton Jones, 1987; Winterhalder, 1996). In addition, the sexual division of foraging labor increases the benefits of trade as well (meat or honey for fruit or tubers) (Milton, 2000). Asynchronous food acquisition and return to camp also increase the payoffs to the delayed exchange of the same foods, so long as scrounging without reciprocation can be limited (Kaplan & Hill, 1985; Winterhalder, 1986; Ofek, 2001). Whether due to in-kind reciprocity, trade, or scrounging, one consequence of CPP among the Hadza, and other foragers, is the camp-wide food sharing commonly observed (Kelly, 1995).

Table 14.3 lists some other possible causes and consequences of CPF and CPP. CPP entails the same costs (further travel) and benefits (greater safety while sleeping) that CPF does but adds to these the costs and benefits of taking food back. Therefore, the same causes that apply to CPF also apply to CPP, which suggests CPF may usually be ancestral to CPP. One important cause of CPF appears to be safety while sleeping (Anderson, 1998). One important consequence, therefore, is probably reduced mortality. The need to sleep in a safe place might lead individuals to adopt CPF if the central place is a particularly protected site, and it would need to be protected to offset the disadvantage of predators knowing where to find those individuals (Reichard, 1999). Alternatively, if the sleeping group is large enough to defend against predators, no central place is required unless the group breaks up into smaller parties during foraging and needs to know where to reunite for sleeping together, which is why it is necessary to distinguish between fission-fusion and cohesive groups in Table 14.1. As noted, the median population of Hadza camps is 21. While there is little risk of predation at night with this many people in camp, there would be a risk if Hadza slept in groups the size of their

Table 14.3. *Possible causes and consequences of Central Place Foraging (CPF) and Central Place Provisioning (CPP) in human foragers*

	Causes	Consequences
CPF	(a_1) Benefit of sleeping at safe site; If fission-fusion foraging, need a place to reunite for sleeping	(a_2) Reduced predation
	(b_1) Benefit of being near water or shelter	(b_2) Day range length increased
	(c_1) Benefit of parking weanlings **Weanlings spend more time in camp than others**	(c_2) Alloparents stay in camp to baby-sit and help mother, but they must be provisioned, which requires surplus production
CPP	(d_1) Benefit of parking sick and injured	(d_2) Sick and injured recover, lower adult mortality, selection for > longevity
	(e_1) Sexual division of labor means fission-fusion so males and females can optimally target different resources—this means couples need a place to reunite each night to share foods	(e_{2a}) Inability to guard one's mate all day (e_{2b}) Taking food back to camp increases opportunity for scrounging **More food transfers in camp than out of camp**

In the top three rows are factors that could lead to CPF but not necessarily CPP, while those in the bottom two rows would result in CPP. Letters with subscript$_1$ are Causes that lead to Consequences with subscript$_2$.

foraging parties. Thus, one reason for sleeping together at a central place, even today, may be to avoid predation. Certainly, sleeping in large groups could have been important for earlier hominins. In more open habitats there will be fewer appropriate sleeping trees nearby when it is time to stop foraging, making CPF more likely. It is interesting that chimpanzees appear to reuse nests more often and more frequently in dryer habitats (Sept, 1992). As already observed, Hadza camps are tethered to sources of water, so in addition to safe sleeping sites, the need to be near water on a daily basis could also lead to CPF (Binford, 1980).

As already noted, most of the causes for CPF may also apply to CPP, but the reverse is not true, and Table 14.3 lists two other reasons for CPP. Among the Hadza, in addition to weanlings, the sick and injured or the very old stay in the central place, and they too need to be provisioned. Rest for and recovery of the sick and injured should result in lower adult mortality rates (Sugiyama, 2004). The fact that females in most mammalian species gestate, lactate, and carry infants may have paved the way for a division of labor in foraging among our ancestors and this suggests another reason for CPP. When males specialize in hunting and females in gathering, as the Hadza do, they cannot forage together effectively since they are targeting different prey. This requires that they fission in the day and fuse at night. Without a central place, it would be difficult to reunite at night, especially for foragers like the Hadza, who have long day range lengths leading in different directions. One consequence of the fissioning between males and females while foraging is that they cannot stay within sight of their mates. This inability to guard mates should make pair bonds less stable. However, household provisioning or trade of male and female foods may strengthen bonds. Compared to Hadza women without infants, those who have nurslings spend less time foraging, bring back less food, and acquire less food per hour of foraging time, while their husbands (unless they are the infants' stepfathers) bring more food back to camp, suggesting men compensate for their wives reduced productivity (Marlowe, 2003). If sharing different types of food within the household is one payoff for the hunting versus gathering specialization, then CPP is favored so that couples have a place to reunite to share foods and provision offspring.

I suggest that the most parsimonious scenario consistent with the fossil and archeological record comprises the following sequence: (1) Our ancestors lived in a more open habitat with a threat from predation greater than that experienced by extant chimpanzees, and this favored sleeping together in sizable groups at central places (CPF). (2) Use of tools such as the digging stick for tuber acquisition and scavenging led to surplus production. (3) Surplus production enabled earlier weaning and longer juvenility, which meant females benefited from leaving weanlings with anyone who would stay in camp to watch them. (4) Leaving weanlings and alloparents in camp meant food needed to be taken back, which was the beginning of CPP. (5) The increasing specialization in hunting or scavenging versus gathering required a sexual fission-fusion, and the central place allowed for the trade of hunted foods for gathered foods. (6) As a consequence of living in large social groups and taking food back to camp for offspring, alloparents, and mates, tolerated scrounging led to camp-wide food sharing.

The search for hearths, middens, tools and lithic debris by paleo-archeologists may turn up the earliest evidence that central places were being used by hominins for CPP. However, it would be wrong to think that such evidence showed when CPF was first adopted; CPF could have begun at the very inception of the hominin clade without leaving any more of an archeological signal than is left by savanna baboons. Central places would have had important consequences even if used only for CPF; one of these consequences, however, would have been to increase the likelihood of CPP evolving, with its even greater implications for human social organization.

Acknowledgments

I thank Pastory Bushozi, Revocatus Bugumba, Daniel Ngumbuke, and Hokki Manonga for gathering data in the field, Dr. Mabulla for assistance in Tanzania, and Tucker Capps, Ben Clouette, and Claire Porter for assistance with data entry and analysis, COSTECH for permission to conduct research in Tanzania, and NSF (grant #0242455) for funding. I appreciate comments from Bill McGrew, Brian Wood, Claire Porter, Cheryl Knott, and three anonymous reviewers.

References

Anderson, J. R. (1998). Sleep, sleeping sites, and sleep-related activities: awakening to their significance. *American Journal of Primatology*, **46**, 63–75.

Asa, C. S. (1997). Hormonal and experiential factors in the expression of social and parental behavior in canids. In *Cooperative Breeding in Mammals*, ed. N. G. Solomon & J. A. French, pp. 129–49. Cambridge: Cambridge University Press.

Binford, L. R. (1980). Willow smoke and dogs' tails: hunter-gatherer settlement systems and archaeological site formation. *American Antiquity*, **45**, 4–20.

Bird, D. & Bliege Bird, R. (1997). Contemporary shellfish gathering strategies among the Meriam of the Torres Strait Islands, Australia: testing predictions of a central place foraging model. *Journal of Archaeological Science*, **24**, 39–63.

Blumstein, D. T. (1998). Quantifying predation risk for refuging animals: a case study with golden marmots. *Ethology*, **104**, 501–16.

Blurton Jones, N. G. (1987). Tolerated theft: suggestions about the ecology and evolution of sharing, hoarding, and scrounging. *Social Science Information*, **26**, 31–54.

Blurton Jones, N. G., Hawkes, K. & O'Connell, J. F. (1989). Modelling and measuring costs of children in two foraging societies. In *Comparative Socioecology: The Behavioural Ecology of Humans and Other Mammals*, ed. V. Standen & R. Foley, pp. 367–90. London: Basil Blackwell.

(2005). Older Hadza men and women as helpers. In *Hunter-Gatherer Childhoods: Evolutionary, Developmental and Cultural Perspectives*, ed. B. S. Hewlett & M. E. Lamb, pp. 214–36. New Brunswick: Transaction Publishers.

Boesch, C. & Boesch-Achermann, H. (2000). *The Chimpanzees of the Tai Forest: Behavioural Ecology and Evolution*. Oxford: Oxford University Press.

Boinski, S., Treves, A., & Chapman, C. A. (2000). A critical evaluation of the influence of predators on primates: effects on group travel. In *On the Move: How and Why Animals Travel in Groups*, ed. S. Boinski & P. A. Garber, pp. 43–72. Chicago, IL: University of Chicago Press.

Bunn, H. T., Bartram, L., & Kroll, E. M. (1988). Variability in bone assemblage formation from Hadza hunting, scavenging, and carcass processing. *Journal of Anthropological Archaeology*, **7**, 412–57.

Chapman, C. A., Chapman, L. J., & McLaughlin, R. L. (1989). Multiple central place foraging by spider monkeys: travel consequences of using many sleeping sites. *Oecologia*, **79**, 506–11.

Clutton-Brock, T. H., Russell, A. F., Sharpe, L. L., Young, A. J., Balmforth, Z., & Mcilrath, G. M. (2002). Evolution and development of sex differences in co-operative behavior in meerkats. *Science*, **297**, 253–6.

Cooper, W. E. (1997). Escape by a refuging prey, the broad-headed skink (*Eumeces laticeps*). *Canadian Journal of Zoology-Revue Canadienne de Zoologie*, **75**, 943–7.

Durant, S. M., Kelly, M., & Caro, T. M. (2004). Factors affecting life and death in Serengeti cheetahs: environment, age, and sociality. *Behavioral Ecology*, **15**, 11–22.

Fruth, B. & Hohmann, G. (1994). Nests: living artefacts of recent apes? *Current Anthropology*, **35**, 310–11.

Gallant, D., Berube, C. H., Tremblay, E., & Vasseur, L. (2004). An extensive study of the foraging ecology of beavers (*Castor canadensis*) in relation to habitat quality. *Canadian Journal of Zoology-Revue Canadienne de Zoologie*, **82**, 922–33.

Garber, P. A. (1987). Foraging strategies among living primates. *Annual Review of Anthropology*, **16**, 339–64.

Goodall, J. (1986). *The Chimpanzees of Gombe*. Cambridge: Belknap Press of Harvard University Press.

Hamilton, W. J. & Watt, K. E. F. (1970). Refuging. *Annual Review of Ecology and Systematics*, **1**, 263–86.

Hawkes, K., Blurton Jones, N. G., & O'Connell, J. F. (1995). Hadza children's foraging: juvenile dependency, social arrangements, and mobility among hunter-gatherers. *Current Anthropology*, **36**, 688–700.

Hawkes, K., O'Connell, J. F., & Blurton Jones, N. G. (1991). Hunting income patterns among the Hadza: big game, common goods, foraging goals and evolution of the human diet. *Philosophical Transactions of the Royal Society of London, B*, **334**, 243–51.

  (1997). Hadza women's time allocation, offspring provisioning, and the evolution of long postmenopausal life spans. *Current Anthropology*, **38**, 551–77.

  (2001). Hadza meat sharing. *Evolution and Human Behavior*, **22**, 113–42.

Isaac, G. (1978). The food-sharing behavior of protohuman hominids. *Scientific American*, **238**, 90–108.

Kaplan, H. & Hill, K. (1985). Food sharing among Ache foragers: tests of explanatory hypotheses. *Current Anthropology*, **26**, 223–46.

Kaplan, H., Hill, K., Lancaster, J., & Hurtado, A. (2000). A theory of human life history evolution: diet, intelligence, and longevity. *Evolutionary Anthropology*, **9**, 156–85.

Kappeler, P. M. (1998). Nests, tree holes, and the evolution of primate life histories. *American Journal of Primatology*, **46**, 7–33.

Kelly, R. L. (1995). *The Foraging Spectrum*. Washington, DC: Smithsonian Institution Press.

Kummer, H. (1995). *In Quest of the Sacred Baboon*. Princeton, NJ: Princeton University Press.

Lima, S., Valone, T. J., & Caraco, T. (1985). Foraging efficiency – predation risk tradeoff in the grey squirrel. *Animal Behaviour*, **33**, 155–65.

Lovejoy, O. (1981). The origin of man. *Science*, **211**, 341–50.

Mabulla, A. (2003). Archeological implications of Hadzabe forager land use in the Eyasi Basin, Tanzania. In *East African Archaeology: Foragers, Potters, Smiths and Traders*, ed. C. M. Kusimba & S. B. Kusimba, pp. 33–58. Philadelphia, PA: University of Pennsylvania Museum of Archaeology and Anthropology.

Markman, S., Pinshow, B., Wright, J., & Kotler, B. P. (2004). Food patch use by parent birds: to gather food for themselves or for their chicks? *Journal of Animal Ecology*, **73**, 747–55.

Marlowe, F. W. (2003). A critical period for provisioning by Hadza men: implications for pair bonding. *Evolution and Human Behavior*, **24**, 217–29.

   (2004). What explains Hadza food sharing? *Research in Economic Anthropology*, **23**, 67–86.

   (2005). Who tends Hadza children? In *Hunter-Gatherer Childhoods: Evolutionary, Developmental and Cultural Perspectives*, ed. B. S. Hewlett & M. E. Lamb, pp. 177–90. New Brunswick: Transaction Publishers.

   (2005). Hunter-gatherers and human evolution. *Evolutionary Anthropology* **14**, 54–67.

Milton, K. (2000). Quo Vadis? Tactics of food search and group movement in primates and other animals. In *On the Move: How and Why Animals Travel in Groups*, ed. S. Boinski & P. A. Garber, pp. 375–417. Chicago, IL: University of Chicago Press.

O'Connell, J. F. (1997). On plio/pleistocene archeological sites and central places. *Current Anthropology*, **38**, 86–8.

Ofek, H. (2001). *Second Nature: Economic Origins of Human Evolution*. Cambridge: Cambridge University Press.

Orians, G. H. & Pearson, N. E. (1979). On the theory of central place foraging. In *Analysis of Ecological Systems*, ed. D. J. Horn, R. D. Mitchell & G. R. Stairs, pp. 154–77. Columbus, OH: Ohio State University Press.

Potts, R. (1987). Reconstructions of early hominid socioecology: a critique of primate models. In *The Evolution of Human Behavior: Primate Models*, ed. W. G. Kinzey, pp. 28–47. Albany, NY: State University of New York Press.

   (1994). Variables versus models of early Pleistocene hominid land use. *Journal of Human Evolution*, **27**, 7–24.

Reichard, U. (1999). Sleeping sites, sleeping places, and presleep behavior of gibbons (*Hylobates lar*). *American Journal of Primatology*, **46**, 35–62.

Rose, L. & Marshall, F. (1996). Meat-eating, hominid sociality, and home bases revisited. *Current Anthropology*, **37**, 307–38.

Ross, C. (2001). Park or ride? Evolution of infant carrying in primates. *International Journal of Primatology*, **22**, 749–71.

Scarratt, A. M. & Godin, J. G. J. (1992). Foraging and antipredator decisions in the hermit-crab *Pagurus acadianus (benedict)*. *Journal of Experimental Marine Biology and Ecology*, **156**, 225–38.

Sept, J. (1992). Was there no place like home? New perspective on early hominid archeological sites from the mapping of chimpanzee nests. *Current Anthropology*, **33**, 187–207.

(1998). Shadows on a changing landscape: comparing nesting patterns of hominids and chimpanzees since their last common ancestor. *American Journal of Primatology*, **46**, 85–101.

Sugiyama, L. (2004). Illness, injury, and disability among Shiwiar forager-horticulturalists: implications of health-risk buffering for the evolution of human life history. *American Journal of Physical Anthropology*, **123**, 371–89.

Swedell, L. (2002). Ranging behavior, group size and behavioral flexibility in Ethiopian hamadryas baboons (*Papio hamadryas hamadryas*). *Folia Primatologica*, **73**, 95–103.

Terborgh, J. & Janson, C. H. (1986). The socioecology of primate groups. *Annual Review of Ecology and Systematics*, **17**, 111–35.

Ulene, A. (1995). *The Nutribase Guide to Carbohydrates, Calories and Fat in Your Food*. Garden City Park: Avery.

Vincent, A. S. (1985). Plant foods in savanna environments: a preliminary report of tubers eaten by the Hadza of northern Tanzania. *World Archaeology*, **17**, 131–47.

Washburn, S. L. & DeVore, I. (1961). Social behavior of baboons and early man. In *Social Life of Early Man*, ed. S. L. Washburn, pp. 91–105. Chicago, IL: Aldine.

Whiten, A. (1992). Commentary: Was there no place like home? A new perspective on early hominid archeological sites from the mapping of chimpanzee nests. *Current Anthropology*, **33**, 200–1.

Winterhalder, B. (1986). Diet choice, risk, and food sharing in a stochastic environment. *Journal of Anthropological Archaeology*, **5**, 369–92.

(1996). Marginal model of tolerated theft. *Ethology and Sociobiology*, **17**, 37–53.

Woodburn, J. (1964). *The Social Organization of the Hadza of North Tanganyika*. Unpublished Ph.D. thesis, Cambridge University.

(1968). Stability and flexibility in Hadza residential groupings. In *Man the Hunter*, ed. R. B. Lee & I. DeVore, pp. 103–10. Chicago, IL: Aldine.

Wrangham, R. W. (1992). Commentary: Was there no place like home? A new perspective on early hominid archeological sites from the mapping of chimpanzee nests. *Current Anthropology*, **33**, 201–2.

Part III

Analyzing nutritional ecology:

Picking up the pace: nutritional ecology as an essential research tool in primatology

Introduction

KATHARINE MILTON

The initial trajectory of primate field studies

The first successful systematic field study of wild primates was carried out by C. Ray Carpenter on Barro Colorado Island (BCI), Panama in the early 1930s. Carpenter, who was jointly sponsored by the National Research Council's Committee for Research in Problems of Sex and Robert M. Yerkes's Department of Psychobiology at Yale, spent some 9 months in total on BCI observing the behavior of wild howler monkeys, and produced a detailed monograph of unusually high quality (Carpenter, 1934). With no precedents to guide his research, he viewed his study as an opportunity "to collect data to answer hundreds of questions . . . on all possible characteristics of . . . a primate living in an undisturbed habitat" (Carpenter, 1965, p. 255). Though Carpenter collected considerable information on the howler diet, at that time there was no comparative framework within which to place it, and his background, training, and interests lay more with behavioral data that could perhaps assist in interpreting results of laboratory research on primates and illuminate aspects of human behavior.

Primate field studies then fell into abeyance for the next 20 or so years, but finally began to pick up speed again in the late 1950s. In 1965, the first collection of papers derived largely from primate field studies, Irven DeVore's *Primate Behavior*, was published. This collection, as its title implies, is filled with descriptions of primate behavior – patterns of social organization, dominance hierarchies, communication calls, mating behaviors, and the like. Diet gets hardly a nod. In fact, the final chapter of the collection, *The Implications of Primate Field Research* written by Washburn & Hamburg (1965) does not even mention the word *diet* until the Conclusion, and here it simply appears as part of a long list of topics warranting investigation.

The next published collection, *Primates: Studies in Adaptation and Variability*, stemmed from a symposium on primate social behavior with the

Feeding Ecology in Apes and Other Primates. Ecological, Physical and Behavioral Aspects, ed. G. Hohmann, M. M. Robbins, and C. Boesch. Published by Cambridge University Press.

emphasis on data from primate field studies. This collection, edited by P. Jay, was published in 1968. The topic of diet is little in evidence. It is rather amazing to read long papers on the behavior of particular primate species in the wild that discuss group size and composition, home range size, activity patterns and so on but hardly mention, much less discuss, diet. Hall's (1968) paper on the behavior and ecology of wild patas monkeys does devote almost four pages to dietary information – but this, in an 88-page paper, is not a great deal. One striking exception to the general non-ecological approach in the Jay volume, however, is provided by Gartlan & Brain (1968) in their paper comparing ecology and social variability in free-ranging vervet (*Cerco-pithecus aethiops*) and blue monkeys (*Cercopithecus mitis*). This paper is modern in its approach and repeatedly stresses that primate behavior needs to be integrated with ecology, not examined in isolation: "Social behavior can be seen to be a function of the interaction of the population with the environment . . . Social behavior is deeply rooted in the evolutionary history of the species; it does not exist in a vacuum, but is the means by which species are adapted to the efficient exploitation of particular niches in just the same way as the evolution of physical structures permits this." (Gartlan & Brain, 1968, p. 282, p. 290). Another pioneer in this area, was Hans Kummer, whose 1971 book *Primate Societies* examined "group techniques of ecological adaptation" for free-ranging hamadryas baboons.

After this point, as the number of primate field studies rapidly increases and training begins to broaden in scope, interest in ecology, and particularly dietary ecology, picks up dramatically By 1986, with the publication of another book titled *Primate Societies*, this one a collection of papers edited by Smuts *et al.*, there is no longer any doubt about the importance of ecological influences on primate social organization and social behavior and, in particular, the importance of diet (e.g., see papers in *Primate Societies* by Oates, 1986; Silk, 1986; Waser, 1986; Wrangham, 1986). No field primatologist today would likely try and examine a behavioral question without obtaining a solid understanding of the diet and food distribution patterns of the species under study.

Since its inception, the ultimate goal of field primatology has repeatedly been stated to be to understand the array of factors giving rise to the wide diversity of primate social systems. However, getting off initially on a heavily behaviorally based research footing, one that largely ignored the role of ecological factors in shaping primate social organization and social behavior, seems to have impeded rather than assisted attainment of this stated goal. Indeed, even today it would seem that full appreciation of the overwhelming influence of diet on almost all aspects of any primate's behavior, morphology, and physiology has yet to be fully realized. We were all taught that behavior is the cutting edge of evolution. But it still seems difficult for some to appreciate that the key force driving behavioral change generally relates to dietary pressures. In fact, as is probably now realized by almost everyone, either overtly or at some unconscious level, social and organizational features largely derive, either directly or indirectly, from the food choices and dietary energetics of the primate species under consideration, and for this reason can only be properly understood in integration with them. Two examples may help to clarify this point.

Ateles

According to Jerison's data (1973), the encephalization quotient (EQ – an estimate of cortical complexity) of spider monkeys (*Ateles* spp.) is unusually high. (I am aware that there are more recent sets of calculations that estimate relative brain size, comparative neocortex size, and so on for primates, but Jerison's EQ serves the purpose.) Regardless of the method employed, all seem in general agreement that among monkeys, spider monkeys are notably "brainy." A popular explanation for their considerable cerebral development might be that their large brain or neocortex relates to large group size and concomitant complex social behavior (Byrne & Whiten, 1988; Dunbar, 1992; Barton, 1996). Spider monkeys, however, do not live in particularly large social groups. Average group size for spider monkeys is around 26.6 ± 11.1 individuals (n = 11 groups; see Appendix for references used in calculation). And, as noted by Symington, in their "group size and socionomic sex ratio, spider monkeys appear to lie somewhere near the middle of the range reported for multimale groups of cercopithecine primates" (Symington, 1988, p. 60). Furthermore, due to their unusually strong focus on ripe fruits in the diet, spider monkeys generally are found foraging in small sub-groups of three or four individuals or even alone (fission-fusion pattern) as only in this way, apparently, are they able to obtain sufficient ripe fruits each day without incurring unacceptable travel costs (Milton & May, 1976; Chapman, 1990) or intragroup food competition (McFarland, 1986; Wrangham, 1986; Symington, 1988).

In a paper published in 2000, I noted (as had McFarland, 1986; Wrangham, 1986; Chapman *et al.*, 1995; and others) that the dietary focus of spider monkeys appeared related to their fission-fusion foraging behavior – a foraging pattern suited to their extreme ripe fruit diet but one requiring each adult individual to possess its own data bank of information regarding the types, locations, and travel routes to a wide array of edible fruit sources (see also Milton, 1982, 1988). I hypothesized that because of this dietary focus and its associated foraging pattern, spider monkeys had also been placed under unusually strong selective pressures to develop (1) an enhanced communication matrix, such that they could exchange information with other members of their social unit at a distance as well as when in proximity, and (2) strong recognition, greeting and bonding behaviors to help reunite them when group members came together (Milton, 2000). Most primate species live in relatively cohesive social units that tend to forage with members in close auditory and/or visual contact with one another and here such behaviors likely are not under such strong selection. The dietary focus of the genus *Ateles* appears to relate to its large brain size on several different levels – selection for enhanced food location skills as well as selection for enhanced social communication skills and elaborated recognition and bonding behaviors. We cannot understand the factors related to the large brain of spider monkeys, nor the role that social behavior plays in selection for enhanced cerebral complexity, if we do not simultaneously consider the spider monkey diet, its energetic content, and its means of acquisition.

Chimpanzees

We can use the same model for common chimpanzees, *Pan troglodytes* (Milton, 2000). According to Jerison's (1973) encephalization quotient data (EQ), the EQ of the common chimpanzee is high, perhaps surpassing those of orangutans and gorillas. As maintaining brain tissue requires a continuous expenditure of energy, we can assume that enhanced cerebral complexity in chimpanzees is netting them more than the energetic price they pay for the brain. Chimpanzees live in social units termed *communities*. These vary considerably in size, both within and between study sites, and can range from around 20 individuals to more than 140 (see Appendix). Available data suggest that many communities fall toward the lower end of the range. Using data from 11 sites, mean community size for common chimpanzees has been estimated by Wrangham at 46.9 ± 26.2 individuals (range 19.7–94.0; R. W. Wrangham, pers. comm., 2005). Regardless of community size, it is the case that chimpanzees generally forage in small subgroups where each individual is often in the company of only three or four other individuals at best and, for adult females, often no other adults at all.

Chimpanzees, like spider monkeys, are extreme ripe fruit specialists and, as has often been stressed (e.g., Wrangham, 1986; Symington, 1988), show a fission-fusion pattern of social organization. It would seem that chimpanzees, like spider monkeys, have been placed under strong selective pressures relative to the development of (1) a long as well as short-range social communication matrix, and (2) an enhanced array of greeting and other social recognition and bonding behaviors. These social behaviors go in tandem with selective pressures related to the ability of each chimpanzee to locate a sufficiency of ripe fruits in the tropical forest environment each day. Enhanced cerebral complexity in chimpanzees therefore relates to this entire complex of essentially diet-based selective pressures. (See Milton, 2000 for progressive steps in this model and further examples.) One can study the social behavior of spider monkeys or chimpanzees, including their long-distance vocalizations, greeting, grooming, and other social recognition and bonding behaviors without referring to the dietary environment. But it is easier to understand why selection might favor particularly complex social behaviors in these two taxa *and* how they can afford both their brain and their sociality if we simultaneously address their diet and energetics.

Nutritional ecology

I take the term *nutritional ecology* to mean the study of how an animal deals with the nutritional, spatial, and temporal heterogeneity of the environment to acquire food. Such study could include examination of the digestive physiology of the species in question, including food transit rates and assimilation efficiencies as well as detoxification limitations (Dearing *et al.*, 2000). Ironically, the first study of primate nutritional ecology, like the first primate field study, appears to have been carried out on Barro Colorado Island, Panama. In November 1966, C. M. Hladik and Annette Hladik came to BCI from France to study the diets of four sympatric primates – howler monkeys, spider monkeys, capuchin monkeys, and red-naped tamarins. They collected slightly more than a year of field data on the dietary behavior of these species and also collected and analyzed samples of numerous wild foods. Analytical results were presented in a multiauthored 1971 paper in *Folia Primatologica* in French (Hladik *et al.*, 1971). The Hladiks and their colleagues carried out an impressive range of studies on the composition of wild plant foods of primates, looking not only at the foods of neotropical primates but also those of Gabon and Sri Lanka (A. Hladik, 1978). They examined not only macronutrient concentrations but also individual amino acids and cellulose, as well as some important minerals and alkaloids (Hladik *et al.*, 1971; A. Hladik, 1978; C. M. Hladik, 1978). These nutritional ecology studies clearly were well ahead of their time and for some years, few others followed their example.

Howler monkey nutritional ecology

In the mid-to-late 1970s, Barro Colorado Island was a popular crossroad for visiting tropical scientists from a wide range of disciplines. Students or post-docs such as myself, working on BCI during this period received broad exposure to a wealth of new ideas and were often able to form research collaborations. By then, I had already compiled data on the dietary behavior of BCI howlers (Milton, 1978, 1980). This material provided the foundation for expansion into the area of howler monkey nutritional ecology. As it might be useful for students to see a range of topics that can be addressed for a single primate species, here is a list of those we examined: (1) basal metabolism of howler monkeys (Milton *et al.*, 1979); (2) free-ranging metabolism of howler monkeys (Nagy & Milton, 1979a); (3) estimates of nutrient intake in howler monkeys (Nagy & Milton, 1979b); (4) quality and anti-quality components of howler foods (Milton, 1979); (5) efficiency of protein and fiber digestion in howlers (Milton *et al.*, 1980); (6) gut passage rates and fermentation returns of howlers eating natural items of diet (Milton, 1981a; Milton & McBee, 1983; Milton, 1998; Milton & Demment, unpubl. data); (7) nitrogen to protein conversion factors for wild plant parts (Milton & Dintzis, 1981); (8) pectin estimates for wild plant parts (Milton, 1991); and (9) vitamin C content of wild plant parts (Milton & Jenness, 1987). I also carried out work on attributes of the nutritional ecology of black-handed spider monkey (*Ateles geoffroyi*), a species sympatric with howler monkeys on BCI (Milton, 1981b). This work explored the question of whether internal features (gut morphology and/or physiology) might set limits to niche breadth for these similar-sized plant-eating monkeys, facilitating their coexistence. This leads me to address the question of just how plastic primate species are in terms of food choice in the natural environment (and indeed, in terms of their social organization and social behaviors).

How plastic are the diets (and social behaviors) of wild primates?

Dietary quality is a much used (and abused) term in the primate literature. The term relates (I believe) to the amount of digestible material relative to the amount of indigestible or unavailable material present in any food item – it is the animal's potential net from eating that item. Though quality in the most general sense relates to the nutritional value of the food to the animal, it can have a more specific sense when placed in the context of animal requirements and availability in the environment. So protein may be more important to a growing animal than to a mature one. Dietary quality therefore is not a constant but a moving target and differs from species to species and even from individual to individual. It depends on many variables – the metabolic body size of the consumer, its sex and age, its digestive physiology, the nutrient mix consumed at any given time and so on. Just like the sweepingly broad (and therefore largely useless) terms *omnivore, folivore* or *frugivore* (see Danish *et al.*, Chapter 18, this volume), the term *dietary quality* needs to be understood as something that must be determined for each primate species.

Analysis of the quality and anti-quality components of the wild plant foods a given primate selects, its net from eating them, *and* the factors underlying its pattern of food selection continue to be only superficially explored for most primate species. Lacking information on factors that may set limits to dietary breadth for a particular species and constrain its range of food choices in the natural environment, many primatologists may overestimate the plasticity of wild primates —viewing them as capable of altering their behavior, including their dietary behavior, to fit almost any environmental circumstance. There is no doubt that primates, with their large brain-to-body ratios and fairly generalized morphology, exhibit considerable plasticity on many different levels and many primate species seem equally at home in a variety of different habitats. However, all primates appear to show species-specific dietary patterns. If we look at primate species, particularly when sympatric – we see that they choose different foods, or eat the same foods but at different stages of maturation, or consume different proportions of a few or many of the same foods while all living together in the same forest and even using many of the same arboreal pathways and food trees. Yet if we bring these same sympatric species into captivity, we often find that many can do equally well on the same diet – often, for monogastric primates, a diet composed of manufactured primate chow and water, although each species or, within species, each sex or age class may consume different quantities of chow, depending on body size and other considerations. This fact suggests, perhaps, that many primate species in the wild have sufficient dietary flexibility to subsist on one another's diets.

But in the natural environment, primates do not have unlimited access to nutritionally fortified, highly digestible monkey chow. In the wild, each primate species has an array of different factors – environmental, behavioral (including social behaviors), morphological and physiological – that appear to constrain it to a particular dietary niche. As has been remarked, the most profitable way to approach an understanding of any animal species may be to view it as a type of "natural experiment," predicated on securing some portion of the always finite dietary resources available on the planet at any one time. The natural experiment we refer to as *Alouatta* or *Ateles* is *a food-acquisition design* worked out over eons of selection. Certainly some degree of dietary plasticity should be possible for any primate species. Some primates, such as savanna baboons (*Papio* spp.), are dietary generalists showing considerable dietary plasticity. But even though they may eat hundreds of different items, they still must take in items of high enough quality to sustain themselves. Just as spider monkeys, wherever studied behave socially and organizationally like spider monkeys (Milton, 1993), I would maintain they must eat like spider monkeys too. One can provide a set of tentative hypotheses as to why a spider monkey cannot eat like a howler monkey (food passage rates, dentition, etc.) or a capuchin monkey (different body size, dentition, lack of manual dexterity, etc.). The question of why a spider monkey cannot eat like a woolly monkey is more difficult to answer but a partial focus on different fruit chemistries (perhaps relating to different digestive or physiological traits) and the deliberate use of animal prey by woolly monkeys are suggested possibilities (Di Fiore, 2004; Dew, 2005). Overall, I continue to view the dietary (and social) plasticity attributed to wild primates as likely over-rated, or to say it another way, at times no more or perhaps even less plastic than the dietary behavior of a number of other non-primate species under similar circumstances. We need to remember that most animals require the same basic set of nutrients. Any readily available nutritious and digestible food source in the natural environment surely can be utilized by a wide array of opportunistic consumers if they can perceive it as potential food.

Analyzing nutritional ecology

The five contributions presented here provide a good cross-section of papers indicating the scope and promise of this area of research. Two of the papers (Chapters 15 and 16) discuss methodologies for estimating dietary intake and food composition for wild primates, two others (Chapters 17 and 18) are "case studies" which explore energy or protein questions for particular primate species, while the final paper (Chapter 19) examines data on the sensory modalities involved in primate food selection.

In Chapter 15, Ortmann *et al.* provide a critical review of methods for estimating the quality and composition of plant foods in wild primate diets, including several newer methods. This compendium will be useful for those contemplating research questions involving some type of dietary analysis and needing orientation in terms of potential methodologies. Of particular value are the authors' detailed comments on the collection and preservation of wild plant foods. The methodologies presented in Chapter 15 are from various sources but many derive directly from livestock studies and grass and herb analysis. As most primates are small- to moderate-sized arboreal animals, some of these analytical techniques, techniques derived largely for forages consumed by large-bodied terrestrial ruminants, may need re-thinking and modification before being applied to the foods and phy-siologies of monogastric (or polygastric) primates consuming tropical tree parts.

In the final portion of their paper, Ortmann *et al.* discuss some potentially promising new analytical methods for dietary analysis. The analysis of differ-ent plant parts and accurate determination of the fate of their chemical constituents in the digestive tract of any animal is a difficult process, fraught with potential errors. Many decades of thoughtful research have gone into the development, refinement and calibration of the analytical methods cur-rently in wide use today and it is likely that newer methods will require time and further study to become as well understood and reliable in terms of results produced.

Chapter 16, by Mayes, examines the applicability of novel marker methods for estimating dietary intake and nutritive value in primate field studies. As Mayes notes, "the quality of data relating to the nutrition of wild primates is generally rather poor." To assist in improving this situation, he discusses several new and relatively straightforward analytical techniques for estimat-ing the composition and digestibility of plant-based diets in the natural environment. Most attention is focused on the use of long-chain fatty acids of plant wax, particularly n-alkanes (also discussed in Chapter 15). Individual n-alkanes in plant wax differ widely from species to species and these different patterns can be used to determine diet or plant mixture composition. To perform this determination, however, one needs food as well as fecal samples. This method has now been demonstrated to work well for herbivores eating diets composed of only a few plant species. Reliability may decline, however, with increasing dietary heterogeneity, though Mayes offers several suggestions for addressing this problem. The cuticular wax of insects and other invertebrates likewise contains saturated branch-chain hydrocarbons which differ significantly between taxonomic groups and have the potential to serve as markers. Mayes also offers possible techniques for fecal identification of ingested honeys or clays.

The use of baits labelled with even-chain alkanes or other marker com-pounds is suggested as a possible technique for monitoring food intake and gut passage rates in free-ranging primates. It likely would not be difficult to get most fruit-eating primates to ingest an alkane-labeled bait as they often show instant affinity for bananas. This thoughtful paper provides a number of interesting new ideas and techniques for primatologists to consider and begin to test in field situations.

In Chapter 17, Conklin-Brittain *et al.* use findings from nutritional ecology to address the question of whether the energetic intake of wild orangutans shows higher variance than that of wild chimpanzees. In their examination, they use data from an array of sources – actual food items and dietary intakes for both species, analysis of the quality and anti-quality components of wild plant foods, fermentation efficiency data derived from captive chimpanzees and so on. For each ape species, energy calculations for each food item are estimated and then total energy intake per month for a 9-month period is obtained by summing the products of the grams of each food consumed for that month and the metabolizable energy content per gram per food. The authors are careful in their assumptions and even though they are working with a variety of different data sets, they appear to have done an excellent first approximation of estimating energy intake for the two apes. Their paper well illustrates how one can take various components and bring them to bear on a specific and important question. Others interested in similar diet-based questions would do well to carefully study the methodologies used in this paper. Conklin-Brittain *et al.* also provide a useful glossary to guide the reader through the various terms employed in their text.

To address a question raised by Conklin-Britten *et al.* in their text, in digestion trials with captive chimpanzees, M. Demment and K. Milton used wheat bran with a mean particle size of 726 μm as the fiber source (Milton & Demment, 1988). Wheat bran fiber derives from a grass and for this reason likely differs in many attributes relative to the fiber constituents of the woody tree and vine parts eaten by wild primates. This could affect estimates presented in Chapter 17. Judith Caton's research on gut passage rates in captive orangutans (Caton *et al.* 1999) may be of interest to readers wishing to know more about the digestive kinetics of apes

Danish *et al.* (Chapter 18) examine sugar concentrations in the diets of sympatric redtail monkeys and red colobus monkeys. Redtails are monogas-trics and red colobus are polygastrics. The authors find a surprisingly high sugar content in some foliage eaten by both species and note that, at times, there is considerable sugary fruit in the diet of red colobus. Ultimately the authors make several important and original points and raise some valuable questions. The analytical data they present suggest that primatologists may be misled by blindly accepting as fact, certain broad generalizations regarding dietary limitations for "colobines" as "folivores" or "frugivores." If, for example, a colobine has an unusually rich source of dietary nitrogen available to it in leaves or other food sources, at such times, it may be able to tolerate a higher sugar content in the diet from fruits.

As Danish *et al.* stress, until more work is done on the digestive physiology of particular primate species, factors facilitating congeneric sympatry in many colobine and cercopithecine species may remain unclear. In future studies, we need to bear in mind, that colobines are not necessarily analogous to small ungulates. Many small ungulates possess primitive rumens and are able to pass some food (e.g., sugary fruits) directly into the acid portion of the stomach, bypassing the fermentation chamber (Demment & Van Soest 1985). The question of the presence of an esophageal shunt in colobines remains unresolved (Milton 1998).

The final paper, Chapter 19, by Dominy *et al.* discusses cognitive processes used by primates to mediate fruit selection in the natural environment. This question of what cues primates use to select edible fruits is a little explored area in primate nutritional ecology and one deserving more attention. In their comparative examination, the authors draw on data from two sites, Pasoh Forest Reserve, Peninsular Malaysia and Kibale National Park, Uganda.

For readers interested in expanding their knowledge in this area, the elegant experimental studies of M. Laska and his associates (e.g., Laska & Seibt 2002a, 2002b) may be of utility. These studies reveal the considerable olfactory sensitivity of some primate species, both Old and New World, to specific aliphatic esters and alcohols involved in the fruit ripening process. These aromatic compounds may alert primates to the maturation state of wild fruit crops and as such could be useful cues mediating fruit selection.

Before, as Dominey *et al.* suggest, we get into study of the integration of different sensory modalities in the central nervous system of primates, perhaps more work is needed on the question of what, besides "texture" and fruit sugars, is actually serving to cue or signal wild primate species as to what is or what is not an appropriate item of diet. Hopefully, increasing interest in this important topic will lead to further experimental work in this area of nutritional ecology.

Appendix

1. Average size for spider monkey groups is around 26.6 ± 11.1 individuals per group (n = 11 groups). Sources of data on group sizes include: 15, Ahumada, 1989; 21, Nunes, 1995; 27 & 20, Klein, 1972; 37 & 40, Symington, 1988; 42, Chapman, 1990; 16 & 41, Ramos-Fernandez & Ayala-Orozco, 2003; 18, van Roosmalen, 1985; 16, Suarez, 2003).

2. Here I provide some estimates for community size for common chimpanzees. Community size for two communities of common chimpanzees at Gombe in the 1970s was reported to be 37 and 19, respectively (Wrangham, 1977). More recently, community size for three communities at Gombe was estimated at 25, 40–45 and 60–80,

respectively (Greengrass, 2000). The largest Gombe community crashed and may now number less than 20 individuals (Greengrass, 2000). A community of common chimpanzees at Bossou, Guinea was reported to contain 20 individuals (Shimada, 2000); a community at Kahuzi, Democratic Republic of Congo was estimated at 23 individuals (Basabose, 2005); Clark & Wrangham (1994) report a community of 27 chimpanzees at Kanywara; C. Stanford, pers. comm., 2005, reports a community of 25 at Bwindi; at Mahale in the late 1960s and early 1970s, K-community contained 28 individuals (mean size for years 1967–1973) and M-community 70 (Nishida, 1979); at Ngogo, 144 individuals has been reported for a single community by Watts & Mitani (2001). It is possible that in some instances, human-modified landscapes surrounding forests inhabited by chimpanzees or other factors related to human intervention may affect community size.

Acknowledgments

I thank Richard Wrangham, Colin Chapman, Craig Stanford, and Tony Di Fiori for providing information on group/community size in *Ateles* or *Pan* and Montague Demment for helpful discussion.

References

Ahumada, J. A. (1989). Behavior and social structure of free-ranging spider monkeys (*Ateles belzebuth*) in La Macarena. *Field Studies of New World Monkeys, La Macarena, Colombia*, **2**, 7–31.

Barton, R. A. (1996). Neocortex size and behavioural ecology in primates. *Proceedings of the Royal Society of London, B*, **263**, 173–7.

Basabose, A. K. (2005). Ranging patterns of chimpanzees in a montane forest of Kahuzi, Democratic Republic of Congo. *International Journal of Primatology*, **26**, 33–54.

Byrne, R. W. & Whiten, A. (1988). *Machiavellian Intelligence: Social Expertise and the Evolution of Intellect in Monkeys, Apes, and Humans*. Oxford: Oxford University Press.

Carpenter, C. R. (1934). A field study of the behavior and social relations of howling monkeys. *Comparative Psychology Monographs*, **10**, 1–168.

    (1965). The howlers of Barro Colorado Island. In *Primate Behavior: Field Studies of Monkeys and Apes*, ed. I. DeVore, pp. 250–91. New York, NY: Holt, Rinehart & Winston.

Caton, J. M., Humen, I. D., Hill, D. M., & Harper, P. (1999). Digesta retention in the gastro-intestinal tract of the orangutan (*Pongo pygmaeus*). *Primates*, **40**, 551–8.

Chapman, C. A. (1990). Ecological constraints on group size in three species of neotropical primates. *Folia Primatologica*, **55**, 1–9.

Chapman, C. A., Wrangham, R. W., & Chapman, L. J. (1995). Ecological constraints on group size: an analysis of spider monkey and chimpanzee subgroups. *Behavioral Ecology and Sociobiology*, **36**, 59–70.

Clark, A. P. & Wrangham, R. W. (1994). Chimpanzee arrival pant-hoots: do they signify food or status? *International Journal of Primatology*, **15**, 185–205.

Dearing, M. D., Mangione, A. M., & Karasov, W. H. (2000). Diet breadth of mammalian herbivores: nutrient versus detoxification constraints. *Oecologia*, **123**, 397–405.

Demment, M. W. & Van Soest, P. J. (1985). A nutritional explanation for body-size patterns of ruminant and non-ruminant herbivores. *The American Naturalist*, **125**, 641–72.

DeVore, I. (1965). *Primate Behavior: Field Studies of Monkeys and Apes*. New York, NY: Holt, Rinehart & Winston.

Dew, J. L. (2005). Foraging, food choice and food processing by sympatric ripe-fruit specialists: *Lagothrix lagothricha poeppigii* and *Ateles belzebuth belzebuth*. *International Journal of Primatology*, **26**, 1107–35.

Di Fiore, A. (2004). Diet and feeding ecology of woolly monkeys in a western Amazonian rain forest. *International Journal of Primatology*, **25**, 767–801.

Dunbar, R. I. M. (1992). Neocortex size as a constraint on group size in primates. *Journal of Human Evolution*, **20**, 469–93.

Gartlan, K. S. & Brain, K. C. (1968). Ecology and social variability. In *Cercopithecus aethiops* and *C. mitis*. In *Primate Ecology: Studies in Adaptation and Variability*, ed. P. C. Jay, pp. 293–312. New York, NY: Holt, Rinehart & Winston.

Greengrass, E. (2000). The sudden decline of a community of chimpanzees at Gombe National Park. *Pan Africa News*, 7.

Hall, K. R. L. (1968). Behaviour and ecology of the wild patas monkey, *Erythrocebus patas*, in Uganda. In *Primate Ecology: Studies in Adaptation and Variability*, ed. P. C. Jay, pp. 32–120. New York, NY: Holt, Rinehart & Winston.

Hladik, A. (1978). Phenology of leaf production in rain forests of Gabon: distribution and composition of food for folivores. In *The Ecology of Arboreal Folivores*, ed. G. G. Montgomery, pp. 51–72. Washington, DC: Smithsonian Institution Press.

Hladik, C. M. (1978). Adaptive strategies of primates in relation to leaf-eating. In *The Ecology of Arboreal Folivores*, ed. G. G. Montgomery, pp. 373–96. Washington, DC: Smithsonian Institution Press.

Hladik, C. M., Hladik, A., Bousset, J., Valdebouze, P., Veroben, G., & Delort-Laval, J. (1971). Le regime alimentaire des primates de l'ile de Barro Colorado (Panama). *Folia Primatologica*, **16**, 85–122.

Jay, P. C. (1968). *Primates: Studies in Adaptation and Variability*. New York, NY: Holt, Rinehart & Winston.

Jerison, H. J. (1973). *Evolution of the Brain and Intelligence*. New York, NY: Academic Press.

Klein, L. L. (1972). *The Ecology and Social Behavior of the Spider Monkey*, Ateles belzebuth. Unpublished Ph.D. Thesis, University of California-Berkeley.

Kummer, H. (1971). *Primate Societies*. Chicago, IL: Aldine.

Laska, M. & Seibt, A. (2002a). Olfactory sensitivity for aliphatic alcohols in squirrel monkeys and pigtail macaques. *Journal of Experimental Biology*, **205**, 1633–43.

Laska, M. & Seibt, A. (2002b). Olfactory sensitivity for aliphatic esters in squirrel monkeys and pigtail macaques. *Behavioural Brain Research*, **134**, 165–74.

McFarland, M. J. (1986). Ecological determinants of fission-fusion sociality in *Ateles* and *Pan*. In *Primate Ecology and Conservation*, ed. J. G. Else & P. C. Lee, pp. 181–90. Cambridge: Cambridge University Press.

Milton, K. (1978). Behavioral adaptations to leaf-eating by the mantled howler monkey. In *The Ecology of Arboreal Folivores*, ed. G. G. Montgomery, pp. 535–50. Washington, DC: Smithsonian Institution Press.

(1979). Factors influencing leaf choice by howler monkeys: a test of some hypotheses of food selection by generalist herbivores. *The American Naturalist*, **114**, 362–78.

(1980). *The Foraging Strategy of Howler Monkeys: A Study in Primate Economics*. New York, NY: Columbia University Press.

(1981a). Food choice and digestive strategies of two sympatric primate species. *The American Naturalist*, **117**, 476–95.

(1981b). Diversity of plant foods in tropical forests as a stimulus to mental development in primates. *American Anthropologist*, **83**, 534–48.

(1982). The role of resource seasonality in density regulation of a wild primate population. In *Ecology of a Tropical Forest*, ed. A. S. Leigh, A. S. Rand, & D. M. Windsor, pp. 273–89. Washington, DC: Smithsonian Institution Press.

(1988). Foraging behavior and the evolution of primate cognition. In *Machiavellian Intelligence: Social Expertise and the Evolution of Intellect in Monkeys, Apes and Humans*, ed. A. Whiten & R. Byrne, pp. 285–305. Oxford: Oxford University Press.

(1991). Pectin content of neotropical plant parts. *Biotropica*, **23**, 90–2.

(1993). Diet and social behavior of a free-ranging spider monkey population: the development of species-typical behaviors in the absence of adults. In *Juvenile Primates: Life History, Development and Behavior*, ed. M. Pereira & L. A. Fairbanks, pp. 173–81. Oxford: Oxford University Press.

(1998). Physiological ecology of howlers (*Alouatta*): energetic and digestive considerations and comparison with the Colobinae. *International Journal of Primatology*, **19**, 513–48.

(2000). Quo vadis? Tactics of food search and group movement in primates and other animals. In *On the Move: How and Why Animals Travel in Groups*, ed. S. Boinski & P. Garber, pp. 375–418. Chicago, IL: University of Chicago Press.

Milton, K. & Demment, M. (1988). Digestive and passage kinetics of chimpanzees fed high and low fiber diets and comparison with human data. *Journal of Nutrition*, **118**, 1–7.

Milton, K. & Dintzis, F. (1981). Nitrogen-to-protein conversion factors for tropical tree samples. *Biotropica*, **13**, 177–81.

Milton, K. & Jenness, R. (1987). Ascorbic acid content of neotropical plant parts available to wild monkeys and bats. *Experientia*, **43**, 339–42.

Milton, K. & May, M. (1976). Body weight, diet and home range in primates. *Nature*, **259**, 459–62.

Milton, K. & McBee, R. H. (1983). Structural carbohydrate digestion in a new world primate, *Alouatta palliata gray*. *Comparative Biochemistry and Physiology*, **74**, 29–31.

Milton, K., Casey, T. M., & Casey, K. K. (1979). The basal metabolism of mantled howler monkeys. *Journal of Mammalogy*, **60**, 373–6.

Milton, K., Van Soest, P. J., & Robertson, J. (1980). Digestive efficiencies of wild howler monkeys. *Physiological Zoology*, **53**, 402–9.

Nagy, K. & Milton, K. (1979a). Energy metabolism and food consumption by howler monkeys. *Ecology*, **60**, 475–80.

  (1979b). Aspects of dietary quality, nutrient assimilation and water balance in wild howler monkeys. *Oecologia*, **30**, 249–58.

Nishida, T. (1979). The social structure of chimpanzees of the Mahale Mountains. In *The Great Apes*, ed. D. A. Hamburg & E. R. McCown, pp. 73–122. Menlo Park: Benjamin/Cummings Publishing Co.

Nunes, A. (1995). Foraging and ranging patterns in white-bellied spider monkeys. *Folia Primatologica*, **65**, 85–99.

Oates, J. F. (1986). Food distribution and foraging behavior. In *Primate Societies*, ed. B. B. Smuts, D. L. Cheney, R. M. Seyfarth, R. W. Wrangham, & T. T. Struhsaker, pp. 197–209. Chicago, IL: University of Chicago Press.

Ramos-Fernandez, G. & Ayala-Orozco, B. (2003). Population size and habitat use of spider monkeys at Punta Laguna, Mexico. In *Primates in Fragments: Ecology and Conservation*, ed. L. K. March, pp. 191–209. New York, NY: Kluwer Academic/Plenum Publishers.

Shimada, M. K. (2000). A survey of the Nimba Mountains, West Africa from three routes: confirmed new habitat and ant-catching wand use of chimpanzees. *Pan Africa News*, 7.

Silk, J. B. (1986). Social behavior in an evolutionary perspective. In *Primate Societies*, ed. B. B. Smuts, D. L. Cheney, R. M. Seyfarth, R. W. Wrangham, & T. T. Struhsaker, pp. 318–29. Chicago, IL: University of Chicago Press.

Smuts, B. B., Cheney, D. L., Seyfarth, R. M., Wrangham, R. W., & Struhsaker, T. T. (1986). *Primate Societies*. Chicago, IL: University of Chicago Press.

Suarez, S. (2003). *Spatio-temporal Foraging Skills of White-bellied Spider Monkeys (Ateles belzebuth belzebuth) in the Yasuni National Park, Ecuador*. Unpublished Ph.D. thesis, State University of New York-Stony Brook.

Symington, M. (1988). Demography, ranging patterns and activity budgets of black spider monkeys (*Ateles paniscus chamek*) in the Manu National Park, Peru. *American Journal of Primatology*, **15**, 45–67.

Van Roosmalen, M. G. M. (1985). Habitat preferences, diet, feeding strategy and social organization of the black spider monkey (*Ateles paniscus paniscus* Linnaeus 1758) in Surinam. *Acta Amazonica*, **15**, 1–238.

Waser, P. M. (1986). Interactions among primate species. In *Primate Societies*, ed. B. B. Smuts, D. L. Cheney, R. M. Seyfarth, R. W. Wrangham, & T. T. Struhsaker, pp. 210–26. Chicago, IL: University of Chicago Press.

Washburn, S. L. & Hamburg, D. A. (1965). The implications of primate research. In *Primate Behavior*, ed. I DeVore, pp. 607–22. New York, NY: Holt, Rinehart & Winston.

Watts, D. P. & Mitani, J. C. (2001). Boundary patrols and intergroup encounters in wild chimpanzees. *Behaviour*, **138**, 299–327.

Wrangham, R. W. (1977). Feeding behaviour of chimpanzees in Gombe National Park, Tanzania. In *Primate Ecology*, ed. T. H. Clutton-Brock, pp. 504–38. London: Academic Press.

(1986). Evolution of social structure. In *Primate Societies*, ed. B. B. Smuts, D. L. Cheney, R. M. Seyfarth, R. W. Wrangham, & T. T. Struhsaker, pp. 282–96. Chicago, IL: University of Chicago Press.

15 Estimating the quality and composition of wild animal diets – a critical survey of methods

SYLVIA ORTMANN, BRENDA J. BRADLEY, CAROLINE
STOLTER, AND JÖRG U. GANZHORN

Introduction

Feeding is a fundamental interaction between an animal and its environment
and several aspects of wildlife ecology are related to nutrition, e.g., population

Feeding Ecology in Apes and Other Primates. Ecological, Physical and Behavioral Aspects, ed.
G. Hohmann, M. M. Robbins, and C. Boesch. Published by Cambridge University Press.
© Cambridge University Press 2006.

dynamics and regulation, mating systems, habitat use, and predator–prey interactions. Nutrients, and particularly energy may set constraints within which animals and populations must operate, and most animal species have to cope with regular, particularly seasonal changes in forage composition, nutrient supply, and the spatial distribution of forage items. A variety of methods and techniques are available for estimating forage quality in natural habitats and evaluating the amount and pattern of dietary intake and assimilation. Although these methods have primarily been developed to quantify the impact of herbivorous livestock on their environment and habitat, they may also be useful for wildlife nutrition studies. The aim of this chapter is to briefly introduce and discuss the application of key methods and to provide a guide to sample collection and treatment in order to optimize the collaboration between field researchers and nutrition laboratories. Since methods continue to be developed and as new, exciting techniques become available to answer old questions, we will discuss two methods in more detail: the "near infrared spectroscopy" (NIRS) for the analysis of diet quality and the "analysis of plant and animal DNA in fecal material" to assess diet composition. We do not intend to discuss the design and analysis of research studies, and instead recommend the respective literature (Cochran, 1977; Mead, 1991; Manly, 1992).

Sampling and storing

To determine seasonal changes in forage availability and forage quality, samples must be collected at regular intervals throughout the year from line transects or sampling plots in the research area. Sampling day as well as sampling time is of vital importance since nutrient composition of fruits and herbs may change with time of day or date and season. The concentration of carbohydrates, e.g., as the main product of photosynthesis, increases during the day and drops again during the night and due to plant metabolism (photosynthesis and respiration). It has also been demonstrated that in perennial ryegrass the content of sucrose increases 3-fold within 6–7 hours after sunrise (Cairns, 2003). In the field it is often not feasible to always collect samples at the same time after sunrise, nor to restrict sampling to a certain time window, e.g., in the morning. But whatever the chosen sampling regime, it is essential to record time of day and date of sampling.

Samples of diet items should consist of those parts of the herbs or fruits that are known or believed likely to be eaten. In addition to young leaves and the pith of terrestrial herbs, primates prefer the soft pulp of fruits rather than the solid outer skin or seeds. However, some fruits are not processed before eating but swallowed whole, and should therefore be collected and analyzed as a whole. After picking, samples should be transferred to the field camp as soon as possible, and if necessary, dissected immediately. Fresh weight and size of the fruit or herb as well as its state of maturity (ripeness of fruits) should be determined before dissection. Weighing and measuring the circumference and length of 10–12 representative fruits or weighing herbs and leaves provide average values for the particular plant species and should be performed each time this species is collected. Fruits and stems must be carefully opened, dissecting those parts known or believed to be eaten. Squeezing of fruits should be avoided to minimize the loss of juice, or in the case of very juicy fruits, a quantitative collection of juice should be attempted. Alternatively, very watery fruits may be cut and dried step by step without great loss of juice by first cutting them in half, letting the halves dry a little bit, then bisecting the halves, allowing them to dry a little, and then continuing to bisect and dry until the whole fruit is cut into small pieces. Plant or fruit parts intended for separate analysis should be stored and dried separately. Immediate drying or freezing stops plant metabolism and further breakdown of nutrients or alteration of cell fractions. For storing in liquid nitrogen, samples must be placed in Cryo-tubes and then stored in Cryo-containers filled with the nitrogen. Evaporative losses of liquid nitrogen can be minimized by opening the containers only once a day and by filling all tubes at once. Frozen samples may also be stored at about -20 °C in a refrigerator if available in the field: storage at this temperature merely slows down but does not completely stop metabolic processes.

In most cases it is neither possible nor necessary to freeze fruits and herbs for storage. Instead, the samples may be dried on site using sun-drying or a light-bulb-heated box, or by circulating warm air from an oven around them. Plants should be laid on or hung from racks and stirred and turned over several times a day. Drying temperature should be above ambient temperature and high enough to allow the samples to dry rapidly and completely within a few days, especially in a humid environment where the onset of molding and rotting may begin very quickly. At ambient temperature the living tissue of the drying plant still metabolizes cell solubles and thus cell wall content would be overestimated (Robbins, 1983) if drying took place under those conditions. However, too much heat may unintentionally cause evaporation of volatile components of food items, for example terpenes, or may induce chemical reactions. For instance, tannins may bind to plant proteins or other molecules, thus forming insoluble complexes that show up as fibers. Thermal degradation (heat damage) of nutrients, known as the "Maillard reaction" or "non-enzymatic browning reaction," occurs in the presence of amines or amino acids and water; and carbohydrates are especially susceptible to this degradation. Since water increases the rate of this reaction, plant samples must be dried very carefully to prevent the loss of carbohydrates. Nonetheless, drying at temperatures below 50 °C should be safe provided it is accomplished rapidly (Van Soest, 1994). The most convenient and safest way of preserving plant samples, however, is to lyophilize them in a freeze-drier. After drying, samples should be weighed again and stored in closed plastic bags or jars with a bag of silica gel or some other agent that attracts water vapor. Prior to nutrient analysis all samples need to be ground in a mill. Since all nutrient analyses need to be performed at least in duplicate, the minimum amount of dry matter required is about 10 g. The high water content of some fruits must be taken into account when estimations are made of the amount of forage needed to complete biochemical analysis. It may be necessary to collect large amounts of fresh material to get a sufficient amount of dry matter (for 10 g dry matter, up to 100 g of fresh material may be required).

Forage quality: analysis of nutrients

To determine the major chemical components of forage, various laboratory methods have been developed that involve drying, burning, and chemical procedures. Wet chemistry procedures based on sound chemical and bio-chemical principles are most widely used and accepted and have been applied to many different animal foods, thus permitting a direct comparison of results across studies. A conventional "proximate analysis" includes the following aspects (in % dry matter): dry matter content (wet weight minus moisture content), crude protein (measured as total nitrogen), ether extract (lipids and fats), ash (mineral content), and crude fiber (cellulose and some lignin). Based on these measurements, the "proximate analysis" then estimates: nitrogen-free extract (sugars, starch, and some of the hemicellulose and lignin) and digestible energy. However, the conventional "proximate analysis" has limitations and some of its aspects have fallen out of favor. In the 1960s a different method of fiber analysis was developed and gained widespread acceptance for forage and food analysis. This is the detergent fiber analysis, also referred to as "Van Soest fiber analysis" after the scientist who first developed it (Goering & Van Soest, 1970; Soest, 1994). By taking advantage of both approaches, most typical forage analyses include the dry matter, crude protein, ether extract, and ash procedures from the "proximate analysis" and the detergent fiber analysis after Van Soest.

Dry matter and ash

Dry matter (DM) is the percentage of forage that is not water, and the basis and prerequisite for comparing different forages. Dry matter and water content, respectively, vary tremendously between plants and even between plant parts, which becomes obvious when comparing, for example, a watermelon with a nut. Dry matter content is determined by weighing a representative sample before and after placing it in an oven at about 105 °C, usually for 24 hours or at least until weight remains constant. Ash is the overall inorganic matter content of dry matter and is determined by placing a sample of the dried forage in a muffle furnace for at least 2 hours at 550 °C. All organic material is burned off and only inorganic matter remains. The relative amount of ash is then calculated by weight difference. Organic compounds can be further classified as proteins, lipids, carbohydrates, vitamins, and other components.

Crude protein

Protein is an important nutrient and extremely important for describing the nutritional ecology of primates. The quality of protein depends on its amino acid composition and differs between and within plant and animal sources (Robbins, 1983; Martinez del Rio, 1994; Milton, 1999). Young leaves, for example, show a higher protein content than fruits, and protein content and amino acid profile change during maturation in leaves and ripening in fruits. Since animals either fail to produce some so-called essential amino acids per se or produce them quickly enough to meet their requirements, they must take up these amino acids or their precursors in their diet. This explains both why free-ranging primates may be very choosy about what they feed on and why they eat fruits and leaves from different sources each day or supplement a fruit-dominated diet with leaves and animal material (insects, prey).

Several well-accepted methods are available to determine the protein content of forage, all varying in procedure accuracy, costs, and duration: individual amino acid analysis (accurate, expensive, time consuming); total ninhydrin protein (accurate, quicker); the Biuret method (single amino acids and di-peptides are ignored); the Bradford method (only intact proteins); and the Lowry method (accurate only on animal products) (Lowry *et al.*, 1951; Aurand *et al.*, 1987; Jones *et al.*, 1989; Barbehenn, 1995). Both the standard Kjeldahl procedure (Association of Official Analytical Chemists, 1984;

Chang, 1994) and the Dumas combustion method accurately measure the total nitrogen content of a sample and are reproducible; however, a conversion factor is then needed to calculate crude protein.

For Kjeldahl analysis, samples are digested in a mix containing sulfuric acid and a catalyst. A concentrated sodium hydroxide solution is then added, followed by a steam distillation. The distillate is finally titrated with HCl. The amount of acid neutralized is an estimate of the amount of nitrogen in the sample. For Dumas combustion, gas chromatography is used, measuring the nitrogen gas liberated from a sample that is entirely combusted at extremely high temperatures (about 950 °C). An alternative method that may provide a quick and cheap alternative to these standard methods is the near infrared spectroscopy (NIRS), which will be discussed in detail later.

Protein conversion factor

Since protein from grain or animal tissue contains, on average, 16% nitrogen, a conversion factor of 6.25 (100% /16%) has been established and is commonly used to estimate protein content from the measured nitrogen content of the sample.

$$\text{Crude protein CP}(\%DM) = 6.25 \times N(\%DM)$$

However, this conversion factor, originally derived from animal protein studies and livestock feed has been heavily disputed in the literature (Jones, 1931; Milton & Dintzis, 1981; Herbst, 1986; Dintzis et al., 1988; Handley et al., 1989; Izhaki, 1993; Yeoh & Wee, 1994; Conklin-Brittain et al., 1999; Levey et al., 2000). Studies on tropical forage clearly indicate that crude protein is overestimated by using the 6.25 conversion factor and a conversion factor of 4.3 is recommended instead (Conklin-Brittain et al., 1999), at least for wild tropical forage, or a conversion factor of 5.64 for wild fruit (Levey et al., 2000). For humans, as the closest relatives to great apes, the FAO (Food and Agriculture Organization of the United Nations) recommends using food-specific conversion factors, if available (FAO, 2003). When the specific factor is not known, the general factor of 6.25 should be used. Nonetheless, converting the measured nitrogen into crude protein content is just a mathematical calculation and values from the literature using the traditional factor 6.25 can easily be recalculated by using another selected factor, e.g., the already mentioned 4.3 or 5.64.

Ether extract

Herbivorous diets are normally quite low in lipid because most plant sources (except seeds) contain such low amounts. Lipids are a group of substances that are generally soluble in ether, chloroform, and other organic solvents but are relatively insoluble in water. Lipids can be divided into storage compounds (chiefly triglycerides), leaf lipids (galactolipids and phospholipids), and miscellaneous (waxes, carotenoids, chlorophyll, essential oils, and other ether-soluble substances; Van Soest, 1994). The standard method for estimating lipid content of natural forage is to extract an oven-dried sample with dry ethyl ether, calculating the difference in weight before and after extraction. However, the validity of the lipid analysis of forage depends on many factors, including proper sampling and preservation of the sample, thorough drying, and choice of an appropriate solvent.

Carbohydrates

Carbohydrates are saccharid compounds and the main storage of photosynthetic energy in plants. In contrast to animal tissue, which is relatively poor in carbohydrates, they constitute roughly 50%– 80% of the dry matter of forages and cereals. Plant carbohydrates can be divided into nonstructural and structural (Figure. 15.1). Nonstructural carbohydrates include mono- and disaccharides such as glucose, fructose, and sucrose, and they store energy from photosynthetic reactions. Fruits are especially high in mono- and disaccharides, with glucose predominating in most wild fruits (Milton, 1999). Due to

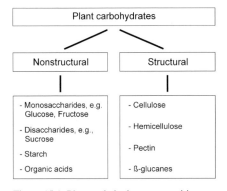

Figure 15.1. Plant carbohydrate composition.

their sugar content, fruits are sweet and palatable and provide an easy to access energy source. Non-structural carbohydrates also include polysaccharides such as starch, which is the primary energy storage compound in seeds and roots. Non-structural carbohydrates can be determined enzymatically with commercialized test assay kits.

The three primary structural carbohydrates are cellulose, hemicellulose, and pectin. Cellulose is the most abundant carbohydrate in the world and is similar to starch in that both are glucose polymers, although they differ in glucose linkage. Cellulose is resistant to any enzymes produced by vertebrates, and thus not digestible by wildlife whose digestive tracts lack microorganisms. Hemicellulose is a diverse and poorly understood group of different polymers made up of five- and six-carbon sugars, and its composition varies from one plant species to another. Hemicelluloses are hydrolyzed by both acid and alkaline solutions and are thus partially digestible by vertebrates. Pectin is a soluble fiber common in fruits and dicotyledonous plants but less common in grasses. It is readily digested by microorganisms but resistant to any enzyme produced by vertebrates. Cellulose and hemicellulose content can be determined by detergent fiber analysis (Van Soest, 1994).

Detergent fiber analysis

The general idea of the detergent fiber analysis, or Van Soest analysis, is to divide plant cell material into cell solubles and cell wall. Most cell solubles (sugars, starch, pectins, lipids, soluble carbohydrates, protein, nonprotein nitrogen, water-soluble vitamins and minerals) are highly digestible, although they also include plant secondary compounds such as tannins. The cell wall fraction consists of cellulose, hemicellulose, lignin, and silica and is generally not digestible by mammals without assistance from symbiotic microorganisms such as bacteria and protozoa (Van Soest, 1996). One of the big advantages of the detergent fiber analysis method is that fiber is divided into chemically meaningful classes, a categorization that does not occur as part of the traditional "proximate analysis."

The stepwise procedure for detergent fiber analysis is as follows (Figure. 15.2):

The weighed dry sample is first boiled in a neutral detergent solution (pH 7.0), filtered, dried to constant weight, and then weighed again. The difference in weight is due to neutral detergent solubles or cell contents, which dissolve in the solution. The residue that remains is termed Neutral Detergent Fiber (NDF) and includes structural carbohydrates, lignin, cutin-suberin, and ash.

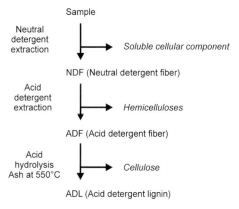

Figure 15.2. Detergent fiber analysis procedure (Van Soest, 1994).

In a second step, the NDF is boiled in an acid detergent solution that dissolves hemicellulose. Again, the amount of hemicellulose is determined by the weight difference of dried samples before and after treatment. This residue is called Acid Detergent Fiber (ADF) and consists of cellulose, lignin, cutin-suberin, and ash.

At this stage of analysis, either the cellulose or the lignin can be removed. The standard procedure is to remove the cellulose by treating the ADF fraction with 71% sulfuric acid followed by ashing the residue at 550 °C in order to estimate the amount of lignin, the only component now remaining in the ADL fraction. Another method is to treat the ADF with potassium permanganate to remove lignin and leave cellulose and cutin-suberin. This residue can then be treated with sulfuric acid to remove cellulose, then ashed to determine the amount of cutin-suberin. Again, the amount of the components is calculated based on the difference in weight of the samples prior to and after treatment.

Energy

The standard method of quantifying total or "gross energy" content of forage is bomb calorimetry, i.e., burning a sample of dry matter in pure oxygen atmosphere and measuring the heat produced in kJ/g DM. However, no animal is able to use 100% of gross forage energy and losses due to digestion, absorption, and utilization must be considered (Figure.15.3). Depending on digestibility of the forage, between 20% and 30% of the gross energy can be lost in fecal material, i.e., the portion of forage that is not digested, and

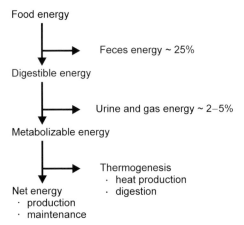

Figure 15.3. Energy losses during digestion, absorption, and utilization.

"digestible energy" is defined as gross energy minus energy in feces. "Metabolizable energy" equals gross energy minus feces energy minus energy losses via urine and combustible gases, the latter being particularly important in ruminants producing significant amounts of methane due to microbial fermentation in the rumen. "Net energy" is finally the energy remaining for maintenance and production (growth, lactation) after subtracting body heat losses from metabolizable energy. A second method for quantifying total or "gross energy" is estimating energy intake from the nutrient contents of the forage, a procedure discussed by Conklin-Brittain *et al.* in Chapter 17 of this volume.

Near infrared spectroscopy for the analyses of nitrogen and fiber fractions

Standard chemical analyses as described above are costly, wasteful in terms of chemicals and the amount of food material required, and also time consuming. Near infrared spectroscopy (NIRS) might provide an option that not only reduces costs but that can also process and analyze a large number of samples (Foley *et al.*, 1998; Stolter *et al.*, 2006). NIRS is based on the absorption of energy by C–H, N–H, and O–H bonds. The absorption (or reflection or transmission) of incidental light at different wavelengths is very characteristic for specific components but can also be used to characterize properties such as digestibility or degradation. The method is nondestructive and does not produce chemical waste. An experienced person can analyze about 100 samples per day. Since many properties (or concentrations of different chemicals) of the sample are represented in the spectrum, a single spectrum can replace several standard chemical analyses.

NIRS has been used successfully in agricultural studies to estimate diet quality, the composition of diets, or digestibility (Lyons & Stuth, 1992; Pearce *et al.*, 1993; Leite & Stuth, 1995; Walker *et al.*, 1998; Xiccato *et al.*, 1999, 2003; DeBoever *et al.*, 2003; Stolter *et al.*, 2005), and to assess properties of decomposition (Gillon *et al.*, 1999; Bouchard *et al.*, 2003). It is an accepted standard method in the Pharmacopöä Europaea (1997) and the USP (United States Pharmacopeia, 2002) for the determination of the quality of pharmaceutical products. It has been successfully used for the determination of soil and peat properties (McTiernan *et al.*, 1998; Chang *et al.*, 2001) and in paleolimnological and limnological studies for sediment and water analyses (Korsman *et al.*, 1992, 1999; Nilsson *et al.*, 1996; Malley *et al.*, 1999; Rosén *et al.*, 2000). Furthermore, it has been used to study food selection criteria and chemical habitat patchiness for marsupial folivores in Australian eucalyptus forests (Lawler *et al.*, 1998, 2000; McIlwee *et al.*, 2001). Application of NIRS might be particularly relevant for assessing spatial and temporal variation in the distribution of plant chemicals (e.g., Ganzhorn & Wright, 1994 ; Ganzhorn, 1995, 2002; Chapman *et al.*, 2003; Stolter *et al.*, 2005). Foley *et al.* (1998) provide an excellent and more extensive review of the method and its applications.

Sampling and sample preparation

Samples are collected and stored as for any other components to be analysed using the wet chemical process, i.e., first ground to a homogenous powder. Since NIR spectra are based on ground raw material, possible sources of error are the difference in grain sizes within and between samples, the water content of the sample, air humidity, and room temperature. To ensure that each sample has the same moisture content all samples can be dried at a given temperature overnight and kept in an exsiccator before scanning. Depending on the components to be analyzed, the appropriate drying temperature is one that will not destroy easily degradable components (a principle that applies to all chemical analyses). For the development of an NIRS model the spectra of the samples are scanned in duplicate. Scans include wavelengths from about 800–2500 nm.

Linking NIR spectra to sample characteristics

Specific plant chemicals (such as nitrogen concentrations) or general properties (such as digestibility) of samples measured conventionally are linked to the spectrum of the sample with commercially available software. The software then searches for properties of the spectra that correspond to the concentration or property of the sample. This model is developed, for example, with partial least squares regression (in our lab, Quant 2-method used in combination with an FT-NIR Spectroscope Vector 22/N, Software Opus NT Version 2.02, Bruker GmbH, Germany). This step is called the calibration procedure.

The models are then verified using cross-validation or test-set-validations. For cross validation one sample (= validation sample x) is removed at a time from the complete data set while the remaining samples are used for the calibration (calibration set = data set − x). This procedure is repeated for the whole data set until each sample had been removed once. For test set validation, the complete data set is divided into two sets. The first is used to develop the model (calibration), the second is used as the validation set (validation; Foley *et al.*, 1998). The "variance explained" by these models is listed as the coefficient of determination (r^2). Ideally, the samples used for the calibration should include the whole range of concentrations occurring in the samples. Since these models are optimized in an iterative way, it is possible that a model with high r^2 values can be created eventually even though the predictive power of the model is low. This might be due to "over-fitting" the NIRS model to the calibration/validation data sets. Optimization of the model for these data sets might have made the model too specific at the cost of a more general applicability. While predictive accuracy is very high as long as we stay with the sample set used for developing the model, accuracy becomes very low when this model is applied to another dataset.

In order to consider these potential problems, it has turned out to be useful to evaluate the optimized models further with yet another set of samples (Stolter *et al.*, 2006) (Figure. 15.4). The failure so far to produce models of general applicability for specific purposes does not rule out the possibility of eventually doing so. But in some cases the time and labor requirements exceed the time required to simply run standard chemical analyses.

Diet composition

Several direct and indirect, invasive and noninvasive techniques are available for studying foraging behavior and diet composition of free-ranging animals. Direct observation of feeding behavior requires well-habituated animals

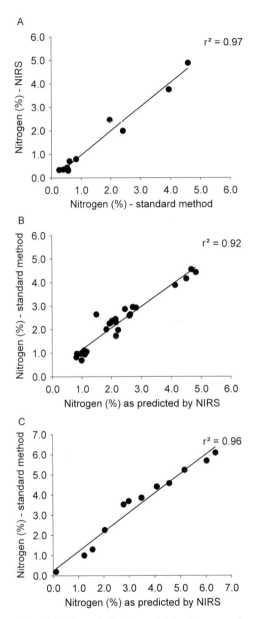

Figure 15.4. Example for the model development and successful application of NIRS. (A) Model developed for the nitrogen concentration in leaves of the bamboo (*Cathariostachys madagascariensis*). (B) Application of the model to a set of independent samples of leaves of *C. madagascariensis*. (C) Application of the model to four different species of bamboo (samples from Ranomafana, Madagascar; collected by C. Tan).

tolerating observers nearby in order to obtain qualitative and quantitative estimates of the selection of plant species and food items, food processing and consumption (Leighton, 1993). However, most studies of primate feeding ecology have to rely on indirect methods simply because the animals forage in tree crowns or thicket, and thus are not easily visible or do not tolerate approaches by observers. Indirect methods include tracing the foraging primate and investigating selected plants and leftovers, or collecting and analyzing feces or hair from nest sites and trails.

Fecal analysis provides broad information about consumed dietary items and can be conducted by visual inspection and biochemical and molecular genetic techniques. Visual inspection of feces, including microscopical techniques, is a cheap approach that allows seeds and matrix, microscopic fragments of plants, insects, spiders, and the bones of birds and mammals to be identified and quantified. Correct identification and classification of the residues using visual inspection require enormous knowledge and experience as well as physical access to a reference collection. Although regional guides are available for some features (e.g., animal hairs and bones), most do not apply to the region where primates still occur today in the wild.

n-alkane-analysis

The analysis of plant wax compounds provides another method for investigating the qualitative and quantitative contribution of certain plants to the primate diet (Dove & Mayes, 1991; Laredo *et al.*, 1991; Mayes *et al.*, 1995; Mayes & Dove, 2000). This technique will be discussed in detail in Chapter 16, but in brief, it makes use of the species-specific composition of plant cuticular wax, e.g., n-alkanes (long chain saturated hydrocarbons). Since plant species differ substantially in their n-alkane profiles, their individual contribution to the profile of a mixture, e.g., a feces sample, can be calculated by statistical methods (least squares optimization algorithm; Dove, 1992; Newman *et al.*, 1995). However, this statistical analysis is limited to the number of potential food plants investigated at any one time (Bugalho *et al.*, 2002). While the n-alkane technique is a powerful tool for investigating diet composition and diet selection in free-ranging herbivorous animals, those plant parts playing a dominant role in primate nutrition (fruit flesh and pith) unfortunately cannot be investigated easily because they lack a cuticular layer. Only fruits that are swallowed whole, or that at least contain fragments of cutin can contribute to the fecal n-alkane profile and be analyzed using this method.

Stable isotope analysis

Another indirect and non-invasive technique used to reconstruct an individual's diet and position in the food web is the analysis of stable isotopes (Griffith, 1991; Gannes *et al.*, 1997). Food items differ in their isotopic signature due to specific metabolic pathways and fractionation effects, and the contribution of different food items to an individual's diet can be calculated from the isotopic signal of animal tissue. Whether animals feed on plant or animal tissue, on C_3 or C_4 plants, whether their diet is terrestrial or marine (DeNiro & Epstein, 1978a, 1978b, 1981; Hobson, 1987, 1995; Sukumar *et al.*, 1987; Smith *et al.*, 1996; Schoeninger *et al.*, 1997, 1998), whether or not they migrate (Fleming *et al.*, 1993; Koch *et al.*, 1995; Hobson *et al.*, 1997; Marra *et al.*, 1998), and what they fed on in prehistoric times (Chisholm *et al.*, 1982; Hilderbrand *et al.*, 1996; Macko *et al.*, 1999a) can be estimated from the analysis of $^{13}C/^{12}C$ ratios and $^{15}N/^{14}N$ ratios of living tissue, bone, and hair (Jones *et al.*, 1981; Nakamura *et al.*, 1982; Minagawa, 1992; Schoeninger *et al.*, 1998; Macko *et al.*, 1999a, 1999b). However, the interpretation of an isotopic profile requires numerous assumptions, careful investigation, and sound statistical techniques. Stable isotopes are analyzed with an IRMS (Isotope Ratio Mass Spectrometer). For most studies, collaboration with an experienced laboratory specializing in isotope analysis is highly recommended since mass spectrometers are very expensive and conducting proper analyses is a challenging task.

DNA-based assessment of diet

Analysis of plant and animal DNA in fecal material offers yet another novel approach to determining the components of wild animal diets. By targeting DNA segments that are highly variable and exhibit consistent sequence differences between taxa, one can identify consumed organisms in the feces simply by their DNA sequences. Plants can be specifically targeted by amplifying and sequencing variable segments of plant nuclear DNA or parts of the chloroplast genome, whereas prey animals can be identified by sequencing fragments of animal nuclear or mitochondrial DNA (Symondson, 2002). DNA sequences obtained from feces can then be compared to those in the publicly accessible GenBank database. This database allows searches for any matches between a submitted sequence and sequences in the database (i.e., by conducting a BLAST search, www.ncbi. nlm.nih.gov/BLAST). As long as the target sequence is sufficiently variable and the database has adequate taxonomic coverage, sample sequences can, in principle, be identified to the genus or species level.

Such analyses have been successfully used to examine fossilized feces (coprolites) preserved in arid caves in the southwestern United States (Hofreiter *et al.*, 2001). These fossilized dung samples have yielded insights into the diet of the extinct Shasta ground sloth (Poinar *et al.*, 1998) as well as information on how that diet changed with corresponding climate changes over the last Glacial Maximum (Hofreiter *et al.*, 2000). Similarly, analysis of 2000-year-old human paleofecal samples from a cave in southwestern Texas revealed that the archaic human inhabitants of the cave ingested a relatively diverse diet of plants and animals (Poinar *et al.*, 2001).

Despite the great potential of this method, broad application in studies of extant animals remains unfulfilled. A few studies (Symondson, 2002) have genetically identified animal prey in the diets of predators (e.g., krill species in diets of pygmy blue whales and Adelie penguins, Jarman *et al.*, 2002). However, to our knowledge genetic analyses have yet to be applied to dietary studies of extant herbivores.

Accuracy, precision, and feasibility

The utility of such a DNA-based approach to dietary analysis clearly depends on the accuracy and precision of the genetic identification as well as the feasibility of obtaining adequate samples and conducting the laboratory analyses within a limited budget.

Confidence in the accuracy and precision of DNA-based dietary analysis requires two types of verification. First, one must verify that obtained DNA sequences are correct and that they accurately represent ingested foods. This can be accomplished by employing adequate precautions against contamination and incorporating numerous negative controls at each stage of the protocol to assess contamination risks. Second, there is a need to confirm that the DNA sequences obtained from fecal samples will be assigned to the proper plant taxa and that the level of taxonomic assignment is informative. This depends largely on the genetic marker targeted and the quality and representation of reference plant sequences in the database.

The accuracy and precision of the taxonomic assignment of the small chloroplast DNA segments (157 bp segments of the ribulose-bisphosphate carboxylase, or *rbcL*, gene) that were targeted in coprolite and ground sloth studies was assessed by applying the method to sequences from 99 herbarium specimens of known taxonomy (Hofreiter *et al.*, 2000). False identifications were rare; only two of 99 sequences were incorrectly identified to the level of family or order. However, all assignments were limited to the family level and approximately 30% of the sequences could not be taxonomically assigned because of a lack of reference sequences in the database. Thus, dietary item identification using this target DNA fragment (of *rbcL*) may be accurate, but can be somewhat imprecise.

Precision may improve with the increasing number of reference sequences that are continually being added to the database, which currently includes over 8000 *rbcL* sequences as compared to some 4000 sequences 4 years ago when Hofreiter *et al.* (2000) conducted their study. Knowledge of local flora or any prior knowledge of diet could potentially make further classification to the genus or species level possible. Similarly, other regions of the plant genome, such as the internal transcribed spacers (*ITS*) between the ribosomal nuclear genes would also be feasible for such analysis and could potentially allow identification to the species level. These multicopy markers are abundant in plant nuclear genomes, and ~200 bp target segments of these markers usually differ between plant species (Baldwin & Markos, 1998). However, the reference collection of sequences for these regions is currently small compared with that of the chloroplast DNA segments targeted in the coprolite studies.

Along with accuracy and precision, the feasibility of a DNA-based dietary study, especially for field-based ecological research where funding is often limited, must also be considered. Obtaining the necessary fecal samples for such dietary studies of wild animals may be relatively simple since many ecological or behavioral field researchers already collect fecal samples for other types of studies. In addition to conducting dietary analyses such as macro-analyses of plant fragments or analyses of plant wax compounds (see above), researchers also collect feces to examine parasite loads on a micro-scopic scale (e.g., Lilly *et al.*, 2002) and viral infections on a molecular scale (e.g., Santiago *et al.*, 2003). In recent years, feces have become the preferred sample material for genetic studies of phylogeography, genetic diversity, and paternity in wild animal populations as they usually yield sufficient quantities of DNA for analysis and can be easily collected without disturbing or even observing the animals (Kohn & Wayne, 1997). Adding a dietary research component to such projects would require only a small amount of fecal material (<0.10 g) or DNA extract (typically <10 μl) from each animal, thereby allowing DNA-based dietary assessments to be easily carried out in tandem with other studies.

This approach is perhaps most valuable for studies whose objective is to gain an overall view of dietary diversity in rare or elusive animals. This would require relatively few samples (5–10), which could be collected over the course of a few days, or over the course of several seasons to evaluate seasonal effects on dietary diversity. Ideally, samples should be collected from different individuals since diets may differ between sex or age categories (e.g., Rose, 1994).

We conservatively estimate the cost at approximately $600 for 10 samples and multiple controls (assuming a target sequence that comprises <500 bp). This price includes all consumables but excludes equipment such as PCR machines and DNA sequencers. Once procedures are established, a full-time person could analyze approximately 10 samples in 2 weeks. Cost and time estimates would not increase linearly with more samples since a project analyzing numerous samples, potentially from different animals, would cost less per sample due to economy of scale.

Ultimately, non-invasive, indirect studies of animal diets can only benefit from a more varied, multidisciplinary approach. Combining several different means of examining dietary quality (e.g., NIRS) and composition (e.g., DNA, n-alkane, and stable isotope analysis) would provide a complementary assessment of the diets of wild animals.

Acknowledgments

We thank Gottfried Hohmann, Martha Robbins, and Christophe Boesch for their invitation to participate in the stimulating meeting on primate feeding ecology, and the Max Planck Institute for Evolutionary Anthropology for the superb hospitality. We are grateful to Heidrun Barleben for excellent technical assistance, and Hendrik Poinar and Linda Vigilant for helpful insights and discussion. We also thank the DFG and the Graduiertenförderung Hamburg for their support to establish the NIRS methodology, as well as the reviewers who provided excellent comments and suggestions on how to improve the manuscript.

References

A. O. A. C. (Association of Official Analytical Chemists) (1984). *Official Methods of Analysis of the Association of Official Analytical Chemists*, ed. S. S. Williams. Arlington: Association of Official Analytical Chemists.

Aurand, L. W., Woods, A. E., & Wells, M. R. (1987). *Food Composition and Analysis.* New York, NY: Van Nostrand Reinhold.

Baldwin, B. G. & Markos, S. (1998). Phylogenetic utility of the external transcribed spacer (ETS) of 18S–26S rDNA: congruence of ETS and ITS trees of Calyca-denia (*Compositae*). *Molecular Phylogenetics and Evolution*, **10**, 449–63.

Barbehenn, R. V. (1995). Measurement of protein in whole plant samples with ninhydrin. *Journal of the Science of Food and Agriculture*, **69**, 353–9.

Bouchard, V., Gillon, D., Joffre, R., & Lefeuvre, J. C. (2003). Actual litter decomposition rates in salt marshes measured using near-infrared reflectance spectroscopy. *Journal of Experimental Marine Biology*, **290**, 149–63.

Bugalho, M. N., Mayes, R. W., & Milne, J. A. (2002). The effects of feeding selectivity on the estimation of diet composition using the n-alkane technique. *Grass and Forage Science*, **57**, 224–31.

Cairns, A. J. (2003). Fructan biosynthesis in transgenic plants. *Journal of Experimental Botany*, **54**, 549–67.

Chang, C. W., Laird, D. A., Mausbach, M. J., & Hurburgh, Jr., C. R. (2001). Near-infrared reflectance spectroscopy – principal components regression analyses of soil properties. *Soil Science Society of America Journal*, **65**, 480–90.

Chang, S. K.C. (1994). Protein analysis. In *Introduction to the Chemical Analysis of Foods*, ed. S. S. Nielsen, pp. 207–19. Boston: Jones and Bartlett.

Chapman, C. A., Chapman, L. J., Rode, K. D., Hauck, E. M., & McDowell, L. R. (2003). Variation in nutritional value of primate foods: among trees, time periods and areas. *International Journal of Primatology*, **24**, 317–33.

Chisholm, B. S., Nelson, D. E., & Schwarcz, H. P. (1982). Stable-carbon isotope ratio as a measure of marine versus terrestrial protein in ancient diets. *Science*, **216**, 1131–2.

Cochran, W. G. (1977). *Sampling Techniques*. New York, NY: John Wiley & Sons.

Conklin-Brittain, N. L., Dierenfeld, E. S., Wrangham, R. W., Norconk, M., & Silver, S. C. (1999). Chemical protein analysis: a comparison of Kjeldahl crude protein and total ninhydrin protein from wild, tropical vegetation. *Journal of Chemical Ecology*, **25**, 2601–22.

DeBoever, J. L., Vanacker, J. M., & DeBrabander, D. L. (2003). Rumen degradation characteristics of nutrients in compounds feeds and the evaluation of tables, laboratory methods and NIRS as predictors. *Animal Feed Science and Technology*, **107**, 29–43.

DeNiro, M. J. & Epstein, S. (1978a). Carbon isotopic evidence for different feeding patterns in two Hyrax species occupying the same habitat. *Science*, **2101**, 906–8.

   (1978b). Influence of diet on the distribution of carbon isotopes in animals. *Geochimica et Cosmochimica Acta*, **42**, 495–506.

   (1981). Influence of diet on the distribution of nitrogen isotopes in animals. *Geochimica et Cosmochimica Acta*, **45**, 341–51.

Dintzis, F. R., Cavins, J. F., Graf, E., & Stahly, T. (1988). Nitrogen-to-protein conversion factors in animal feed and fecal samples. *Journal of Animal Science*, **66**, 5–11.

Dove, H. (1992). Using the n-alkanes of plant cuticular wax to estimate the species composition of herbage mixtures. *Australian Journal of Agricultural Research*, **43**, 1711–24.

Dove, H. & Mayes, R. W. (1991). The use of plant wax alkanes as marker substances in studies of the nutrition of herbivores: a review. *Australian Journal of Agricultural Research*, **42**, 913–52.

Fleming, T. H., Nunez, R. A., & Lobo-Sternberg, L. S. (1993). Seasonal changes in the diets of migrant and nonmigrant nectarivorous bats as revealed by carbon stable isotope analysis. *Oecologia*, **94**, 72–5.

Foley, W. J., McIlwee, A., Lawler, I. R., *et al.* (1998). Ecological applications of near infrared spectroscopy – a tool for rapid, cost-effective prediction of the composition of plant and animal tissues and aspects of animal performance. *Oecologia*, **116**, 293–305.

Food and Agriculture Organization of the United Nations (2003). *Food Energy – Methods of Analysis and Conversion Factors*. Report of a technical workshop, Rome, 3–6 December 2002. FAO Food and Nutrition Paper 77.

Gannes, L. Z., O'Brien, D. M., & Martinez del Rio, C. (1997). Stable isotopes in animal ecology: assumptions, caveats, and a call for more laboratory experiments. *Ecology*, **78**, 1271–6.

Ganzhorn, J. U. (1995). Low level forest disturbance effects on primary production, leaf chemistry, and lemur populations. *Ecology*, **76**, 2084–96.

   (2002). Distribution of a folivorous lemur in relation to seasonally varying food resources: integrating quantitative and qualitative aspects of food characteristics. *Oecologia*, **131**, 427–35.

Ganzhorn, J. U. & Wright, P. C. (1994). Temporal pattern in primate leaf eating: the possible role of leaf chemistry. *Folia Primatologica*, **63**, 203–8.

Gillon, D., Joffre, R., & Ibrahma, A. (1999). Can litter decomposability be predicted by near infrared reflectance spectroscopy? *Ecology*, **80**, 175–86.

Goering, H. K. & Van Soest, P. J. (1970). Forage fiber analysis. In *Agricultural Handbook Number 379*. Washington, DC: United States Department of Agriculture, Agricultural Research Service.

Griffith, H. (1991). Application of stable isotope technology in physiological ecology. *Functional Ecology*, **5**, 254–69.

Handley, L. L., Mehran, M., Moore, C. A., & Cooper, W. J. (1989). Nitrogen-to-protein conversion factors for two tropical C_4 grasses, *Brachiaria mutica* (Forsk) Stapf and *Pennisetum purpureum* Schumach. *Biotropica*, **21**, 88–90.

Herbst, L. H. (1986). The role of nitrogen from fruit pulp in the nutrition of the frugivorous bat *Carollia perspicillata*. *Biotropica*, **18**, 39–44.

Hilderbrand, G. V., Farley, S. D., Robbins, C. T., Hanley, T. A., Titus, K., & Servheen, C. (1996). Use of stable isotopes to determine diets of living and extinct bears. *Canadian Journal of Zoology*, **74**, 2080–8.

Hobson, K. A. (1987). Use of stable-carbon isotope analysis to estimate marine and terrestrial protein content in gull diets. *Canadian Journal of Zoology*, **65**, 1210–13.

   (1995). Reconstructing avian diets using stable-carbon and nitrogen isotope analysis of egg components: patterns of isotopic fractionation and turnover. *The Condor*, **97**, 752–62.

Hobson, K. A., Hughes, K. D., & Ewins, P. J. (1997). Using stable-isotope analysis to identify endogenous and exogenous sources of nutrition in eggs of migratory birds: application to great lakes contaminants research. *The Auk*, **114**, 467–78.

Hofreiter, M., Poinar, H. N., Spaulding, W. G., *et al.* (2000). A molecular analysis of ground sloth diet through the last glaciation. *Molecular Ecology*, **9**, 1975–84.

Hofreiter, M., Serre, D., Poinar, H. N., Kuch, M., & Paabo, S. (2001). Ancient DNA. *Nature Review Genetics*, **2**, 353–9.

Izhaki, I. (1993). Influence of nonprotein nitrogen on estimation of protein from total nitrogen in fleshy fruits. *Journal of Chemical Ecology*, **19**, 2605–15.

Jarman, S. N., Gales, N. J., Tierney, M., Gill, P. C., & Elliott, N. G. (2002). A DNA-based method for identification of krill species and its application to analysing the diet of marine vertebrate predators. *Molecular Ecology*, **11**, 2679–90.

Jones, C. G., Hare, J. D., & Compton, S. J. (1989). Measuring plant protein with the Bradford assay. 1. Evaluation and a standard method. *Journal of Chemical Ecology*, **15**, 979–92.

Jones, D. B. (1931). Factors for converting percentages of nitrogen in foods and feeds into percentages of protein. *USDA Circular Number 183*. Washington, DC: United States Department of Agriculture.

Jones, R. J., Ludlow, M. M., Throughton, J. H., & Blunt, C. G. (1981). Changes in the natural carbon isotope ratios of the hair from steers fed diets of C_4, C_3 and C_4 species in sequence. *Search*, **12**, 85–7.

Koch, P. L., Heisinger, J., Moss C. *et al.* (1995). Isotopic tracking of change in diet and habitat use of African elephants. *Science*, **267**, 1340–3.

Kohn, M. H. & Wayne, R. K. (1997). Facts from feces revisited. *Trends in Ecology and Evolution*, **12**, 223–7.

Korsman, T., Nilsson, M. B., Landgren, K., & Renberg, I. (1999). Spatial variability in the surface sediment composition characterised by near-infrared (NIR) reflectance spectroscopy. *Journal of Paleolimnology*, **21**, 61–71.

Korsman, T., Nilsson, M. B., Öhman, J., & Renberg, I. (1992). Near-infrared reflectance spectroscopy of sediments: a potential method to infer the past pH of lakes. *Environmental Science and Technology*, **26**, 2122–6.

Laredo, M. A., Simpson, G. D., Minson, J. D., & Orpin, C. G. (1991). The potential for using n-Alkanes in tropical forages as a marker for the determination of dry matter by grazing ruminants. *Journal of Agricultural Science*, **117**, 355–61.

Lawler, I. R., Foley, W. J., & Eschler, B. M. (2000). Foliar concentration of a single toxin creates habitat patchiness for a marsupial folivore. *Ecology*, **81**, 1327–38.

Lawler, I. R., Foley, W. J., Eschler, B. M., Pass, D. M., & Handasyde, K. (1998). Intraspecific variation in *Eucalyptura* secondary metabolites determines food intake by folivorous marsupials. *Oecologia*, **116**, 160–9.

Leighton, M. (1993). Modeling dietary selectivity by Bornean orangutans: evidence for integration of multiple criteria in fruit selection. *International Journal of Primatology*, **14**, 257–313.

Leite, E. R., & Stuth, J. W. (1995). Fecal NIRS equations to assess diet quality of free-ranging goats. *Small Ruminant Research*, **15**, 223–30.

Levey, D. J., Bissell, H. A., & O'Keefe, S. F. (2000). Conversion of nitrogen to protein and amino acids in wild fruits. *Journal of Chemical Ecology*, **26**, 1749–63.

Lilly, A. A., Mehlman, P. T., & Doran, D. (2002). Intestinal parasites in gorillas, chimpanzees, and humans at Mondika research site, Dzanga-Ndoki National Park, Central African Republic. *International Journal of Primatology*, **23**, 555–73.

Lowry, O. H., Rosebrough, N. J., Farr, A. R., & Randall, R. J. (1951). Protein measurement with the Folin phenol reagent. *Journal of Biological Chemistry*, **193**, 265–75.

Lyons, R. K. & Stuth, J. W. (1992). Fecal NIRS equations for predicting diet quality of free-ranging cattle. *Journal of Range Management*, **45**, 238–44.

Macko, S. A., Engel, M. H., Andrusevich, V. *et al.* (1999a). Documenting the diet in ancient human populations through stable isotope analysis of hair. *Philosophical Transactions of the Royal Society of London, B*, **354**, 65–76.

Macko, S. A., Lubec, G., Teschler-Nicola, M., Andrusevich, V., & Engel, M. H. (1999b). The Ice Man's diet as reflected by the stable nitrogen and carbon isotopic composition of his hair. *FASEB Journal*, **13**, 559–62.

Malley, D. F., Rönicke, H., Findlay, D. L., & Zippel, B. (1999). Feasibility of using near-infrared reflectance spectroscopy for the analysis of C, N, P and diatoms in lake sediments. *Journal of Paleolimnology*, **21**, 295–306.

Manly, B. F. J. (1992). *The Design and Analysis of Research Studies*. Cambridge: Cambridge University Press.

Marra, P. P., Hobson, K. A., & Holmes, R. T. (1998). Linking winter and summer events in a migratory bird by using stable-carbon isotopes. *Science*, **282**, 1884–6.

Martinez del Rio, C. (1994). Nutritional ecology of fruit-eating and flower-visiting birds and bats. In *The Digestive System in Mammals: Food, Form, and Function*, ed. D. J. Chivers & P. Langer, pp. 103–27. Cambridge: Cambridge University Press.

Mayes, R. W. & Dove, H. (2000). Measurement of dietary nutrient intake in free-ranging mammalian herbivores. *Nutrition Research Reviews*, **13**, 107–38.

Mayes, R. W., Dove, H., Chen, X. B., & Guada, J. A. (1995). Advances in the use of faecal and urinary markers for measuring diet composition, herbage intake and nutrient utilization in herbivores. In *Recent Developments in the Nutrition of Herbivores. Proceedings of the IVth International Symposium on the Nutrition of Herbivores*, ed. M. Journet, E. Grenet, M.-H. Farce, M. Theriez, & C. Demarquilly, pp. 381–406. Paris: INRA editions.

McIlwee, A. M., Lawler, I. R., Cork, S. J., & Foley, W. J. (2001). Coping with chemical complexity in mammal–plant interactions: near infrared spectroscopy as a predictor of Eucalyptus foliar nutrients and the feeding rates of folivorous marsupials. *Oecologia*, **128**, 539–48.

McTiernan, K. B., Garnett, M. H., Mauquoy, D., Ineson, P., & Couteaux, M. M. (1998). Use of near-infrared reflectance spectroscopy (NIRS) in paleoecological studies of peat. *The Holocene*, **8**, 729–40.

Mead, R. (1991). *The Design of Experiments*. Cambridge: Cambridge University Press.

Milton, K. (1999). Nutritional characteristics of wild primate foods: do the diets of our closest living relatives have lessons for us? *Nutrition*, **15**, 488–98.

Milton, K. & Dintzis, F. R. (1981). Nitrogen-to-protein conversion factors for tropical plant samples. *Biotropica*, **13**, 177–81.

Minagawa, M. (1992). Reconstruction of human diet from $\delta^{13}C$ and $\delta^{15}N$ in contemporary Japanese hair: a stochastic method for estimating multi-source contribution by double isotopic tracers. *Applied Geochemistry*, **7**, 145–58.

Nakamura, K., Schoeller, D. A., Winkler, F. J., & Schmidt, H.-L. (1982). Geographical variation in the carbon isotope composition of the diet and hair in contemporary man. *Biomedical Mass Spectrometry*, **9**, 390–4.

Newman, J. A., Thompson, W. A., Penning, P. D., & Mayes, R. W. (1995). Least-squares estimation of diet composition from n-alkanes in herbage and faeces using matrix mathematics. *Australian Journal of Agricultural Research*, **46**, 793–805.

Nilsson, M. B., Dabakk, E., Korsman, T., & Renberg, I. (1996). Quantifying relationships between near-infrared reflectance spectra of lake sediments and water chemistry. *Environmental Science and Technology*, **3**, 2586–90.

Pearce, R. A., Lyons, R. K., & Stuth, J. W. (1993). Influence of handling methods on faecal NIRS evaluations. *Journal of Range Management*, **46**, 272–6.

Pharmacopöa Europaea (1997). *Europäisches Arzneibuch Ph. Eur.* Stuttgart: Deutscher Apotheker Verlag.

Poinar, H. N., Hofreiter, M., Spaulding, W. G. *et al.* (1998). Molecular coproscopy: dung and diet of the extinct ground sloth. *Science*, **281**, 402–6.

Poinar, H. N., Kuch, M., Sobolik, K. D. *et al.* (2001). A molecular analysis of dietary diversity for three archaic Native Americans. *Proceedings of the National Academy of Sciences of the United States of America*, **98**, 4317–22.

Robbins, C. T. (1983). *Wildlife Feeding and Nutrition*. London: Academic Press.

Rose, L. M. (1994). Sex-differences in diet and foraging behavior in white-faced capuchins (*Cebus capucinus*). *International Journal of Primatology*, **15**, 95–114.

Rosén, P., Dabakk, E., Renberg, I., Nilsson, M., & Hall, R. (2000). Near-infrared spectrometry (NIRS): a new tool for inferring past climatic changes from lake sediments. *The Holocene*, **10**, 161–6.

Santiago, M. L., Bibollet-Ruche, F., Bailes, E. *et al.* (2003). Amplification of a complete simian immunodeficiency virus genome from fecal RNA of a wild chimpanzee. *Journal of Virology*, **77**, 2233–42.

Schoeninger, M. J., Iwaniec, U. T., & Glander, K. E. (1997). Stable isotope ratios indicate diet and habitat use in new world monkeys. *American Journal of Physical Anthropology*, **103**, 69–83.

Schoeninger, M. J., Iwaniec, U. T., & Nash, L. T. (1998). Ecological attributes recorded in stable isotope ratios of arboreal prosimian hair. *Oecologia*, **113**, 222–30.

Smith, R. J., Hobson, K. A., Koopman, H. N., & Lavigne, D. M. (1996). Distinguishing between populations of fresh- and salt-water harbour seals (*Phoca vitulina*) using stable-isotope ratios and fatty acid profiles. *Canadian Journal of Fisheries and Aquatic Sciences*, **53**, 272–9.

Stolter, C., Ball, J. B., Lieberei, R., Julkunen-Tiitto, R., & Ganzhorn, J. U. (2005). Winter browsing of moose (*Alces alces*) on two different forms of willows: food selection in relation to plant chemistry and plant response. *Canadian Journal of Zoology*, **83**, 807–19.

Stolter, C., Julkunen-Tiitto, R., & Ganzhorn, J. U. (2006). Application of near infrared reflectance spectroscopy (NIRS) to assess some properties of a sub-arctic ecosystem. *Basic and Applied Ecology*, **7**, 167–87.

Sukumar, R., Bhattacharya, S. K., & Krishnamurthy, R. V. (1987). Carbon isotopic evidence for different feeding patterns in an Asian elephant population. *Current Science*, **56**, 11–14.

Symondson, W. O. C. (2002). Molecular identification of prey in predator diets. *Molecular Ecology*, **11**, 627–41.

United States Pharmacopeia (2002). *USP25–NF20 Supplement, Official Monographs.* Rockville, MD: The United States Pharmacopeial Convention, Inc.

Van Soest, P. J. (1994). *Nutritional Ecology of the Ruminant.* Ithaca, NY: Cornell University Press.

  (1996). Allometry and ecology of feeding behavior and digestive capacity in herbivores: a review. *Zoo Biology*, **15**, 455–79.

Walker, J. W., Clark, D. H., & McCoy, S. D. (1998). Fecal NIRS for predicting percent leafy spurge in diets. *Journal of Range Management*, **51**, 450–5.

Xiccato, G., Trocino, A., Carazzolo, A. *et al.* (1999). Nutritive evaluation and ingredient prediction of compound feeds for rabbits by near-infrared reflectance spectroscopy (NIRS). *Animal Feed Science and Technology*, **77**, 201–12.

Xiccato, G., Trocino, A., DeBoever, J. L. *et al.* (2003). Prediction of chemical composition, nutritive value and ingredient composition of European compound feeds for rabbits by near infrared reflectance spectroscopy (NIRS). *Animal Feed Science and Technology*, **104**, 153–68.

Yeoh, H. H. & Wee, Y.-C. (1994). Leaf protein contents and nitrogen-to-protein conversion factors for 90 plant species. *Food Chemistry*, **49**, 245–50.

16 The possible application of novel marker methods for estimating dietary intake and nutritive value in primates

ROBERT W. MAYES

Introduction

Knowledge of the feeding behavior and nutritional status of animals living in the wild is central to gaining a better understanding of their interaction with their habitats and, for vulnerable species in particular, to developing strategies for protecting their future. Despite considerable observational data on feeding behavior, the extreme timidity and mobility of many animals means that such

Feeding Ecology in Apes and Other Primates. Ecological, Physical and Behavioral Aspects, ed. G. Hohmann, M.M. Robbins, and C. Boesch. Published by Cambridge University Press. © Cambridge University Press 2006.

data are often incomplete. The difficulty in making observations at night exacerbates the problem. Because wild animals have opportunities to choose what, where, and when to eat, and the availability of their food is often transient, feeding behavior can be very variable. Although foods consumed by herbivores are constrained to particular areas, the species composition, availability, and nutritional quality of vegetation may still show drastic variations throughout the year. Visual observations of feeding behavior, either directly or through film or video recording, may give indications of the range of food items in the diet, but are generally inadequate in quantitatively describing the nutritional status of the animals; additional techniques are required to assess nutritional parameters such as dietary intake, digestibility, and the proportions of different food items in the diet. Thus knowledge of intake and utilization of dietary nutrients in wild, free-ranging animals is generally very limited. Information obtained from captive animals can give some indication of their nutritional needs, but rarely reflects the diets consumed by animals in the wild.

Although as a taxonomic group primates encompass a large number of species occupying a wide range of food niches, they are overwhelmingly plant-eating (although other food items may play an important nutritional role) and it is likely that the functioning of their digestive tracts will generally be typical of nonruminant mammalian herbivores. For many primate species, the wealth of long-term studies on habituated social groups has meant that many highly detailed observational data have been obtained on their feeding ecology. Furthermore, some such studies can allow the collection of samples of food, food residues, feces and urine, which have the potential to provide nutritional information about the animals. Despite this, the quality of data relating to the nutrition of wild primates is generally rather poor.

The physical examination of primate feces for residues of different food items has served as a useful complement to direct observation in dietary studies with free-ranging animals. Most of the current techniques used in studying the feeding ecology of primates have provided detailed temporal information on feeding behavior and the range of food items in the diet. Much less information on the gravimetric proportions of different items in the diet, daily nutrient intake and diet digestibility has been gathered, largely because of technique limitations.

In contrast, largely for reasons concerning the profitability of commercial enterprises, the last 150 years has seen the accumulation of considerable knowledge about the nutrient requirements, intake, and feed utilization in domestic farm livestock. While most of this information has been derived from studies with housed animals, a considerable amount of research on the dietary intake and digestibility in grazing ruminant livestock has been carried out using appropriate measurement techniques. Recent developments in the use of fecal and urinary markers have led to improvements in the reliability of intake and digestibility measurements, and also allow quantitative estimates of the plant species composition of the diet to be made. Such techniques are also applicable to nonruminants. This paper summarizes these developments, which have been reviewed more extensively elsewhere (Dove & Mayes, 1991, 1996; Mayes *et al.*, 1995; Oliván *et al.*, 1999; Mayes & Dove, 2000) and explores the potential for them to be used for measuring diet composition, and possibly other parameters, in primates and other wild herbivores.

Markers used in farm livestock feeding studies

For over 130 years, indigestible substances appearing in feces have been used as markers for providing nutritional information on farm livestock, laboratory animals, and humans (Kotb & Luckey, 1972). Fecal markers have allowed convenient means of determining diet digestibility, nutritive value, and gut passage rate, and in grazing animals, dietary intake. Until recently, diet composition investigations with grazing farm animals have been largely confined to simple (two- or three-component) dietary mixtures; little use has been made of fecal markers to characterize complex diets of livestock grazing within heterogenous environments.

Certain metabolites excreted in the urine have been exploited as markers for a few specific purposes in nutritional studies in housed ruminant livestock (Mayes *et al.*, 1995). In particular, excretory products of purine metabolism have been used as urinary markers as indices of the ruminal synthesis of microbial protein, which is an important factor in protein utilization (Chen *et al.*, 1995). Because ingested toxins and other plant compounds produce characteristic urinary excretory products, there is potential for such metabolites to be used to help describe the plant species composition of the diet in free-ranging animals. The main limitation has been the ability to sample and measure the output of urine under field conditions.

Determination of intake in grazing farm ruminants

Methods developed to determine intake in individual grazing animals have generally used the equation:

$$Intake = Fecal\ output/(1 - Digestibility) \tag{16.1}$$

Both fecal output and whole-diet digestibility can be conveniently estimated using fecal markers.

Estimation of fecal output

Daily fecal output can be directly measured in grazing animals using attached bags, but this is labor-intensive and may severely disturb the animals, which could in turn affect intake. A more satisfactory method is to determine fecal output by the dilution of an external marker (absent from the diet), orally dosed once or twice daily using the equation:

$$\textit{Fecal output}(g/day) = \text{Dose of external marker}(mg/day)/ \atop (\textit{External} \text{ marker concentration in feces}(mg/g)) \quad (16.2)$$

Because of its inertness and high fecal recovery, chromium sesquioxide (Cr_2O_3) has been the most widely used marker. Other fecal output markers include titanium dioxide (TiO_2), barium sulfate ($BaSO_4$), and various rare-earth salts, such as ytterbium (Yb) chloride and metal complexes (e.g., Ru-phenanthroline). More recently, the even-chain alkane, hexatriacontane (C_{36}), dosed in a paper pellet or powdered cellulose (Dove & Mayes, 1991), has been used as a fecal output marker. As with most even-chain alkanes, it is virtually absent from the diet.

In order to reduce the disturbance to animals, by avoiding the need for repeated marker dosing, intraruminal controlled-release devices (CRDs) have been developed, which release either a Cr_2O_3 (Furnival *et al.*, 1991) or alkane marker (Dove *et al.*, 2002a) into the rumen at a constant predictable rate for about 3 weeks. An alternative method is to administer a single large dose of marker, and collect a series of fecal samples over the following 7 days (France *et al.*, 1988; Galyean, 1993).

Estimation of diet digestibility

Whether used to measure intake or indicate dietary nutritive value, digestibility estimates require a feed sample that is representative of the material actually eaten by the study animals. The concept of using a natural indigestible component of the diet as an internal marker to determine digestibility was suggested over 130 years ago (see Kotb & Luckey, 1972), using the equation:

$$\textit{Digestibility} = 1 - (\text{Internal marker concentration in diet}/ \atop \text{Internal marker concentration in feces}) \quad (16.3)$$

However, fiber fractions, such as indigestible acid detergent fiber (IADF) or lignin, can sometimes be unreliable as digestibility markers since the analytical methods employed may not measure the same chemical entity in the diet and feces. Various in vitro methods, such as incubation with rumen liquor (Tilley & Terry, 1963), cellulase or bacterial cultures from ruminant feces (e.g., Akter *et al.*, 1999) have thus been widely used to estimate herbage digestibility. Because in vitro digestibility methods may not give valid herbage intake estimates when animals are receiving cereal supplements, and give no indication of between-animal variability, plant-wax compounds were evaluated as digestibility markers.

The long-chain fatty acids of plant wax were first suggested as digestibility markers by Grace & Body (1981). More easily analyzed are the plant-wax hydrocarbons, which are predominantly odd-chain length mixtures of *n*-alkanes (C_{21}–C_{37}). Studies in mainly sheep and cattle (see Dove & Mayes, 1991, 1996) have shown that the fecal recoveries of dietary *n*-alkanes increase with carbon chain length. Although its high recovery (typically 95%) would suggest that C_{35} would make a good digestibility marker, in many plants its concentrations are too low for reliable measurements to be made. However, C_{35} has been effectively used (Dove *et al.*, 1990) where its dietary concentrations are reasonable (>15 mg/kg dry matter).

The use of n-alkanes as markers for estimating dietary intake

Because *n*-alkanes can be used for estimating both fecal output and diet digestibility, it is logical to use separate dosed and natural dietary alkanes concurrently to obtain estimates of intake (Mayes *et al.*, 1986). For dietary alkane, *i*, and dosed even-chain alkane, *j*, with respective fecal recoveries, R_i and R_j:

$$Daily\ Intake = \frac{Daily\ Dose_j}{\frac{Fecal\ content_j \times R_i}{Fecal\ content_i \times R_j} \times Dietary\ content_i - Dietary\ content_j}$$

$$(16.4)$$

This equation has essentially been derived from the equations shown earlier, for intake, fecal output, and digestibility, but with account taken of the levels of the dosed alkane naturally present in the diet. The main advantage of this procedure is that the ratio of the fecal marker concentrations is required, rather than their absolute concentrations. Since the concentrations of both alkanes can be obtained from a single analysis a number of potential sources of error are eliminated. There is an additional advantage in using dosed C_{32} and natural C_{33} alkanes since their fecal recoveries have been found, from a number of studies (Dove & Mayes, 1991), to be virtually the same and thus cancel out. The results of a number of validation studies have been reviewed by Dove & Mayes (1996). With the additional requirement to take dietary samples, the practical procedures for determining intake in grazing ruminants are similar to those described for making marker-based fecal output measurements. Alkane-containing CRDs (Dove *et al.*, 2002a) and procedures involving a single large dose of even-chain alkane, with a series of subsequent feces samples, have been developed for intake determination (Letso, 1995; Giráldez *et al.*, 2004).

When using the alkane method for intake measurement, it is important that the alkane concentrations of analyzed dietary samples are representative of the material ingested by the animals under study; alkane concentrations differ not only among species within a pasture, but also between different plant parts. It was originally recommended that dietary samples should be obtained from animals, surgically prepared with a fistula in the esophagus, but it was later found that, for high-quality (vegetative) grass pastures, samples obtained by hand-clipping were satisfactory (Mayes & Dove, 2000). For animals grazing heterogenous vegetation it becomes more difficult to obtain representative dietary samples. Individual animals can vary widely in the proportions of different plant species and/or plant parts that they select. A logical way of addressing this problem is to estimate diet composition in each animal, allowing the whole-diet alkane concentrations to be quantified, and hence intake to be determined.

Estimation of diet composition in ruminant livestock

Studies on the plant composition of the diet have mainly been with animals feeding in heterogenous vegetation environments, such as extensive hill and upland pastures, scrublands, and rangelands. Methods have been generally based on the visual identification of fragments of material originating from different plant species in feces, rumen contents, or material (extrusa) collected from esophageal-fistulated animals (Holechek *et al.*, 1982). Extrusa samples can provide a good indication of the botanical and plant-part composition of the material ingested over a short period (1–2 hours). Plant fragments in fecal and stomach (rumen) samples may represent a slightly longer ingestion period, but would have been subjected to degradation in the digestive tract; identification of fragments is thus difficult, and there is potential for bias in favor of the least digestible plant species or parts. Furthermore, methods based on visual identification are highly dependent on the skills of the operator. However, despite these limitations, much useful dietary information has been obtained, especially for giving qualitative descriptions (presence or absence of plant species) of the diet selected by free-ranging ruminants. Before the advent of plant-wax methods, there was little use of chemical fecal markers to estimate diet composition in livestock; one example was the estimation of tropical grass:legume dietary ratios by exploiting the different $^{13}C:^{12}C$ isotope ratios in plants using the C_4 photosynthetic pathway (tropical grasses) and C_3 plants (legumes) (Jones *et al.*, 1979).

Plant-wax compounds as markers for estimating diet composition

Long before plant-wax *n*-alkanes were used as digestibility and intake markers, it was known that the patterns of individual *n*-alkanes in plant wax differed widely across different plant species (see Tulloch, 1976). Such differences in alkane patterns can be exploited to determine diet or plant mixture composition by calculating the proportions of individual plant species that contribute to the alkane patterns found in herbivore feces. Diet composition can be calculated using simultaneous equations, in which the numbers of plant components and alkane markers used are the same (see Dove & Mayes, 1996). A more versatile approach that can accommodate more markers than components is to use a least-squares optimization procedure that seeks to obtain a composition estimate that gives the 'best fit' rather than a unique solution. While various least-squares algorithms have been used (Mayes *et al.*, 1994; Dove & Moore, 1995; Newman *et al.*, 1995; Hameleers & Mayes, 1998), all essentially carry out a similar process – searching for dietary component intakes or proportions that produce calculated fecal alkane values that best match (minimal discrepancy sum of squares) the actual (recovery-corrected) fecal values, as follows:
    Minimize:

$$\sum_{alkane_{1-n}} (Calculated\ fecal\ content\ of\ alkane_i - Actual\ fecal\ content\ of\ alkane_i)^2 \qquad (16.5)$$

Although the various algorithms give similar results, they differ in the calculation of the fecal alkane content. Procedures described by Dove& Moore (1995) and Newman *et al.* (1995) are perhaps the most straightforward. For a three-component diet, as an example:

$$Calculated\ fecal\ content\ of\ alkane_i = \alpha \times A_i + \beta \times B_i + \gamma \times C_i \qquad (16.6)$$

where the unknowns α, β, and γ are the respective intakes of dietary components A, B, and C, which, together, would result in one kilogram of feces.

Thus, dietary proportions x, y, and z can be calculated:

$$x = \alpha/(\alpha + \beta + \gamma); y = \beta/(\alpha + \beta + \gamma); z = \gamma(\alpha + \beta + \gamma). \qquad (16.7)$$

Digestibility can also be calculated:

$$Digestibility = (\alpha + \beta + \gamma - 1)/(\alpha + \beta + \gamma). \qquad (16.8)$$

A variant of this approach, when one or more of the component intakes is known, is to use the optimization routine to produce best fit values for the actual fecal output and intakes of the remaining dietary components (Dillon *et al.*, 2002):

$$\begin{aligned} Calculated\ &fecal\ content\ of\ alkane \\ &= (\alpha \times A_i + \beta \times B_i + \gamma \times C_i)/Fecal\ output \end{aligned} \qquad (16.9)$$

With appropriate fecal recovery corrections, the main advantage of this approach is that intake estimates can be obtained without the need to dose animals. This has recently been tested in sheep to determine roughage intake, by feeding known amounts of sunflower meal sprayed with beeswax, to provide a distinct alkane profile (Dove *et al.*, 2002b).

Statistical evaluations of the quality of diet composition estimates obtained from fecal marker patterns are very complex. However, some approximate judgments can be made about factors that may affect the reliability of diet composition values. The best estimates could be expected where the differences among marker patterns in individual diet components are most extreme, but with the total marker concentrations being similar. The reliability of composition estimates is likely to be case-specific since it will depend upon how marker patterns in individual food plants relate to each other. One means of pre-empting the likely reliability of diet composition estimates has been to use multivariate statistical procedures to compare marker patterns in different potential dietary components (Bugalho *et al.*, 2004). Most validation studies to date have compared actual diet compositions with marker-based estimates. These have shown that reliable estimates can be obtained using plant-wax *n*-alkane patterns in feces in simple mixtures of two or three dietary plant species (Dove & Mayes, 1996; Ali, 2003) or plant parts (Dove *et al.*, 1999), but the reliability will likely decline as the number of potential dietary components increases. There is thus a need to find ways of improving the reliability of the n-alkane diet composition method when used for complex diets. Mayes & Dove (2000) suggested three approaches to address this problem: to (1) combine this method with other diet composition techniques; (2) reduce the number of dietary components to be discriminated; and (3) increase the number of markers used in the estimation procedure. Salt *et al.* (1994) combined the alkane marker method with microhistological proced- ures to characterize the complex species composition of the diet of sheep having free access to *Deschampsia flexuosa* – dominant grassland and *Cal- luna vulgaris* – dominant moorland. One way of reducing the number of plant components to be discriminated is to form groups of individual species. In diet composition studies with red deer consuming mainly browse plants, Bugalho *et al.* (2002) grouped grass and forb species into a single 'herbage' component. This method should be reliable as long as differential selection of individual species within a group does not influence the overall diet composition estimate. This can be tested by simulation studies, examining the effects of within-group selection (Bugalho *et al.*, 2002). Another approach could be to use a qualitative technique, such as fecal plant fragment analysis, or (possibly in the future) plant DNA techniques to eliminate plant species that are absent from the diet, prior to using the alkane technique.

Other plant-wax compounds as diet composition markers

Of attempts made to improve the reliability of marker profile methods for use with complex diets, the main emphasis has been on increasing the number of plant markers used in the estimation. In most situations fewer than 10 alkanes are available as diet composition markers. Diet composition markers should ideally possess the following attributes: (1) methods exist or can be developed for their reliable, quantitative analysis; (2) their behavior in the gut is predictably understood (preferably complete recovery); and (3) they exhibit large between-species variation, but small within-species variation in their concentration patterns.

A number of types of plant-wax compounds, in addition to *n*-alkanes, have potential use as markers; most can be analyzed using similar procedures to those adopted for *n*-alkanes (gas chromatography). Mixtures of *n*-alkenes are often present in floral parts of plants and can be quantified at the same time as *n*-alkanes. In grasses these are mainly odd-chain monoenes with C_{29} predominating; for each carbon chain there are two alkenes with the double bond at different positions along the carbon chain (C_7-C_8 and C_9-C_{10}). The leaves of some deciduous trees (e.g., *Salix* and *Betula* spp.) contain both odd- and even-chain *n*-alkenes ($C_{22}-C_{30}$) (Oliván *et al.*, 1999). Although rela- tively low (30–50%), the fecal recoveries of alkenes tend to be similar over the range of chain lengths, and have been shown to be useful diet composition markers (H. Dove & M. Oliván, unpubl. data). Branched-chain alkanes tend to be rare in plant waxes. However, they have been found in the grass *Agrostis*

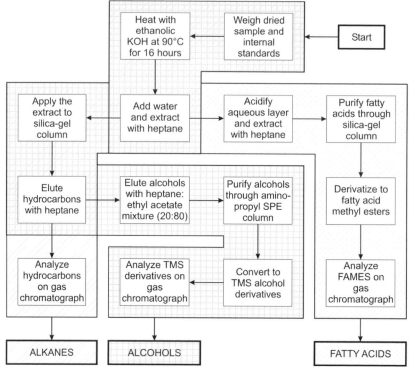

Figure 16.1. Schematic diagram of the method for analyzing alkanes, long-chain fatty alcohols, and fatty acids in plant material and in feces, described by Ali *et al.* (2004).

capillaris, an important food source for ruminants grazing upland areas of Western Europe (Oliván *et al.*, 1999).

The wax of most higher plants contains high proportions of long-chain fatty alcohols and acids of similar range in carbon chain length($C_{20}-C_{34}$), mainly as wax esters. Total (free plus esterified) alcohols and acids can conveniently be analyzed by gas chromatography, from separate fractions obtained as by-products of the analytical procedure for alkanes. Methods of analysis have recently been published (Ali *et al.*, 2004) and are summarized conceptually in Figure 16.1. The patterns of long-chain fatty alcohols differ greatly across different species (Oliván *et al.*, 1999; Ali, 2003). Furthermore, many plant species that contain low levels of *n*-alkanes tend to have reasonable levels of fatty alcohols, including some grasses, clover, and bracken (*Pteridium* spp.). Both fatty alcohols and fatty acids have been evaluated as diet composition markers in sheep. Ali *et al.* (2004) found that, like *n*-alkanes, fecal recoveries of both fatty alcohols and fatty acids increased with carbon chain length. In different sheep fed various three-component dietary mixtures of known composition, Ali (2003) observed that fatty alcohols and *n*-alkanes used together produced better estimates of diet composition than estimates obtained from using only *n*-alkanes. However, despite large between-plant differences in composition, fatty acids did not improve diet composition estimates when used as additional markers.

There remain many as yet uninvestigated fat-soluble compounds in the wax and other plant compartments that may have potential as fecal markers for estimating diet composition. Long-chain aliphatic compounds in plant wax, of similar chain length range to alkanes, alcohols, and fatty acids such as fatty aldehydes and β-diketones, may be good candidates, but there is first a need to develop appropriate methods for their routine quantitative analysis.

The use of fecal markers to measure the rate of passage of material along the digestive tract

The rate of passage of material along the gut is a useful parameter in understanding the factors affecting an animal's ability to extract nutrients from its diet. Passage rate is usually expressed as mean retention time or transit time, determined from the relationship between the time following a single oral dose of an indigestible marker, and its concentration or amount excreted in feces. Polyethylene glycol, CrEDTA, and CoEDTA are effective markers for determining the passage rate of liquid material. However, markers conventionally used to estimate the passage rate of particulate matter can be unreliable since it is difficult to be sure that they exactly follow the passage of particulate feed residues. This can be due to inappropriate densities or particle sizes (Cr_2O_3, TiO_2, Cr-mordanted fiber and plastic particles) or a tendency to move from one particle to another (e.g., rare earths and Ru-Phenanthroline). Passage rate determinations in ruminant livestock have generally been made indoors under controlled conditions, but measurements could be made in free-ranging animals.

Alkanes as passage rate markers

In ruminant gut, dietary *n*-alkanes are associated with the particulate digesta (Mayes *et al.*, 1997) and may be useful as passage rate markers. Because such alkanes are a normal component in the diet, an appropriate marker could be isotopically labeled hydrocarbons in intact plant material, given as a single feed or oral dose. Since ^{14}C and ^3H are equally effective as plant hydrocarbon labels for passage-rate determination (F. J. Giráldez & R. W. Mayes, unpublished data), ^{13}C and ^2H, could probably be used as non-radioactive alternatives. Another approach is to give a single dose (or feed) of plant material, such as cabbage, which has very high levels of natural *n*-alkanes (Hatt *et al.*, 1998). Synthetic even-chain alkanes, absorbed into paper pellets or sprayed onto plant material, have also been used as passage rate markers (Dove & Mayes, 1999). Labeling the wax of separate dietary plant components with different markers allows concurrent passage-rates of these components to be determined. Using grass fractions, separately labeled with ^{14}C and ^3H, and sprayed with different even-chain alkanes, Mayes *et al.* (1997) demonstrated that the leaf, stem, and flower parts had similar passage rates in grass-fed sheep.

Application of fecal marker techniques in non-ruminants and wild herbivores

Although most of the marker techniques described above are in theory equally applicable for use in domestic non-ruminants, it has only been recently that the importance of determining grazed herbage intake and other dietary parameters has been fully recognized. Fecal alkane recovery measurements and intake validation trials have shown the methods to be reliable in pigs (Wilson *et al.*, 1999) and horses (Dove & Mayes, 1996; O'Keefe & McMeniman, 1998). Unlike ruminants, fecal recovery in pigs and horses does not change with alkane chain length. However, from intake validation studies, Letso (1995) showed that in rabbits, alkane recoveries increased with carbon chain length.

Studies to evaluate *n*-alkanes as fecal markers have been carried out in a range of captive wild herbivores. Ruminant species investigated include fallow deer (R. W. Mayes, unpublished data), moose (Mayes *et al.*, 2001), and giraffe (Hatt *et al.*, 1998). Non-ruminant species studied include mountain hares, wombats (Woolnough, 1998), pigeons (Hatt *et al.*, 2001), and Galapagos tortoises (Hatt *et al.*, 2002). A number of intake and diet composition measurements have been made with free-ranging herbivores confined to paddocks, including red deer (Heydon *et al.*, 1993), wapiti (Gedir & Hudson, 2000), elephants (I. J. Gordon, unpublished data), and ostriches (Orr, 1998). These studies suggest that the methodologies using alkanes as fecal markers are universally applicable, but that there may be differences in fecal recovery and alkane dosing procedures between animal species.

Greater challenges exist if studies are to be carried out with wild, free-ranging herbivores. The main problem is to minimize the degree to which the animals are disturbed while not compromising the quality of the measurements. If representative samples of the dietary plant components can be collected (following direct observation of animals or indirectly from plant damage), the determination of diet composition and digestibility do not present a major problem since feces can be collected from the ground without directly disturbing the animals. For example, the diet composition of wild mountain hares in the Scottish Highlands has been measured by collecting feces after trapping the animals for a short period (Hulbert *et al.*, 2001; Rao *et al.*, 2003). Diet composition measurements have also been made in wombats, red deer (Bugalho *et al.*, 2002) and moose by collecting feces from the ground. Since the estimation of intake, fecal output, and gut passage rate require animals to be dosed or fed a known quantity of alkane, these measurements are more difficult to make in wild, free-ranging animals. Intake has been estimated in wild moose using CRDs containing C_{32} alkane (Mayes *et al.*, 2001). To insert the CRDs into the rumen, the animals were anesthetized by darting; at the same time the animals were fitted with collars carrying a Global Positioning System (GPS) device, which after release, allowed them to be located for sampling of feces and vegetation. Such use of CRDs is not possible with wild non-ruminants, but the single-dosing method (described above) could be employed for intake measurement.

The potential for using marker methods in primate feeding studies

Most studies of diet and feeding behavior in free-living primates have consisted of detailed observations of the number of feed items ingested and relative duration and frequency of their consumption over different time periods and differing study sites. By comparison, there is little information on the intake and digestibility of the whole diet and nutrients, such as energy, protein, and minerals. It is feasible that such nutritional information could be obtained through the use of appropriate marker techniques. Mostly in captive animals, a range of conventional markers, such as Cr_2O_3, plastic particles, Cr-mordanted fiber, and CoEDTA have been used in passage rate and digestibility studies in primates (Cabré-Vert & Feistner, 1995; Caton *et al.*, 1996; Campbell *et al.*, 1999; Edwards & Ullrey, 1999; Remis & Dierenfeld, 2004). Since primates are predominantly plant-eaters, it is probable that methods using plant-wax *n*-alkanes and long-chain fatty alcohols, could be applicable. From studies with a range of herbivore species, it can be expected that the behavior of fecal markers in the primate digestive tract will not be greatly different from other mammalian herbivores, and thus allow intake, diet composition, digestibility, and gut passage rate to be determined. However, there are likely to be practical problems to overcome for particular primate species in different scenarios, and these would need to be addressed. Fortunately, the presence of many captive species held under controlled conditions in zoos potentially allows some validation tests to be carried out.

Making quantitative measurements relating to diet and feeding in primates presents a number of challenges that would not normally be encountered in studies with wild and domestic ungulates. Some of these are listed below:

1. Primates as a taxonomic group can consume a very wide range of food items, including a number of nonplant materials.
2. Some primate species ingest a narrow range of food types such as predominantly leaves or fruits, while other primate species are omnivorous, eating plant, insect, vertebrate, and mineral materials.
3. Depending on species, primates can be very highly selective in the food items eaten and the way such items are harvested.
4. Because primates generally live in groups, measurement methods must allow either for individual animal measurements or representative group measurements to be made without influencing the animals' social behaviors or compromising the measurements obtained.
5. Since the ease of collecting feces samples is probably dependent on the type of understory vegetation cover, there could be possible bias in favor of easily visible feces.
6. It is possible that the occurrence of coprophagy in wild primates (sometimes observed in captive animals) may potentially affect the fecal recoveries of dietary markers. Such effects have not been investigated.

While these points may present additional obstacles to the use of dietary marker techniques in primates, there are other general characteristics of primates that may facilitate measurements, for example:

1. Studying habituated groups allows observational data, such as knowledge of actual plant species, plant parts, and other food items selected, to be much more accurate than equivalent data from wild ungulates.
2. The sampling of representative feed items and feces is likely to be straightforward when studying habituated groups.

3. There is potential to persuade individuals to consume a special feed or bait containing a marker, thus allowing intake, fecal output, or passage rate to be determined.

4. The arboreal habitat of many primates may allow measures to be taken so that not only feces and discarded food items can be collected, but also urine.

Addressing the problem of the wide range of food types consumed by primates

Plant materials

In free-ranging primates that eat only leaves it should be feasible to use plant-wax markers to estimate diet composition, digestibility, and (possibly) intake and passage rate, with constraints, such as the number of dietary plant species, similar to those encountered with ungulates. The contributions that flowers, plant buds, and green stems may make to the primate diet should also be measurable using plant-wax markers. However, estimating the dietary proportions of plant parts such as fruits and woody plant stems could be difficult since the waxes tend to be present only on the outer surfaces of plants; the effective concentrations of plant-wax markers would depend on the degree of consumption of fruit skins or stem bark, which could be very variable. Other plant items containing little or no plant wax include roots, edible seed parts, nectar, and plant gums. In order to quantify the contributions made to the diet by food items containing little or no plant wax (including non-plant components), other methods must be used. One obvious approach is to search for suitable markers associated with the individual food items. Such markers could be non-degradable components appearing in feces such as seeds from fruit (e.g., figs), other chemical entities recovered in feces, or if urine can be collected, metabolites arising from the ingestion of specific plant secondary compounds. Furthermore, conventional observational data may have an important role. The use of fecal DNA to identify plant species (Ortmann, *et al*, this volume, Chapter 15), and perhaps animal prey and other nonplant food items, may also help characterize the diet. Having established the composition of the complete diet, digestibility and intake could be estimated, using as markers the alkanes or other compounds from those dietary components that contain plant wax. Because of its widespread occurrence in higher plants, the plant sterol, β-sitosterol, and the products of its hydrogenation in the gut, may have potential as fecal markers to indicate the proportion of plant material in an omnivore's diet. It follows that the effective use of novel markers or other methods contributing to the estimation of diet composition would need to be evaluated in studies under controlled conditions at zoos or other appropriate institutions.

Insects and other invertebrates

Insects, both as adult and larval forms, may make a significant contribution to the diet of many primate species, especially the great apes (Tutin & Fernandez, 1992). Like the higher plants, insects and spiders have an outer waxy cuticle (Blomquist *et al.*, 1987). Insect cuticular wax contains complex mixtures of mainly saturated branched-chain hydrocarbons, which differ greatly between taxonomic groups. Preliminary studies with bats and insectivorous birds (R. W. Mayes & L. Stephen, unpublished data) have shown that the patterns of branched-chain alkanes in feces are similar to those of the insects in the diet. Thus there is potential for using insect hydrocarbons as markers for determining the insect species composition of the diet, which is likely to be an improvement upon the largely qualitative method of identification of hard body parts in feces. The contribution that insects as a group make to the omnivore's diet may also be measurable.

Honey, an occasional food item for some primates, contains appreciable concentrations of hydrocarbons (*n*-alkanes and alkenes) and wax esters with characteristic concentration patterns. These compounds originate from beeswax, which has been shown to be a useful marker for labeling supplementary concentrate feeds for sheep (Dove *et al.*, 2002b).

While the presence in feces of indigestible body parts from invertebrates other than insects and spiders could be used to provide dietary information, little is known about potential marker compounds in these species. Soft-bodied herbivorous invertebrates can contain *n*-alkanes and fatty alcohols typical of plant material, originating from the contents of their digestive tracts (L. Stephen & R. W. Mayes, unpublished data). This may offer a means of determining their contribution to the diet of primates.

Vertebrate prey items

Except for the possibility of using fecal DNA, discriminating at the animal species level, there appears to be little opportunity for using chemical fecal marker methods to quantify vertebrate food items in the primate diet. Thus established methods based on direct observation and the identification of fragments of bone, hair, feathers, and other indigestible body parts in feces will continue to be used.

Fungal material

At present there are no established marker methods for quantifying the proportion of fungal material in the mammalian diet. The unsaturated sterol, ergosterol, is present in both the hyphae and fruiting bodies of fungi, and was tested as a fungal marker in goats, but none was recovered in feces (R. W. Mayes, unpublished data). However, the possibility of finding ergosterol hydrogenation products in the feces was not addressed. Further work would be required to investigate the behavior of fungal sterols in the primate gut.

Soil

Some primates are known to practice geophagy (Knezevich, 1998). Fecal marker methods have been developed to determine soil intake in grazing livestock, which should be directly transferable to primate studies. Titanium has been the most popular marker (Mayland *et al.*, 1975) since typical soil concentrations are high (1000–2000 mg/kg dry soil), yet uptake by plants is extremely low, and is completely recovered in feces. Recent advances in x-ray diffraction (XRD) techniques (Hillier, 2003) have enabled specific clays and other minerals to be quantitatively analyzed in soil samples. The technique may be applicable to fecal analysis. If appropriately validated, this method may enable quantification of particular types of clay ingested by primates in order to study interactions with ingested toxins.

Possible fecal marker substances and food residues appearing in feces that could be used to determine the diet composition in primates have been summarized in Table 16.1.

The potential for measuring intake and gut passage rates in primates

In order to use fecal markers to determine intake, fecal output, or gut passage rate in free-ranging animals, it is necessary to administer a known amount of marker by oral dosing or giving a labeled feed. While repeated marker administration is theoretically possible, it is feasible for intake and fecal output to be more conveniently determined in primates by the 'single dose'

Table 16.1. *Potential fecal and urinary markers or food residues that might be used to determine the composition of the diets of primates*

Food item	Measured parameter	Type[a]	Tests[b]	Other comments
Plants				
Leaves	Plant-wax markers	Qn	E	Should be routine
Stems & buds	Plant-wax markers	Qn	E	Problem if outer surface is discarded
Flowers	Plant-wax markers	Qn	E	Some other compounds in flowers
Roots	Plant-wax markers	Qn	E	Low levels could be a problem
Fruits	Plant-wax markers	Qn	E	Problem if outer skin is discarded
Fruits	Indigestible seeds	?	E	Applicability depends on fruit species
Seeds	Plant-wax markers	Qn	S	Only if outer coating contains waxes
General	Plant fragments	Ql	E	Should be routine
General	Plant DNA	Ql	E	Species, but not plant part-specific
General	Fecal sterols	?	S	Not yet tested
General	Urinary metabolites	?	S	Limited data from ruminants
Invertebrates				
Insects and spiders	Cuticular hydrocarbons	Qn	D	Some bird and bat data
Insects and spiders	Body fragments	Ql	E	Tested in birds
Other hard-bodied	Body fragments	Ql	E	Tested in birds
Other soft-bodied	Plant-wax markers	?	S	Maybe from plant material in guts
Vertebrates	Scales; feathers; hair	Ql	?	Only if parts consumed
	Animal DNA	Ql	D	Animal species specific
Other				
Honey	Alkanes; alcohols; acids	Qn	D	Beeswax tested in sheep
Fungal material	Fecal sterols	?	S	Behavior in gut not yet known
Fungal material	Spores	?	S	Not yet tested
Any life form	DNA	Ql	D	Species specific
Soil	Ti, Sc	Qn	E	Validated in sheep and cattle
Soil	Clay and other minerals	Qn	S	Mineral-specific; not yet tested

Notes:

Food items that are unlikely to contain any known marker types or leave any identifiable fecal residues (e.g., nectar, plant gums and exudates and nut kernels) have not been included.

[a] Indicates whether method is quantitative (Qn) or qualitative (Ql).

[b] Indicates whether the method is established (E), under development (D), or speculative (S);

? indicates uncertainty about whether the marker would be quantitative, qualitative, or both.

method. Backed up by direct observations, individuals may be persuaded to ingest a bait, suitably labeled with even-chain alkane or other marker compound, followed by the collection of a series of fecal samples from that animal. Gut passage rates could be determined using the same procedure if accurate times of defecation can be established.

Collection of fecal samples from individual animals within primate groups

There are likely to be many occasions when it is impossible to use direct observation to ascertain which individual animal produced a particular feces sample. In farm livestock, marker methods have been employed to identify which animals produced individual feces samples. Plastic particles of differing colors dosed to individual cattle have been used for fecal identification (Minson *et al.*, 1960). Another method, yet to be fully tested, could be to dose individuals with differing patterns of even-chain alkanes. Such fecal identification could be obtained concurrently with alkane analyses employed for making other measurements. This concept may be applicable to primate studies. A simpler and more elegant approach, which is likely to have widespread use in the future, would be to use fecal DNA to identify individual animals.

Urinary markers in primate feeding studies

Primates are unusual among wild animal species in that with some long-term habituated groups it is possible, on occasion, to collect urine samples. It is thus feasible that metabolites in primate urine may provide information on their diet and nutrient utilization. Studies in goats (Keir, 2000) have shown that different plant species fed as the sole diet resulted in different complex patterns of phenolic and aromatic acid metabolites of plant secondary compounds appearing in the urine. In sheep, urinary excretion of the simple phenol, orcinol, was found to be linearly related to the intake of heather (*Calluna vulgaris*) (Martin *et al.*, 1983). These limited studies suggest that urinary metabolite patterns could potentially be used to characterize diet composition. However, because differences are likely between the metabolites excreted in ruminant and non-ruminant urine, relevant validation studies would be necessary if this approach were to be used in primate studies. Such urinary markers may be useful for food items, such as fruit pulp, which may contain plant secondary compounds but little or no plant-wax substances.

Conclusions

There appears to be considerable potential for using marker methods, especially those using plant-wax compounds, originally developed for farm livestock, to advance our understanding of the feeding behavior and nutrition of free-ranging primates. With appropriate modifications, and following validation studies with captive animals, these techniques should allow the quantitative measurement of diet composition, digestibility and, possibly, intake and gut passage rate. There is likely to be a need to search for additional markers, especially in food items that contain no wax lipid compounds, to allow the composition of the complete diet to be quantified. There may also be potential for urinary metabolites to provide information on diet and nutrition.

References

Akter, S., Owen, E., Theodorou, M. K., Butler, E. A., & Minson, D. J. (1999). Bovine faeces as a source of micro-organisms for the in vitro digestibility assay of forages. *Grass and Forage Science*, **54**, 219–26.

Ali, H. A. M. (2003). *The Potential Use of Diet Composition Markers in Studies of Nutrition of Free-ranging Herbivores*. Unpublished Ph.D. thesis, University of Aberdeen.

Ali, H. A. M., Mayes, R. W., Lamb, C. S. *et al.* (2004). The potential of long-chain fatty alcohols and long-chain fatty acids as diet composition markers: development of methods for quantitative analysis and faecal recoveries of these compounds in sheep fed mixed diets. *Journal of Agricultural Science*, **142**, 71–8.

Blomquist, G. J., Nelson, D. R., & Derenobales, M. (1987). Chemistry, biochemistry, and physiology of insect cuticular lipids. *Archives of Insect Biochemistry and Physiology*, **6**, 227–65.

Bugalho, M. N., Dove, H., Kelman, W. M., Wood, J. T., & Mayes, R. W. (2004). Plant wax alkanes and alcohols as herbivore diet composition markers. *Journal of Range Management*, **57**, 259–68.

Bugalho, M. N., Mayes, R. W., & Milne, J. A. (2002). The effects of feeding selectivity on the estimation of diet composition using the n-alkane technique. *Grass and Forage Science*, **57**, 224–31.

Cabré-Vert, N. & Feistner, A. T. C. (1995). Comparative gut passage time in captive lemurs. *Dodo, Journal of Wildlife Preservation Trusts*, **31**, 76–81.

Campbell, J. L., Eisemann, J. H., Glander, K. E., & Crissey, S. D. (1999). Intake, digestibility and passage of a commercially designed diet by two propithecus species. *American Journal of Primatology*, **48**, 237–46.

Caton, J. M., Hill, D. M., Hume, I. D., & Crook, G. A. (1996). The digestive strategy of the common marmoset, *Callithrix jacchus*. *Comparative Biochemistry and Physiology, Part A: Physiology*, **114**, 1–8.

Chen, X. B., Mejia, A. T., Kyle, D. J., & Ørskov, E. R. (1995). Evaluation of the use of the purine derivative-creatinine ratio in spot urine and plasma samples as an index of microbial protein supply in ruminants – studies in sheep. *Journal of Agricultural Science*, **125**, 137–43.

Dillon, P., Crosse, S., O'Brien, B., & Mayes, R. W. (2002). The effect of forage type and level of concentrate supplementation on the performance of spring-calving dairy cows in early lactation. *Grass and Forage Science*, **57**, 212–23.

Dove, H. & Mayes, R. W. (1991). The use of plant wax alkanes as marker substances in studies of the nutrition of herbivores: a review. *Australian Journal of Agricultural Research*, **42**, 913–52.

(1996). Plant wax components: a new approach to estimating intake and diet components in herbivores. *Journal of Nutrition*, **126**, 13–26.

(1999). Developments in the use of plant wax markers for estimating diet selection in herbivores. In *Emerging Techniques for Studying the Nutritional Status of Free-ranging Herbivores. Satellite Meeting of the Vth International Symposium on the Nutrition of Herbivores*, ed. H. Dove & S. W. Coleman, CD-ROM. San Antonio.

Dove, H. & Moore, A. D. (1995). Using a least-squares optimisation procedure to estimate botanical composition based on the alkanes of plant cuticular wax. *Australian Journal of Agricultural Research*, **46**, 1535–44.

Dove, H., Mayes, R. W., Lamb, C. S., & Ellis, K. J. (2002a). Factors influencing the release rate of alkanes from an intraruminal controlled release device, and the resultant accuracy of intake estimation in sheep. *Australian Journal of Agricultural Research*, **53**, 681–96.

Dove, H., Milne, J. A., & Mayes, R. W. (1990). Comparison of herbage intakes estimated from in vitro or alkane-based digestibilities. *Proceedings of the New Zealand Society of Animal Production*, **50**, 457–9.

Dove, H., Scharch, C. P., Oliván, M., & Mayes, R. W. (2002b). Using n-alkanes and known supplement intake to estimate roughage intake in sheep. *Animal Production in Australia*, **24**, 57–60.

Dove, H., Wood, J. T., Simpson, R. J. *et al.* (1999). Spray-topping annual grass pasture with glyphosate to delay loss of feeding value during summer. III Quantitative basis of the alkane-based procedures for estimating diet selection and herbage intake by grazing sheep. *Australian Journal of Agricultural Research*, **50**, 475–85.

Edwards, M. S., & Ullrey, D. E. (1999). Effect of dietary fiber concentration on apparent digestibility and digesta passage in non-human primates. I. Ruffed lemurs (*Varecia variegate vauiegata and V. v. rubra*). *Zoo Biology*, **18**, 537–49.

France, J., Dhanoa, M. S., Siddons, R. C., Thornley, J. H. M., & Poppi, D. P. (1988). Estimating the production of faeces by ruminants from faecal marker concentrations. *Journal of Theoretical Biology*, **135**, 383–91.

Furnival, E. P., Ellis, K. J., & Pickering, F. S. (1991). Evaluation of controlled release devices for administration of chromium sesquioxide using fistulated grazing sheep. I. Variation in rate of release from the device. *Australian Journal of Agricultural Research*, **41**, 977–86.

Galyean, M. L. (1993). Technical note: an algebraic method for calculating fecal output from a pulse dose of an external marker. *Journal of Animal Science*, **71**, 3466–9.

Gedir, J. V. & Hudson, R. J. (2000). Estimating dry matter digestibility and intake in wapiti (*Cervus elaphus canadensis*) using the double n-alkane ratio technique. *Small Ruminant Research*, **36**, 57–62.

Giráldez, F. J., Lamb, C. S., López, S., & Mayes, R. W. (2004). Effects of carrier matrix and dosing frequency on digestive kinetics of even-chain alkanes and implications on herbage intake and rate of passage studies. *Journal of the Science of Food and Agriculture*, **84**, 1562–70.

Grace, N. D. & Body, D. R. (1981). The possible use of long-chain (C_{19}-C_{32}) fatty acids in herbage as an indigestible faecal marker. *Journal of Agricultural Science*, **97**, 743–5.

Hameleers, A. & Mayes, R. W. (1998). The use of n-alkanes to estimate herbage intake and diet composition by dairy cows offered a perennial ryegrass/white clover mixture. *Grass and Forage Science*, **53**, 164–9.

Hatt, J-M., Gisler, R., Mayes, R. W. *et al.* (2002). The use of dosed and herbage n-alkanes as markers for the determination of intake, digestibility, mean retention time and diet selection in Galapagos tortoises (*Geochelone nigra*). *The Herpetological Journal*, **12**, 45–54.

Hatt, J-M., Lechner-Doll, M., & Mayes, B. (1998). The use of dosed and herbage *n*-alkanes as markers for the determination of digestive strategies of captive giraffes (*Giraffa camelopardalis*). *Zoo Biology*, **17**, 295–309.

Hatt, J-M., Mayes, R. W., Clauss, M., & Lechner-Doll, M. (2001). Use of artificially applied n-alkanes as markers for the estimation of digestibility, food selection and intake in pigeons (*Columba livia*). *Animal Feed Science and Technology*, **94**, 65–76.

Heydon, M. J., Sibbald, A. M., Milne, J. A., Brinklow, B. R., & Loudon, A. S. I. (1993). The interaction of food availability and endogenous physiological cycles on the grazing ecology of red deer hinds (*Cervus elaphus*). *Functional Ecology*, **7**, 216–22.

Hillier, S. (2003). Quantitative analysis of clay and other minerals in sandstones by X-ray powder diffraction (XRPD). *International Association of Sedimentologists, Special Publication*, **34**, 213–51.

Holechek, J. L., Vavra, M., & Pieper, R. D. (1982). Botanical composition determination of range herbivore diets: a review. *Journal of Range Management*, **35**, 309–15.

Hulbert, I. A. R., Mayes, R. W., & Iason, G. R. (2001). The flexibility of an intermediate feeder: dietary selection by mountain hares measured using faecal *n*-alkanes. *Oecologia*, **128**, 499–508.

Jones, R. J., Ludlow, M. M., Throughton, J. H., & Blunt, C. G. (1979). Estimation of C_3 and C_4 plant species in the diet of animals from the ratio of the natural ^{12}C and ^{13}C isotopes in the faeces. *Journal of Agricultural Science*, **92**, 91–100.

Keir, B. L. (2000). *The Potential Use of Urinary Metabolites of Plant Compounds as Markers for Assessing the Botanical Composition of the Diet of Free-ranging Herbivores.* Unpublished Ph.D. thesis, University of Aberdeen.

Knezevich, M. (1998). Geophagy as a therapeutic mediator of endoparasitism in a free-ranging group of rhesus macaques (*Macaca mulatta*). *American Journal of Primatology*, **44**, 71–82.

Kotb, A. R., & Luckey, T. D. (1972). Markers in nutrition. *Nutrition Abstracts and Reviews*, **42**, 813–45.

Letso, M. (1995). *A Study of the Use of n-alkanes to Determine Dietary Intake and Digestibility in Grazing Rabbits*. Unpublished M.Sc. thesis, University of Aberdeen.

Martin, A. K., Milne, J. A., & Moberly, P. (1983). The origin of urinary aromatic compounds excreted by ruminants. 4. The potential use of urine aromatic acid and phenol outputs to measure voluntary food intake. *British Journal of Nutrition*, **49**, 87–99.

Mayes, R. W., & Dove, H. (2000). Measurement of dietary nutrient intake in free-ranging mammalian herbivores. *Nutrition Research Reviews*, **13**, 107–38.

Mayes, R. W., Beresford, N. A., Lamb, C. S. *et al.* (1994). Novel approaches to the estimation of intake and bioavailability of radiocaesium in ruminants grazing forested areas. *Science of the Total Environment*, **157**, 289–300.

Mayes, R. W., Dove, H., Chen, X. B., & Guada, J. A. (1995). Advances in the use of faecal and urinary markers for measuring diet composition, herbage intake and nutrient utilisation in herbivores. In *Recent Developments in the Nutrition of Herbivores*, ed. M. Journet, M-H. Farce & C. Demarquilly, pp. 381–406. Paris: INRA Editions.

Mayes, R. W., Giráldez, F. J., & Lamb, C. S. (1997). Estimation of gastrointestinal passage rates of different plant components in ruminants using isotopically-labelled plant wax hydrocarbons or sprayed even-chain alkanes. *Proceedings of the Nutrition Society*, **56**, 187A.

Mayes, R. W., Iason, G. R., White, N., & Palo, T. (2001). Measuring diet composition and food intake by moose in the Swedish boreal forest: integrating GPS and faecal marker technologies. In *Tracking Animals with GPS*, ed. A. M. Sibbald & I. J. Gordon, pp. 77–80. Aberdeen: Macaulay Institute.

Mayes, R. W., Lamb, C. S., & Colgrove, P. M. (1986). The use of dosed and herbage *n*-alkanes as markers for the determination of herbage intake. *Journal of Agricultural Science*, **107**, 161–70.

Mayland, H. F., Florence, A. R., Rosenau, R. C., Lazar, V. A., & Turner, H. A. (1975). Soil ingestion by cattle on semi-arid range as reflected by titanium analysis of feces. *Journal of Range Management*, **28**, 448–452.

Minson, D. J., Tayler, J. C., Alder, F. E. *et al.* (1960). A method for identifying the faeces produced by individual cattle or groups of cattle grazing together. *Journal of the British Grassland Society*, **15**, 86–8.

Newman, J. A., Thompson, W. A., Penning, P. D., & Mayes, R. W. (1995). Least-squares estimation of diet composition from n-alkanes in herbage and faeces using matrix mathematics. *Australian Journal of Agricultural Research*, **46**, 793–805.

O'Keefe, N. M., & McMeniman, N. P. (1998). The recovery of natural and dosed n-alkanes from the horse. *Animal Production in Australia*, **22**, 37.

Oliván, M., Dove, H., Mayes, R. W., & Hoebee, S. E. (1999). Recent developments in the use of alkanes and other plant wax components to estimate herbage intake and diet composition in herbivores. *Revista Portuguesa de Zootecnia*, **6**, 1–26.

Orr, A. (1998). *The intake of Herbage, Supplementary Feed and Performance of growing Ostriches* (Struthio camelus) *Given two Different Feeding Regimens in North East Scotland.* Unpublished M.Sc. thesis, University of Aberdeen.

Rao, S. J., Iason, G. R., Hulbert, I. A. R., Mayes, R. W., & Racey, P. A. (2003). Estimating diet composition for mountain hares in newly established native woodland: development and application of plant-wax faecal markers. *Canadian Journal of Zoology*, **81**, 1047–56.

Remis, M. J. & Dierenfeld, E. S. (2004). Digesta passage, digestibility and behavior in captive gorillas under two dietary regimens. *International Journal of Primatology*, **25**, 825–45.

Salt, C. A., Mayes, R. W., Colgrove, P. M., & Lamb, C. S. (1994). The effects of season and diet composition on the radiocaesium intake by sheep grazing on heather moorland. *Journal of Applied Ecology*, **31**, 125–36.

Tilley, J. M. A., & Terry, R. A. (1963). A two-stage technique for the in vitro digestion of forage crops. *Journal of the British Grassland Society*, **18**, 104–11.

Tulloch, A. P. (1976). Chemistry of waxes of higher plants. In *Chemistry and Biochemistry of Natural Waxes*, ed. P. E. Kollattukudy, pp. 235–87. Amsterdam: Elsevier.

Tutin, C. E. G. & Fernandez, M. (1992). Insect-eating by sympatric lowland gorillas (*Gorilla gorilla gorilla*) and chimpanzees (*Pan troglodytes troglodytes*) in the Lope Reserve, Gabon. *American Journal of Primatology*, **28**, 29–40.

Wilson, H., Sinclair, A. G., Hovell, F. DeB., Mayes, R. W., & Edwards, S. A. (1999). Validation of the n-alkane technique for measuring herbage intake in sows. *Proceedings of the British Society of Animal Science*, **171**.

Woolnough, A. P. (1998). *The Feeding Ecology of the Northern Hairy-nosed Wombat,* Lasiorhinus krefftii (Marsupiala: vambatidae). Unpublished Ph.D. thesis, Australian National University.

17 Energy intake by wild chimpanzees and orangutans: methodological considerations and a preliminary comparison

NANCY LOU CONKLIN-BRITTAIN, CHERYL D. KNOTT, AND RICHARD W. WRANGHAM

Feeding Ecology in Apes and Other Primates. Ecological, Physical and Behavioral Aspects, ed. G. Hohmann, M. M. Robbins, and C. Boesch. Published by Cambridge University Press.
© Cambridge University Press 2006.

Introduction

Energy intake and energy balance are critical for successful ovulation and conception in great apes, including humans (Lipson & Ellison, 1996; Knott 1998, 1999, 2001; Ellison, 2001). Energy is also the principal currency used in optimal foraging theory (Schoener, 1987). For these reasons, energy intake is an important variable for research in primate ecology, but it has rarely been estimated in the wild. Even in the controlled conditions of a laboratory, energy intake is difficult to measure accurately, and even more so in the wild (Dasilva, 1992; Nakagawa, 1997, 2000; Wasserman & Chapman, 2003).

There are two methods for determining the energy content of a food. The first, and most direct, requires experimental control of feeding; it is also the most time-consuming and expensive. It begins by using bomb calorimetry to determine the gross energy (GE) value of a food item (see glossary) (Schneider & Flatt, 1975). Since the gross energy value includes calories that are not digestible, the indigestible fraction must then be estimated. This can be done by feeding the food in question to several individuals and using bomb calorimetry to determine the energy content of the resulting feces. Subtraction yields the digestible energy (DE) (gross energy minus fecal energy). A more precise version of energy intake, metabolizable energy (ME), can be obtained in a similar way, by subtracting the energy in urine. Net energy (NE) subtracts heat and gaseous losses from ME (Schneider & Flatt, 1975), but requires special equipment, and is usually not measured for non-ruminants. While the above procedures are potentially precise, they are not normally practical for the study of wild foods. We therefore used the second method, consisting of calculated estimates, which will be described in the Methods section.

An initial estimate of energy intake can be calculated from two main variables, the amount of time spent feeding on different food items, and the concentration of the macronutrients in those items. One aim of this paper, therefore, is to use such data to do a preliminary comparison of the daily energy intake of wild chimpanzees (*Pan troglodytes schweinfurthii*) and orangutans (*Pongo pygmaeus abelii*). Our data come from 8 and 9 months of observation, respectively, for the two species, including periods of both high and low fruit availability. They include simple activity budgets that allow us to assess energy usage. Following the hypothesis that orangutans are adapted to larger variance in fruit availability than most primates (Knott, 1998, 1999, 2005) we predict that variation in energy intake between months is greater for orangutans than for chimpanzees.

Our second aim is to evaluate the problems associated with energy intake estimates. In particular, we consider the critical but poorly quantified role of the fermentation of fiber in the hindgut. We rely heavily on the fiber digestion coefficient determined by Milton & Demment (1988) because there are no others available. Following this we rely on three assumptions: (1) fiber in wild foods chemically resembles the fiber in the captive diet fed to the chimpanzees during the digestion trial; (2) chimpanzees in captivity have a capacity for fiber fermentation similar to that of those in the wild; and (3) body size differences between chimpanzees and orangutans will not interfere with our use of the same digestion coefficient for both. These and additional sources of error will be further discussed.

We use only neutral-detergent fiber (NDF) in our calculations rather than acid-detergent fiber (ADF) or crude fiber (see glossary). This is because NDF is the only one of these fiber measures that includes hemicellulose (van Soest, 1994; N. R. C., 2003). Since hemicellulose is partially fermentable by chimpanzees (and presumably by orangutans) (Milton & Demment, 1988), NDF is the best index of total insoluble fiber and of energy available from fiber. We do not recommend the use of ADF or crude fiber.

Methods

Field methods

Chimpanzees

Methods for observing chimpanzees were described by Wrangham *et al.* (1998). Briefly, chimpanzees were studied in the Kanyawara sector of Kibale National Park, Uganda in 1992 and 1993. They were habituated sufficiently to allow uninterrupted observation when they were in trees. When on the ground, most adult males allowed observers to sit about 5 m away, but even so, thick vegetation made viewing of terrestrial foods difficult at times. Chimpanzees could not always be located and followed predictably. Therefore, chimpanzees were observed whenever they could be found.

Feeding observations were recorded by instantaneous, focal-animal sampling (Wrangham *et al.*, 1998). Focal observations were rotated among all available individuals in the party, changing targets every 10 minutes. At the end of a 10-minute focal-animal session, the observer located a new target individual whose behavior had not been recorded for at least 20 minutes. After a target was selected, it was observed for 60 s, at the end of which its instantaneous behavior was recorded. During those 60 s, a feeding rate of bites/minute was recorded whenever possible. The monthly number of 10-minute plant-feeding observations was 271 ± 106 (mean ± s/d per month) for the 8 months considered in this study, i.e., an average of 45.2 plant-feeding hours per month.

Orangutans

Methods for recording orangutan data were described in Knott (1998). The orangutans were studied in the Cabang Panti Research Site in Gunung Palung National Park, West Kalimantan, Indonesia between 1994 and 1996. They were also well habituated, but could not always predictably be located and followed, and so were also observed whenever they could be found.

Feeding observations were recorded by a combination of continuous recording of a focal-animal throughout the day and instantaneous, focal-animal sampling within each feeding bout. The start of a feeding bout was recorded and then every 2 minutes, when possible, the observer counted the number of fruits or leaves, etc. the individual put in its mouth during 60 s (feeding rate), for a total of 3 minutes per record. When possible, five feeding rates were collected for each bout and if the bout lasted longer than 10 minutes, subsequent feeding rates were collected every 5 minutes until the bout ended. The monthly number of plant-feeding minutes was 7459 ± 3331 (mean ± standard deviation per month) for the 9 months included in this study, an average of 124.3 plant-feeding hours per month.

Definition of feeding

For both the chimpanzees and orangutans, feeding was defined as reaching, picking, handling, or chewing a food item, and these actions were included in the feeding rate. Plant food items were categorized as: fruit (ripe or unripe; pulp, husk, or skin), seed (ripe or unripe; with or without seed coat), leaf (young or mature; petiole, bud, or blade), flower (age and part, e.g., sepal, petal, buds), pith, root, bark, cambium, and wood.

We characterized food items by the average weight eaten per minute. For small items, we estimated intake rate using the number of bites per minute and the size of bites, e.g., one fruit per bite, 10 leaflets per bite, half a leaf per bite, etc. Some fruits eaten by orangutans required more than 1 minute to consume. For those large items, we calculated intake rate using the period between picking one fruit, eating it, and picking the next.

Sample processing

We determined the fresh weights of individual items eaten as soon as possible after harvesting them. The fresh weights of 5–20 individual specimens per food type were recorded to determine the average wet (fresh) and dry weights for a food. To process chimpanzee fruits, we weighed the whole fruit, then cut the pulp off the seed or seeds and weighed the individual seed or group of seeds. The weight of the fresh fruit pulp was determined by difference (whole fruit weight minus seed(s) weight, less other undigested parts such as skins). For orangutan fruits, each plant part was weighed separately after weighing the whole fruit. Fruits were dissected within the weighing dishes to avoid loss of pulp liquid while processing. We recorded the number of seeds per fruit. In addition, for nutrient analysis, about 100 g of fresh weight of the same food was collected, weighed, and processed in bulk. We reconstructed pith or bark quantities by examining feeding remains and determining how many centimeters of pith within a plant's stem or bark had been eaten and harvesting similar sized pieces to determine the fresh and dry weights consumed over a measured amount of time.

The plant samples were then air-dried in the shade. Analyses necessary for the calculation of energy content are not compromised by air-drying. For a description of the chimpanzee food dryer, see Conklin-Brittain *et al.* (1998). We constructed racks with legs and multiple shelves made of wire mesh, which were then set over a heat source (kerosene lamps). The orangutan foods were dried in a large metal, kerosene drying oven with wire racks inside the oven. In both cases, trays or dishes made of aluminum foil worked well to conduct heat and allow fast drying.

The most important characteristic of a drying arrangement, in addition to maintaining a 30–40 °C temperature, is to dry samples quickly to prevent both fermentation and molding, which can change the chemical composition of a sample (Harborne, 1984). The critical period in drying is the first several hours, when the water content is the highest and the samples are sitting in a warm environment. To facilitate the escape of water we initially reduced large items of pulp or pith to pieces, sometimes as small as 5–10 mm^2 in area. We stirred the drying samples frequently, and examined them several times daily for mold. Samples that became moldy (yellow or grey fuzz) or fermented (blackened) were discarded; we re-collected that sample, sliced it more thinly and dried it at a higher temperature (maximum 50 °C). In a few cases, the new samples also turned black; nevertheless we accepted such items in our analyses. We recorded the dry weights as the "field-dried" weight.

In Kibale, a comparatively dry environment, we stored the dried samples in paper envelopes in a large, covered basket suspended from the ceiling in the drying room. In Gunung Palung, we stored dried samples in heat-sealed plastic bags, and then placed these bags in a larger bag containing silica gel to prevent humid air from rehydrating the samples.

Lab methods

We assayed nutrient composition of each food using standard chemical methods, and used conventional estimates of the energy content of different nutrients to assess total energy content (Atwater physiological fuel values: see glossary) (Watt & Merrill, 1963; N. R. C., 1981, 2003). Chemical values are reported as a percentage of organic matter instead of dry matter (see below and glossary).

Protein and fat

For chemical analysis we ground the field-dried samples through a #20-mesh screen in a Wiley Mill. We determined crude protein (CP) by using the Kjeldahl procedure for total nitrogen and multiplying by 6.25 (Pierce & Haenisch, 1947). The digestion mix contained Na_2SO_4 and $CuSO_4$. The distillate was collected in 4% boric acid and titrated with 0.1N HCl. We measured lipid content using petroleum ether extraction for four days at room temperature, a modification of the method of the Association of Official Analytical Chemists (A.O.A.C., 1984).

Fiber and carbohydrates

We used the Detergent System of Fiber Analysis (Goering & van Soest, 1970) as modified by Robertson & van Soest (1980) to determine the neutral-detergent fraction (NDF), also referred to as the insoluble, structural, plant cell-wall fiber fraction (see glossary). We determined total nonstructural carbohydrates (TNC), i.e., the digestible carbohydrates, by difference (see glossary):

$$\%TNC = 100 - \%lipid - \%crude\ protein$$
$$- \%total\ ash - \%NDF \tag{17.1}$$

This fraction (TNC) is sometimes referred to as the nitrogen-free extract (NFE).

As stated in the introduction, if the acid-detergent fiber (ADF) was used in this equation instead of NDF, the hemicellulose fraction would be included in the TNC and subsequently treated as if it were a highly digestible carbohydrate, which it is not. Hemicellulose is an insoluble fiber (see glossary) and insoluble fibers are not digested, except to a limited extent by fermentation, e.g., in the hindgut. Likewise, the hemicellulose and lignin that the crude fiber extraction misses would be included in the TNC and mistakenly treated as though they were both digestible (van Soest, 1994).

Dry matter, total ash, and organic matter

We determined the final dry matter (DM) correction coefficient (A.O.A.C., 1984) by drying a small subsample of the field-dried sample at 100 °C for 8 h and weighing it hot (= g 100 °C DM/g field-dried sample). Total ash (g ash/g field-dried sample) was measured by ashing (burning) the same DM subsample at 520 °C for 8 h and then weighing it at 100 °C. We calculated the organic matter (OM) correction coefficient of the sample as:

$$\text{OM correction coefficient(g OM/g field-dried sample)} = (1 - \text{total ash}) \times \text{DM correction coefficient} \quad (17.2)$$

These DM and OM correction coefficients are applied to the field-dried DM value to determine the grams of OM in the fresh food:

$$\text{g OM/g fresh food} = \text{field DM coefficient}$$
$$\times \text{ final DM correction coefficient}$$
$$\times \text{ OM correction coefficient}$$

The ash values for the foods analyzed here varied between 3% and 20%, which are probably typical values for wild plants. Since ash (total inorganic minerals, see glossary) does not contribute energy to a food, and since dry matter (DM) includes the ash, we express nutrient values as a percentage of OM instead of DM.

Physiological fuel values

Following conventional practice, we calculated energy content using standard conversion factors (i.e., physiological fuel values): 4 kcal/g (16 kJ/g) carbohydrate (TNC), 4 kcal/g (17 kJ/g) crude protein (CP) and 9 kcal/g (37 kJ/g) lipid (N. R. C., 1980, 1981, 1989, 2003). We also included a fourth conversion factor for the potentially available energy in NDF. A fiber digestion coefficient for chimpanzees (see glossary) has been published by Milton & Demment (1988), and in that study, chimpanzees digested 54.3% of the NDF fed to them in the form of a primate biscuit containing 34% NDF (meaning the digestion coefficient is 0.543). Conveniently, reported wild diets consumed by both the chimpanzees and the orangutans averaged about 34% NDF (Conklin-Brittain *et al.*, 1998; Knott, 1999).

The second consideration in the fiber conversion factor is as follows: since the fibers hemicellulose and cellulose are carbohydrates, the conversion factor of 4 kcal/g (16 kJ/g) could theoretically be applied to them, except that the anaerobic microbes doing the digestion/fermentation keep about 1 kcal/g (4 kJ/g) of fiber for their own growth (of the 36 moles ATP in glucose, they keep as much as 6 moles). This leaves up to 3 kcal/g (12 kJ/g) fiber for the host animal (30 moles/36 moles × 4 kcals) (Conklin & Wrangham, 1994; van Soest, 1994). Thus the physiological fuel value from fiber is $3 \times 0.543 = 1.6$ kcal/g (6.7 kJ/g).

Plant fibers vary in their ratios of hemicellulose, cellulose, and lignin, even if the overall fiber value (NDF) is the same. For example, the lignin level in the orangutans' diet was 16%, twice as high as the average chimpanzee lignin level (Conklin-Brittain & Knott, unpublished data), although both wild diets averaged 34% NDF. Lignin is a polyphenol, not a carbohydrate, and essentially unfermentable. Consequently, we used a lower fiber conversion factor for the orangutans, assuming a lower digestion coefficient of 0.181 because of the high lignin level. We therefore used a physiological fuel value for orangutan fiber of $3 \times 0.181 = 0.543$.

Energy calculations

We calculated the energy per food item in three ways, depending on how much metabolizable energy (ME) was assumed to be derived from fermentation of NDF:

1. Zero-fermentation Metabolizable Energy (ME_O) assumes that there is no energy obtained from fiber fermentation:

$$ME_O \text{ kcal}/100 \text{ g OM} = (4 \times \%TNC) + (4 \times \%CP) + (9 \times \%lipid) \tag{17.3}$$

2. Low-fermentation Metabolizable Energy (ME_L) assumes partial NDF fermentation, with a fiber digestion coefficient of 0.181:

$$ME_L \text{ kcal}/100 \text{ g OM} = (4 \times \%TNC) + (4 \times \%CP) + (9 \times \%lipid) + (0.543 \times \%NDF) \tag{17.4}$$

3. High-fermentation Metabolizable Energy (ME_H) assumes maximal NDF fermentation, with a fiber digestion coefficient of 0.543:

$$ME_h \text{ kcal}/100 \text{ g OM} = (4 \times \%TNC) + (4 \times \%CP) + (9 \times \%lipid) + (1.6 \times \%NDF) \tag{17.5}$$

These calculations give a range of estimates for the energy content for 100 g of dry food. To obtain the energy content per gram of fresh weight, we converted to total kcals (or Kj) by restoring the water. For example:

Energy content of a fresh food $= (\text{ME}/100\,\text{g OM})$
$$\times (\text{g OM/g of the fresh food})$$
$$(17.6)$$

The total energy intake per month by the chimpanzees was calculated as the sum of the products of the grams of each food consumed per month and the energy content per gram of that food.

$$\sum(\text{g food}_1/\text{month}) \times (\text{energy content food}_1/\text{g}) + (\text{g food}_2/\text{month})$$
$$\times (\text{energy content food}_2/\text{g}) + \ldots + \text{g food}_n/\text{month})$$
$$\times (\text{energy content food}_n/\text{g})$$
$$(17.7)$$

The total energy intake per month by the orangutans was calculated as the sum of the products of the grams of each food consumed per feeding bout and the energy content per gram of that food. This was then summed for all feeding bouts per month.

$$\sum(\text{g food}_1/\text{bout}) \times (\text{energy content food}_1/\text{g}) + (\text{g food}_2/\text{bout})$$
$$\times (\text{energy content food}_2/\text{g}) + \ldots + \text{g food}_n/\text{bout})$$
$$\times (\text{energy content food}_n/\text{g})$$
$$(17.8)$$

For some foods we lacked one or more data points, specifically no feeding rates and/or nutrient values (17% of the chimpanzee feeding minutes, 10% of the orangutan feeding minutes). For chimpanzees we treated these foods as if they had the average values for all foods that month. For orangutans, we treated these foods as if they had the average values for foods of the same plant part eaten by that animal on that day.

We used energy expenditure values previously reported for Kibale chimpanzees by Pontzer & Wrangham (2004) and for Gunung Palung orangutans by Knott (1999). The energy intake calculations did not differentiate between males and females, in keeping with the very preliminary nature of this report.

Results

Chimpanzees

We evaluated data for 8 months, including three major non-fig fruit seasons (each dominated by a single, highly preferred, fruit species, based on time spent feeding), one minor non-fig fruit season (dominated by several, less-preferred fruits) and one fig season (Table 17.1). Figs were eaten throughout the year but the fig season comprised about eight different fig species, and was a 3-month period when very few non-fig fruit species were eaten. Table 17.2 shows the energy values calculated for the adult and sub-adult chimpanzees of both sexes combined, using Equations 1 and 3 described in the Methods.

Using the energy expenditures by males and lactating female chimpanzees, as estimated by Pontzer & Wrangham (2004), and examining the differences between the two different methods of energy intake calculation (Figure 17.1), the data in Tables 17.1 and 17.2 suggest that without including the potential contribution from fiber, the energy from ME_O was enough to satisfy maintenance, walking, and climbing energy expenditures for only a few months.

As can be seen in Figure 17.1, if the energy from fiber fermentation were not included, lactating chimpanzees would frequently experience inadequate energy intake. However, if they could ferment fiber at the rate that captive chimpanzees can (Milton & Demment, 1988), both males and females would have surplus energy. For the eight months of this study, although there appeared to be fairly dramatic differences between ME_O and ME_H energy intake (Figure 17.1), energy intake was not significantly correlated with fruit availability, as measured by monthly phenological observations (Wrangham *et al.*, 1998) ($r^2 = 0.12$ for ME_O, NS; $r^2 = 0.05$ for ME_H; NS). The same was

Table 17.1. *Feeding ecology summary: source of data and sample sizes for chimpanzee feeding observations*

	Dominant fruit available	Total no. feeding obsv's	No. obsv's used in energy calc.	% of obsv's included	No. food types eaten	No. food types used in calc.
Month						
September 92	*Pseudospondias*	267	213	80	25	12
October 92	Misc. fruit	205	176	86	13	7
January 93	*Mimusops*	73	64	89	10	4
February 93	*Mimusops*	277	238	87	29	14
March 93	Figs	332	277	83	40	19
April 93	Figs	242	193	80	35	13
May 93	Figs	424	311	73	44	15
June 93	*Uvariopsis*	351	302	90	28	7
Total/mean		2171	1774	83	28	11

Total number of feeding observations comes from 10-minute samples. Food types are a combination of plant species and plant part. Season refers to the dominant fruit species eaten. Obsv's = observations; calc.= calculation.

Table 17.2. *Estimated daily energy intake by chimpanzees*

Month	Dominant fruit	ME_H Kcal	kJ	ME_O Kcal	kJ	% ME_H from fiber
September 92	*Pseudospondias*	1806	7558	1206	5046	33.2
October 92	Misc. fruit	1955	8179	1540	6443	21.2
January 93	*Mimusops*	2603	10892	2002	8376	23.1
February 93	*Mimusops*	3333	13947	2535	10606	23.9
March 93	Figs	2009	8406	1381	5778	31.3
April 93	Figs	2116	8853	1456	6092	31.2
May 93	Figs	2286	9563	1601	6699	30.0
June 93	*Uvariopsis*	2616	10947	1912	8000	26.9

Season refers to the dominant fruit species eaten.

ME_H = the total estimated metabolic energy intake in kilocalories (kilojoules) assuming a high level of fiber fermentation.

ME_O = the total estimated metabolic energy intake in kilocalories (kilojoules) assuming no fiber fermentation.

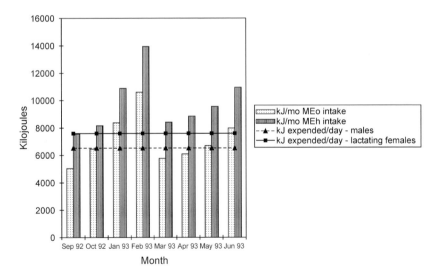

Figure 17.1. Chimpanzee energy intake compared to energy expenditure. Female chimpanzees were lactating. There was no measure available for monthly variation in energy expenditure.

true for the percentage of energy derived from fiber (correlation with fruit availability, $r^2 = 0.08$, NS.).

Orangutans

We evaluated data from 9 months, including 3 months of a mast (a super-abundant fruiting season that occurs on an unpredictable basis every few years), and 3 months of extremely low fruit availability (ELFA), which frequently follows a mast period. We also included 3 months of intermediate fruit availability (Table 17.3). Table 17.4 presents energy values calculated using Equations 1, 2, and 3 for the orangutans and includes both adults and sub-adults of both sexes combined.

Using the maximum and minimum energy expenditures by male and female orangutans, as calculated by Knott (1999), averaged for the year, we graphed energy intake with energy expenditure (Figure 17.2). As can be seen in Figure 17.2, during the mast it does not matter which energy conversion factor is used; both males and females are more than adequately fed. Likewise, during ELFA, it does not matter which calculation is used because food is so scarce and of such low quality (high fiber) that energy intake is inadequate. During the intermediate fruit availability months, however, our preliminary conclusion is that the fiber digestion coefficient becomes important for the accurate estimate of energy obtained from fiber.

Using ANOVA, we compared energy intake in the three periods. The mast months were, as expected, significantly higher in energy intake than both the intermediate months ($p < 0.01$), and the ELFA months ($p < 0.02$) regardless of the method used to calculate energy intake. The intermediate months were not quite significantly higher in energy intake than the ELFA months ($p \geq 0.1$).

To compare the relative importance of energy from fiber we summarized the values obtained with ME_L (Table 17.4). The statistical results were similar for ME_H. Comparing the percentage of kcals derived from fiber to food availability, the mast months' energy intake contained significantly less energy from fiber than either the intermediate or the ELFA months (mean 3.5%, 15.2%, and 13.7%, respectively, $p < 0.001$), but the intermediate and ELFA months were not significantly different ($p = 0.21$). However, when comparing the actual energy derived from fiber, the reverse was true. The mast months and the ELFA months were not significantly different (mean 202 kcal vs. 147 kcal respectively, $p = 0.42$). The mast months and the intermediate months were almost significantly different (mean 202 kcals vs. 344 kcals respectively, $p = 0.07$). The intermediate and ELFA months were significantly different (mean 344 kcals vs. 147 kcals, respectively, $p = 0.02$).

Table 17.3. *Feeding ecology summary: source of data and sample sizes for orangutan feeding observations*

Month	Fruit availability	Total no. feeding obsv's	No. obsv's used in energy calc.	% of obsv's included	No. food types eaten	No. food types used in calc.
Dec 94	mast – high	11453	10695	93	48	36
Jan 95	mast – high	7041	6236	89	36	24
Feb 95	mast – high	5804	5467	94	34	23
April 95	ELFA – low	4155	4033	97	30	25
June 95	ELFA – low	6909	5491	80	48	30
Aug 95	ELFA – low	7033	5857	83	45	29
Nov 95	intermediate	5241	4362	83	41	30
Sept 96	intermediate	5110	4811	94	32	23
Oct 96	intermediate	14390	13278	92	64	48
Total/mean		67135	60230	90	42	30

Obsv's = observations; calc. = calculation; ELFA = extremely low fruit availability.
Total number of feeding observations means 1-minute points. Food types are a combination of plant species and plant part.

Table 17.4. *Estimated daily energy intake by orangutans*

Month	Fruit availability	ME$_H$		ME$_L$		ME$_O$		%ME$_H$ from fiber	%ME$_L$ from fiber
		Kcal	kJ	Kcal	kJ	Kcal	kJ		
Dec 94	mast – high	4279	17903	3990	16696	3842	16076	11.4	3.7
Jan 95	mast – high	7827	32749	7183	30054	6866	28728	14.0	4.4
Feb 95	mast – high	6346	26552	6075	25416	5935	24833	6.9	2.3
Apr 95	ELFA – low	1186	4962	935	3913	807	3375	47.0	13.7
Jun 95	ELFA – low	787	3291	598	2504	502	2099	56.8	16.1
Aug 95	ELFA – low	1783	7462	1363	5702	1147	4798	55.4	15.8
Nov 95	intermediate	2550	10667	1989	8322	1701	7118	49.9	14.5
Sep 96	intermediate	3634	15205	2826	11824	2411	10087	50.7	14.7
Oct 96	intermediate	3419	14307	2778	11624	2449	10245	39.6	11.8

ME$_H$ = the total estimated metabolic energy intake in kilocalories (kilojoules) assuming a high level of fiber fermentation.
ME$_L$ = the total estimated metabolic energy intake in kilocalories (kilojoules) assuming a low level of fiber fermentation.
ME$_O$ = the total estimated metabolic energy intake in kilocalories (kilojoules) assuming no fiber fermentation.

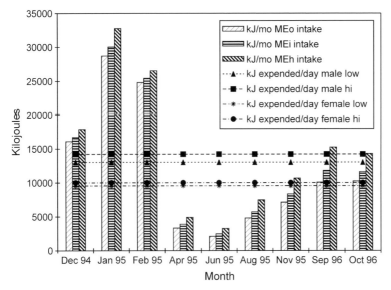

Figure 17.2. Orangutan energy intake compared to energy expenditure. Females used included all reproductive classes. The maximum and minimum ranges represent annual differences in energy expenditure for males and females. More details on energy expenditure are available in Knott (1999).

In sum, the total energy intake during the ELFA months was low, and energy from fermenting fiber did not appear sufficient to balance energy expenditure. The food was of such poor quality and the quantity of food was so low, that the animals apparently catabolized major amounts of body fat, as shown by Knott (1998) by the presence of ketones in the urine. However, during months with intermediate food availability orangutans could take advantage of the energy from fiber fermentation.

To more directly compare chimpanzee and orangutan energy intake, Figure 17.3 combines Figures 17.1 and 17.2. The chimpanzee intake levels were consistent with the low or intermediate intake levels of the orangutans ($ME_O p = 0.56$, $t = 0.60$; $ME_H p = 0.81$, $t = 0.25$). The differences in monthly energy intakes, however, were greater for the orangutans than for the chimpanzees ($ME_O p = 0.003$, $t = -5.3$; $ME_H p = 0.001$, $t = -6.3$).

Discussion

Our calculations of energy intake generally conform to expected patterns. In particular, variation in energy intake across months was higher for

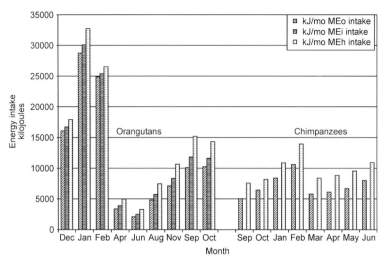

Figure 17.3. The chimpanzee energy intake range compared to the orangutan energy intake range.

orangutans than chimpanzees. The lack of correlation between energy intake and fruit availability for the chimpanzees was contrary to expectation. The energy intake estimates for the low fruit availability months were perhaps overestimates, a result of difficulties collecting pith intake data.

There are many aspects that need further refinement in the energy calculation process. We will address four of the major points. First, the captive primate diet used to obtain the 0.543 digestion coefficient had a very low lignin content (about 2.5%), even though it was 34% neutral-detergent fiber (NDF). Lignin is one of the three principal insoluble fibers (together with hemicellulose and cellulose, see glossary), but is virtually unfermentable compared with the other two. As a result, fiber in the captive diet would have contained a fermentable NDF value of $34 - 2.5 = 31.5\%$. Meanwhile, the lignin content of wild chimpanzee foods can range from 1% to 38% lignin. The wild chimpanzee diet averaged about 8% lignin annually (Conklin-Brittain *et al.*, 1998) so the average fermentable NDF would have been $34 - 8 = 26\%$. Consequently, wild foods probably had lower digestion coefficients than the captive diet. This problem can be investigated by determining digestibility coefficients for fiber in wild foods in vitro, using fresh fecal microbes from adults consuming wild foods. The protocols for this procedure have been worked out for humans (Barry *et al.*, 1995).

The second caveat for wild foods is that the 4, 4, 9 conversion factors are average metabolizable energy values for low-fiber mixed human diets of meat, vegetables, grains etc. (Watt & Merrill, 1963; N.R.C., 1980). They are undoubtedly overestimates for the availability of these nutrients in a high-fiber, wild diet (Miles *et al.*, 1988). This problem can in theory be investigated by in vitro determination of enzymatic digestion coefficients for TNC, CP, and lipid in high fiber diets of varying NDF levels, using commercially available enzymes. These protocols have been reviewed (Moughan, 1999), but in vivo values for primates are needed to calibrate the in vitro results.

The third point refers specifically to the carbohydrate value of 4 kcal (16 kJ)/g, which may be too high because we did not differentiate between starch (the only complex carbohydrate directly digestible by mammalian enzymes) and all of the nonstarch polysaccharides, also known as soluble fiber (see glossary). Soluble fiber is fermented and can be given the conversion factor of 3 kcal (12 kJ)/g, in conjunction with a digestion coefficient, which is often fairly high (Bourquin *et al.*, 1996; Livesey, 2001). We used 4 kcal (16 kJ)/g in our calculation because the Detergent System of Fiber analysis (see glossary) only measures insoluble rather than soluble fiber. This problem can be ameliorated by determining the soluble fiber content (Lee *et al.*, 1992), a procedure commonly used in human nutrition.

Broadly speaking, resolving these three issues would lower all the energy intake values in Figures 17.1 and 17.2 somewhat. Whether the corrections would lower the lines uniformly, leaving the statistical relationships intact, needs to be further investigated (Charrondiere *et al.*, 2004).

The question of which conversion factor to use for fiber leads us to discuss a fourth caveat, that concerning passage rate. One might argue that male orangutans should have a higher digestion coefficient than male chimpanzees simply because of their larger body size and the increased gut capacity that size probably provides (Demment & van Soest, 1985). However, during times of high-food abundance, the passage rate of food through the gut will probably be faster than usual because of the extreme increase in food intake. During the mast, orangutans consume approximately 4.5 times as many grams of food dry matter compared with the ELFA months (Knott, 1999). This large difference in total food intake results in a fiber dry matter intake that is 2.5 times as many grams in the mast compared with the ELFA months, in spite of the fact that the ELFA foods are individually higher in fiber. Faster passage rates lower digestion coefficients of fiber and of all the nutrients (Milton & Demment, 1988; van Soest, 1994). During times of extremely low fruit availability, when orangutans eat leaves and bark, the lignin content of the fiber increases, as already mentioned, which would lower the digestion coefficient of fiber even further. Overall, there is reason to use a lower conversion factor, perhaps a range of 0.5 to 1.0 kcal (2–4 kJ)/g, for both the orangutans and the chimpanzees. Determining passage times and digestion coefficients for fiber at varying intake levels would assist in resolving this fourth problem. The marker used for passage time of the fiber fraction should be one that chemically bonds to the food fiber, e.g., chromium-mordanted fiber; plastic pellets do not work well (van Soest *et al.*, 1983).

A small number of other researchers have also reported preliminary energy intake values (Dasilva, 1992; Nakagawa, 1997, 2000; Knott, 1998, 1999; Wasserman & Chapman, 2003). The caveats described here apply to these studies as well. In addition, Wasserman & Chapman (2003) and Dasilva (1992) used a single digestion coefficient from Watkins *et al.* (1985), just as we used a single one from Milton & Demment (1988). Ideally we should have all used a different digestion coefficient unique for each and every food item eaten. Nakagawa (1997, 2000) did not include any estimate for fiber digestion. Perhaps the most pragmatic approach at this time is to use maximum/minimum ranges. We suggest a range of 0.5–1.0 kcal (2–4 kJ)/g, in conjunction with energy expenditure estimates collected simultaneously with energy intake.

As research questions become more detailed and complicated, it is important to improve energy intake determinations. For the moment, we can make progress by calculating approximate energy intakes, especially for questions regarding the maintenance of adults. We strongly recommend that the energy calculation process be refined, as we have outlined, in order to assess energy intake for increasingly sensitive analyses, such as questions regarding growth, reproduction, and optimal foraging.

Acknowledgments

We thank the Uganda National Research Council, Forestry Department, and National Parks Board, for permission to work in the Kibale National Park and study the chimpanzees. Facilities were provided by Makerere University Biological Field Station and the Wildlife Conservation Society. The Department of Zoology, Makerere University also assisted. Funding was from the NSF (DEB-9120960), USAID, and MacArthur Foundation.

We also thank the Indonesian Ministry of Forestry (Direktorat PHKA), the Indonesian Institute of Research Science (LIPI) and the Center for Research and Development in Biology (PPPB) for their sponsorship. Grants from the National Geographic Society, the Leakey Foundation, the US Fish and

Wildlife Service, the Conservation, Food and Health Foundation, NSF, and the Wenner-Gren Foundation made the orangutan work possible. Thanks, also, to the many field assistants and managers at both sites who collected data for this study.

References

A. O. A. C. (Association of Official Analytical Chemists) (1984). *Official Methods of Analysis of the Association of Official Analytical Chemists*, 14th edn., ed. S. S. Williams. Arlington: Association of Official Analytical Chemists.

Barry, J. L., Hoebler, C., Macfarlane, G. T. *et al.* (1995). Estimation of the fermentability of dietary fiber *in-vitro* – a European interlaboratory study. *British Journal of Nutrition*, **74**, 303–22.

Bourquin, L. D., Titgemeyer, E. C., & Fahey, Jr., G. C. (1996). Fermentation of various dietary fiber sources by human fecal bacteria. *Nutrition Research*, **16**, 1119–31.

Charrondiere, U. R., Chevassus-Agnes, S., Marroni, S., & Burlingame, B. (2004). Impact of different macronutrient definitions and energy conversion factors on energy supply estimations. *Journal of Food Composition and Analysis*, **17**, 339–60.

Conklin, N. L., & Wrangham, R. W. (1994). The value of figs to a hind-gut fermenting frugivore: a nutritional analysis. *Biochemical Systematics and Ecology*, **22**, 137–51.

Conklin-Brittain, N. L., Dierenfeld, E. S., Wrangham, R. W., Norconk, M., & Silver, S. C. (1999). Chemical protein analysis: a comparison of Kjeldahl crude protein and total ninhydrin protein from wild, tropical vegetation. *Journal of Chemical Ecology*, **25**, 2601–22.

Conklin-Brittain, N. L., Wrangham, R. W., & Hunt, K. D. (1998). Dietary response of chimpanzees and cercopithecines to seasonal variation in fruit abundance. II. Macronutrients. *International Journal of Primatology*, **19**, 971–98.

Dasilva, G. L. (1992). The western black-and-white colobus as a low-energy strategist: activity budgets, energy expenditure and energy intake. *Journal of Animal Ecology*, **61**, 79–91.

Demment, M. W., & van Soest, P. J. (1985). A nutritional explanation for body size patterns of ruminant and nonruminant herbivores. *The American Naturalist*, **125**, 641–72.

Ellison, P. T. (2001). *On Fertile Ground*. Cambridge, MA: Harvard University Press.

Goering, H. K. & van Soest, P. J. (1970). Forage fiber analysis. In *Agricultural Handbook Number 379*. Washington, DC: United States Department of Agriculture, Agricultural Research Service.

Harborne, J. B. (1984). *Phytochemical Methods: A Guide to Modern Techniques of Plant Analysis*, 2nd edn. London: Chapman and Hall.

Knott, C. D. (1998). Changes in orangutan caloric intake, energy balance, and ketones in response to fluctuating fruit availability. *International Journal of Primatology*, **19**, 1061–79.

(1999). *Reproductive, Physiological and Behavioral Responses of Orangutans in Borneo to Fluctuations in Food Availability.* Unpublished Ph.D. thesis, Harvard University.

(2001). Female reproductive ecology of the apes: implications for human evolution. In *Reproductive Ecology and Human Evolution*, ed. P. Ellison, pp. 429–63. New York, NY: Aldine de Gruyter.

(2005). Energetic responses to food availability in the great apes: implications for hominin evolution. In *Primate Seasonality: Implications for Human Evolution*, ed. D. K. Brockman & C. P. van Schaik. Cambridge: Cambridge University Press.

Lee, S. C., Prosky, L., & De Vries, J. W. (1992). Determination of total, soluble, and insoluble dietary fiber in foods – enzymatic – gravimetric method, MES-TRIS buffer: collaborative study. *Journal of the AOAC International*, **75**, 395–416.

Lipson, S. F. & Ellison, P. T. (1996). Comparison of salivary steroid profiles in naturally occurring conception and non-conception cycles. *Human Reproduction*, **11**, 2090–6.

Livesey, G. (2001). A perspective on food energy standards for nutrition labeling. *British Journal of Nutrition*, **85**, 271–87.

Miles, C. W., Kelsay, J. L., & Wong, N. P. (1988). Effect of dietary fiber on the metabolizable energy of human diets. *Journal of Nutrition*, **118**, 1075–81.

Milton, K. & Demment, M. W. (1988). Chimpanzees fed high and low fiber diets and comparison with human data. *Journal of Nutrition*, **118**, 1082–8.

Milton, K. & Dintzis, F. R. (1981). Nitrogen-to-protein conversion factors for tropical plant samples. *Biotropica*, **13**, 177–81.

Moughan, P. J. (1999). In vitro techniques for the assessment of the nutritive value of feed grains for pigs: a review. *Australian Journal of Agricultural Research*, **50**, 871–9.

Nakagawa, N. (1997). Determinants of the dramatic seasonal changes in the intake of energy and protein by Japanese monkeys in a cool temperate forest. *American Journal of Primatology*, **41**, 267–88.

(2000). Foraging energetics in patas monkeys (*Erythrocebus patas*) and tantalus monkeys (*Cercopithecus aethiops tantalus*): implications for reproductive seasonality. *American Journal of Primatology*, **52**, 169–85.

N. R. C. (National Research Council) (1980). *Recommended Dietary Allowances (RDA)*, 9th edn. Washington, DC: National Academy Press.

(1981). *Nutritional Energetics of Domestic Animals & Glossary of Energy Terms*, 2nd revised edn. Washington, DC: National Academy Press.

(1989). *Recommended Dietary Allowances (RDA)*, 10th edn. Washington, DC: National Academy Press.

(2003). *Nutrient Requirements of Nonhuman Primates*, 2nd revised edn. Washington, DC: National Academy Press.

Pierce, W. C. & Haenisch, E. L. (1947). *Quantitative Analysis*, 3rd edn. New York, NY: John Wiley & Sons.

Pontzer, H. & Wrangham, R. W. (2004). Climbing and the daily energy cost of locomotion in wild chimpanzees: implications for hominoid locomotor evolution. *Journal of Human Evolution*, **46**, 317–35.

Robertson, J. B. & van Soest, P. J. (1980). The detergent system of analysis and its application to human foods. In *The Analysis of Dietary Fiber in Food*, ed. W. P. T. James & O. Theander, pp. 123–58. New York, NY: Marcel Dekker.

Schneider, B. H. & Flatt, W. P. (1975). *The Evaluation of Feeds through Digestibility Experiments*. Athens, GA: The University of Georgia Press.

Schoener, T. (1987). A brief history of optimal foraging ecology. In *Foraging Behavior*, ed. A. C. Kamil, J. R. Krebs, & H. R. Pulliam, pp. 5–67. New York, NY: Plenum Press.

van Soest, P. J. (1994). *Nutritional Ecology of the Ruminant*, 2nd edn. Ithaca, NY: Comstock Publishing Associates.

van Soest, P. J., Uden, P., & Wrick, K. F. (1983). Critique and evaluation of markers for use in nutrition of humans and farm and laboratory animals. *Nutrition Reports International*, **27**, 17–28.

Wasserman, M. D. & Chapman, C. A. (2003). Determinants of colobine monkey abundance: the importance of food energy, protein and fibre content. *Journal of Animal Ecology*, **72**, 650–9.

Watkins, B. E., Ullrey, D. E., & Whetter, P. A. (1985). Digestibility of a high-fiber biscuit-based diet by black and white colobus (*Colobus guereza*). *American Journal of Primatology*, **9**, 137–44.

Watt, B. K. & Merrill, A. L. (1963). Composition of foods: raw, processed, prepared. In *Agricultural Handbook 8*. Washington, DC: United States Department of Agriculture, Agricultural Research Service.

Wrangham, R. W., Conklin-Brittain, N. L., & Hunt, K. D. (1998). Dietary response of chimpanzees and cercopithecines to seasonal variation in fruit abundance. I. Antifeedants. *International Journal of Primatology*, **19**, 949–70.

Appendix 1. *List of chimpanzee plant foods used in energy calculations*

Chimpanzee foods eaten	Plant part	ME$_O$ Kcals/ 100 g OM	ME$_H$ Kcals/ 100 g OM	Months eaten
Acanthus pubescens	pith	194	259	2, 3, 4, 5
Aframomum spp.	ripe pulp	256	308	9, 1, 2, 3, 4, 5, 6
Blighia unijugata	ripe pulp	627	644	5
Celtis africana	ripe pulp	240	277	10
Celtis africana	young leaf	220	272	9, 10, 1, 2, 3, 4, 6
Celtis durandii	ripe pulp	511	549	10
Celtis durandii	unripe pulp	274	318	10
Ficus brachylepis	ripe pulp	246	305	2, 3
Ficus brachylepis	unripe pulp	221	288	3, 4
Ficus capensis	ripe pulp	167	255	2, 3
Ficus capensis	unripe pulp	181	269	3, 5
Ficus cyathistipula	ripe pulp	209	287	2, 3
Ficus dawei	ripe pulp	173	260	2, 3, 4
Ficus exasperata	ripe pulp	226	290	9, 4, 5, 6
Ficus exasperata	unripe pulp	241	299	4, 5, 6
Ficus exasperata	young leaf	205	265	9, 4, 5
Ficus natalensis	ripe pulp	134	236	2, 3, 4, 5
Ficus natalensis	unripe pulp	127	233	3, 5
Ficus stipulifera	ripe pulp	199	275	2, 3, 5
Ficus urceolaris	ripe pulp	314	351	2, 3, 4, 5
Linociera johnsonii	ripe pulp	207	284	9
Linociera johnsonii	unripe pulp	170	261	9
Marantochloa spp.	pith	93	196	9, 10, 2, 3, 4, 5
Mimusops bagshawei	ripe pulp	226	293	10, 1, 2
Mimusops bagshawei	unripe pulp	178	265	10
Musa x. *paradisiaca*	pith	280	326	3
Pennisetum purpurea	pith	58	183	9, 2, 3, 4, 5
Pseudospondias microcarpa	ripe pulp	164	254	9
Pseudospondias microcarpa	unripe pulp	167	256	9
Tabernaemontana holstii	ripe pulp	315	345	3
Tabernaemontana holstii	unripe pulp	298	338	3
Trichilia splendida	young leaf	152	240	9, 2, 3, 4, 5, 6
Uvariopsis congensis	ripe pulp	187	260	5, 6

Appendix 2. *Partial list of orangutan foods used in energy calculations*

Orangutan foods eaten	Plant part	ME_O Kcals/ 100 g OM	ME_L Kcals/ 100 g OM	ME_H Kcals/ 100 g OM	Months eaten
Aromadendron sp.	seed	273	298	346	10
Artabotrys sp.	seed	158	199	280	2, 6, 8, 9, 10
Artocarpus sp.	flowers	140	177	250	12, 6
Artocarpus fulvicortex	pulp	277	297	336	1, 10
Artocarpus fulvicortex	seed	219	245	296	12, 9
Artocarpus sp.	pulp	290	310	350	12, 2, 4, 6, 8, 11, 10
Baccaurea sp.	seed	284	308	353	12, 1, 2, 8, 10
Baccaurea angulata	seed	298	319	361	1
Blumeodendron sp.	seed	308	324	355	8, 11, 9
Castanopsis sp.	seed	284	300	330	12, 1, 2, 10
Chaetocarpus sp.	seed	443	469	519	12
Dillenia sp.	flowers	183	213	272	12, 4, 10
Diospyros confertiflora	seed	81	124	208	12, 8, 11, 10
Diospyros phillipenensis	pulp and seed	77	121	206	9, 10
Dipterocarpus sublamellatus	seed	369	376	390	12, 1, 2
Durio sp.	new leaves	157	191	258	4, 6, 8, 11, 9, 10
Durio sp.	mature leaves	133	172	246	10
Durio sp.	seed	214	240	289	12, 1, 2, 10
Durio sp.	pulp	273	293	331	12, 1, 2, 10
Ficus binnendykii	whole fruit	100	141	223	6, 8, 10
Ficus dubia	pulp and skin	243	270	324	4, 10
Ficus kerkhovenii	pulp and skin	83	126	211	6
Ficus punctata	pulp and skin	204	235	295	4, 6, 10
Ficus stupenda	whole fruit	137	174	246	12, 4, 6, 8, 11, 9, 10
Ficus subgelderi	whole fruit	159	194	261	4, 6, 8, 11, 10
Ficus sycidium	whole fruit	158	193	263	4, 6, 8, 11, 10
Garcinia cowa	pulp and seed	287	309	351	10, 1, 2
Gironierra sp.	whole fruit	208	240	302	9
Gnetum sp.	skin	157	191	257	12, 4, 6, 8, 11, 10
Grewia sp.	whole fruit	102	143	225	9, 10
Hydnocarpus sp.	pulp	302	316	343	12, 9, 10

Appendix 2. (*cont.*)

Orangutan foods eaten	Plant part	ME_O Kcals/ 100 g OM	ME_L Kcals/ 100 g OM	ME_H Kcals/ 100 g OM	Months eaten
Irvingia sp.	seed	397	408	429	12
Lithocarpus sp.	seed	319	331	353	4, 6, 9, 10
Neesia sp.	seed	590	596	606	2, 4, 6
Palaquium sp.	pulp	249	272	316	4, 6, 11, 9, 10
Pandanus *epiphyticus*	Pith	185	215	273	1, 2, 4, 6, 8, 11, 9, 10
Pandanus sp.	seed	255	286	346	10
Polyalthia sp.	seed	62	110	203	4, 11, 10
Pternandra sp.	whole fruit	174	206	267	8
Santiria sp.	whole fruit	234	260	310	2, 6
Scaphium sp.	seed	331	345	374	12, 1, 2
Scutinanthe sp.	pulp	390	405	434	12, 1
Sindora sp.	seed	326	341	369	12, 1, 2, 8, 11, 10
Sterculia sp.	seed	98	140	222	12
Syzygium sp.	skin and pulp	106	146	224	12, 1, 4, 11
Willughbeia sp.	seed	284	301	333	12, 1, 2, 4, 6, 8, 9, 11, 10
Willughbeia sp.	pulp	309	326	359	12, 1, 2, 4, 6, 8, 9, 11, 10

Glossary (van Soest, 1994; N. R. C., 2003)

Ash.   The mineral content of food, it can also contain possible soil or dust contamination. A small sub-sample is placed in a furnace and burned at 500–550 °C. This removes the organic matter and leaves the inorganic mineral as a residue.

Bomb calorimetry.   A method for determining energy content in organic material. A sub-sample is placed in the burn chamber, which is surrounded by a second chamber full of water. The sample is ignited and the increase in temperature of the water is measured; energy is calculated from that increase and reported as kilocalories or kilojoules.

Crude fiber (CF).   A profoundly antiquated method for determining fiber content, it principally measures cellulose, detecting varying, small amounts of the other two important fibers, lignin and hemicellulose.

Crude protein (CP).   See Kjeldahl crude protein below.

Detergent System of Fiber Analysis.

1. *Neutral-detergent Fiber* (NDF). Total insoluble fiber, contains the three principal insoluble fibers hemicellulose, cellulose, and lignin. A neutral pH detergent (ND) solution is used to dissolve the soluble cell contents, as well as soluble fibers, leaving the insoluble fibers of the plant cell wall. Also referred to as structural, plant cell wall fibers.

2. *Acid-detergent Fiber* (ADF). An intermediate step in the process of separating the three principal fibers, it is not meant to be used as if it were a fiber fraction itself. An acidic detergent solution (AD) is used to:

    (i)   remove the soluble cell contents and the hemicellulose if this is a nonsequential extraction, or

    (ii) remove only the hemicellulose if used sequentially after a pre-extraction with ND. Both cellulose and lignin remain in the AD residue.

3. *Hemicellulose* (HC). A general term for a group of similar, insoluble carbohydrate fibers. The acid stomach can break the hemicellulose strands into smaller pieces, making them more available to microbial attack, but HC is not digested by mammalian enzymes, requiring microbial fermentation to break it down. The quantity of HC present in a sample is determined by difference: %NDF minus %ADF = % HC. Hemicellulose is not closely related to cellulose, despite the name. It is a heterogenous group of carbohydrates rich in 5-carbon sugars.

4. *Lignin* (Ls or ADL). An indigestible and essentially unfermentable group of similar and insoluble polyphenolic fibers. A common method for determining lignin quantity dissolves the cellulose using sulfuric acid and leaves the lignin as residue (Ls or ADL). The lignin residue is weighed and the cellulose (Cs) is determined by difference: %ADF minus % Ls = % Cs. Other methods exist (e.g., potassium permanganate; Van Soest, 1994).

5. *Cellulose* (Cs). An insoluble carbohydrate fiber, chemically very similar to starch except that the glucose (a 6-carbon sugar) units are linked by beta-linkages instead of the alpha-linkages in starch. This renders cellulose indigestible by mammalian enzymes but fermentable by microbial enzymes. See *Lignin* for how cellulose is quantified analytically.

Dry matter (DM).   All of the nutrients of a food are contained within the dry matter. The water in fresh food can interfere with chemical evaluation and interpretation, so nutrient analyses are performed on the dry material after the water has been evaporated. Nutrients are then expressed as a percentage of dry matter.

1. *DM digestion coefficient.* Is the percentage of the total dry matter of the whole food that is digested. This is in contrast to, for example, a fiber digestion coefficient or a protein digestion coefficient.

Energy.   One kilocalorie = 4.184 kilojoules.

1. *Gross energy* (GE). The total energy released by a food item when burned in a bomb calorimeter. To determine nutritionally available energy, the feces and urine resulting from an ingested food are collected and also bombed to determine their GE value.
2. *Digestible energy* (DE). = GE minus fecal energy.
3. *Metabolizable energy* (ME). = DE minus urinary energy. There are additional, minor energy losses from the body, nevertheless, ME is generally used for simple-stomached animals like apes and humans.

Fermentation.   The microbial digestion of food, including fiber. Adequate amounts of free simple sugars and moisture can result in a variety of end-products, e.g., alcohols and acids. The common end-products resulting from fermentation in the gastrointestinal (GI) tract are volatile fatty acids. These acids are absorbed into the bloodstream of the host animal and used as energy sources.

Fiber digestion coefficient.   Defines the fraction of ingested fiber that is digested and thus not measured in feces. If 100 g of fiber (NDF) were ingested and you determined that 45.7 g of NDF escaped fermentation and came out in the feces, that means that 54.3 g (or 54.3%) stayed in the body and were digested through fermentation. The digestion coefficient is 0.543. This is a physical separation, not a metabolic or energetic separation.

1. *Fiber conversion factor.* The GI tract microbes keep about 1 kcal/g for their own growth, leaving 3 kcal/g for the host in the form of volatile fatty acids. However, you cannot apply the 3 kcal/g conversion factor to all 100 g eaten above, only to the 54.3 g actually digested. Thus, the real, applicable conversion factor for our example is $3 \times 0.543 = 1.6$ kcal/g of the original amount of fiber consumed.

Hot weighing.   An alternative to the process of weighing samples using a desiccator. Samples or extracts are weighed directly out of an oven.

Kjeldahl crude protein (CP).   Kjeldahl is the most commonly used and accurate method for determining nitrogen content of a food sample. There is disagreement over which conversion factor should be used to convert %N into %protein (6.25 or 4.3, etc.) (Milton & Dintzis, 1981). An alternative solution for determining nutritionally available nitrogen/protein is suggested by Conklin-Brittain *et al.* (1999).

Lipid.   Lipid and fat are synonymous. Lipids are hydrophobic organic compounds and need to be extracted with organic solvents, e.g., petroleum ether. Room temperature extraction removes less of the indigestible waxes or latexes than does extraction at higher temperatures, probably giving a more nutritionally available estimate of fat content.

Mineral content.   See "Ash".

Organic matter.   There are two major fractions present within dry matter (see above), organic compounds and inorganic (mineral) compounds. If the ash values from the various foods in a diet vary by more than 5 percentage points, report the nonmineral nutrients as a percentage of organic matter instead of as a percentage of dry matter. The amount of ash in dry matter can be too variable and potentially confounding.

1.   *OM correction coefficient.* = (1 – ash coefficient) × DM coefficient.

Physiological fuel values.   4 kcal/g (16 kJ/g) TNC, 4 kcal/g (17 kJ/g) CP, 9 kcal/g (37 kJ/g) lipid.

Soluble fibers and pectin.   See also TNC. Most soluble fibers are found in the cell sap, except for pectin, which is found in the cell wall along with the insoluble fibers. Other examples of soluble fibers in the human diet are beta-glucans in oats and barley, fructosans in Jerusalem artichokes (a tuber), and mannans in seaweeds and nuts.

Total Dietary Fiber (TDF).   An enzymatic method that partitions insoluble and soluble fibers. It gives a total insoluble fiber value, roughly equivalent to NDF, and a total soluble fibers value.

Total Non-structural Carbohydrates (TNC).   Includes monosaccharides (e.g., glucose), disaccharides (e.g., sucrose), starch, and soluble fibers (also known as nonstarch polysaccharides, see also Soluble Fibers). The complex carbohydrates are referred to as "water-soluble" but water is not sufficient to dissolve them into their individual sugar units; at most they become soft or gelatinous, and enzymes are needed to break them down further.

18 The role of sugar in diet selection in redtail and red colobus monkeys

LISA DANISH, COLIN A. CHAPMAN, MARY BETH HALL, KARYN D. RODE, AND CEDRIC O'DRISCOLL WORMAN

Introduction

When considering the distribution of nutrients in tropical forests, the traditional view has been that fruits tend to be high quality, provide easily digested forms of carbohydrates but low levels of fiber, usually contain

Feeding Ecology in Apes and Other Primates. Ecological, Physical and Behavioral Aspects, ed. G. Hohmann, M.M. Robbins, and C. Boesch. Published by Cambridge University Press. © Cambridge University Press 2006.

few secondary compounds, but also provide little protein. In contrast, leaves offer more protein but have higher levels of fiber, little energy, and are more likely to include undesirable secondary compounds (Waterman, 1984; Milton, 1993; Janson & Chapman, 1999). From an evolutionary perspective, there are two ways to cope with this distribution of nutrients (Ganzhorn, 1989; Milton, 1993). The first is to have a digestive tract that allows the structural polysaccharides of fiber to be metabolized via a fermentative digestion process involving the use of microbes. This is the foraging strategy used by folivorous primates, including *Alouatta* spp., colobines, and some lemurs (Milton, 1998). The second means of coping is the adoption of behavioral strategies that will obtain foods of the highest quality. The tendency for species to be large in size is associated with morphological changes that allow animals to survive on lower-quality plant parts. Larger animals can obtain adequate nourishment by taking in less energy per unit of body mass and thus can meet their energy requirements on lower-quality foods (Milton, 1993; Robbins, 1993; Cork, 1996; McNab, 2002; Remis, 2002). Thus, large-bodied primates tend to be folivorous, consuming a low-quality diet (high fiber/low energy), while many small-bodied primates typically consume a largely frugivorous diet, presumably due to the high energy provided by sugars (Ungar, 1995; Cork, 1996).

However, viewing diet selection in such a dichotomous fashion may curtail studies that examine the importance of particular nutrients (e.g., sugars) and may bias our initial perception of dietary preferences. For example, despite expectations based on body size differences, gorillas and chimpanzees exhibit similar food preferences (Remis, 2002; male gorillas are almost triple the weight of male chimpanzees, Rowe, 1996). Both species prefer foods high in non-starch sugars and low in fiber, and do not avoid foods containing tannins. Furthermore, the diets of different populations of species traditionally classed as frugivores or folivores can be highly variable with respect to the amount of fruit and leaves they consume (Chapman & Chapman, 1990; Norconk & Conklin-Brittain, 2004), suggesting that care must be taken in accepting any generalization with respect to fiber/sugar use and primate digestive strategies.

The objective of this study was to contrast the use of sugars by the redtail monkeys (*Cercopithecus ascanius*) and red colobus (*Piliocolobus tephrosceles*) of Kibale National Park, Uganda. Red colobus are anatomically adapted for digesting leaves, while redtail monkeys are not. Therefore, we had three specific goals: to (1) provide a comparable description of the diet of these two species; (2) contrast the sugar content of plant foods each species eats; and (3) evaluate if either of these species is selecting for or against food items based on the food's sugar content.

Redtail monkeys (adult male 4.2 kg, female 3.3 kg) and other guenons are generally considered frugivorous (Conklin-Brittain *et al.*, 1998; Chapman *et al.*, 2002a). Fruit makes up an average of 47% of the diet of redtail monkeys, while leaves and insects each make up an average of 24%. However, there is a great deal of variation among redtail populations, particularly in the consumption of fruit and leaves. For example, the consumption of fruit has been documented as ranging from 13%–61% and the consumption of leaves ranges from 7%–74% across populations (Chapman *et al.*, 2002a). Redtail monkeys do not have specialized stomachs (Lambert, 2002), and as a result, cannot effectively metabolize the structural carbohydrates found in leaves.

Unlike guenons, the colobines, including the red colobus (adult male 10.5 kg, female 7.0 kg), are considered folivorous (Oates, 1994; Milton, 1998). For example, leaves make up an average of 62.4% of the diet of the red colobus (Oates, 1994). Fruit makes up an average of 27.9% of their diet, but the consumption of fruit can be as low as 5.7%. Colobine monkeys use forestomach fermentative digestion to metabolize the structural carbohydrates found in leaves (Bauchop, 1978; Chivers, 1994). It is believed that colobines avoid foods that are high in sugar since these foods can lower forestomach pH, which can result in a decrease in fermentation efficiency or cause acidosis, and in extreme cases can result in death (Kay & Davies, 1994; Milton, 1998). Thus, the ingestion of ripe fruit has been viewed as incompatible with a diet containing significant amounts of leaf material (Kay & Davies, 1994). Furthermore, given the long retention time in colobines (Lambert, 1998), they may not be able to obtain sufficient quantities of other nutrients (e.g., protein) on a diet with large amounts of fruit.

Based on anatomical adaptations, body size differences, and the fact that high levels of sugar can cause acidosis in folivores, the expectation is that redtail monkeys will frequently feed on foods that have high sugar contents, while red colobus monkeys will not. However, the large variation among populations of these species in the amount of time they spend eating fruit suggests that the consumption of simple sugars needs to be carefully evaluated.

Methods

Plant samples were gathered in Kibale National Park (795 km^2) in western Uganda (0°13′–0°41′N and 30°19′–30°32′E) near the foothills of the Rwenzori Mountains (Chapman & Chapman, 2002; Chapman *et al.*, 2002b). Mean rainfall in the region is 1741 mm (1990–2003). Rainfall is bimodal with the two rainy seasons generally occurring from March to May and September to November. The mean daily minimum temperature is 15.5 °C and the mean daily maximum temperature is 23.7 °C.

Information on the diets of red colobus and redtail monkeys was derived from Chapman *et al.* (2002b) and Rode *et al.* (in press). For redtails, six groups were observed from May 2001–May 2002. A preliminary field season conducted from May–July 2000 identified six study groups: three in the unlogged area (Kanyawara K-30 Forestry Compartment) and three in logged areas (Mikana, K-15). Preceding the beginning of data collection in May 2001, the four less habituated groups were followed every month for 4 months to increase habituation. To examine seasonal variation in diet, a single redtail group (primary group) was followed in each area for 6 days each month. The two other groups in each area (secondary groups) were followed 3 days every other month to determine if diets and behaviors observed in the primary group in each area were characteristic of both logged and unlogged habitats. An average of 418 hours of observation was collected on each of the two primary groups and 134 hours on each of the four secondary groups for a total of 1372 hours.

For red colobus, behavioral observations were made in the Kanyawara area of Kibale between July 1994 and June 1999 for a total of 2425 hours. This area (K-30, elevation = 1500 m) consists of 282 ha that have not been commercially harvested. However, prior to 1970, a few large stems (0.03 to 0.04 trees/ha) were removed by pitsawyers. This extremely low level of extraction seems to have had very little impact on the structure and composition of the forest (Struhsaker, 1997). In both of these studies, during each half-hour the observer was with the group, five point samples were made of different adults. If the animal was feeding, the species and plant part (e.g., fruit, young leaf, and leaf petiole) were recorded. We made an effort to avoid repeatedly sampling particularly conspicuous animals by moving throughout the group when selecting subjects and by sampling animals that were both in clear view and those that were more hidden. Groups were followed from approximately 700–1700 hours, with scans being conducted every 15 minutes.

Interspecific dietary overlap between redtail and red colobus monkeys was calculated using the following formula: $D = \Sigma S_i$ where D = dietary overlap and S_i = percentage of diet shared between two species, evaluated on a plant species and part basis. This formula was first used by Holmes & Pitelka (1968) and has become a standard means of expressing dietary overlap for primates (Struhsaker, 1975; Struhsaker & Oates, 1975; Chapman, 1987; Maisels *et al.*, 1994).

Plant samples were obtained using a tree-pruning pole to cut down a limb from individual trees that had reached mature size (Chapman & Chapman, 2002). Items were processed in the same manner as that used by the study animals, and only those parts selected by the study animals were collected. Samples were dried in the field by sun drying, using a dehydrator that circulated warm air past the samples, or using a light-bulb heated box containing a series of racks (Chapman *et al.*, 2003). Samples were dried thoroughly to avoid mold. We assured that all samples were dried below 50 °C by placing max/min thermometers with the drying samples. When samples were dried in the drying oven, the oven was set at its lowest heat setting (37 °C). Dried samples were stored in sealed plastic bags until they could be transported to the University of Florida for analysis. Dried samples were ground finely enough to pass through a 1-mm mesh screen in a Wiley mill. Dry matter was determined by drying a portion of each sample overnight at 105 °C. Due to variations in nutritional parameters over time and between individual trees, four different samples were analyzed for each species, provided that there were a sufficient number of samples.

The primary components of plant carbohydrates that are easily digestible by mammalian enzymes were quantified using a phenol-sulfuric acid assay method that requires extraction (80% ethanol) and digestion with colorimetric analysis of filtrates (Dubois *et al.*, 1956; Hall *et al.*, 1999; Hall, 2001). Standards were made using sucrose and absorbance was read from a spectrometer at 490 nm. This procedure allows an assessment of all simple sugars lumped together monosaccharides- (glucose, fructose), disaccharides- (sucrose, lactose), and oligosaccharides. Oligosaccharides include α-galactosides (raffinose, stachyose) and short chain fructans (fructooligosaccharides). These oligosaccharides are not considered digestible; as a result, the level of sugar in primate diets may be overestimated (Asp, 1993). However, it is not known whether the foregut (ruminal) fermentation of the red colobus monkeys allows oligosaccharides to be digested. The concentration of these oligosaccharides in primate foods is also not known; α-galactosides are generally found in leguminous seeds, while fructans are found in several vegetables, including onions and artichokes. For ease of discussion we refer to all of these compounds as an evaluation of sugar content.

We present a standard description of foraging efforts (per cent of foraging scans devoted to particular items) for each species and contrast species and plant part using t-tests. To determine whether the monkeys select for foods based only on the sugar content we correlate foraging effort and sugar content using data presented in Tables 18.1a and b. We also conduct partial correlation analyses to statistically control for the effects of tree density and size (DBH).

Results

Red colobus spent an average of 2.7% (sd = 4.60) of their monthly foraging time eating fruit; however, this value ranged from 0–31.8% (Figure 18.1). In contrast, redtail monkeys fed on fruit much more often and spent an average of 34.1% (sd = 10.6) of their monthly foraging time eating fruit. Over the course of the observation months, the proportion of the foraging time they devoted to fruit ranged from 20.1%–52.4% (Figure 18.2).

In contrast, red colobus spent an average of 74.3% (mature 8.10, sd = 8.46 / young 66.2, sd = 14.28; also petioles 11.24, flowers 2.22) of their monthly foraging time eating leaves; however, young leaf consumption ranged from 47.8–100%. Redtail monkeys fed on leaves much less often and they spent an average of 14.9% (mature 0.23, sd = 0.74, young 10.66, sd 4.73, petioles 4.65, sd 4.41) of their monthly foraging time eating leaves. Over the course of the observation months, the proportion of their foraging time devoted to leaves ranged from 4.8%–26.9% (Figure 18.2).

The mean sugar content of fruits consumed by red colobus monkeys (13.65%) did not differ significantly from that of fruits consumed by redtail monkeys (16.10%, t = 0.555, p = 0.584). Fruits consumed by redtail monkeys were expected to be high in sugar; however, the folivorous red colobus monkeys also ate fruits with high levels of sugar (maximum for both species = 38.89%).

The mean sugar content of leaves consumed by red colobus monkeys (7.71%) was not significantly lower than that of leaves consumed by redtail

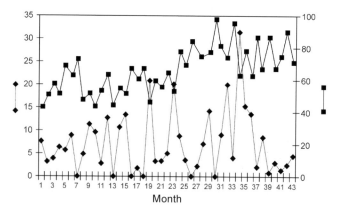

Figure 18.1. Monthly variation in red colobus foraging efforts (percent of feeding scans) devoted to eating young leaves (squares with solid line) and fruits (diamonds with dashed line) in Kibale National Park, Uganda.

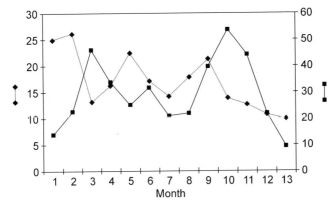

Figure 18.2. Monthly variation in redtail monkeys foraging efforts (percent of feeding scans) devoted to eating leaves (young, old, and petioles: squares with solid line) and fruits (diamonds with dashed line) in Kibale National Park, Uganda.

monkeys (8.49%, t = 0.456, p = 0.652). Given the amount of time typically spent eating leaves in a single day, the red colobus ingest substantial amounts of sugars in their daily diet.

For both red colobus (t = 2.146, p = 0.041) and redtails (t = 2.363, p = 0.027) the sugar content of fruits eaten was higher than that of leaves. However, a number of fruit species have sugar contents similar or even lower than the mean sugar level of the leaves that were eaten (Table 18.1). Thus, it is evident that leaves are a potentially significant source of sugars.

There was no evidence to suggest that either species was selecting food items based only on the sugar content (i.e., there was no correlation with foraging effort – percent of the feeding scans – and sugar content). This held true for both leaves and fruits, and regardless of whether we statistically controlled for the effects of tree density or size (the lowest probability in any of these relationships was 0.347).

While in this study we collected the data on foraging and nutritional content of red colobus and redtail foods at different times, previously we have observed the association patterns and foraging behavior of these two species simultaneously (Chapman & Chapman, 1996; 2000), but did not in these cases collect items for nutritional analysis. The data collected in these previous studies indicates that these species were in association between 23.0 and 52.5% of the time, depending on the year of study, and that their diets overlapped by 25.1%. They were often observed feeding in the same tree on the same food item when in association.

Table 18.1. *The foraging effort (% of all feeding scans) and sugar content (mean and range, if multiple samples could be analyzed) of young leaves (1a) and fruits (1b) consumed by redtail and red colobus monkeys*

(a)

Leaves	% Sugar	Range	RC	RT All foods	RT Plant foods
Apocynaceae					
Funtumia latifolia (YL)	8.98%	7.72–11.10	0.018	0.017	0.030
Bignoniaceae					
Markhamia lutea (YL)	4.83%	2.80–8.80	2.964	0.474	0.826
Ebenaceae					
Diospyros abyssinica (YL)	15.00%	11.50–16.70	0.054	0.090	0.157
Euphorbiaceae					
Macaranga sp. (YL)	6.24%	6.02–6.47	0.029	0	0
Leguminosae					
Albizia grandibracteata (YL)	7.81%	5.40–9.78	5.643	0.011	0.019
Moraceae					
Trilepisium madagascariense (YL)	7.00%	5.85–9.04	3.036	7.490	13.059
Ficus brachylepsis (YL)	8.37%	8.37	0.643	0	0
Ficus exasperata (YL)	1.73%	0.84–2.93	0.768	0.022	0.038
Olacaceae					
Strombosia scheffleri (YL)	3.32%	1.77–4.79	4.143	0	0
Oleaceae					
Olea welwitschii (YL)	17.39%	15.48–20.67	0.643	0.33	0.575
Rosaceae					
Parinari excelsa (YL)	9.38%	7.60–11.11	4.429	0	0
Prunus africana (YL)	13.31%	10.75–17.36	4.750	0.022	0.038
Prunus africana (ML)	11.15%	9.37–13.13	2.786	0	0
Rubiaceae					
Rothmannia urcelliformis (YL)	3.43%	1.75–4.81	0	2.63	4.586
Sapotaceae					
Pouteria altissima (YL)	6.11%	6.11	0.768	0	0
Chrysophyllum sp. (YL)	5.06%	5.06	0.875	0	0
Mimusops bagshawei (YL)	3.22%	1.33–4.21	1.643	0	0
Sterculiaceae					
Dombeya kirkii (YL)	3.63%	2.51–5.10	4.429	0	0
Ulmaceae					
Celtis africana (YL)	4.03%	2.02–5.27	8.768	0.045	0.078
Celtis durandii (YL)	9.97%	6.89–13.97	10.696	0.338	0.589
Rubiaceae					
Rothmannia urcelliformis (YL)	3.43%	1.75–4.81	0	2.63	4.586

(b)

Fruit	% Sugar	Range	RC	RT All foods	RT Plant foods
Annonaceae					
Uvariopsis congensis (RF)	5.47%	4.35–6.80	0	4.66	8.125
Uvariopsis congensis (UF)	7.80%	4.19–11.40	0	1.18	2.057

Table 18.1. (*cont.*)

Fruit	% Sugar	Range	RC	RT All foods	RT Plant foods
Boraginaceae					
Cordia millenii (RF)	34.65%	34.65	0	0.32	0.558
Ebenaceae					
Diospyros abyssinica (RF)	26.55%	16.24–36.20	Obsv'd	0.520	0.907
Diospyros abyssinica (UF)	13.35%	9.19–19.20	Obsv'd	0.63	1.098
Euphorbiaceae					
Bridelia micrantha (RF)	17.17%	17.17	Obsv'd	0.090	0.157
Leguminosae					
Millettia dura (RF)	6.66%	6.66	Obsv'd	0	0
Moraceae					
Ficus brachylepsis (RF)	12.21%	11.31–13.10	0.089	1.55	2.703
Ficus natalensis (RF)	7.80%	6.20–9.40	Obsv'd	0	0
Oliniaceae					
Psidium guajava (RF)	3.37%	2.49–4.24	Obsv'd	Obsv'd	Obsv'd
Rosaceae					
Prunus africana (RF)	38.89%	38.89	0.036	0.650	1.133
Rubiaceae					
Rothmannia urcelliformis (RF)	33.81%	33.81	0	0.022	0.038
Vangueria apiculata (RF)	4.71%	4.71	0	0.135	0.235
Sapotaceae					
Mimusops bagshawei (RF)	20.85%	15.10–28.50	0	1.250	2.179
Ulmaceae					
Celtis africana (UF)	1.34%	1.30–1.38	0.018	1.420	2.476
Celtis durandii (RF)	10.00%	2.75–12.80	1.204	1.420	2.476
Celtis durandii (UF)	12.90%	4.95–22.60	3.125	1.860	3.243
Chaetacme aristata (RF)	22.72%	22.72	0	0.022	0.038

For redtail monkeys, foraging effort is presented with respect to all foods (i.e., including insects) and all plant foods (i.e., excluding insects). Obsv'd indicates that animals have been seen eating that species/part, but it was not recorded in the scan data; this includes instances when neighboring groups ate this species/part (abbreviations RC = red colobus, RT = redtails, UF = unripe fruit, RF = ripe fruit, ML = mature leaves, YL = young leaves).

Discussion

Although the sugar content of leaves is less than that of fruits, it is evident that leaves have the potential to be an important source of sugars. As a result, some leaves can be considered high-quality food since they are a source of protein (Chapman & Chapman, 2002), minerals (Rode *et al.*, 2003), and sugars (this study). Thus, since redtail monkeys are frugivorous and small-bodied, it is not surprising that their diet includes leaves whose high sugar content helps meet their heavy metabolic demands. By consuming leaves, redtail monkeys can obtain protein and minerals that they do not typically get from fruits, while still consuming a significant quantity of sugar. The high sugar content of leaves may make them a more valuable food source for frugivores than has typically been thought. Additionally, the availability of leaves with high sugar content may be important as fall-back foods for frugivorous primates during periods of fruit scarcity.

Interpreting the dietary strategy of the folivorous red colobus with respect to sugars is more difficult. Some of the leaves they feed on have considerable amounts of sugar and thus are likely a significant source of energy. There is evidence that this is also the case for other folivores. For example, Koenig *et al.* (1998) documented that two out of the three key resources for a population of Hanuman langurs (*Presbytis entellus*) contained significantly higher levels of sugar than other foods. Even the mature and young leaves of the most preferred food plant contained over 9.7% sugar on a dry matter basis. On average, the red colobus of Kibale did not feed on fruits to any great extent. However, during some months fruits did constitute a large proportion of their diet. For example, in one month the study group spent 21.2% of their foraging time eating fruit. During this month the most frequently eaten fruit was *Celtis durandii*, which has a relatively high sugar content (10%). Koenig *et al.* (1998) similarly documented that Hanuman langurs can feed extensively on fruits that contain high sugar levels (e.g., *Dillenia pentagyna*, 27.6% sugar on a dry matter basis). Other small-bodied animals with fermentative digestion systems have also been reported to consume foods high in sugars. For example, the fruits eaten by duiker (*Cephalophus* spp.) vary in their sugar content from 0.2%–15.7% (there are no data regarding the sugar content of the leaves consumed by duikers, nor comparable information on their efforts to forage for these fruits; Dierenfeld *et al.*, 2002).

Studies that have examined the effects of sugars on foregut fermentation and apparent nutrient supply in dairy cattle offer some insights into the use of sugars by colobines. Sudden introduction of sucrose and other sugars have been used to induce low ruminal pH (Krehbiel *et al.*, 1995); however, this detrimental effect was not seen in animals used to consuming sugars (Broderick *et al.*, 2002). Reduced microbial fermentation of fiber that is not related to low pH has been noted with sugar consumption by cattle when nitrogen degradable by microbes was limiting (Heldt *et al.*, 1999) or was apparently due to proteinaceous inhibitors produced by glucose utilizing microbes in vitro (Piwonka & Firkins, 1996). In contrast, improvement in fiber digestion has been reported with sugar consumption by cattle when microbially degradable protein was not limiting (Heldt *et al.*, 1999). In terms of the net effect of sugars on nutrient supply from fermentation, reduction in nutrient yield from fiber digestion may be balanced by nutrients from fermented sugars, whereas an increase in fiber digestion could increase the overall supply. Additionally, some portion of the sugars may be stored as glycogen by microbes (Thomas, 1960), offering a source of digestible glucose if it passes to the small intestine.

In cattle used to consuming higher dietary concentrations of sugar (10%–13% of dry matter), sugars supported high levels of milk production relative to starch, suggesting that the nutrient supply was not compromised and may have been enhanced (Broderick *et al.*, 2002). However, in both studies, efficiency of nitrogen utilization was decreased for animals on the specially prepared sugar diets. That may be a function of the lower yield of microbial protein from sucrose than from starch (Hall & Herejk, 2001; Sannes *et al.*, 2002), but may also be related to decreases in ammonia and branch chain volatile fatty acid concentrations (Sannes *et al.*, 2002) – components that are essential to protein production by fiber digesting bacteria. Some studies have reported either decreased milk production in dairy cattle (Sannes *et al.*, 2002) or no change (McCormick *et al.*, 2001) with sugar supplementation. In the latter case, the ryegrass forage that formed the basal diet contained approximately 14%–17% nonfiber carbohydrates, of which more than half would likely have been classified as sugars. These studies suggest that the effect of adding a food item that is high in sugars to the diet will depend on the level of protein and sugar in the basal diet and the animal's previous exposure to the diet. Given the importance of protein to folivorous primates (Ganzhorn, 1992; Chapman & Chapman, 2002; Chapman *et al.*, 2004), it seems likely that colobus will only eat foods high in sugars when protein acquisition will not be compromised. In the month that the red colobus fed the most on fruit (21.2% of feeding scans), they also fed extensively on the young leaves of *Celtis durandii*, which has the highest protein to fiber ratio of any plant part eaten; in fact it is 35% higher than the protein/fiber content of the next species/part. Norconk & Conklin-Brittain (2004) found that lipids were as important in a similar fashion for white-faced saki (*Pithecia pithecia*) as proteins are for colobines: namely that the saki select foods high in lipids even when lipids are negatively correlated with sugars and positively correlated with lignin.

Several studies have shown that the composition of sugars can affect both food selection and fermentation processes. A study of frugivorous bats showed that even in a mammal with a non-fermenting digestive system, thresholds for sugar intake existed and differed with respect to the types of sugars ingested (Herrera *et al.*, 2000). Additionally, the composition of volatile fatty acids in the rumen of sheep differed depending on what type of sugar was added to the diet (Chamberlain *et al.*, 1993). Thus, it is possible that red colobus and redtail monkeys consume fruits and leaves that differ in specific types of sugars. For example, in sheep, sucrose had less of a negative effect on microbial fermentation than did fructose (Chamberlain *et al.*, 1993). Similar effects could occur in colobines, which would be an interesting area for future studies.

It is apparent that we do not yet fully understand the fermentative digestion system. This study presents data that contradict the historically accepted idea that high levels of sugar cannot be consumed by primates with fermentative digestion systems, and this challenge of the historical view is supported by investigations of other non-primates with similar digestive systems. As opposed to being harmful, sugars may simply alter the digestion efficiency of other nutrients, particularly protein and fiber. It is possible that this decrease in efficiency is a reasonable exchange for the benefit of consuming sugars, but only under certain conditions (e.g., when protein intake is high). Clearly, more research is needed to understand the physiological limitations of fermentative digestive systems. It seems reasonable to hypothesize that there is a threshold ratio of sugar to fiber for colobus at which diet digestibility declines, and that compared with colobus, redtails could tolerate a much higher sugar to fiber ratio. Furthermore, the effect of sugar on digestibility may be dependent on sugar type. To address these questions, feeding trails are desperately needed and would be a fascinating direction for future research.

Acknowledgment

Funding for this research was provided by the Wildlife Conservation Society, McGill University, the National Science Foundation (grant number SBR-990899, SBR 0342582), and the Leakey Foundation. Permission to conduct this research was given by the National Council for Science and Technology, and the Uganda Wildlife Authority. Lauren Chapman and Tom Gillespie provided helpful comments on this research.

References

Asp, N.-G. (1993). Nutritional importance and classification of food carbohydrates. In *Plant Polymeric Carbohydrates*, ed. F.D. Meuser, J. Manneis, & W. Seibel, pp. 121–6. Cambridge: Royal Society of Chemistry.

Bauchop, T. (1978). Digestion of leaves in vertebrate arboreal folivores. In *The Ecology of Arboreal Folivores*, ed. G.G. Montgomery, pp. 193–204. Washington, DC: Smithsonian Institution Press.

Broderick, G.A., Luchini, N.D., Radloff, W.J., Varga, G.A., & Ishler, V.A. (2002). Effect of replacing dietary starch with sucrose on milk production in lactating dairy cows. In *U.S. Dairy Forage Research Center 2000–2001 Research Report*, pp. 116–18. Madison, WI: United States Department of Agriculture Agricultural Research Service.

Chamberlain, D.G., Robertson, S., & Choung, J.J. (1993). Sugars versus starch as supplements to grass-silage: effects on ruminal fermentation and the supply of microbial protein to the small-intestine, estimated from the urinary-excretion of purine derivatives, in sheep. *Journal of the Science of Food and Agriculture*, **63**, 189–94.

Chapman, C.A. (1987). Flexibility in diets of three species of Costa Rican primates. *Folia Primatologica*, **49**, 90–105.

Chapman, C.A. & Chapman, L.J. (1990). Dietary variability in primate populations. *Primates*, **31**, 121–8.

(1996). Mixed species primate groups in the Kibale Forest: ecological constraints on association. *International Journal of Primatology*, **17**, 31–50.

(2000). Interdemic variation in mixed-species association patterns: common diurnal primates of Kibale National Park, Uganda. *Behavioral Ecology and Sociobiology*, **47**, 129–39.

(2002). Foraging challenges of red colobus monkeys: influence of nutrients and secondary compounds. *Comparative Biochemistry and Physiology Part A: Molecular and Integrative Physiology*, **133**, 861–75.

Chapman, C.A., Chapman, L.J., Cords, M. *et al.* (2002a). Variation in the diets of *Cercopithecus* species: differences within forests, among forests, and across species. In *The Guenons: Diversity and Adaptation in African Monkeys*, ed. M. Glenn & M. Cords, pp. 325–50. New York, NY: Plenum Press.

Chapman, C.A., Chapman, L.J., & Gillespie, T.R. (2002b). Scale issues in the study of primate foraging: red colobus of Kibale National Park. *American Journal of Physical Anthropology*, **117**, 349–63.

Chapman, C.A., Chapman, L.J., Naughton-Treves, L., Lawes, M.J., & McDowell, L.R. (2004). Predicting folivorous primate abundance: validation of a nutritional model. *American Journal of Primatology*, **62**, 55–69.

Chapman, C.A., Chapman, L.J., Rode, K.D., Hauck, E.M., & McDowell, L.R. (2003). Variation in the nutritional value of primate foods: among trees, time periods, and areas. *International Journal of Primatology*, **24**, 317–33.

Chivers, D.J. (1994). Functional anatomy of the gastrointestinal tract. In *Colobine Monkeys: Their Ecology, Behavior, and Evolution*, ed. A.G. Davies & J.F. Oates, pp. 205–28. Cambridge: Cambridge University Press.

Conklin-Brittain, N.L., Wrangham, R.W., & Hunt, K.D. (1998). Dietary response of chimpanzees and cercopithecines to seasonal variation in fruit abundance. II. Macronutrients. *International Journal of Primatology*, **19**, 971–98.

Cork, S.J. (1996). Optimal digestive strategies for arboreal herbivorous mammals in contrasting forest types. Why koalas and colobines are different. *Australian Journal of Ecology*, **21**, 10–20.

Dierenfeld, E.S., Mueller, P.J., & Hall, M.B. (2002). Duikers: native food composition, micronutrient assessment, and implications for improving captive diets. *Zoo Biology*, **21**, 185–96.

Dubois, M., Gilles, K.A., Hamilton, J.K., Rebers, P.A., & Smith, F. (1956). Colorimetric method for determination of sugars and related substances. *Analytical Chemistry*, **28**, 350–6.

Ganzhorn, J.U. (1989). Niche separation of seven lemur species in the eastern rainforest of Madagascar. *Oecologia*, **79**, 279–86.

(1992). Leaf chemistry and the biomass of folivorous primates in tropical forests: test of a hypothesis. *Oecologia*, **91**, 540–7.

Hall, M.B. (2001). *Neutral Detergent-Soluble Carbohydrates: Nutritional Relevance and Analysis, a Laboratory Manual*. Gainsville: University of Florida Extension Bulletin 339.

Hall, M.B. & Herejk, C. (2001). Differences in yields of microbial crude protein from in vitro fermentation of carbohydrates. *Journal of Dairy Science*, **84**, 2486–93.

Hall, M.B., Hoover, W.H., Jennings, J.P., & Miller-Webster, T.K. (1999). A method for partitioning neutral detergent-soluble carbohydrates. *Journal of the Science of Food and Agriculture*, **79**, 2079–86.

Heldt, J.S., Cochran, R.C., Stokka, G.L. *et al.* (1999). Effects of different supplemental sugars and starch fed in combination with degradable intake protein on low-quality forage use by beef steers. *Journal of Animal Science*, **77**, 2793–802.

Herrera, L.G., Leblanc, D., & Nassar, J. (2000). Sugar discrimination and gustatory thresholds in captive-born frugivorous Old World bats. *Mammalia*, **64**, 135–43.

Holmes, R.T. & Pitelka, F.A. (1968). Food overlap among coexisting sandpipers on northern Alaskan tundra. *Systematic Zoology*, **17**, 305–18.

Janson, C.H. & Chapman, C.A. (1999). Resources as determinants of primate community structure. In *Primate Communities*, ed. J.G. Fleagle, C.H. Janson, & K.E. Reed, pp. 237–67. Cambridge: Cambridge University Press.

Kay, R.N.B. & Davies, A.G. (1994). Digestive physiology. In *Colobine Monkeys: Their Ecology, Behavior, and Evolution*, ed. A.G. Davies & J.F. Oates, pp. 229–50. Cambridge: Cambridge University Press.

Koenig, A., Beise, J., Chalise, M.K., & Ganzhorn, J.U. (1998). When females should contest for food – testing hypotheses about resource density, distribution, size, and quality with Hanuman langurs (*Presbytis entellus*). *Behavioral Ecology and Sociobiology*, **42**, 225–37.

Krehbiel, C.R., Britton, R.A., Harmon, D.L., Wester, T.J., & Stock, R.A. (1995). The effects of ruminal acidosis on volatile fatty acid absorption and plasma activities of pancreatic enzymes in lambs. *Journal of Animal Science*, **73**, 3111–21.

Lambert, J.E. (1998). Primate digestion: interactions among anatomy, physiology, and feeding ecology. *Evolutionary Anthropology*, **7**, 8–20.

(2002). Resource switching and species coexistence in guenons: a community analysis of dietary flexibility. In *The Guenons: Diversity and Adaptation in African Monkeys*, ed. M. Glenn & M. Cords, pp. 309–23. New York, NY: Plenum Press.

Maisels, F., Gautier-Hion, A., & Gautier, J.-P. (1994). Diets of two sympatric colobines in Zaïre: more evidence on seed-eating in forests on poor soils. *International Journal of Primatology*, **15**, 681–701.

McCormick, M.E., Redfearn, D.D., Ward, J.D., & Blouin, D.C. (2001). Effect of protein source and soluble carbohydrate addition on rumen fermentation and lactation performance of Holstein cows. *Journal of Dairy Science*, **84**, 1686–97.

McNab, B.K. (2002). *The Physiological Ecology of Vertebrates: A View from Energetics.* Ithaca, NY: Cornell University Press.

Milton, K. (1993). Diet and primate evolution. *Scientific American,* August, 86–93.

(1998). Physiological ecology of howlers (*Alouatta*): energetic and digestive considerations and comparison with the Colobinae. *International Journal of Primatology,* **19**, 513–48.

Norconk, M.A. & Conklin-Brittain, N.L. (2004). Variation on frugivory: the diet of Venezuelan white-faced sakis. *International Journal of Primatology,* **25**, 1–26.

Oates, J.F. (1994). The natural history of African colobines. In *Colobine Monkeys: Their Ecology, Behavior, and Evolution,* ed. A.G. Davies & J.F. Oates, pp. 75–128. Cambridge: Cambridge University Press.

Piwonka, E.J. & Firkins, J.L. (1996). Effect of glucose fermentation on fiber digestion by ruminal microorganisms in vitro. *Journal of Dairy Science,* **72**, 2196–206.

Remis, M.J. (2002). Food preferences among captive western gorillas (*Gorilla gorilla gorilla*) and chimpanzees (*Pan troglodytes*). *International Journal of Primatology,* **23**, 231–49.

Robbins, C.T. (1993). *Wildlife Feeding and Nutrition.* New York, NY: Academic Press.

Rode, K.D., Chapman, C.A., Chapman, L.A., & McDowell, L.D. (2003). Mineral resource availability and consumption by colobus in Kibale National Park, Uganda. *International Journal of Primatology,* **24**, 541–73.

Rode, K.D., Chapman, C.A., McDowell, L.R., & Strickler, C. The role of nutrition in population: a comparison of redtail monkeys' diets and densities across habitats and logging intensities. *Biotropica* (in press).

Rowe, N. (1996). *The Pictorial Guide to the Living Primates.* East Hampton: Pogonias Press.

Sannes, R.A., Messman, M.A., & Vagnoni, D.B. (2002). Form of rumen-degradable carbohydrate and nitrogen on microbial protein synthesis and protein efficiency of dairy cows. *Journal of Dairy Science,* **85**, 900–8.

Struhsaker, T.T. (1975). *The Red Colobus Monkey.* Chicago, IL: University of Chicago Press.

(1997). *Ecology of an African Rain Forest: Logging in Kibale and the Conflict between Conservation and Exploitation.* Gainesville, FL: University Press of Florida.

Struhsaker, T.T. & Oates, J.F. (1975). Comparison of the behavior and ecology of red colobus and black-and-white colobus monkeys in Uganda: a summary. In *Socioecology and Psychology of Primates,* ed. R.H. Tuttle, pp. 103–24. The Hague: Mounton Publishers.

Thomas, G.J. (1960). Metabolism of the soluble carbohydrates of grasses in the rumen of the sheep. *Journal of Agricultural Science,* **54**, 360–72.

Ungar, P.S. (1995). Fruit preferences of four sympatric primate species at Ketambe, Northern Sumatra, Indonesia. *International Journal of Primatology,* **16**, 221–45.

Waterman, P.G. (1984). Food acquisition and processing as a function of plant chemistry. In *Food Acquisition and Processing in Primates,* ed. D.J. Chivers, B. A. Wood, & A. Bilsborough, pp. 177–211. New York, NY: Plenum Press.

19 *Primate sensory systems and foraging behavior*

NATHANIEL J. DOMINY, PETER W. LUCAS, AND NUR
SUPARDI NOOR

Introduction

Fruits are a key food for many primates. However, finding edible fruits in
tropical forests is a task complicated by a high diversity of fruiting species,
the variable quality of fruits, and the spatial and temporal complexity of fruit
production. How primates solve this challenging task is a poorly understood
process. The representational paradigm of neuroscientists assumes that the
nervous system forms relatively complex internal representations of the
environment, and that primates use these global representations to make
complex foraging decisions (Gallistel, 1990). Yet few behavioral ecologists

Feeding Ecology in Apes and Other Primates. Ecological, Physical and Behavioral Aspects, ed.
G. Hohmann, M.M. Robbins, and C. Boesch. Published by Cambridge University Press.
© Cambridge University Press 2006.

accept the notion of decisive behavior, the implication that animals make conscious choices or appreciate the computational structure underlying the problem to be solved. Instead, they assume that simple processes mediate apparently complex behaviors (Wehner, 1997).

Behavioral studies of primates support both paradigms to different extents. Primates appear to form complex spatial representations of their environment, or mental maps (Garber, 2000; Janson, 2000), but they appear unable to anticipate the availability and/or edibility of fruits. As Byrne (1996, p. 114) points out, "the job of feeding does not cease once a food is located." Fruit trees require monitoring (van Roosmalen, 1985) and the fruits themselves are subject to discrete special-purpose behaviors, or subroutines, each responsible for a particular sensory task (Figure 19.1). For example, spider monkeys often "inspect fruits by sniffing or biting them, since the external properties of the fruit (like colour) do not give a decisive answer on the stage of maturity" (van Roosmalen, 1985, p. 87). Similarly, moloch gibbons (*Hylobates moloch*) "examine food (by smell or taste) before it is ingested" (Kappeler, 1984, p. 228), and chimpanzees (*Pan troglodytes*) "inspect individual food items by sight, touch, or smell" (Wrangham, 1977, p. 510).

If the outcome of a subroutine is to reject a fruit, a primate will target another fruit and repeat its particular idiosyncratic series of behaviors (cf. Byrne 1996, 1999). The iterative process of sequential behaviors has been termed *handlungskette* (Tinbergen, 1953) or a behavioral chain (van Loon & Dicke, 2001). We define a behavioral chain as a broad synergism of behavioral elements all operating to achieve a specific goal. For primates, fruit selection is not a fixed succession of subroutines, but a mix of sensory behaviors (Figure 19.1). The output of each behavior is combined in the central nervous system to support the perception of fruit edibility.

Acceptance (deglutition) of a fruit is often based on an oral signal, usually the amount of soluble sugar (Figure 19.1F). In food preference trials, gorillas and chimpanzees preferred sugar-rich foods (Remis, 2002; Remis & Kerr, 2002). More generally, sugars are regarded as the key reward of vertebrate-dispersed fruits (Baker *et al.*, 1998). In some cases lipids appear to be the primary reward, particularly in arillate fruits and fruits in the family Lauraceae (Snow, 1981; Pannell & Koziol, 1987). In primates, the mechanisms of lipid perception are insufficiently understood, but lipids do at least appear to evoke positive sensations (Lermer & Mattes, 1999; Levine *et al.*, 2003; Verhagen *et al.*, 2004). Thus lipids, sweet proteins, e.g., thaumatin (Glaser *et al.*, 1978), and sweet amino acids, e.g., aspartic acid (Haefeli *et al.*, 1998), are all potential gustatory signals. Their effectiveness on primates depends on taste sensitivity, which varies with primate taxonomy and body size (Hladik & Simmen, 1997; Simmen & Hladik, 1998).

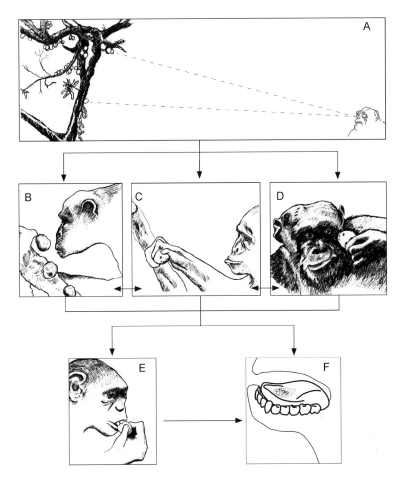

Figure 19.1. The sequential transitions between actions exhibited by a chimpanzee foraging on *Ficus brachylepis* (= *F. sansibarica*). Foraging can be divided into consecutive phases: searching and contact-testing. The searching phase (A) may involve random or oriented movement aided by spatial memory. Visual signals can be effective up to 30 m (Janson & Di Bitetti, 1997) and olfactory signals might be effective up to 100 m (Murlis, 1997). The searching phase ends when contact with a food item is established. Contact-testing can involve olfaction (B), or haptic behaviors, such as digital palpation (C). Visual scrutiny for defects or color can also occur during this stage (D), as can repeated contact with the lips and incisors (E), which are mechanosensitive (Agrawal & Lucas, 2003). Contact-testing ends with food acceptance (deglutition), when mechanical, gustatory, and olfactory information have become available in the oral cavity (F). The decision to accept or reject a food item is based not only on the properties of the food itself, but also on a primate's physiological state (satiety and reproductive status) and information from previous experiences. These factors are integrated in the central nervous system and in concert determine food acceptance and preference. (B and D are based on photographs by N.J.D.; C is based on a photograph by A. Houle).

Evolution of the primate sensory toolkit

The existence of behavioral chains in primates implies a learned association between the taste of a fruit and a variety of outward traits. Given that foraging efficiency depends on such traits, it stands to reason those traits requiring fine discrimination have shaped the evolution of the primate sensory toolkit. However, understanding how cognitive processes mediate fruit selection is a difficult problem, requiring several independent sets of data. Here we provide two data sets. First, we provide an extensive quantitative comparison of fruit traits across and within plant species from South-East Asia. These data assess if, and to what extent, fruit traits might contribute to the perception of nutritional quality. Second, we integrate spatial, physical, and chemical variables into an analysis of primate fruit selection in Kibale Forest, Uganda. These data assess if primates use specific fruit traits consistently to form internal representations of fruit quality.

Methods

Study sites

South-East Asia
We collected fruits opportunistically while conducting fieldwork in South-East Asia. In January 2002, we collected specimens of vertebrate-dispersed fruits in the Pasoh Forest Reserve, Peninsular Malaysia (2°59'N, 102°18'E). A minor mast-fruiting event occurred during this period (Numata *et al.*, 2003). Approximately 2500 ha in size, Pasoh consists of primary lowland dipterocarp forest; mean annual rainfall is *c.* 1800 mm (1983–1995; Noguchi *et al.*, 2003). Plant taxonomy follows Kochummen (1997). In August 2003, we collected specimens of vertebrate-dispersed fruits in the Bukit Timah Nature Reserve, central Singapore (1°21'N, 103°47'E). Approximately 225 ha in size, the reserve consists of primary lowland dipterocarp and secondary forest; annual rainfall is *c.* 2600 mm (Corlett, 1990). Plant taxonomy follows Turner (1995a, 1995b).

East Africa
In January–October 1999, we collected fruits in Kibale National Park, Uganda (0°13'N–0°41'N and 30°19'E–30°32'E). Kibale is organized on a 56-km north-south axis, comprising nearly 795 km² (Struhsaker, 1997). Rainfall is bimodal in distribution, occurring in two distinct rainy seasons: March–May and August–November. Annual rainfall is *c.* 1750 mm (1990–2001; Chapman

et al., 2003). The forest is characterized as a moist evergreen forest (closely related to moist montane forest) and lowland tropical rain forest with affinities to montane rain forest and mixed tropical deciduous forest (Struhsaker, 1997). Plant taxonomy follows Hamilton (1991).

We observed four species of catarrhine primate: *Cercopithecus ascanius* (red-tailed monkey), *Colobus guereza* (black-and-white colobus), *Pan troglodytes* (chimpanzee), and *Piliocolobus badius* (red colobus). A different species was followed daily, in rotational order, and focal animal data were recorded. Focal animals were selected opportunistically, their behavior being recorded continuously for 10 minutes, after which a new focal animal was selected.

We analyzed phenological patterns from K-30, a relatively undisturbed site. Twelve 200 × 10 m-transects were established along existing trails in January 1990, producing a total sampling area of 2.4 ha (Chapman *et al.*, 1999). All trees with a diameter at breast height (dbh) >10 cm and within 5 m of the trail were identified and monitored monthly. A total of 1171 trees (67 species) were tagged. The presence and maturity of fruit on all marked trees was determined, and their abundance was ranked on a relative scale of 0–4 (Chapman *et al.*, 1994).

Measurement of fruit properties

Fruit specimens were obtained either in the context of primate feeding observations (Kibale) or opportunistically (Bukit Timah and Pasoh). In Kibale, most fruits were collected in situ by ascending the trees (Dominy & Duncan, 2001). Some specimens were dropped or mishandled by foraging primates. In Bukit Timah and Pasoh, we collected a broad array of fleshy, vertebrate-dispersed fruits. The majority of specimens were collected from the forest floor. A stout slingshot was used to detach fruits <25 m overhead. Obviously decaying fruits or those judged to be overripe were not collected, as the traits are not representative of fruits consumed by vertebrates.

Physical measurements

Reflectance spectra were measured with an Ocean Optics S2000 spectrometer (Dunedin, FL, USA) fitted with diffraction grating No. 2, admitting light from 200–850 nm. Within four hours of collection, fruit specimens were mounted in a custom-built chamber and illuminated with a 12-volt, 3100-Kelvin tungsten halogen lamp (LS-1; Ocean Optics; range: 350–700 nm; Lucas *et al.*, 2001). Light was focused onto specimens through a 400-μm diameter patch fiber-optic cable and lens (Ocean Optics). Light was collected with a second lens and transmitted through a 200-μm fiber cable to the diffraction grating of the S2000. Specimens were placed 45° to the incident illumination with 20-degree axes separating the illuminating and recording lenses. Spectra were referenced to $BaSO_4$. The data are available at http://anthro.ucsc.edu/faculty/njdominy/spectral/

Mechanical properties were measured with a portable universal tester (Darvell *et al.*, 1996). Radial samples of fruit flesh were cut orthogonal to the outer surface and shaped with a 4-mm cork borer into right cylinders, *c.* 5 mm high. The modulus of elasticity, or Young's modulus, E, of fruit flesh was determined from tests on short cylinders in compression. Fracture toughness, R, was determined with a 15° angle wedge driven into small rectangular specimens cut from the fruit wall. Toughness was approximated by dividing the area beneath the force-deformation curve by the product of crack depth (i.e., wedge displacement) and initial specimen width (Lucas *et al.*, 2001). To account for some of the anisotropic variation within a fruit, mechanical measurements were taken from each hemisphere and averaged. The energetic equivalent of the critical stress intensity factor, K_{IC}, was determined by calculating the square root of $E \times R$ (Agrawal & Lucas, 2003). K_{IC} represents the perception of fruit hardness during incisory evaluation (Vincent *et al.*, 2002).

Spectral modeling

Reconstructing the color of a primate food item requires modeling. This is achieved by multiplying the reflectance spectrum of the object by an illuminant (Figure 19.2). Attenuation of short-wavelengths by the lens and neural retina was not modeled as in other studies (Regan *et al.*, 2001). The quantum catch (Q) of S-, M-, and L-cone classes was calculated by multiplying the stimulus radiance and cone absorption spectra and integrating the resulting spectrum over wavelength (Figure 19.2). Chromaticity coordinates analogous to MacLeod–Boynton (1979) coordinates may be graphed by plotting a y value of $Q_S/(Q_L + Q_M)$, which defines yellow-blueness (yellow low, blue high), against an x value of $Q_L/(Q_L + Q_M)$, which defines green-redness (green low, red high). These coordinates correspond to the physiological subsystems of catarrhines, the phylogenetically recent green-red subsystem (subserved by midget ganglion cells) and the S-cone mediated yellow-blue subsystem (subserved by small bistratified ganglion cells).

Chemical measurements

High performance liquid chromatography (HPLC) was used to measure concentrations of sucrose, fructose, and glucose (Lee, 1990). Extractions of 0.1 g fruit flesh in 5 ml 1:1 $dH_2O:CH_3OH$ were made with a Tissue Tearor (BioSpec, Bartlesville, OK, USA). Ethanol concentrations were determined

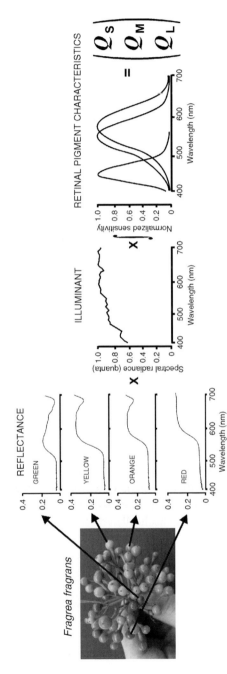

Figure 19.2. Estimation of catarrhine photoreceptor responses. The quantal catches (Q) of S-, M-, L-cone photoreceptors are determined from the fruits of *Fagrea fragrans* (Loganiaceae). Each reflectance spectrum is multiplied by an open sky illuminant (Endler, 1993) to produce an adjusted radiance spectrum. The product is then multiplied by Smith–Pokorny cone sensitivities (Wyszecki & Stiles, 1982) and integrated over wavelength to give $Q_{S,M,L}$. These quantal catches are used to calculate post-receptive channels.

from vapor pressure measurements at 24 °C using a modified electrochemical sensor (PAS Systems, Fredericksburg, VA, USA) calibrated against solutions of known concentration (Dudley, 2004). Fruit acidity was measured with a solid-state pH meter (model 1001, Sentron BV, Roden, the Netherlands). Protein and phenolic measurements follow Lucas *et al.* (2001).

Statistical analyses

The assumptions of all statistical tests were examined, and data transformed to meet the assumptions before analysis proceeded; in nearly all cases, a natural log-transformation of the original data produced conformity with the assumptions of homoskedasticity and normality of residuals.

Correlation analyses

To determine the extent to which sensory and nutritional variables are correlated across vertebrate-dispersed fruits, we performed a pair-wise correlation on data representing 31 species of paired fruits collected from individual trees in Pasoh Forest. Table 19.1 provides select examples; each pair represents an anthropogenic judgment of distinctly different developmental stages. Significance was determined using a sequential Bonferroni adjustment on the significance probabilities (Sokal & Rohlf, 1995).

Because pair-wise correlations across fruit species might mask developmental trends within a fruit species, we performed a t-test for paired comparisons on the above data set to determine the extent to which traits change during fruit development.

Multiple regression analyses

For tree species used by Kibale primates (Table 19.2), the properties of edible fruits and measures of primate visitation were averaged over all plant individuals measured. Each monthly set of averages for a tree species constituted one datum in a linear multiple regression analysis; this was performed with JMP 5.0.1 for Macintosh. Natural-log transformations of both independent and dependent variables ensured homoskedasticity of the residuals for the measures of primate feeding (proportion of trees visited) against each of the independent variables, which were chosen to represent discrete sensory modalities: trichromatic vision (red-green and yellow-blue chromaticity values), texture (fracture toughness and Young's modulus), and taste (total soluble sugars). Because some of the potentially explanatory variables were correlated with each other, violating an assumption of the model, an iterative step-wise regression model was used to reduce the

Table 19.1. *The sensory traits of fruits at two developmental stages*

Species	Early-intermediate stage of development						Late stage of development					
	Chromaticity[a]		Texture[b]		Sweetness[c]		Chromaticity[a]		Texture[b]		Sweetness[c]	
	S/(M+L)	L/(M+L)	E	R	Sucrose	Fructose	S/(M+L)	L/(M+L)	E	R	Sucrose	Fructose
Annonaceae												
Polyalthia jenkinsii	0.223	0.496	3.6	272	0.0	18.0	0.113	0.558	0.1	34	0.0	129.0
P. sclerophylla	0.188	0.495	0.6	137	0.0	3.0	0.397	0.577	0.5	71	0.0	10.0
Popowia tomentosa	0.039	0.941	5.3	756	1.0	3.0	0.618	0.495	0.7	132	1.0	5.0
Unidentified sp. 1	0.092	0.512	–	134	57.0	83.0	0.079	0.538	–	11	0.0	212.0
Clusiaceae												
Garcinia forbesii	0.225	0.506	1.1	555	0.0	89.0	0.227	0.537	0.2	17	0.0	209.0
G. parvifolia	0.323	0.507	3.9	491	0.0	62.0	0.231	0.544	0.1	23	80.0	69.0
G. prainiana	0.129	0.503	0.2	140	21.0	26.2	0.386	0.614	0.2	31	167.0	130.0
Moraceae												
Artocarpus elasticus	0.156	0.540	0.7	196	0.0	62.6	0.158	0.553	0.0	71	0.0	215.0
A. kemando	0.170	0.525	–	–	0.0	112.0	0.200	0.524	0.1	13	242.0	211.0
A. lanceifolius	0.160	0.536	1.9	413	7.0	58.8	0.136	0.577	0.0	5	298.0	198.0
A. nitidus var. *griffithii*	0.124	0.523	1.5	398	0.0	71.0	0.182	0.531	1.3	406	0.0	193.0
Sapindaceae												
Nephelium costatum	0.146	0.520	0.2	52	–	–	0.207	0.599	0.1	5	126.0	114.0
Pometia pinnata	0.226	0.531	0.2	0	0.0	295.0	0.296	0.575	0.1	0	140.0	296.0

Notes:

Specimens were collected in Pasoh Forest and paired from the same tree or liana. Selected taxonomic families and genera are those that produce primate fruits (Leighton & Leighton, 1983; Leighton, 1993).

[a] Chromaticity calculations are described in the text.

[b] Young's modulus, E, units are MPa. Fracture toughness, R, units are joules (J) m^{-2}.

[c] Sugar concentration units are mM.

Table 19.2. *Fruiting trees in Kibale National Park monitored for phenology and modeled using multiple regression analysis*

Family and species	Tree density (individuals ha^{-1})
Apocynaceae	
Funtumia latifolia	33.8
Rauvolfia oxyphylla	0.4
Araliaceae	
Polyscias fulva	0.8
Ebenaceae	
Diospyros abyssinica	40.0
Moraceae	
Ficus brachylepis	1.7
F. exasperata	3.8
F. natalensis	0.4
Olacaceae	
Strombosia scheffleri	12.5
Rosaceae	
Prunus africana	0.4
Rutaceae	
Fagaropsis angolensis	2.5
Teclea nobilis	17.1
Sapotaceae	
Mimusops bagshawei	3.3
Ulmaceae	
Celtis africana	4.2
C. durandii	47.1
Chaetacme aristata	17.1

regression model to a smaller number of explanatory variables. For all regressor variables, p-to enter the forward step-wise model was set to p $\leq$ 0.05 and p-to remove was set to p $\geq$ 0.10 (Sokal & Rohlf, 1995).

Results

Correlations between fruit traits

Pair-wise correlations across fruits collected from Pasoh Forest showed total sugar concentration to be correlated with three variables: fracture toughness (R), Young's modulus (E), and the critical stress intensity factor (K_{IC}), all of which correlate strongly with each other (Table 19.3). Fruit width was positively correlated with mass, but negatively correlated with phenolics. Phenolics in fruits were positively correlated with protein content. These

Table 19.3. *Pair-wise correlation coefficients of the independent variables measured to predict sensory and nutritional variables in a sample of 62 fruits (n = 31 species) collected in the Pasoh Forest Reserve*[a]

| | Sensory and nutritional variables | | | | | | | Chemical traits | | | |
| | Physical traits | | | | | | | | | | |
	S/(M+L)	L/(M+L)	E	R	K_{IC}	Mass	Width	Sugars	pH	Protein	Phenolics
S/(M+L)	1.00							0.20	0.01	0.19	0.17
L/(M+L)	−0.39	1.00						0.03	0.13	−0.13	−0.20
Young's modulus – E	−0.16	−0.11	1.00					**−0.58**	0.18	0.34	0.24
Toughness – R	−0.23	−0.11	**0.84**	1.00				**−0.63**	0.23	0.26	0.16
Intensity Factor – K_{IC}	−0.24	−0.12	0.92	**0.96**	1.00			**−0.64**	0.09	0.27	0.24
Fruit mass	−0.21	0.20	−0.31	−0.20	−0.25	1.00		0.31	−0.25	−0.22	−0.40
Fruit width	−0.26	0.20	−0.38	−0.20	−0.28	**0.99**	1.00	0.29	−0.19	−0.29	**−0.47**
Sugars								1.00	−0.01	−0.01	−0.20
pH									1.00	0.27	−0.01
Protein										1.00	**0.72**
Phenolics											1.00

Notes:

Significance (p ≤ 0.05) was determined using a Bonferroni adjustment on the significance probabilities and are highlighted bold.

[a] Species used in the analysis: *Aglaia ridleyi, Aidia wallichiana, Artocarpus elasticus, Artocarpus lanceifolius, Clidemia hirta. Dissochaeta* sp., *Durio griffithii, Dysoxylum cauliflorum, Ficus schwarzii, Ficus fulva, Ficus glandulifera, Garcinia forbesii, Garcinia parvifolia, Garcinia prainiana, Gironniera parvifolia, Gnetum* cf. *cuspidatum, Ixora grandifolia* var. *lancifolia, Lithocarpus conocarpus, Nephelium costatum, Polyalthia jenkinsii, Polyalthia sclerophylla, Pometia pinnata, Popowia tomentosa, Pternandra coerulescens, Pternandra echinata, Trema angustifolia, Urophyllum glabrum, Vitex pinnata, Vitex* sp., unknown Annonaceae, unidentified species A.

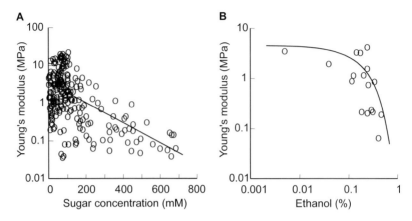

Figure 19.3. The relationship between total sugar concentration and Young's modulus in 225 specimens collected from all field sites ($r^2 = 0.346$; P < 0.001) (B). The x-axis is untransformed because proportional changes in sugar concentration did not produce linear responses in Young's modulus (Sokal & Rohlf, 1995). The relationship between Young's modulus and ethanol content was inversely related for 17 fruits collected in Bukit Timah (A), indicating rapid ethanol production during the softening process ($r^2 = 0.459$; P < 0.01). These relationships suggest that primates could use the olfactory or textural properties of a fruit to predict sugar content.

results suggest that fruit texture is a broadly informative cue of nutritional quality. A regression between Young's modulus and total soluble sugars for 225 fruit specimens collected at all field sites was significant (Figure 19.3A), as was a regression between Young's modulus and the ethanol content of fruits from Bukit Timah (Figure 19.3B).

The results of a t-test for paired comparisons within fruits showed the more developed, or riper, of paired fruits to be yellower, softer, less tough, larger, heavier, sweeter, and richer in protein and phenolics (Table 19.4). Only redness, sucrose, and acidity did not differ between paired fruits. However, for some fruit species, such as *Uvariopsis congensis* from Kibale Forest, redness did predict sugar content (Figure 19.4B). The effectiveness of a particular color signal is therefore taxon dependent, as sugars can be equally high in fruits that do not undergo developmental changes in red-green chromaticity (Figure 19.4D).

Primate fruit selection

Results of regressions on the monthly proportion of trees > 10 cm dbh fed in by *C. ascanius* ($r^2_{adj} = 0.33$; $F_{[5,23]} = 3.28$; p = 0.03), *C. guereza* ($r^2_{adj} = 0.64$; $F_{[5,9]} = 4.21$; p = 0.09), *P. troglodytes* ($r^2_{adj} = 0.72$; $F_{[5,6]} = 4.08$; p = 0.36),

Table 19.4. *Results of a t-test for paired comparisons in a sample of 62 fruits (n = 31 species) collected in Pasoh Forest*

Sensory and nutritional variables	F_S	p-value
Physical traits		
S/(M+L)	10.382	0.006
L/(M+L)	0.310	0.587
Young's modulus – E	15.807	0.001
Toughness – R	36.722	<0.001
Intensity Factor – K_{IC}	28.215	<0.001
Fruit mass	8.379	0.012
Fruit width	8.876	0.010
Chemical traits		
Sugars	48.970	<0.001
Sucrose	0.172	0.685
Fructose	42.075	<0.001
Glucose	36.717	<0.001
pH	1.422	0.253
Protein	5.567	0.033
Phenolics	4.670	0.049

The F_S-ratio is the squared value of t[s]. The more developed, or riper, of two fruits tended to be yellower, softer, less tough, larger, heavier, sweeter, and richer in protein and phenolics.

and *P. tephrosceles* ($r^2_{adj} = -0.07$; $F_{[5,21]} = 0.76$.; p = 0.59) are presented in Table 19.5. Models were significant only for *C. ascanius*. None of the explanatory variables were statistically significant (Table 19.5.), but some correlated with each other in a different data set (Table 19.3), potentially violating assumptions of the model and possibly resulting in a Type 1 error, rejecting the null hypothesis of a zero slope. Accordingly, an iterative step-wise regression model was performed and some variables were retained. Total soluble sugars was the only predictor variable of visitation for two species: fruits with high sugar content predicted increased visitation by *C. ascanius*, ($R^2_{adj} = 0.40$; $F = 16.48$), and *P. troglodytes* ($R^2_{adj} = 0.81$; $F = 26.97$). For *C. guereza*, toughness and yellow-blueness were key predictor variables of visitation: fruits with high toughness and low yellowness predicted increased visitation ($R^2_{adj} = 0.65$; $F = 9.19$).

Discussion

The function of animal cognition is not to represent reality, it is to allow the animal to act in an adaptive way.

Konrad Lorenz (Vaas, 2002).

Table 19.5. *Results of regressions on the monthly proportion of trees > 10 cm dbh fed in by Cercopithecus ascanius, Colobus guereza, Pan troglodytes, and Piliocolobus badius*

Fruit sensory trait	Cercopithecus ascanius		Colobus guereza		Pan troglodytes		Piliocolobus badius	
	slope	p-value	slope	p-value	slope	p-value	slope	p-value
Chromaticity								
Yellow-blue - S/(M+L)	-1.167	0.435	-1.121	0.106	0.570	0.372	0.291	0.277
Green-red - L/(M+L)	-10.387	0.315	-4.980	0.348	-29.508	0.422	-2.257	0.312
Texture								
Fracture toughness	-0.048	0.695	-0.301	0.181	1.193	0.414	-0.054	0.601
Young's modulus	-0.185	0.473	0.575	0.288	0.476	0.410	-0.105	0.424
Sweetness								
Sugars	-0.187	0.462	-0.781	0.235	-2.154	0.357	-0.216	0.304

Slopes are the regression coefficients and give the expected change (in standard deviations) of the proportion of trees visited for each standard deviation change in a given fruit trait. Probability levels test the null hypothesis that the slope equals zero.

Here we provide a quantitative comparison of traits across and within fruit species from South-East Asia. A trait may function as a cue or a signal to a primate. Hasson (1994) defines a cue as any trait that can be used to guide behavior. A trait is a signal if it is adaptive for the sender and if it results from mechanisms attributable to natural selection (Williams, 1992; Maynard Smith & Harper, 2003). However, the effectiveness of a signal is subject to the sensory physiology of the intended receiver. The distinction between a cue and a signal is best illustrated by an example. A chimpanzee feeding on *Ficus brachylepis* uses texture to guide selection (Figure 19.1). In this case, fruit texture is a cue for the primate, but it is an unlikely signal from the plant. Softening is a complex process in fruits (Brady, 1987), and because it is uncoupled with the production of external pigments (Brady, 1987), it is an improbable signal to seed-dispersing frugivores.

The value of texture as a cue depends on how it correlates with sugars. Our results show that total sugar concentration is strongly correlated with textural variables across fruit species: softer, less tough fruits had more sugar. Interestingly, the sensitivity of primate digits tends to correlate with the proportion of fruits in the diet (Hofmann *et al.*, 2004). Softer fruits were also characterized by a higher ethanol content, a compound that could function as a cue (Janzen, 1977) or a signal to primates (Dominy, 2004). The role of ethanol as an olfactory signal requires further study (Dudley, 2004). Within plant species, a riper fruit differed from a less developed one in a variety of traits: it was larger, softer, yellower, and possessed more sugar. These results suggest that a primate seeking sugars could use visual signals to determine which fruits within a tree are more developed.

While these results provide conceptual data for the types of traits primates could use to select fruits efficiently, they are impractical for understanding how primates determine which trees to visit. In a multiple regression model that controls for the availability of fruits, we show that *C. ascanius* and *P. troglodytes* select fruits rich in sugars (cf. Janson *et al.*, 1986; Leighton, 1993). However, the variables that predict how primates select sugar-rich fruits are not immediately clear. Observations suggest that primates vary the use of particular sensory modalities according to the species being evaluated. For example, some species of fruits, e.g., *U. congensis*, produce a strong red signal perceptible to trichromatic primates. When selecting fruits from this species, primates likely give priority to the chromatic signal. For other species, in which color is an insufficient signal, other modalities receive priority, e.g., haptic senses when foraging on *F. brachylepis* (Figure 19.4). This view is consistent with recent developments in behavioral ecology; specifically, the notion that a complex task such as fruit selection can be supported by neural mechanisms devoted to subroutines (Wehner, 1997).

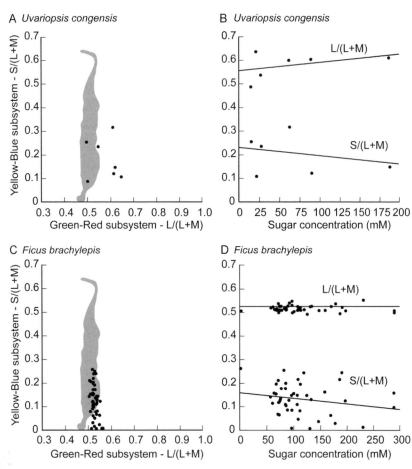

Figure 19.4. Chromaticity diagrams analogous to that of MacLeod Boynton (1979). Solid circles represent the chromaticities of individual fruits. The shaded region depicts the chromatic domain of mature foliage (n = 461 specimens), the background against which fruits are seen. Fruits of a species with prominent chromatic changes on the green-red subsystem, *Uvariopsis congensis* (Annonaceae) (A), are compared to a species lacking such changes, *Ficus brachylepis* [= *Ficus sansibarica*] (Moraceae) (C). Increasing increments on the green-red subsystem attend increasing sugar concentrations in *Uvariopsis congensis* (B), but not *Ficus brachylepis* (D), which yields higher sugar concentrations at equivalent edible stages.

Subroutines are combined to provide a global output. The output represents an integration of different sensory modalities in the central nervous system and supports the ecological problem needing to be solved, whether the problem is navigation or food recognition.

However, given the complexity of tropical forests, it would be surprising if primates did not have a means of categorizing fruits or remembering how fruit species vary during development. Although evidence that primates recognize plants at the specific taxonomic level is unavailable, the cognitive skill is implied from studies of auto-medication (Huffman, 1997). Perhaps more important, however, is the role of social learning. Primates have experience with the fruits in their environment, having learned from conspecifics those that are edible. Because ecologically naïve, ex-captive orangutans struggle to identify suitable food items in their natural habitat (Russon, 2003), sensory subroutines can be insufficient for assessing fruit edibility. Rijksen (1978) reported that wild orangutans do not accept edible cultivated fruits that are unfamiliar unless they observe others eating them (cf. Glander, 1977; Milton, 1980). Thus, recognizing and selecting a food item is a challenging task that integrates learned associations, memory, and sensory subroutines.

In this chapter we have described the traits of fruits and outlined how primates might accomplish the task of selecting edible ones. We avoided the tendency to focus on a single sensory modality because we believe the approach suffers from an intellectual inertia that distracts from the goal of understanding how different sensory inputs are integrated by the living, breathing animal. No single sensory modality allows a primate to predict precisely how many calories are available in a particular fruit, but subroutines ensure that edible fruits within a patch or territory are identified. We believe this approach has the potential to shed light on the selective pressures underlying the sensory and neural mechanisms of primates. Guiding mechanisms, evolutionary processes, and importance of context are important themes in the study of primate behavior, and provide important caveats and new directions for researchers interested in the evolution of primate senses.

Acknowledgments

We thank C. Boesch, G. Hohmann, M. Robbins, four anonymous reviewers, and the participants of the Foraging Ecology in Apes and Other Primates conference. We also thank S. Ahmad, L. Alport, K.Y. Ang, B. Balyeganira, Y.C. Chan, Y.-Y. Chen, P.Y. Cheng, P. Chiyo, H. Corke, P. Kagoro, M. LaBarbera, S. Latif, J. Magnay, M. Musana, D. Osorio, L. Ramsden, J.J. Socha, I.-F. Sun, H.T.W. Tan, R.W. Wrangham, and S.J. Wright for their contributions. We are particularly grateful to C.A. Chapman for access to his phenological data. Research permits were granted by the Forest Research Institute of Malaysia, the Makerere University Biological Field Station, the National Parks Board of Singapore (permit no. RP328A), the Ugandan

National Council for Science and Technology, and the Uganda Wildlife Authority. Funding was received from the Croucher Foundation, the Explorer's Club, the National Geographic Society (nos. 6584–99 and 7179–01), the Raffles Museum of Biodiversity, the Research Grants Council of Hong Kong (no. 7241/97M), a Ruth L. Kirschstein National Service Research Award, and Sigma Xi.

References

Agrawal, K.R. & Lucas, P.W. (2003). The mechanics of the first bite. *Proceedings of the Royal Society of London, Series B*, **270**, 1277–82.

Baker, H.G., Baker, I., & Hodges, S.A. (1998). Sugar composition of nectars and fruits consumed by birds and bats in the tropics and subtropics. *Biotropica*, **30**, 559–86.

Brady, C.J. (1987). Fruit ripening. *Annual Review of Plant Physiology*, **38**, 155–78.

Byrne, R.W. (1996). The misunderstood ape: cognitive skills of the gorilla. In *Reaching into Thought: The Minds of the Great Apes*, ed. A.E. Russon, K.A. Bard, & S.T. Parker, pp. 111–30. Cambridge: Cambridge University Press.

(1999). Cognition in great ape ecology: skill-learning ability opens up foraging opportunities. *Symposia of the Zoological Society of London*, **72**, 333–50.

Chapman, C.A., Chapman, L.J., Rode, K.D., Hauck, E.M., & McDowell, L.R. (2003). Variation in the nutritional value of primate foods: among trees, time periods, and areas. *International Journal of Primatology*, **24**, 317–33.

Chapman, C.A., Wrangham, R.W., & Chapman, L.J. (1994). Indices of habitat-wide fruit abundance in tropical forests. *Biotropica*, **26**, 160–71.

Chapman, C.A., Wrangham, R.W., Chapman, L.J., Kennard, D.K., & Zanne, A.E. (1999). Fruit and flower phenology at two sites in Kibale National Park, Uganda. *Journal of Tropical Ecology*, **15**, 189–211.

Corlett, R.T. (1990). Flora and reproductive phenology of the rain-forest of Bukit Timah, Singapore. *Journal of Tropical Ecology*, **6**, 55–63.

Darvell, B.W., Lee, P.K.D., Yuen, T.D.B., & Lucas, P.W. (1996). A portable fracture toughness tester for biological materials. *Measurement Science and Technology*, **7**, 954–62.

Dominy, N.J. (2004). Fruits, fingers, and fermentation: the sensory cues available to foraging primates. *Integrative and Comparative Biology*, **44**, 295–303.

Dominy, N.J. & Duncan, B.W. (2001). GPS and GIS methods in an African rain forest: applications to tropical ecology and conservation. *Conservation Ecology*, **5**, 537–49.

Dudley, R. (2004). Ethanol, fruit ripening, and the historical origins of human alcoholism in primate frugivory. *Integrative and Comparative Biology*, **44**, 315–23.

Endler, J.A. (1993). The color of light in forests and its implications. *Ecological Monographs*, **63**, 1–27.

Gallistel, C.R. (1990). *The Organization of Learning*. Cambridge, MA: MIT Press.

Garber, P.A. (2000). Evidence for use of spatial, temporal, and social information by primate foragers. In *On the Move: How and Why Animals Travel in Groups*, ed. S. Boinski & P.A. Garber, pp. 261–98. Chicago, IL: University of Chicago Press.

Glander, K.E. (1977). Poison in a monkey's Garden of Eden. *Natural History*, **86**, 34–41.

Glaser, D., Hellekant, G., Brouwer, J.N., & van der Wel, H. (1978). The taste responsiveness in primates to the proteins thaumatin and monellin and their phylogenetic implications. *Folia Primatologica*, **29**, 56–63.

Haefeli, R.J., Solms, J., & Glaser, D. (1998). Taste responses to amino acids in common marmosets (*Callithrix jacchus jacchus*, Callitrichidae) a non-human primate in comparison to humans. *Food Science and Technology*, **31**, 371–6.

Hamilton, A. (1991). *A Field Guide to Uganda Forest Trees*, 2nd edn. Kampala: Makerere University Printery.

Hasson, O. (1994). Cheating signals. *Journal of Theoretical Biology*, **167**, 223–38.

Hladik, C.M. & Simmen, B. (1997). Taste perception and feeding behavior in non-human primates and human populations. *Evolutionary Anthropology*, **5**, 58–71.

Hoffmann, J.N., Montag, A.G., & Dominy, N.J. (2004). Meissner corpuscles and somatosensory acuity: the prehensile appendages of primates and elephants. *The Anatomical Record, Part A*, **281**, 1138–47.

Huffman, M.A. (1997). Current evidence for self-medication in primates: a multidisciplinary perspective. *Yearbook of Physical Anthropology*, **40**, 171–200.

Janson, C.H. (2000). Spatial movement strategies: theory, evidence, and challenges. In *On the Move: How and Why Animals Travel in Groups*, ed. S. Boinski & P.A. Garber, pp. 165–203. Chicago, IL: University of Chicago Press.

Janson, C.H. & Di Bitetti, M.S. (1997). Experimental analysis of food detection in capuchin monkeys: effects of distance, travel speed, and resource size. *Behavioral Ecology and Sociobiology*, **41**, 17–24.

Janson, C.H., Stiles, E.W., & White, D.W. (1986). Selection on plant fruiting traits by brown capuchin monkeys: a multivariate approach. In *Frugivores and Seed Dispersal*, ed. A. Estrada & T.H. Fleming, pp. 83–92. Dordrecht: Dr. W. Junk Publishers.

Janzen, D.H. (1977). Why fruits rot, seeds mold, and meat spoils. *The American Naturalist*, **111**, 691–713.

Kappeler, M. (1984). Diet and feeding behaviour of the moloch gibbon. In *The Lesser Apes: Evolutionary and Behavioural Biology*, ed. H. Preuschoft, D.J. Chivers, W.Y. Brockelman, & N. Creel, pp. 228–41. Edinburgh: Edinburgh University Press.

Kochummen, K.M. (1997). *Tree Flora of Pasoh Forest*. Kepong: Forest Research Institute of Malaysia.

Lee, Y.C. (1990). High-performance anion-exchange chromatography for carbohydrate analysis. *Analytical Biochemistry*, **189**, 151–62.

Leighton, M. (1993). Modeling dietary selectivity by Bornean orangutans: evidence for integration of multiple criteria in fruit selection. *International Journal of Primatology*, **14**, 257–313.

Leighton, M. & Leighton, D.R. (1983). Vertebrate responses to fruiting seasonality within a Bornean rain forest. In *Tropical Rain Forest: Ecology and Resource*

Management, ed. S.L. Sutton, T.C. Whitmore, & A.C. Chadwick, pp. 181–96. Oxford: Blackwell Sciente.

Lermer, C.M. & Mattes, R.D. (1999). Perception of dietary fat: ingestive and metabolic implications. *Progress in Lipid Research*, **38**, 117–28.

Levine, A.S., Kotz, C.M., & Gosnell, B.A. (2003). Sugars and fats: the neurobiology of preference. *Journal of Nutrition*, **133**, 831S–4S.

Lucas, P.W., Beta, T., Darvell, B.W., *et al.* (2001). Field kit to characterize physical, chemical, and spatial aspects of potential primate foods. *Folia Primatologica*, **72**, 11–25.

MacLeod, D.I.A. & Boynton, R.M. (1979). Chromaticity diagram showing cone excitation by stimuli of equal luminance. *Journal of the Optical Society of America*, **69**, 1183–6.

Maynard Smith, J. & Harper, D. (2003). *Animal Signals*. Oxford: Oxford University Press.

Milton, K. (1980). *The Foraging Strategy of Howler Monkeys: A Study in Primate Economics*. New York, NY: Columbia University Press.

Murlis, J. (1997). Odor plumes and the signal they provide. In *Insect Pheromone Research: New Directions*, ed. R.T. Cardé & A.K. Minks, pp. 221–31. New York, NY: Chapman & Hall.

Noguchi, S., Rahim Nik, A., & Tani, M. (2003). Rainfall characteristics of tropical rainforest at Pasoh Forest Reserve, Negeri Sembilan, Peninsular Malaysia. In *Pasoh: Ecology of a Lowland Rain Forest in Southeast Asia*, ed. T. Okuda, N. Manokaran, Y. Matsumoto, K. Niiyama, S.C. Thomas, & P.S. Ashton, pp. 51–8. Tokyo: Springer.

Numata, S., Yasuda, M., Okuda, T., Kachi, N., & Supardi Noor, N. (2003). Temporal and spatial patterns of mass flowerings on the Malay Peninsula. *American Journal of Botany*, **90**, 1025–31.

Panell, C.M. & Koziol, M.J. (1987). Ecological and phytochemical diversity of arillate seeds in Aglaia (*Meliaceae*): a study of vertebrate dispersal in tropical trees. *Philosophical Transactions of the Royal Society of London, Series B*, **316**, 303–33.

Regan, B.C., Julliot, C., Simmen, B. *et al.* (2001). Fruits, foliage and the evolution of primate colour vision. *Philosophical Transactions of the Royal Society of London, Series B*, **356**, 229–83.

Remis, M.J. (2002). Food preference among captive western gorillas (*Gorilla gorilla gorilla*) and chimpanzees (*Pan troglodytes*). *International Journal of Primatology*, **23**, 231–49.

Remis, M.J. & Kerr, M.E. (2002). Taste responses to fructose and tannic acid among gorillas (*Gorilla gorilla gorilla*). *International Journal of Primatology*, **23**, 251–61.

Rijksen, H.D. (1978). *A Fieldstudy on Sumatran Orangutans (Pongo pygmaeus abelii Lesson 1827)*. Wageningen, the Netherlands: H. Veenman & Zonen.

Russon, A.E. (2003). Developmental perspectives on great ape traditions. In *The Biology of Traditions: Models and Evidence*, ed. D.M. Fragaszy & S. Perry, pp. 329–64. Cambridge: Cambridge University Press.

Simmen, B. & Hladik, C.M. (1998). Sweet and bitter taste discrimination in primates: scaling effects across species. *Folia Primatologica*, **69**, 129–38.

Snow, D.W. (1981). Tropical frugivorous birds and their food plants: a world survey. *Biotropica*, **13**, 1–14.

Sokal, R.S. & Rohlf, F.J. (1995). *Biometry*, 3rd edn. New York: W. H. Freeman.

Struhsaker, T.T. (1997). *Ecology of an African Rainforest: Logging in Kibale and the Conflict between Conservation and Exploitation.* Gainesville, FL: University of Florida Press.

Tinbergen, N. (1953). *Social Behaviour in Animals with Special Reference to Vertebrates (Reprint 1962).* London: Methuen.

Turner, I.M. (1995a). A catalogue of the vascular plants of Malaya. *Gardens Bulletin Singapore*, **47**, 1–346.

(1995b). A catalogue of the vascular plants of Malaya. *Gardens Bulletin Singapore*, **47**, 347–757.

Vaas, R. (2002). Consciousness and its place in nature. *Journal of Consciousness Studies*, **9**, 69–78.

van Loon, J.J.A. & Dicke, M. (2001). Sensory ecology of arthropods utilizing plant infochemicals. In *Ecology of Sensing*, ed. F.G. Barth & A. Schmid, pp. 253–70. Berlin: Springer-Verlag.

van Roosmalen, M.G.M. (1985). Habitat preferences, diet, feeding strategy and social organization of the black spider monkey (*Ateles paniscus paniscus* Linnaeus 1758) in Surinam. *Acta Amazonica*, **15** (Suppl. 3/4), 1–238.

Verhagen, J.V., Kadohisa, M., & Rolls, E.T. (2004). Primate insular/opercular taste cortex: neuronal representations of the viscosity, fat texture, grittiness, temperature, and taste of foods. *Journal of Neurophysiology*, **92**, 1685–99.

Vincent, J.F.V., Saunders, D.E.J., & Beyts, P. (2002). The use of stress intensity factor to quantify 'hardness' and 'crunchiness' objectively. *Journal of Texture Studies*, **33**, 149–59.

Wehner, R. (1997). Sensory systems and behaviour. In *Behavioural Ecology: An Evolutionary Approach*, 4th edn., ed. J.R. Krebs & N.B. Davies, pp. 19–41. Oxford: Blackwell Science.

Williams, G.C. (1992). *Natural Selection: Domains, Levels and Challenges.* Oxford: Oxford University Press.

Wrangham, R.W. (1977). Feeding behaviour of chimpanzees in Gombe National Park, Tanzania. In *Primate Ecology: Studies of Feeding and Ranging Behaviour in Lemurs, Monkeys, and Apes*, ed. T.H. Clutton-Brock, pp. 503–38. London: Academic Press.

Wyszecki, G. & Stiles, W.S. (1982). *Color Science: Concepts and Methods, Quantitative Data and Formulae*, 2nd edn. New York, NY: John Wiley & Sons.

Index